W9-BQY-547

Springer Advanced Texts in Life Sciences

David E. Reichle, Editor

Photo by Jan Hahn, courtesy of Woods Hole Oceanographic Institution.

Ivan Valiela

Marine Ecological Processes

With 220 Figures

Springer-Verlag
New York Berlin Heidelberg Tokyo

Ivan Valiela
Boston University Marine Program
Marine Biological Laboratory
Woods Hole, Massachusetts 02543
U.S.A. and
Department of Biology
Boston University
Boston, Massachusetts 02215 U.S.A.

Series Editor:
David E. Reichle
Environmental Sciences Division
Oak Ridge National Laboratory
Oak Ridge, Tennessee 37830 U.S.A.

Library of Congress Cataloging in Publication Data
Valiela, Ivan.
Marine ecological processes.
(Springer advanced texts in life sciences)
Bibliography: p.
Includes index.
1. Marine ecology. I. Title. II. Series.
QH541.5.S3V34 1984 574.5′2636 83-20037

Typeset by Bi-Comp, Incorporated, York, Pennsylvania.
Printed and bound by Halliday Lithograph Corporation, West Hanover, Massachusetts.
Printed in the United States of America.

9 8 7 6 5 4 3 2 1

ISBN 0-387-90929-X Springer-Verlag New York Berlin Heidelberg Tokyo
ISBN 3-540-90929-X Springer-Verlag Berlin Heidelberg New York Tokyo

Preface

This text is aimed principally at the beginning graduate or advanced undergraduate student, but was written also to serve as a review and, more ambitiously, as a synthesis of the field. To achieve these purposes, several objectives were imposed on the writing. The first was, since ecologists must be the master borrowers of biology, to give the flavor of the eclectic nature of the field by providing coverage of many of the interdisciplinary topics relevant to marine ecology. The second objective was to portray marine ecology as a discipline in the course of discovery, one in which there are very few settled issues. In many instances it is only possible to discuss diverse views and point out the need for further study. The lack of clear conclusions may be frustrating to the beginning student but nonetheless reflects the current—and necessarily exciting—state of the discipline. The third purpose is to guide the reader further into topics of specialized interest by providing sufficient recent references—especially reviews. The fourth objective is to present marine ecology for what it is: a branch of ecology. Many concepts, approaches, and methods of marine ecology are inspired or derived from terrestrial and limnological antecedents. There are, in addition, instructive comparisons to be made among results obtained from marine, freshwater, and terrestrial environments. I have therefore incorporated the intellectual antecedents of particular concepts and some non-marine comparisons into the text.

The plan of this book is to present information on specifics about physiological and populational levels of biological organization in Chapters 1-7. Notions of how populations relate to each other, and their environment, are documented (Chapters 8–9) and so community ecology is introduced. This is followed by Chapters 10–12, where major aspects of the chemistry of organic matter and nutrients in marine ecosystem are developed, based on much of the material of previous chapters. Then, having provided the essentials for understanding the working of various processes in marine ecosystems, the final chapters (Chapters 12–15) dwell on how the structure of marine communities and ecosystems may be maintained over space and time.

Although I am responsible for whatever errors remain, this book has been greatly improved by many people. I have to thank my colleages in Woods Hole, especially John Teal and John Hobbie, for many years of discussion and exchange of ideas. One or more chapters were criticized by Randy Alberte, Karl Banse, Judy Capuzzo, Hal Caswell, Jon Cole, Joseph Connell, Tim Cowles, Werner Deuser, Bruce Frost, Joel Goldman, Charles Greene, Marvin Grosslein, Loren Haury, John Hobbie, Robert Howarth, Michael Landry, Cindy Lee, Jane Lubchenco, Kenneth Mann, Roger Mann, Scott Nixon, Mark Ohman, Bruce Peterson, Donald Rhoads, Amy Schoener, Sybil Seitzinger, Charles Simenstad, and Wayne Sousa.

The graduate students associated with my laboratory during the writing of this book have served as a critical sounding board, and have substantially contributed in many ways. I therefore have to acknowledge the contributions of Gary Banta, Donald Bryant, Robert Buchsbaum, Nina Caraco, Charlotte Cogswell, Joseph Costa, Cabell Davis, William Dennison, Kenneth Foreman, Rod Fujita, Anne Giblin, Jean Hartman, Brian Howes, Alan Poole, Armando Tamse, Christine Werme, David White, and John Wilson. All of them have helped in some fashion with this text, especially Kenneth Foreman and Anne Giblin, who read and criticized most of the chapters. Virginia Valiela did much of the work on the index. Sarah Allen provided technical help throughout the writing of this book, and Jean Fruci was invaluable in helping put together the final manuscript. Lastly, I want especially to thank Virginia, Luisa, Cybele, and Julia Valiela for putting up with me while I was writing this book and my parents for providing a learning environment long ago.

Most of this text was written at the Boston University Marine Program, Marine Biological Laboratory, in Woods Hole. Arthur Humes and Richard Whittaker, Directors of BUMP, were always helpful and provided the time and academic environment in which to put this book together. Dorothy Hahn, Mark Murray-Brown, and Dale Leavitt patiently converted my endless sheets of illegible scribbles into neat piles of readable word processor output. I owe thanks also to Jane Fessenden and her staff, especially Lenora Joseph and Judy Ashmore, at the MBL Library for ever-ready help. The drafting skill of Laurie Raymond is obvious in the illustrations, and her sharp eye for errors was invaluable.

A necessary and stimulating stint of writing took place during a leave of absence at the Department of Oceanography, University of Washington. Karl Banse, Bruce Frost, Mike Landry, Amy Schoener, and the oceanography graduate students were hospitable and provided stimulation for my writing.

Contents

Part V Structure of Marine Communities

Part I

Primary Producers in Marine Environments

A coccolithophorid, *Syracosphaera syrausa*. Photo courtesy of Susumo Honjo.

All that takes place biologically in marine ecosystesm—75% of the earth's surface—is based on the activity of organisms that use light or chemical energy, plus carbon and other essential elements to produce organic matter. These organisms are the primary producers, the main subject of Part I. The major kinds of producers, the processes by which production takes place, the methods by which production may be measured, and comparisons of rates of production in various environments are included in Chapter 1. The rate of primary production varies over time and space; the major factors that directly influence rates of primary production are the subject of Chapter 2. This second chapter concludes with an examination of how the physically determined movements of water mediate the effects of limiting factors and effectively establish the regional patterns of primary production.

Chapter 1

Producers and Processes Involved in Primary Production

1.1 The Kinds and Amounts of Primary Producers in the Sea

There are many kinds of marine organisms that fix inorganic carbon into organic compounds using external sources of energy. There are many thousands of producer species with widely different phyletic origins (Taylor 1978), including single-celled phytoplankton and bottom-dwelling benthic algae, large many-celled macroalgae or seaweeds, symbiotic producers such as corals, and higher plants such as seagrasses.

The many species of phytoplankton may contribute 95% of marine primary production (Steeman Nielsen 1975). This importance of phytoplankton is largely due to the very large area of the earth that is covered by the open sea. In shallow water near coasts, attached single-celled algae, larger multicellular algae, and vascular plants make considerably more important contributions to marine primary production.

1.11 Phytoplankton

1.111 Principal Taxonomic Groups

The principal taxa of planktonic producers over most of the world's oceans are diatoms, dinoflagellates, coccolithophorids, silicoflagellates, and blue-green and other bacteria.

Diatoms (Fig. 1-1, top row) have cell walls of silica and pectin, often with complex sculptured surfaces and range in size from 0.01 to 0.2 mm. They occur floating in the water column or attached to surfaces, singly or

Figure 1-1

as chains of cells. Diatoms are major contributors to production, especially, as discussed below, in coastal waters. Diatoms occur everywhere in the sea, but are most abundant in colder, nutrient-rich waters. Division occurs by fission, and each fission is accompanied by a reduction in the size of the cell and silicious cell wall; after several divisions the cells reach some lower limit in size. At that point the diatoms leave their silicious cell wall and the bare cells grow in size before forming a new cell wall.

Dinoflagellates (Fig. 1-1, left bottom) occur as single cells, either as naked cells or within cellulose cell walls, and range in size from 0.001 to 0.1 mm. They are planktonic, drifting with the water, although many species use flagella to move. Dinoflagellates are only second to diatoms in contributing to primary production, and are widespread over the sea. Metabolically they are very versatile: they may engage in photosynthesis, be parasitic or symbiotic, absorb dissolved organic matter, or ingest particles of organic matter. Reproduction in dinoflagellates is by division, and daughter cells grow to the size of the parent before dividing.

Coccolithophorids (Fig. 1-1, bottom center) occur as single cells and are protected by ornate calcareous plates (coccoliths) embedded in a gelatinous sheath that surrounds the cell. They are most abundant in warm, open-ocean waters, although sometimes abundant nearer shore. Coccolithophorids photosynthesize and may also absorb organic matter. Individuals may form cysts, from which spores develop to produce new individuals. Many species are flagellated.

Silicoflagellates (Fig. 1-1, right bottom) occur as single small (0.06-mm), flagellated cells. The silicoflagellates typically secrete a silicious outer skeleton. These organisms are photosynthetic, but some may consume organic matter. They are common in cold, nutrient-rich water, and reproduce by simple division.

Blue-green bacteria are photosynthetic procaryotes with cell walls made of chitin; they occur as single, very small cells or as longer filaments (Fig. 1-4, bottom). The latter may occur in bundles of many cells. Other groups of bacteria may be photosynthetic but are often restricted to waters with low oxygen content. The various types of bacteria may exist as single freefloating cells, as single cells attached to surfaces, or as long filaments composed of many cells. Most marine bacteria are motile and gram-negative, and divide by fission.

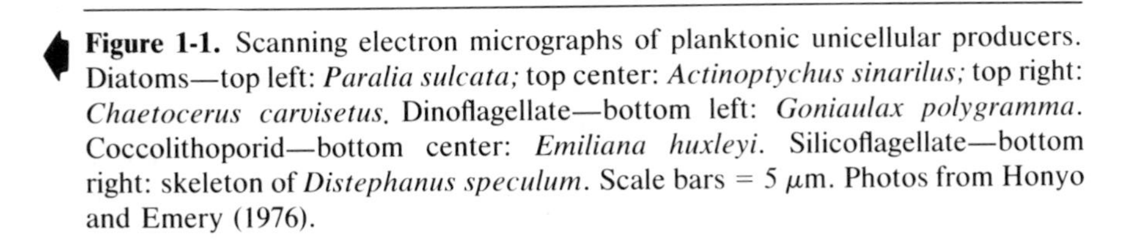

Figure 1-1. Scanning electron micrographs of planktonic unicellular producers. Diatoms—top left: *Paralia sulcata;* top center: *Actinoptychus sinarilus;* top right: *Chaetocerus carvisetus.* Dinoflagellate—bottom left: *Goniaulax polygramma.* Coccolithoporid—bottom center: *Emiliana huxleyi.* Silicoflagellate—bottom right: skeleton of *Distephanus speculum.* Scale bars = 5 μm. Photos from Honyo and Emery (1976).

1.112 Sizes of Producers in the Plankton

The size of cells of bacteria and algae is an important feature of marine plankton, as will be seen below, and diameters of cells vary over several orders of magnitude. Dussart (1965) and Sieburth et al. (1978) devised convenient subdivisions within the range in sizes of phytoplankton. Producers smaller than 2 μm—mainly bacteria and blue-greens—are ultrananoplankton. Small phytoplankton between 2 and 20 μm are nanoplankton—including diatoms, coccolithophores, and silicoflagellates. Cells 20–200 μm in diameter are microplankton, and diatoms and dinoflagellates are common in this size class.

The role of the various size fractions in planktonic production needs further study and is discussed further in Chapter 15, but recent work has pointed out that the smallest size classes may contribute significantly to primary production. For example, in the eastern tropical Pacific particles less than 1 μm in diameter contribute 25 to 90% of the biomass and 20 to 80% of the carbon fixation (Li et al. 1983).

1.113 Abundance of Various Groups of Phytoplankton

The taxonomic composition of the phytoplankton of a given body of seawater varies over time and space. There are many instances of localized short-term blooms, where one taxon may temporarily be very abundant such as in the so-called red tides of estuaries and coastal waters, caused by rapid growth of red-pigmented dinoflagellates (*Gonyaulax*, *Gymnodinium*, and others).

There are, however, some notable and regular regional patterns. Phytoplankton abundance decreases along coastal to oceanic gradients. Den-

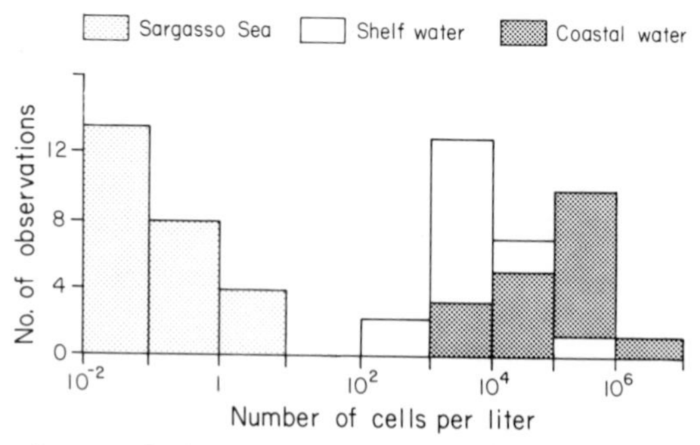

Figure 1-2. Abundance of algal microplankton cells in near-surface waters of three areas of the Atlantic [from data of Hulburt (1962)]. The Sargasso Sea represents truly oceanic waters, while the other two sites show the situations over the continental shelf and near-shore coastal waters. The bars are not plotted cumulatively. The Sargasso Sea is located within the northern semitropical gyre of the Atlantic Ocean.

sities of microplankton reaching several thousand algal cells per milliliter are common near shore and on the continental shelf, while in the ocean it is more typical for algal microplankton to average a few to 100 cells ml^{-1} (Figs. 1-2 and 1-3).

The fertility of seawater also affects taxonomic composition. Diatoms, for example, often dominate nutrient-rich near-shore stations, while the relative abundance of coccolithophorids and dinoflagellates increases in nutrient-poor oceanic waters offshore (Fig. 1-3). The relative dominance of one or another taxon in the water column has many important consequences, since, for instance, phytoplankton dominated by coccolithophores provides only small particles for grazing animals or may produce a calcareous ooze in the sediments on the seafloor below. A diatom-dominated phytoplankton, in contrast, may furnish larger particles and a silicious ooze below. These are just two of many possible consequences.

In oceanic waters, colonial or at least aggregated phytoplankton cells can be found. Visible clumps of cells of the blue green bacterium *Oscilla-*

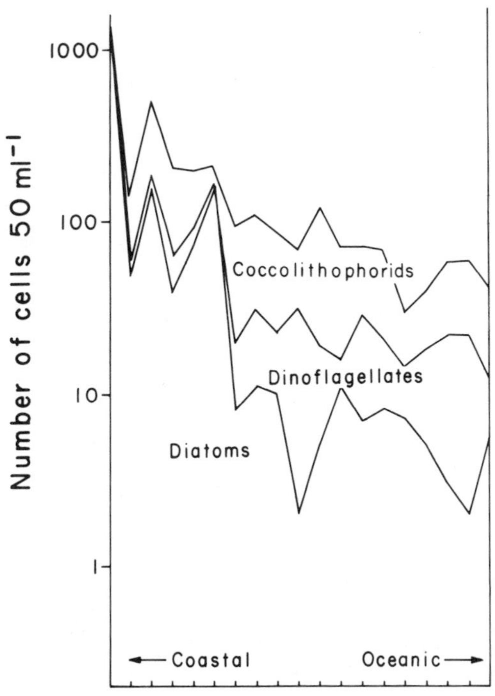

Figure 1-3. Change in composition of microphytoplankton in a series of stations (shown as tics) in a transect from the coast of Venezuela out to the Caribbean Sea. Note that abundance is expressed logarithmically and that the data are graphed cumulatively, so that abundance of each of the three groups is given between the lines. From data of Hulburt (1962).

Figure 1-4. Top: Water surface of the Sargasso Sea. The light-colored rafts are made up of the brown alga *Sargassum* and associated organisms. Bottom: Tufts of filaments of the blue-green *Oscillatoria*, common in the Sargasso Sea. The width of the filaments is about 10 μ. Photo by Natalie Pascoe.

toria (also called *Trichodesmium*) are common in surface waters of the Sargasso Sea, Caribbean, and other nutrient-poor areas (Figs. 1-4, bottom). Mats of the diatom *Rhizosolenia* are frequent in the tropical Atlantic (Carpenter et al. 1977). Larger multicellular algae—referred to as macroalgae—may be locally abundant in the plankton. Species of the brown alga *Sargassum* are common in surface waters of the tropical Atlantic (Fig. 1-4, top). In certain estuaries, free-floating fronds of the brown macroalga *Ascophyllum nodosum* may be dense (Chock and Mathieson 1976). Rafts of a red alga (*Antithamnion*) are found off the coast of Australia (Womersley and Norris 1959).

1.12 Benthic Producers

1.121 Sand and Mud Bottoms

Mixtures of algae and vascular plants can often be found growing in or anchored to sand and mud bottoms in shallow waters. Unicellular or filamentous algae may be very abundant on the surface of sediments or rocks and on fronds and leaves of other producers. Diatoms are often dominant in algal mats common in shallow water environments (up to 40 $\times 10^6$ diatoms cm^{-2}) (Grontved 1960) and may occur as single cells or in colonial arrangements (Fig. 1-5). Green, red, and brown algae, flagellates, and blue-green and photosynthetic bacteria can also be very abundant.

Marine vascular plants, with only about 60 species, are far less diverse taxonomically than phytoplanktonic taxa, but are widespread in coastal sands and muds. In cold and temperate waters, stands of *Zostera* (eelgrass), *Phyllospadix* (surfgrass), *Ruppia* (widgeon grass), and other species can cover extensive subtidal and intertidal areas (Fig. 1-6). *Thalassia* (turtle grass), *Posidonia*, and a variety of other genera occur in soft sediments of warmer waters. The taxonomy and natural history of sea grasses are reviewed in Den Hartog (1970).

Salt marshes fringe the intertidal zone of the muddy or sandy coasts of estuaries and protected shores in temperate and cold latitudes. The primary producers in salt marshes vary regionally; for example, in the coasts of the northeastern North America (Fig. 1-7), the major taxa are vascular plants in the genera *Spartina*, *Distichlis*, *Puccinellia*, *Carex*, *Juncus*, and *Salicornia*. The lowest part of the intertidal belt in these marshes is characteristically populated by attached or loose fronds of the brown algae *Ascophyllum* or *Fucus*. Tall (up to 3 m) plants of *Spartina alterniflora* are found slightly higher in the intertidal range. Shorter plants of *S. alterniflora* are present farther up the tidal range, while higher zones are dominated by other grasses (*Spartina patens* and *Distichlis spicata*). Elsewhere in the world, the species present and the vertical zonation of plants

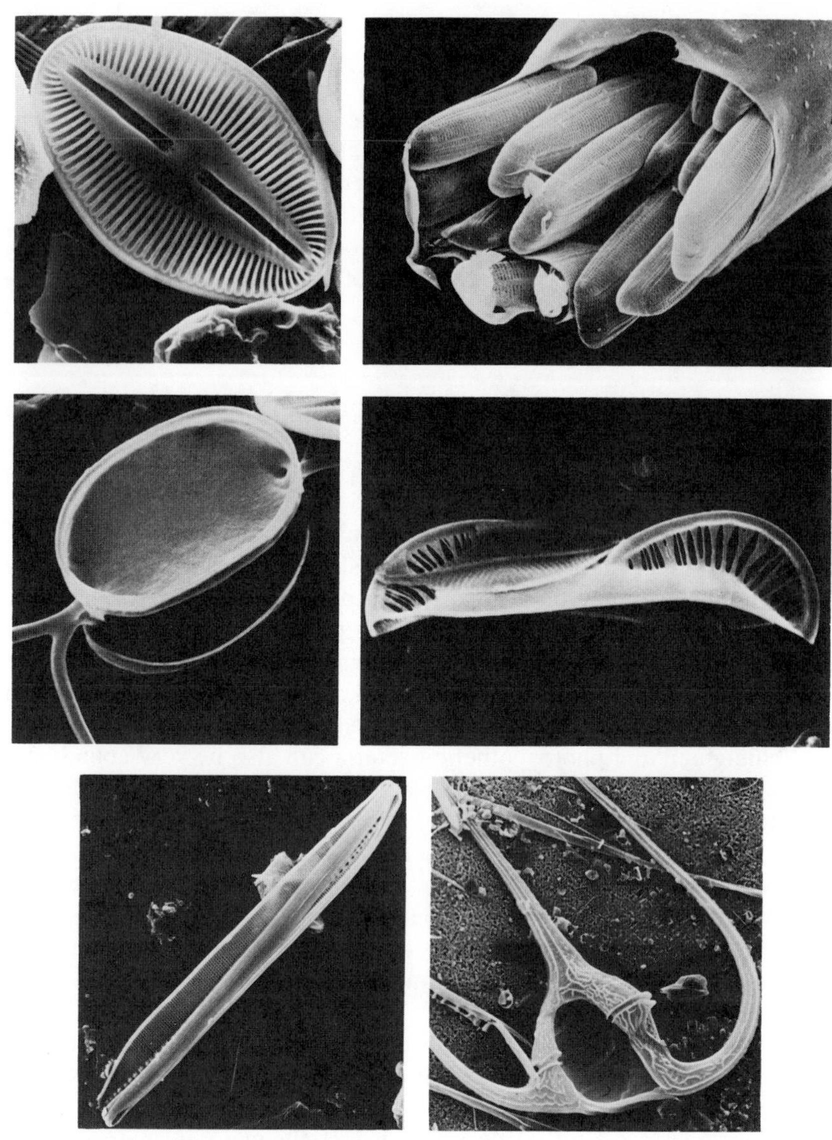

Figure 1-5. Scanning electron microphotographs of unicellular producers associated with the benthos: Diatoms (top left, *Navicula heufleri*, ×4300; top right, *Amphipleura elegans*, each cell about 5 μm wide; center left, *Chaetoceras*; center right, *Amphiprora*, ×5100; bottom left, *Nitzschia*, ×4300) and dinoflagellate (bottom right, *Ceratium* skeleton). Such algae are not obligate benthic dwellers and may be found in the water column; *Ceratium*, for example, may be more commonly planktonic, and skeleton or spores may sink to the benthos. *A. elegans* is a common colonial diatom found in salt marshes and lives within tubes large enough to be visible. Photos by Stjepko Golubic and his students, except that for *A. elegans* by Rod Fujita.

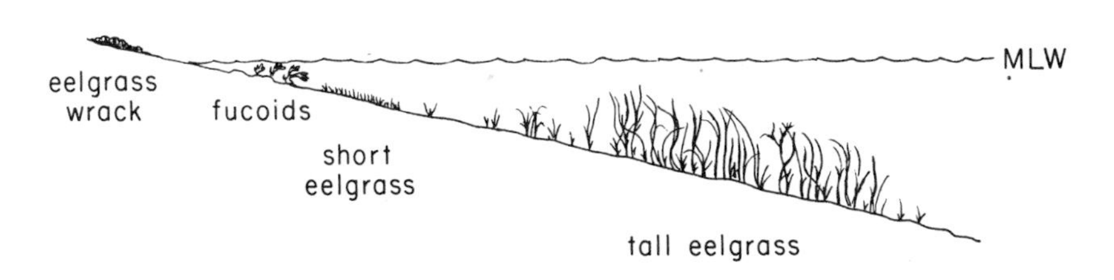

Figure 1-6. Top: Intertidal seagrass (*Zostera marina*) flat in Izembek Lagoon, Alaska. Photo by William Dennison. Middle: Underwater view of *Z. marina* stand. Photo by R. Phuillips. Bottom: Profile of subtidal *Zostera* bed in Nova Scotia, Canada. Adapted from Harrison and Mann (1975). © Canadian Journal of Fisheries and Aquatic Sciences, reprinted by permission.

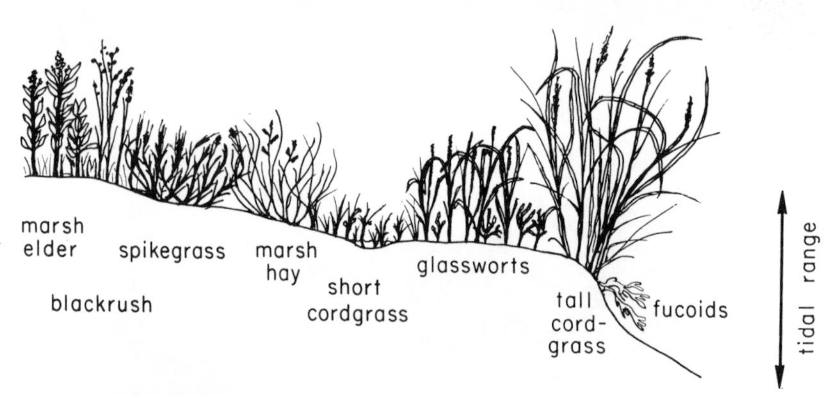

Figure 1-7. Salt marshes of New England, United States. Top: Tidal creek through an area dominated by the cordgrass (*Spartina alterniflora*). Bottom: Diagrammatic section through a New England salt marsh, corresponding to a transect from the upland on the left top of the photo to the tidal creek. The species included are (in the same sequence as in the diagram, left to right): *Iva frutescens, Juncus gerardi, Distichlis spicata, Spartina patens, Salicornia europaea, Spartina alterniflora*. The fucoids include *Fucus* and *Ascophyllum*. Drawing by Margery Taylor.

Figure 1-8. Mangrove swamps on the coast of Florida, United States. Top: Seaward edge of the mangrove, showing prop roots of the red mangrove (*Rhizophora mangle*). Bottom: Diagrammatic profile of a mangrove swamp, including, in sequence, *Conocarpus, Avicennia* (with respiratory organs, the pneumatophores, emerging vertically from the roots), and *Rhizophora*. The viviparous seedlings of the red mangrove have a long root and are shown hanging from the parent trees in the lower figure. The seedlings fall to the water and usually float off to establish new stands. The diagram has been extensively modified from Davis (1940).

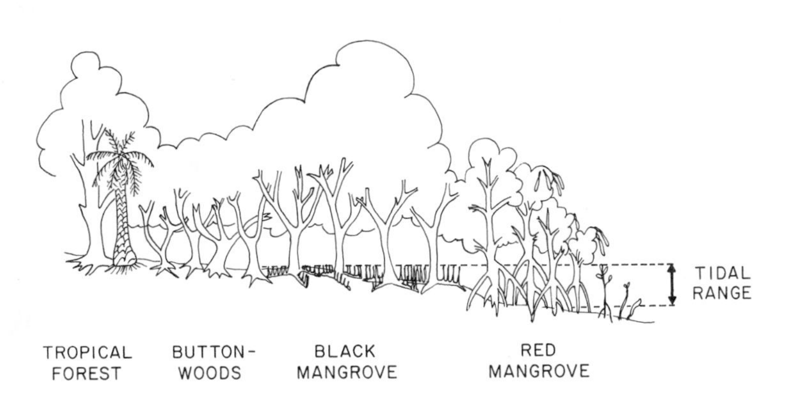
TIDAL
RANGE
TROPICAL
FOREST
BUTTON-
WOODS
BLACK
MANGROVE
RED
MANGROVE

vary markedly; Chapman (1977) and Frey and Basan (1978) review some of these aspects of salt marsh vegetation.

Mangrove swamps (Fig. 1-8) replace salt marshes in protected coastlines with soft sediments in tropical latitudes (Chapman 1977). In the American continent, the mangroves are mainly *Rhizophora* (red mangrove) and *Avicennia* (black mangrove). There are often well-defined horizontal belts of these species, with *Rhizophora* usually found lower in the intertidal range than *Avicennia* or other mangrove species. Mangrove swamps are best developed in Indomalasia, where canopies 10–20 m high are not unusual.

1.212 Rocky Bottoms and Reefs

On rocky coastlines, we have the greatest variety of macroalgae found in the sea as well as many kinds of unicellular benthic algae. As in the case of mangrove swamps and salt marshes, there is usually a distinct zonation of producers in the subtidal and intertidal distribution of species. This zonation is extremely variable from one geographical location to another, but within a given site the zones are clearly demarcated (Fig. 1-9). Brown algae are prominent in the various belts of vegetation, notably kelp such as *Macrocystis*, *Nereocystis*, *Laminaria*, and *Eklonia*, and fucoid rockweeds such as *Fucus* and *Ascophyllum*. There are also many red (*Chondrus*, *Gracilaria*, among many others) and green algae (*Enteromorpha*, *Ulva*, and others). Fucoids dominate the intertidal zone. The subtidal vegetation is usually dominated by kelp attached by holdfasts to the rocky bottom; the stipes and fronds of many kelp extend upward, often reaching the sea surface. These kelp may be found from the low tide mark to a depth of about 30 m. In gently sloping shores, this results in broad bands of kelp parallel to the coastline. These bands are referred to as kelp beds or forests (Fig. 1-10) and occur on rocky coasts* where seawater is relatively cold (Mann 1973).

Coral reefs are found in shallow depths in tropical regions (Fig. 1-11). Reef-building corals are primary producers, with photosynthesis carried out by symbiotic dinoflagellates called zooxanthellae living within the tissues of the coral (cf. Chapter 5). In coral reefs and elsewhere, there are also heavily calcified red coralline algae (Dawson 1966) that at times are responsible for a major portion of the reef-building activity. Coral reefs also show zonation, and the common occurrence of such stratification of species of coastal communities suggests that there are very powerful and ubiquitous ecological processes—to be discussed later—that determine where certain species may exist.

---

* The rocky coasts of Antarctica are uniquely devoid of kelp (Laminariales); they are replaced there by kelp-like macroalgae of the Desmarestiales, which form thickets rather than forests (Moe and Silva 1977).

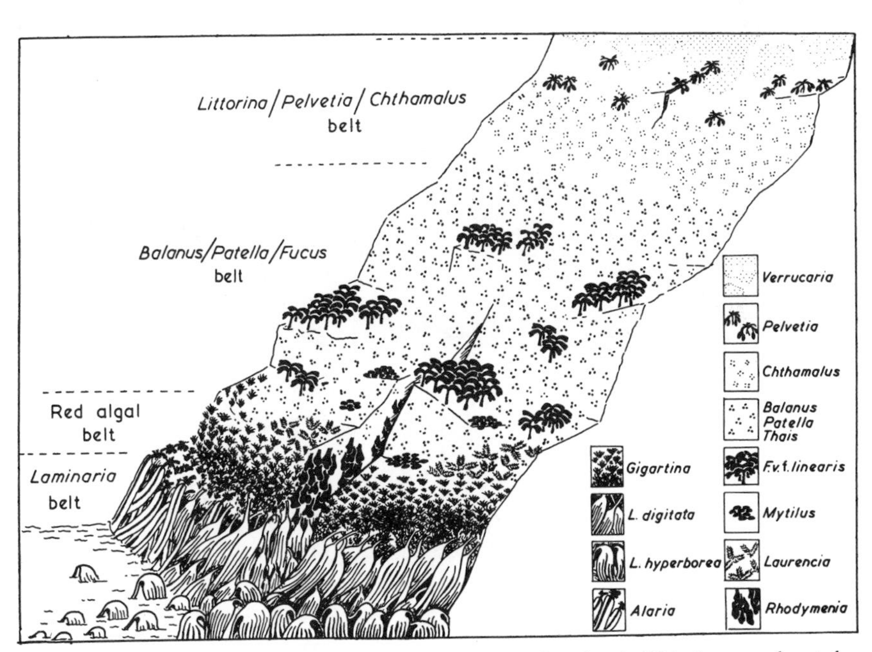

Figure 1-9. Top: Rocky shore on Sound of Jura, Scotland. This is a moderately exposed shore. Bottom: Diagram of rocky shore common in Scotland and Northern Ireland. Both producers and some animals are included in the diagram. (F.v.f: *Fucus vesiculosus* form *linearis*). The tidal range is about 3-4 m. Adapted from Lewis (1964).

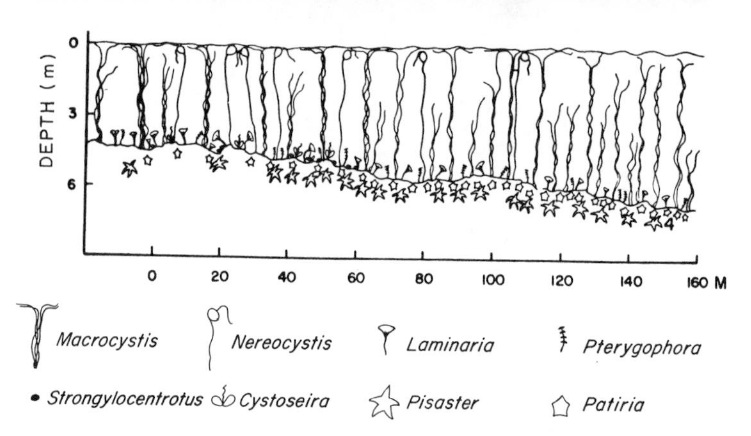
DEPTH (m)
0
3
6
0
20
40
60
80
100
120
140
160 M
Macrocystis
Nereocystis
Laminaria
Pterygophora
Strongylocentrotus
Cystoseira
Pisaster
Patiria

Figure 1-10. Kelp forests. Top: Surface view of a *Macrocystis* stand near Oudekraal, South Africa. Photo by John Field. Middle: View of a kelp forest off Catalina Island, California, United States. The canopy is formed by *Macrocystis pyrifera* with an understory of *Eisenia arborea*. Photo by M. S. Foster. Bottom: Diagrammatic section through kelp forest off Point Santa Cruz, California. Adapted from Yellin et al. (1977).

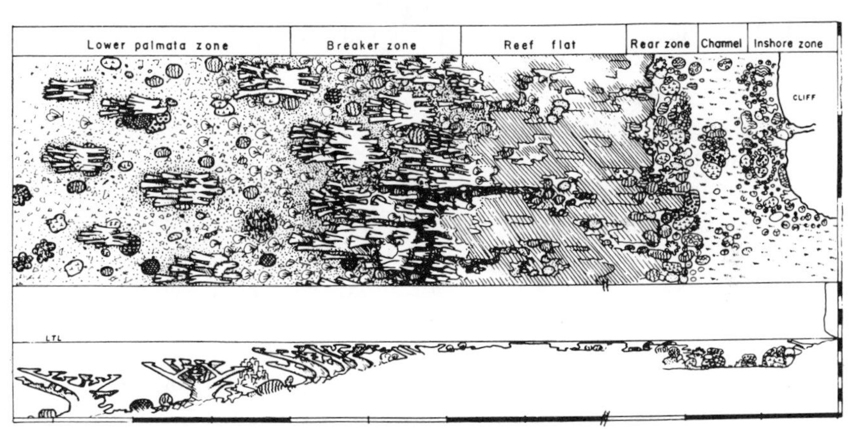

Figure 1-11. Coral reefs in the Caribbean. Top left: The coral *Acropora palmata* on the reef slope where waves approach from the left. Note the high degree of relief and percentage of the area that is covered by the coral. Photo by James W. Porter, Discovery Bay, Jamaica. Top right: The branching coral *A. palmata* grows over the massive head coral *Montastrea annularis*. The shading results in death of the head coral. Photo by James W. Porter, Guadelupe. Bottom: Diagrammatic plan and lateral view of coral reef, Ocho Rios, Jamaica. Size of corals not to scale. Horizontal distances are marked in 10 m intervals; vertical distance in 1 m intervals; vertical distance in 1 m intervals. The different symbols refer to different species. From Gorean (1959). © Ecological Society of America, reprinted by permission.

1.2 Production: The Formation of Organic Matter

Autotrophic organisms, including algae, some bacteria, and plants, produce energy-rich organic compounds from water and carbon dioxide. Autotrophs that use light as the energy source with which to fix carbon are termed photosynthetic organisms, while organisms that use energy stored in inorganic compounds, such as H_2S, methane, ammonia, nitrite, sulfur, hydrogen gas, or ferrous iron, are termed chemosynthetic.

1.21 Photosynthesis

Carbohydrates and other organic compounds are synthesized by photosynthetic autotrophs through the activity of the photosynthetic apparatus and associated biochemical pathways. The basic photosynthetic equation can be expressed as

$$CO_2 + 2H_2A \xrightarrow[\text{pigments}]{\text{light}} CH_2O + 2A + H_2O. \qquad (1\text{-}1)$$

The actual details of photosynthesis are far more complicated than those that appear in Eq. (1-1) (see review by Govindjee 1976). Photosynthesis is a multistep process comprised of two independent series of reactions. The "light" reactions take place only when light is available and depend on the capture of photons by the photosynthetic pigments. In this process an electron donor, H_2A, is split, liberating two electrons. In the case of oxygenic photosynthesis, the electron donor is water, and two H_2O are split to form an O_2 molecule and four protons. The energy in the excited electrons released in this reaction is transferred by a series of oxidation-reduction reactions involving various electron "carriers," to produce ATP and a strong reductant, $NADPH_2$.

Photosynthetic bacteria carry out anoxygenic photosynthesis in the anoxic environments in which they live. These organisms are relics of the time when the ancient atmosphere of the earth lacked oxygen; light is the source of energy but the electron donors are reduced inorganic compounds such as hydrogen sulfide or thiosulfate (in purple and green sulfur bacteria) and organic compounds (in the purple nonsulfur bacteria) (Table 1-1) (Fenchel and Blackburn 1979).

The ATP and $NADPH_2$ provided by the light reactions is used to reduce CO_2 into complex organic molecules. These reactions are not light dependent and hence are called the "dark" reactions.

There are two major pathways of carbon fixation during the dark reactions. In the "C_3" pathway the first product of carbon fixation is a 3-carbon compound called 3-phosphoglyceric acid. Primary producers that manufacture this 3-carbon compound as an end product of carbon fixation are referred to as C_3 organisms. In the "C_4" pathway a 4-carbon com-

Table 1-1. Major Electron Donors, Acceptors, and End Products for the Three Major Types of Primary Production[a]

	Electron donor (reductants)	Electron acceptor (oxidants)	Oxidized end products
Photosynthesis			
Oxygenic	H_2O	CO_2[b]	O_2
Anoxygenic	H_2S, H_2	CO_2[b]	S^0, SO_4^{2-}
Chemosynthesis			
Nitrifying bacteria	NO_2, NH_4^+, NH_2OH	O_2, NO_3^-	NO_3^-, NO_2
Sulfur bacteria[c]	H_2S, S^0, $S_2O_3^-$	O_2	S^0, SO_4^{2-}
Hydrogen bacteria[c]	H_2	O, SO_4	H_2O
Methane bacteria[c]	CH_4	O_2	CO_2
Iron bacteria[c]	Fe^{2+}	O_2	Fe^{3+}
Carbon monoxide bacteria[c]	CO	H_2	CH_4

[a] From Fenchel and Blackburn (1979) and Parsons et al. (1977). There are many other possible chemosynthetic reactions and end products (see Tables 10-7, 10-8).
[b] Takes place if light furnishes the large amounts of energy needed to reduce the CO_2.
[c] These groups may also live heterotrophically, using a variety of organic compounds manufactured by other organisms as sources of energy (or electron donors), and with CO_2, H_2O, or more oxidized organic compounds as the end products.

pound (malic or aspartic acid) is the first end product of fixation. The malate is decarboxylated, and the CO_2 thus released is fixed through the C_3 pathway (Hatch and Black 1970). There are important physiological differences associated with C_3 and C_4 metabolism, many of which have ecological significance (cf. Section 6.32). C_4 metabolism has been demonstrated in vascular plants but not in phytoplankton.

The organic compounds produced by photosynthesis may be stored or used immediately. The energy contained in the organic compounds is made available by a series of oxidative reactions. This oxidation is called dark respiration, and is the process that provides energy to sustain metabolic needs. The complete oxidation of glucose to carbon dioxide and water, for example, yields 36 ATP:*

$$C_6H_{12}O_6 + 6O_2 \rightarrow 6CO_2 + 6H_2O + 36ATP. \quad (1\text{-}2)$$

* Respiration of course occurs in all organisms. In animals or many microbes, ingested or absorbed carbon compounds serve as the principal substrate for respiration (cf. Sections 7.3 and 10.1232). Organisms whose metabolism is based on organic compounds fixed by autotrophs are called heterotrophs. This is in contrast to autotrophs, defined as organisms able to use CO_2 by reductive assimilation to supply carbon requirements.

Such oxidation-reduction reactions are not light dependent and in primary producers provide the sole source of energy during periods when photosynthesis is not active. Glucose is initially broken down by the process of glycolysis to produce NADH and ATP. The electrons in these reduced carriers and those produced by the Krebs cycle are donated to the electron transport chain where they are coupled to phosphorilation of ribulose. The amount of dark respiration by producers relative to photosynthesis varies and is difficult to measure in the sea because rates are low and also because bacteria and small zooplankton also respire in the seawater where the producers are suspended.

In C_3 plants there is an additional light-dependent respiration activity—referred to as photorespiration—that consumes oxygen and produces carbon dioxide (Tolbert 1974; Harris 1980). Under high partial pressures of oxygen, the apparent rate of photosynthesis by plant cells declines. When the intracellular concentrations of O_2 are high, one of the enzymes responsible for CO_2 fixation in the C_3 dark reactions—ribulose-bisphosphate carboxylase—can also catalyze the oxidation of a Calvin cycle intermediate, initiating a series of reactions that produce CO_2 and glycine as end products. Hence, when the splitting of water during the light reaction furnishes oxygen, photorespiration takes place. It has proven difficult to ascertain the degree to which these reactions occur, and there is much contradictory information on the relative rates of photorespiration and photosynthesis by marine algae and angiosperms (Banse 1980).

The generalized equation for photosynthesis [Eq. (1-1)] is incomplete, because producers require a variety of inorganic nutrients to provide the building blocks for the synthesis of the many compounds present in cells. A more inclusive equation would be

$$\begin{aligned}&1{,}300 \text{ kcal light energy} + 106 \text{ mole } CO_2 + 90 \text{ mole } H_2O\\&\quad + 16 \text{ mole } NO_3 + 1 \text{ mole } PO_4 + \text{small amounts of mineral}\\&\quad \text{elements} = 3.3 \text{ kg biomass} + 150 \text{ mole } O_2 + 1{,}287 \text{ kcal heat.}\end{aligned} \tag{1-3}$$

The biomass yield of Eq. (1-3) would contain on the average 13 kcal of energy, 106 g C, 180 g H, 46 g O, 16 g N, 1 g P, and 825 g of mineral ash. These values are based on the average contents of phytoplankton cells (Odum 1971), but there is considerable variation in these elemental ratios. Although more comprehensive than Eq. (1-1), Eq. (1-3) still gives a very simplified picture of biomass production by producers. For example, phytoplankton may use ammonium rather than nitrate to satisfy their nitrogen requirement, and in the case of diatoms, substantial amounts of silica are required for growth since this element is a major component of their cell wall. Despite the intricacy of photosynthetic processes, the efficiency in capture of light energy is low. Only 0.4 to 0.9% of incident light energy is converted into energy stored in cells (Platt and Subba Rao 1975).

1.22 Chemosynthesis

Chemosynthesis is carried out by bacteria that obtain energy from simple inorganic compounds (Table 1-1). The biochemical pathways of different kinds of chemosynthetic bacteria are very diverse (Fenchel and Blackburn 1979; Parsons et al. 1977). The basic reaction of chemosynthesis is is

$$nH_2A + nH_2O \rightarrow nAO + 4n[H^+ + e^-], \quad (1\text{-}4)$$

where H_2A represents a relatively reduced inorganic compound. Dehydrogenase converts this reduced compound to the oxidized end product, AO, and the reducing power gained is represented as $[H^+ + e^-]$. Some of the reducing power gained through Eq. (1-4) is devoted to energy production such as synthesis of ATP via transfer through the cytochrome system to O_2. Most chemosynthetic bacteria thus require free oxygen to function as the electron acceptor. Obligate anaerobic or facultative bacteria can use oxygen bound in nitrate or sulfate for this purpose. The rest of the reducing power obtained through Eq. (1-4) may be used to reduce NAD to $NADPH_2$. The ATP and $NADPH_2$ can then be used to assimilate CO_2 and make organic matter:

$$12NADH_2 + 18ATP + 6CO_2 \rightarrow C_6H_{12}O_6 + 6H_2O + 18ADP + 18P + 12NAD. \quad (1\text{-}5)$$

Chemosynthetic bacteria thus obtain both energy and reducing power from inorganic compounds (Table 1-1). Since the inorganic compounds used as electron donors are not assimilated by the bacteria, these reactions have been called dissimilative reactions, in contrast to assimilative reactions whose purpose is the uptake of elements or compounds.

The organic matter resulting from the carbon fixed as CO_2 by the chemosynthetic bacteria can be considered as truly primary production. The energy obtained from inorganic compounds and elements by chemosynthesizers was, however, fixed originally by photosynthesis in other organisms. The process of chemosynthesis therefore may not yield completely new primary production.

In this book, photosynthetic processes have been given the spotlight. This is justified in view of the extent to which photosynthesis today produces organic matter and so determines many of the dynamics of contemporary food webs. The book might have been organized differently, giving precedence to bacterial chemosynthesis. This latter approach would be justified because of the pervasive biogeochemical influence of chemosynthesis, as will be seen in Chapters 10 and 11, and from an evolutionary standpoint. Chemosynthetic transformations took place very early in the history of the earth. Photosynthesis, in fact, may be thought of as a derived process carried out by organisms that have access to light to provide the large amounts of energy needed to reduce CO_2, and have belatedly had O_2 to provide the needed electrons to complete the process of fixing CO_2 and energy.

1.3 The Measurement of Producer Biomass and Primary Production

1.31 Phytoplankton

1.311 The Measurement of Algal and Microbial Biomass

The most frequently used method to estimate the standing crop of phytoplankton is to extract and measure the amount of chlorophyll *a* in seawater samples containing algae (Strickland and Parsons 1968; Holm-Hansen and Rieman 1978). The chlorophyll concentration can be used directly or, if desired, the algal standing crop can be obtained using a calculated ratio between weights of chlorophyll and biomass. This ratio averages about 62 and varies between 22 and 154. The species composition of the sample, extent, and variability of light adaptation, age, and nutritional state of the cells all affect these ratios.

More recently the content of ATP of phytoplankton has been used to estimate standing crop (Holm-Hansen and Booth 1966). The ATP values are extrapolated to total carbon by multiplying by a derived factor of 250, although the ATP content of cells is variable due to the influence of several environmental and growth factors (Holm-Hansen 1970; Banse 1977; Banse and Mosher 1980).

Hobbie et al. (1972) compared several methods of estimating microbial and algal biomass in seawater obtained from nutrient-poor and nutrient-rich stations in the Atlantic from the surface to a depth of 700 m. There were relatively high correlations among measurements of phytoplankton and organic carbon, volume of particulates, concentrations of ATP, DNA, and chlorophyll *a*. Comparisons among these correlated variables, however, are not straightforward, since, for example, not only algal but also fungal and bacterial biomass are included in the organic carbon, particulate volume, ATP, and DNA values. Organic carbon, particulate volume, and DNA may include nonliving materials. Knowledge of the ratio of algal abundance to that of other microbes or detritus would be needed to calculate phytoplankton biomass using these other variables. Hobbie et al. (1972) made an effort to obtain samples of water whose nutrient content ranged from rich to very poor. Their values of standing crops therefore varied over a broad range with high correlations over that range. In any one study of a more local nature, the range of concentration of organisms would likely be smaller, variation of the samples would be proportionally greater, and the correlation between methods could be considerably reduced.

All indirect methods to estimate phytoplankton biomass need to be used cautiously, but if the limitations are kept in mind and if they are applied with discrimination, these procedures can be useful.

1.312 Measurements of Production *in Situ*

The rate of primary production might be measured by determining the change over time of any of the components of Eq. (1-2), for example, the disappearance of phosphate or nitrate from the water column over a time interval. One major problem with this approach is that no natural water body is static and inhabited solely by producers. It is therefore difficult to ensure that the observed changes in nutrient concentrations are not due to movement of water in or out of the study area, or to uptake and release by bacteria or animals. Further, nutrients may be quickly released by algal cells after uptake; phosphate, for example, may be retained only a few minutes in phytoplankton cells (Pomeroy 1960). This rapid turnover makes it difficult to determine exactly what the actual rate of disappearance of the specific nutrient actually was.

The changes in oxygen dissolved in water due to photosynthesis and respiration over a given period of time—including night and day—can be used to obtain net production* rates after correction for the diffusion of oxygen at the sea surface. This method has been used in places where there are well-defined water masses such as in fjords (Gilmartin 1964), certain places in the Mediterranean (Minas 1970), in coral reefs (Marsh and Smith 1978), and elsewhere (Johnson et al. 1981).

Carbon dioxide dissolved in water is the most frequent carbon source for marine photosynthesis and diffuses much more slowly than oxygen in seawater. The lowered concentrations due to consumption by photosynthesis thus tend to be measurable over periods of hours, so that CO_2 concentrations provide a relatively stable tracer of photosynthetic activity. Production of CO_2 by respiration of microorganisms and animals is unavoidably included in results obtained by this method (Johnson et al. 1981).

The rate of appearance of new algal biomass over time is a direct estimate of net primary production, and measurement of this is discussed in Section 1.3133. To apply this approach to the sea, the losses of cells through grazing and sinking need to be measured. These are difficult to do, so this method has not been widely used to measure production.

Yet another *in situ* method involves estimating the amount of chlorophyll *a* and relating this to net production using a ratio of production rate

* Gross production minus respiration is referred to as net production.

to chlorophyll content. This ratio, called the coefficient of assimilation or the assimilation number, has been taken to be 3–4, but is not a constant (Curl and Small 1965), and varies from less than 1 to over 20 (Glover 1980). Hence, it is necessary to conduct preliminary studies to estimate the coefficient for the population being studied. Estimates of rates of production can then be calculated routinely from the chlorophyll values using equations that consider the effect of light attenuation in the water column (Ryther and Yentsch 1957). Although subject to all the problems of conversions using ratios, this procedure is attractive because of the simplicity of the chlorophyll *a* determination (Strickland and Parsons 1968). The development of ideas based on these equations has resulted in the development of mathematical models to predict primary production rates under a variety of conditions (Hall and Moll 1975; Kiefer 1980).

1.313 Methods Using Containers

1.3131 Light and Dark Bottle Oxygen Technique

The methods that have received the most attention involve placing seawater in small containers, some of which are transparent and some dark, and measuring the amount of oxygen dissolved in the water before and after some suitable time period has elapsed (Gaarden and Gran 1927). The oxygen method has in the past only been applicable in waters where there was considerable photosynthetic activity. The sensitivity of a standard Winkler titration for oxygen is about ± 0.02 mg O_2 liter^{-1} (Steeman Nielsen 1975), too low to measure changes in oxygen content produced by oceanic phytoplankton within a short period (2–4 hr) using the standard oxygen method. If longer periods of incubation are used, "bottle effects" result, with the walls of the container providing an attachment surface for microorganisms and stimulating growth and oxygen consumption in some as yet undefined fashion. Oxygen electrodes have been used if changes in oxygen due to production by algae are large (Lazplewski and Parker 1973; Kanwisher et al. 1974). Recently very sensitive methods to measure oxygen have been developed so that it is now possible to measure production rates of oceanic phytoplankton using changes in oxygen (Tijssen 1979).

The increase in dissolved oxygen in the light bottle is a measure of photosynthetic activity (P_a) by the algae minus the respiration by algae (R_a), bacteria (R_b), and whatever zooplankton (R_z) were present in the sample; it is not net photosynthesis ($P_n = P_a - R_a$), since it includes R_b and R_z. The decrease in oxygen concentration in the dark bottle is a measure of respiration by all the aerobic organisms present:

$$\text{change in } O_2 \text{ in light bottle} = P_a - R_a - R_b - R_z, \tag{1-6}$$

$$\text{change in } O_2 \text{ in dark bottle} = R_a + R_b + R_z. \tag{1-7}$$

By adding the change in O_2 in the light bottle and the change in O_2 in the dark bottle we get P_g, the gross photosynthesis by the algae over the period of incubation.

The rates obtained from such light and dark bottle measurements can, if desired, be converted to amounts of CO_2 assimilated (Strickland and Austin 1960). The photosynthetic ratios, i.e., the relation between O_2 released and CO_2 assimilated, vary. For example, where the CO_2 is devoted for synthesis of carbohydrates the ratio is 1, while for lipid synthesis it is 1.4. The ratio is 1.05 for protein synthesis if the source of nitrogen is ammonium, and 1.6 or higher if nitrate is the source of nitrogen used to make protein. This difference reflects the fact that the nitrate has to be reduced intracellularly to ammonium prior to incorporation into amino acids. For fieldwork a value of about 1.2–1.4 is usually applied.

1.3132 Carbon-14 Method

The most common way to measure algal production is to measure the rate of uptake of ^{14}C, a radioactive isotope of carbon. A known, small amount of ^{14}C, usually in the form of dissolved $NaH^{14}CO_3$, is added to seawater samples contained in transparent and opaque 100- to 300-ml bottles. The seawater is incubated for a period that may vary from 0.5 to 24 hr, after which the sample is filtered, cells are killed, and the quantity of ^{14}C assimilated is determined by using a liquid scintillation counter to measure the beta-radiation emitted by the ^{14}C incorporated into the cells. The carbon uptake can be calculated as carbon uptake = ^{14}C incorporated into organic form/total ^{14}C added × available inorganic carbon × 1.05.

The available inorganic carbon has to be measured by chemical means and the value of 1.05 is a factor to account for the fact that $^{14}CO_2$ is taken up at a somewhat slower rate than the lighter $^{12}CO_2$. The rates of carbon uptake are subtracted from those of the light bottle if uptake in the light is the desired measurement. Strickland and Parsons (1968) provide further details of the method and Peterson (1980) and Carpenter and Lively (1980) evaluate the present status of the method.

The ^{14}C is taken up by cells; this uptake can be considered gross primary production. During the same time interval, however, the phytoplankton cells respire some of the ^{14}C. In addition, phytoplankton cells release carbon continually, and in seawater many heterotrophic organisms are also usually present in the water sample, and these take up ^{14}C released by phytoplankton. The respiration of these heterotrophs releases ^{14}C, so most actual measurements include respiration by phytoplankton and heterotrophs, and therefore do not exactly measure gross primary production but something closer to net production.

The ^{14}C method has been the most widely used of all the available procedures, and knowledge of the rates of production by phytoplankton over the oceans is basically derived from ^{14}C measurements. At present a critical reassessment of the ^{14}C method is taking place, due to various independent observations that suggest that this method underestimates primary production.* Perhaps the underestimation may be an artifact of the procedure, as suggested by Gieskes et al. (1979), who found that even short (4 hr) incubations in small (300 ml) bottles yielded ^{14}C uptake rates 5–15 times smaller than obtained in a 4-liter bottle.† Most ^{14}C measurements have been done using 300-ml bottles or smaller. Further testing of the effect of container size and other factors is needed. It seems likely that after the current controversy has subsided we may find that in nutrient-rich water, where organisms are less sensitive to manipulation, and production rate is high, the ^{14}C method is adequate. In nutrient-poor oceanic waters, it might be that ^{14}C somewhat underestimates production, but it is not yet possible to say positively whether the reduction is significant.

1.3133 Particle Counting Method

The advent of electronic particle counters, devices that quickly and accurately assay the abundance of suspended particles in a fluid and also record their sizes, has made possible another way to measure primary production. The number of particles in a relatively small sample of seawater is determined for the size classes of living cells in the sample. The samples are then incubated in light and dark bottles. After a few hours, a new measurement of the abundance of the relevant size classes is made. Since most particles in seawater are detrital (Table 1-2), considerable arithmetical manipulations are necessary so that only living cells are included in the measurement. Even so, the method attempts to measure changes of living cells within a very large and variable background of nonliving particles, except perhaps under bloom conditions, when live cells may outnumber other particles (Table 1-2). This is a relatively elaborate method that requires considerable diligence (Cushing and Nichols

* These observations include (a) measurements of changes in O_2, CO_2, and dissolved organic carbon that suggest that CO_2 consumed or O_2 released exceeds primary production as measured by ^{14}C [Shulenberger and Reid (1981) and studies summarized in Johnson et al. (1981)]; (b) measurements of formation of particulate carbon exceeding rates of ^{14}C primary production (Postma and Rommets 1979); (c) estimates of rates of O_2 formation in water masses whose age has been established by tritium and ^{3}He dating (Jenkins 1977), where estimates of O_2 formed exceed rates obtained by ^{14}C; (d) measurement of high primary production (11-56% of total water column production) associated with fragile aggregates that are seldom sampled by usual water sampling methods (G. A. Knauer, personal communication). All of these observations need confirmation and evaluation.

† The bottle size effect has been claimed to be due to enhanced mortality of algal cells through contact with bottle walls, nutrient depletion within the container, toxic effects of trace contaminants in the chemicals used in the ^{14}C method (Fitzwater et al. 1982), changes in the species composition within the container, enhanced bacterial activity, and inhibition of photosynthesis due to constant exposure to high light intensity (Venrick et al. 1977; Gieskes et al. 1979; J. Goldman personal communication). All these factors are more serious in small containers, and all these postulated mechanisms need further study.

Table 1-2. Composition of the Particulate Matter Suspended in Seawater in Selected Areas of the Sea[a]

	Particulate organic matter (mg C m^{-3} or μg C liter^{-1})	Percentage of total particulate organic matter			
		Phyto-plankton	Zoo-plankton	Bac-teria	Detri-tus[b]
Sea of Azov	750–1500	5–10	3–10	0.3–7	80–92
Arabian Sea	100–250	1–31	—	—	—
Black Sea	200–250	0.2–1	5–20	0.4	78–95
Tropical Atlantic					
15th Meridian	450–600	0.5–1.3	0.6	—	98–99
16th Parallel	100–250	0.6–1.3	0.7	—	98–99
Upwelling off					
S.W. Africa	70–900	30–43	4–14	—	9–14
Hudson River Estuary	660–2250	2–72			40–93
New York Bight	200–840	12–51			38–90
Western Baltic Sea	492–505	23–27	33–35	—	41–43
Chesapeake Bay	11.5–84[c]	23	—	—	77
English Channel	950–2500	15–17			
Aberdeen Bay,					
Scotland	200–3400	8–10			
Wadden Sea,					
Netherlands	1000–4000	10–25			
Akkeshi Bay,		9.7	1.7		
Hokkaido			0.1[d]		

[a] Adapted from Finenko and Zaika (1970), Hobson (1971), Chervin (1978), Lenz (1977), Van Valkenberg et al. (1978), Jorgensen (1962), and Hogetsu et al. (1977).
[b] Detritus include attached fungi and bacteria.
[c] Particles ml^{-1} × 10^4. Totals and percents do not include zooplankton.
[d] Microplankton.

1966; Sheldon and Parsons 1967; Parsons et al. 1969; Sheldon 1979), and whose value has not been fully assessed.

1.314 Comparisons Among Methods of Measuring Primary Production in Phytoplankton

Comparisons of different *in situ* methods were obtained off the Peruvian coast by Ryther et al. (1971), who followed drogues that drift with the water mass and so could sample the same parcel of water repeatedly. The algae (primarily diatoms) increased in abundance in the nutrient-rich water, as evidenced by the increase in particulate carbon, phosphorus, and chlorophyll during the first 3 days (Fig. 1-12A).

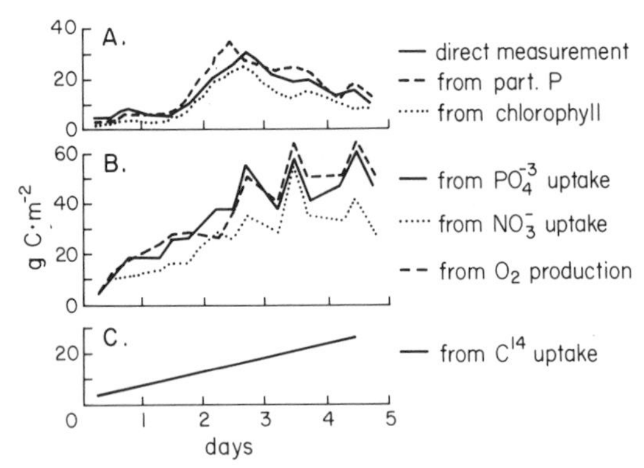

Figure 1-12. Comparison of carbon obtained from various production estimates from 0.50 m in depth in recently upwelled water off Peru. A and B are integrated from the surface to 50 m in depth while C refers to the entire euphotic layer. Adapted from Ryther et al. (1971).

The conversion of chlorophyll and particulate phosphorus measurements to carbon (Fig. 1-12A) provided results that agreed roughly with the carbon measurements. The particulate carbon value derived from chlorophyll measurements are uniformly lower than carbon values measured directly. The conversion is of course dependent on the value of the ratio of chlorophyll to carbon that was used. If the Chl : C ratio used had been 50 : 1, the chlorophyll-derived curve would virtually overlie the carbon curve. Thus it is important, as mentioned before, to procure a good estimate of the conversion ratio before computations are done. The point to note in Fig. 1-12A, however, is that the pattern over time of the three curves is roughly similar.

Ryther and colleagues also calculated the change in carbon standing stock by using stoichiometric ratios similar to those of Eq. (1-2). The ratios were used to convert observed changes in dissolved oxygen, phosphate, and nitrate in the water to estimates of the amount of carbon fixed during the time of the study. The estimates of changes in organic standing crop over time computed from changes in nutrient concentrations in the water were much more irregular (Fig. 1-12B) than those for particulate measurements (Fig. 1-12A). This is not surprising; microbes, for example, could have consumed nitrate and ammonia at different rates (cf. Section 11.2). The stocks of carbon calculated on the basis of nutrient uptake predicted that carbon would remain high for the 5 days of the study (Fig. 1-12B), but the measured carbon declined after the third day. This suggests that grazers or sinking removed part of the carbon produced. It is also likely that the algae lost dissolved organic matter during growth. Ryther and colleagues also calculated the carbon yields implied by the measured increase in O_2 in the water. The rate of O_2 consumed to CO_2 produced was estimated using a photosynthetic quotient of 1.25. The relation of O_2 produced to CO_2 consumed is partly a function of the relative amounts of lipid, protein, and fat produced by the phytoplankton and of the redox state of the nitrogen source used. Use of nitrate, rather than ammonia, produces two extra O_2 molecules during reduction in cells ($NO_3 + 2H_2O = NH_3 + OH^- + 2O_2$). The ratio used by Ryther et al. (1971) is applicable if algae use ammonium as the nitrogen source; where, as in Ryther's study, nitrate is the principal source of nitrogen, the photosynthetic ratio is larger, perhaps up to 2 (Williams et al. 1979). If the higher ratio is used in the comparison, the discrepancy between the estimates based on ^{14}C and O_2 methods (Fig. 1-12B, C) is reduced.

Comparisons of different methods of measuring production were also made by McAllister et al. (1961) and Antia et al. (1963). In these studies, rather than follow a body of water, seawater was enclosed in large free-floating plastic bags suspended below the surface in a protected bay in the coast of British Columbia. The bags were filled with filtered seawater and then inocula of natural seawater were added. The water was sampled for nutrient analysis and growth of algae was assayed using oxygen and ^{14}C methods (Fig. 1-13). There was a lag of some 10 days until the algal bloom began. The nutrients, especially nitrate, were rapidly depleted by algal uptake. The estimates of gross production are provided by the light–dark oxygen bottles. The production rates obtained by the ^{14}C method were lower than the rates calculated using the oxygen method. The difference

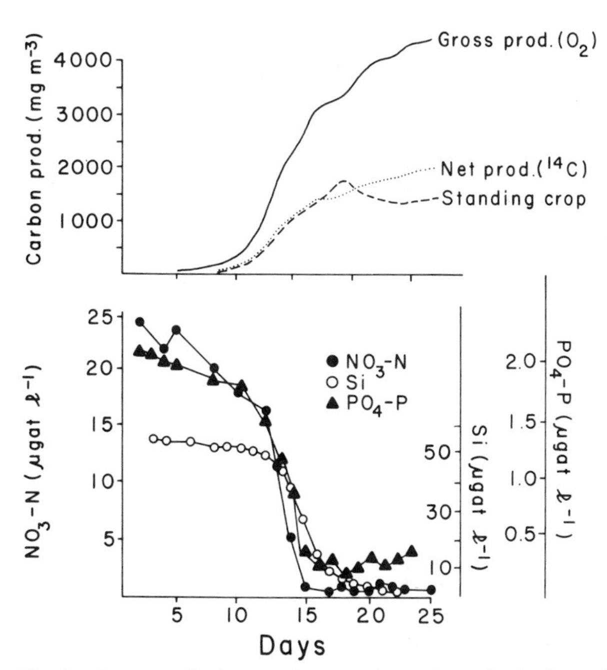

Figure 1-13. Production, particulate carbon, and nutrients in the large bag experiment, Saanich Inlet, British Columbia. Adapted from Antia et al. (1963).

between the O_2 and the ^{14}C methods bring up the matter of what it is that ^{14}C methods measure. Peter Le B. Williams (personal communication) has done many experiments in which there is good agreement between O_2 and ^{14}C determinations where heterotrophic respiration is low. Thus, ^{14}C *can* approach a measure of gross production. This agreement indicates that generally respiration by phytoplankton is low relative to photosynthesis. The ostensible discrepancy between estimates of "net" and "gross" production seen in many studies may be partly the result of calculation using values of photosynthetic quotients that are too low (Williams et al. 1979).

The large bag experiment started with filtered seawater, so that the particulate organic carbon (POC), was phytoplankton that grew during the experiment. The actual observed POC measured, however, was only about 65% of measured net production. The loss of carbon may be the result of release of dissolved organic carbon (DOC) by cells. We will return to such losses in DOC from living cells in Chapter 10.

These two experiments show some of the problems and the insights obtainable by some of the available methods used to measure primary production. No one method is free from drawbacks, but each has specific applications. The suite of available procedures should be scrutinized when planning a research program so as to choose the method best suited to answer the hypotheses guiding the research.

1.32 Benthic Microalgae

The production of algal cells growing attached to sediments is best measured by adapting the O_2 or ^{14}C method. There are various procedures available (Gargas 1970; Hargrave 1969; Pomeroy 1959; Van Raalte et al. 1974; Leach 1970; Skauen et al. 1971).

1.33 Production by Macroalgae and Vascular Plants

Various methods exist to estimate the production of macroalgae and vascular plants (Wiegert and Evans 1964; Williams and Murdoch 1972; Hopkinson et al. 1980). The most used approach involves repeated harvest of standing crop and calculation of the growth increment during each time period. This procedure has often been used in salt marshes, but care must be taken to include the parts of the plant that die during the growth period (Valiela et al. 1975). Such methods provide estimates of net aboveground production; Hopkinson et al. (1980) compare various alternative methods applied to salt marsh production.

Leaf or frond marking methods have also been used. Growth and production in kelp have been measured by punching small holes at the base of a frond and measuring the rate at which the hole moves away from the stipe as the frond grows (Mann 1973; Chapman and Craigie 1977). Production in seagrasses (Sand-Jensen 1975) and mangroves (Onuf et al. 1977) has also been measured using leaf-marking techniques. Rates of elongation or of addition of tissues are then converted into production by using approximate conversion factors.

In marshes and seagrass beds, much of the plant production is translocated into sediments. Estimation of such below-ground "production" by harvest techniques is difficult and laborious (Valiela et al. 1976), since the living tissues are difficult to separate from the sediments.

Measurement of gas exchanges by a macroalgae or vascular plant contained in a bell jar or gas-proof tent is an accurate method to obtain total production (Kanwisher 1966). In water, the changes in O_2 can be used to measure production. In air or water, a gas chromatograph or an infrared CO_2 analyzer can be used to estimate consumption of CO_2. Measurements under light and dark conditions can allow the calculation of gross and net production (Teal and Kanwisher 1961; Lugo et al. 1975). Estimates of net production obtained using CO_2 exchange measurements agree well with estimates obtained by harvest at short intervals, at least in salt marshes (Valiela et al. 1976).

Estimates of potential production have been obtained based on the amount and interception of light and pigments in mangrove forests (Bunt et al. 1979) and in intertidal macroalgae (Brinkhuis 1977). Such an approach may provide a fairly easy way to obtain production estimates. These methods are useful where light rather than nutrients is the dominant limiting factor. Careful measurements of local conditions are needed to make these methods reliable. Comparisons with other better established procedures should be carried out before extensive use of light-based calculations is made.

In certain situations where water traverses a stand of subtidal vegetation in a well-defined direction, as in some kelp forests, it is possible to estimate production by macroalgae by measuring the changes in O_2 and CO_2 in the seawater entering the stand compared to that leaving the stand (McFarland and Prescott 1959). Such measurements must be interpreted carefully since microbial and animal respiration may be important.

1.34 Production by Chemosynthetic Bacteria

Chemosynthetic activity can be roughly measured by ^{14}C uptake measurements done in the dark. Uptake of CO_2 by photosynthetic organisms may take place in the dark, so these ^{14}C uptake values are not only due to chemosynthesis. Dark uptake of CO_2 by phytoplankton may reach at most 5% of the light uptake, so the error in measurement of chemosynthetic activity may be small.

1.4 Production Rates by Marine Primary Producers

1.41 Rates of Production in Photosynthetic Producers

Many measurements of net primary production in various marine environments have been made using the various methods sketched above. At any one site and at any one time the rates of production may vary greatly; in some environments there may be short-lived bursts of intense production. To make some sort of comparison among various marine producers, it is convenient therefore to express production in terms of amounts of carbon fixed per unit area per year.

The ranges of reported values of annual production by marine producers (Fig. 1-14, top) show that attached macroalgae, corals, and vascular plants are generally more productive than single-celled algae. Al-

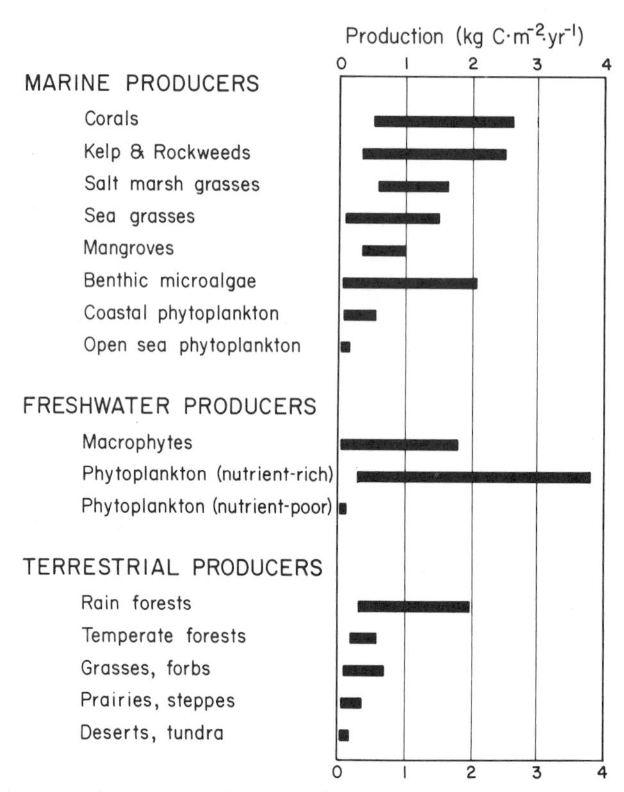

Figure 1-14. Annual net production rates in a variety of producers within marine, freshwater, and terrestrial environments. Modified from many sources and compilations by Margalef (1974), Bunt (1975), Wetzel (1975), Van Raalte (1975), Sournia (1977), and De Vooys (1979).

though attached microalgae can be very productive, phytoplanktonic production is considerably lower than production by macroalgae. Production by oceanic phytoplankton is particularly low.

Reported values of production by diverse marine organisms lie within the same ranges measured in freshwater (Fig. 1-14, middle) and terrestrial (Fig. 1-14, bottom) environments. Production in rainforests is comparable to that of corals and macroalgae, while production in deserts is similar to oceanic production.

Average values of production by whole assemblages of producers in marine and terrestrial habitats are summarized in Table 1-3. The average annual production rates of estuaries, algal beds, and coral reefs are about four times those of shelf or upwelled water and more than an order of magnitude greater than those of the open ocean (Table 1-2). The open sea, however, comprises over 90% of the area of the world oceans, so that even though oceanic production rates are much lower than those of other marine environments, the area of open ocean is so large compared to coastal waters that over 60% of marine production takes place in oceanic waters (Table 1-3).

The differences in standing crops of producers between terrestrial and ocean environments are far larger than the differences in production rates. The average producer biomass in the open ocean (Table 1-3) is several orders of magnitude lower than the average biomass per unit area on land and one order of magnitude smaller than those of freshwater environments. Even the much more productive coastal shelf waters have

Table 1-3. Average Primary Production and Biomass, Turnover Time, and Chlorophyll in Major Environments[a]

	Area (10^6 km^2)	Net production (g m^{-2} yr^{-1})	Biomass (kg m^{-2})	Turnover time (P/B, yr^{-1})	Chlorophyll (g m^{-2})
Open ocean	332	125	0.003	42	0.03
Upwellings	0.4	500	0.02	25	0.3
Continental shelf	27	300	0.001	300	0.2
Algal beds and reef	0.6	2,500	2	1.3	2
Estuaries (excl. marsh)	1.4	1,500	1	1.5	1
Total marine	361	155	0.01		0.05
Terrestrial environments	145	737	12	0.061	1.54
Swamp and marsh	2	3,000	15	0.2	3
Lakes and streams	2	400	0.02	20	0.2
Total continental	149	782	12.2	0.064	1.5

[a] The more productive habitats can have considerable more biomass and production than the average shown in this table (cf. Table 15-1 in Whittaker and Likens, 1975). Production and biomass expressed as dry matter. Terrestrial environments include all land habitats. Adapted from Whittaker and Likens (1975).

less biomass than most freshwater systems. Producer biomass in estuaries and marine algal beds is higher but does not match those of freshwater swamps and marshes. This contrast means that the standing crop of marine plants turns over faster than that of terrestrial environments (Table 1-3). Terrestrial plants, because of the amount of nongrowing tissues needed for structural support, show very slow turnover times, as do plants in freshwater swamps and marshes, and to some extent estuarine producers and macroalgal beds in the sea. The most striking comparison, though, is that between phytoplankton and multicellular producers. Whether in freshwater or seawater, phytoplankton turn over 20–40 times a year, probably faster in oligotrophic waters. Rates of turnover vary seasonally, but there are times of year in which phytoplankton biomass turns over daily. This is in sharp contrast to the 1-2 times a year that most vascular plants achieve.

1.42 Production by Bacteria

1.421 Anoxygenic Photosynthesis

Anoxygenic photosynthesis demands anaerobic conditions and light. In the sea this combination is rare; for example, in the Black Sea the anaerobic layers lie below the depth to which light penetrates (Fig. 1-15, right), so bacterial photosynthesis is not likely to be high.

In seasonally stratified lakes with an anaerobic lower layer, production by photosynthetic bacteria may be 20–85% of total annual production (Takahashi and Ichimura 1968; Cohen et al. 1977a,b). In Solar Lake, a stratified coastal pond with very high salinities, anoxygenic photosynthesis is 91% of total production (Cohen et al. 1977b).

In shallow coastal mud and sandflats, light penetrates the thin oxic surface layers of sediment, and some photosynthesis by anaerobic bacteria takes place. A thin plate of photosynthetic purple sulfur bacteria is often present just below the anoxic boundary. In such situations high rates of chemosynthetic production may take place.

Given the right circumstances, anoxygenic photosynthesis can therefore be locally important, and can take place at high rates. In general, it is not important for aerobic water columns.

1.422 Chemosynthesis

Chemosynthesis occurs in environments where anaerobic and aerobic conditions exist side by side, since oxidized reactants as well as anaerobic conditions are needed (cf. Table 1-1). Where such conditions are available, chemosynthetic activity can be substantial: in a small bay in Japan chemosynthetic rates exceed photosynthetic rates during spring, and amount to about 20% of annual carbon fixation (Seki 1968).

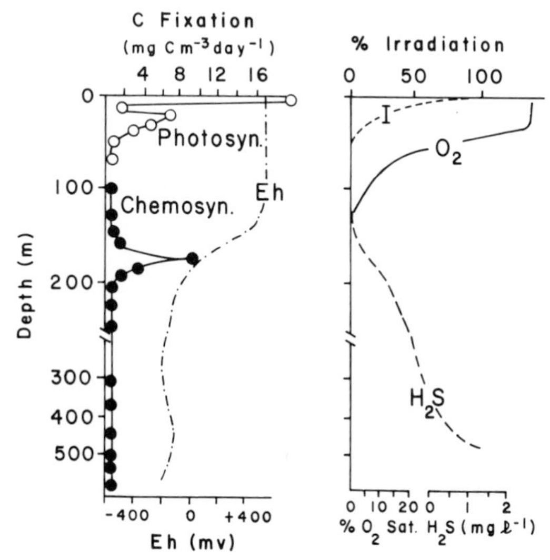

Figure 1-15. Profile of producer activity and certain physicochemical properties of the Black Sea. Eh is the redox potential; I is the % of incident radiation. Adapted from Sorokin (1964).

The boundary between aerobic and anaerobic conditions is the most favorable site for chemosynthetic bacteria, and there is often a layer or plate of these organisms in such boundaries. The presence of an aerobic-anaerobic boundary depends on the balance of conditions that supply or deplete oxygen in the adjoining habitats.

The presence of oxygen is a function of renewal of water-borne O_2, and of consumption of O_2 by organisms. Oxygen-poor water may occur where the circulation of water is impaired. In the Black Sea and in fjords, for example, fairly shallow sills at the exits of these bodies of water prevent mixing below the depth of the sills. Any newly diffused or advected* oxygen that reaches the lower depths is quickly consumed by organisms. In the water column of the central Black Sea, aerobic photosynthesis due to algae takes place near the surface (Fig. 1-15, left). Below about 150 m, reducing conditions are found (note the lowered Eh at 150–200 m), dissolved O_2 disappears, and H_2S is present (Fig. 1-15, right) (cf. Section 11.3). At or below the interphase between the oxidized and the reduced zones of the water column, chemosynthetic bacteria show considerable carbon-fixing activity (Fig. 1-15, left).

Chemosynthetic activity is also important in anaerobic sediments. Where primary production is high, large amounts of organic matter are

* Advection and turbulent (or eddy) diffusion can both be agents of transport of substances. Advection refers to the mass movement of a parcel of water. Diffusion in seawater tends to occur primarily due to small turbulent eddies rather than to molecular diffusion. In sediments molecular diffusion is more important.

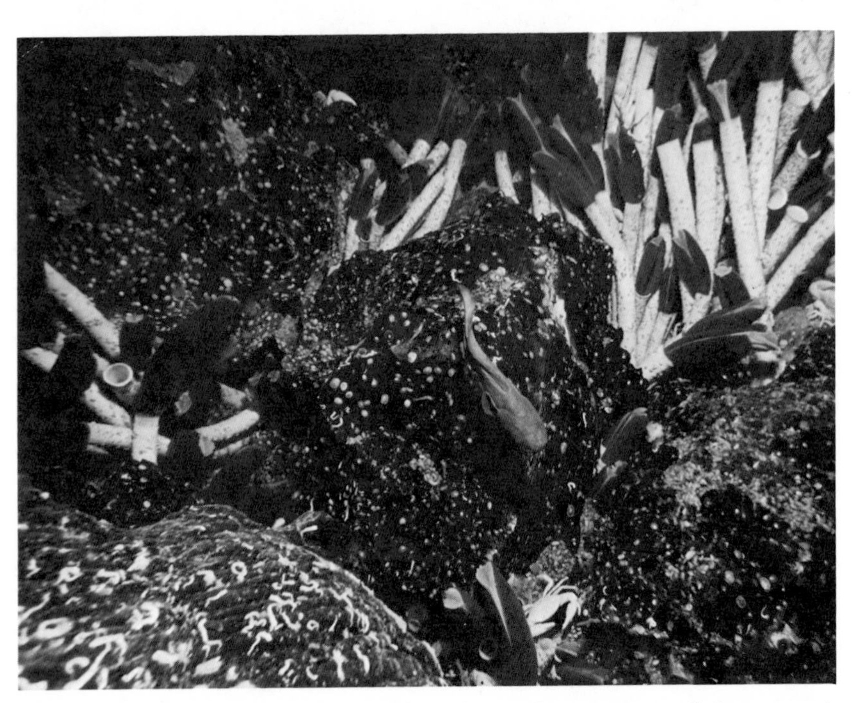

Figure 1-16. Views of the communities of organisms—many of them new to science—associated with vents on the Galapagos rift zones. Photos taken from the research vessel *Alvin*. Top: View of water flowing through some of the vents. Some of the prominent animals are the large bivalves (up to 30 cm in length),

released into and onto sediments. The decay of these organic substrates consumes oxygen in excess of renewal by water exchange (cf. Section 10.2) and leads to anoxic conditions. Such a situation is common in sediments below many coastal waters, including especially regions of upwelling, mudflats, salt marshes, and mangrove swamps. The delivery of organic matter to deeper sediments is usually considerably less than that to coastal sediments (cf. Section 10.1121), so that deeper sediments are seldom anoxic; chemosynthesis is therefore more active in shallower sediments.

A situation where chemosynthetic bacteria are responsible for all the production—and carry out truly primary production—is in the vicinity of hydrothermal vents in the deep sea, where the seafloor is spreading (Fig. 1-16). Geothermally reduced sulfur compounds are emitted from such vents, and chemosynthetic bacteria use the energy in these reduced sulfur compounds for the reduction of carbon dioxide to organic matter (Jannasch and Wirsen 1979; Karl et al. 1980). The high chemosynthetic bacterial production near the vents serves as the food base for a complex community of heretofore unknown organisms (Lonsdale 1977).

It should be emphasized again that the oxygen content in water of most marine environments is adequate for aerobic producers and therefore algal photosynthesis is the principal means by which carbon dioxide is converted into organic matter in marine environments. It is only in certain situations where the physicochemical conditions are such that bacterial photosynthesis or chemosynthesis may make more than trivial contributions to the production of organic matter.

Calyptogena magnifica, a new genus; the pogonophoran worms (left bottom), *Riftia pachypttila*, a new family; and the white crab *Dythograea thermydron*, a member of a new family. Photo by H. Sanders, courtesy Woods Hole Oceanographic Institution. Bottom: View of area at 2,800 m populated by *R. pachypttila*, whose tubes can reach up to 1.6 m in length, and limpets, crabs, worms, and an unidentified fish. Photo by John Edmond, courtesy Woods Hole Oceanographic Institution.

Chapter 2

Factors Affecting Primary Production

The rate of primary production of a parcel of a marine environment depends on light and on the chemical conditions provided by the physics of water masses. There is thus a complex coupling of physics, chemistry, and biology in marine environments. In this chapter we start in reductionist fashion by examining how light, nutrients, and temperature affect primary producers. We focus on two major variables—light and nutrients—and their role in determining primary production. Grazing and sinking, the other major factors affecting producers, are discussed in Chapters 5 and 10. We end this chapter with some examples of how the motion of water masses affects production in the ocean through its effect on availability of light, nutrients, and temperature.

2.1 Light

2.11 Amount of Light

2.111 Availability of Light

The amount of solar radiation reaching the earth's surface is strongly influenced by absorption by water vapor, carbon dioxide, oxygen, and ozone present in the atmosphere (Fig. 2-1). The absorption by ozone of ultraviolet (UV) light is especially important to biological systems because UV light is harmful to organisms, particularly because of the absorption of these wavelengths by nucleic acids such as DNA. In fact, it was not until the evolution of oxygen-yielding photosynthetic cyanobacteria that the reducing atmosphere of the primitive earth was provided

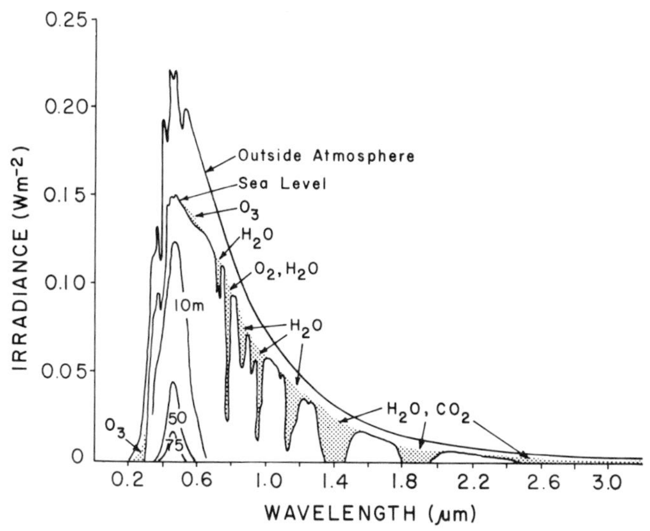

Figure 2-1. Absorption of solar radiation in the atmosphere and at depths of 10, 50, and 75 m in the sea. Adapted from Valley (1965) and Jerlov (1951). Note that little radiation in the ultraviolet radiation (below about 0.3 μm) reaches the surface of the sea and that long wave radiation (beyond 0.6 μm) is very effectively absorbed in the very upper layers of the water column. Marine organisms are left with a rather narrow window of wavelengths to use.

with oxygen and hence ozone. The ozone intercepted much of the UV radiation and its presence facilitated the subsequent evolution of aerobic organisms.

Passage through the atmosphere reduces the intensity of solar radiation, but absorption by the various compounds of the atmosphere over the broad spectrum of energy is very uneven (Fig. 2-1), as shown by the very irregular curve of irradiance that arrives at the surface of the sea (Fig. 2-1). Ozone, oxygen, water, and carbon dioxide are principal absorbers in the nonvisible ranges. Clouds may reduce radiation reaching the sea, and some solar energy is lost by scattering and reflection at the sea surface. When the sun is low in the sky or during rough seas, over 30% of radiation may be lost; on calm, clear days when the sun is high in the sky loss via these mechanisms is only a few percent of incident radiation (Von Arx 1962).

There is further absorption in the water itself so that only small amounts of light penetrate the deeper layers of the sea (Fig. 2-1). Light is absorbed by water, suspended particles, and dissolved matter (Fig. 2-2). Absorption by algae peaks below the surface, as can be expected from the vertical distribution of chlorophyll or primary production (cf. Fig. 2-4). Deeper in the water column, water itself is responsible for most of the absorption. The amount of available light used by algae is directly proportional to the amount of photosynthetically active pigments in the photic

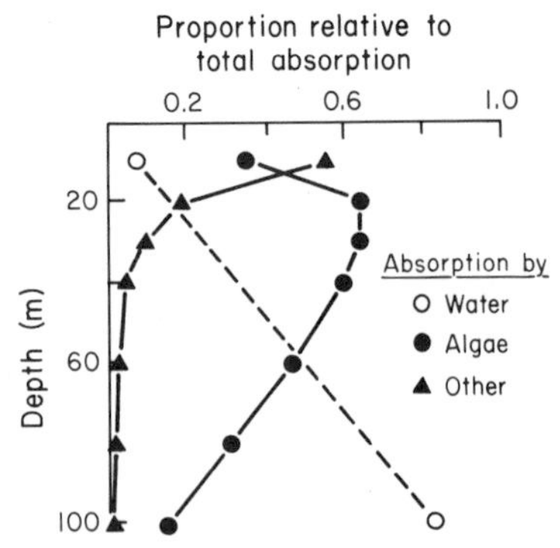

Figure 2-2. Relative absorption of light by water, algae, and organic materials in the water column. The "other" category refers to suspended particulate and dissolved matter. Adapted from Lorenzen (1976). © W. B. Saunders Co, reprinted by permission.

zone. In a representative measurement in coastal water the photic zone was 10 m deep, the algae were dense and absorbed 56% of the light energy reaching the water surface. In an oceanic situation algal densities were low, the photic zone extended to 100 m, and only 1% of the energy available was absorbed by algae (Lorenzen 1976).

The total amount of light entering the water column from the surface and penetrating to any depth z can be described by Beer's Law:

$$I_z = I_0 e^{-kz}, \tag{2-1}$$

where I_z is the intensity of light at z, the depth of interest, I_0 is the intensity at the surface, and k is the extinction coefficient of water. The extinction coefficient varies from one place to another and is wavelength specific. When converted to logarithms and plotted, Eq. (2-1) becomes linear (Fig. 2-3). The vertical dashed lines indicate thresholds for various processes. One important feature of Fig. 2-3 is that the lower limit for photosynthesis (and hence for growth of phytoplankton) is very near the surface. This lower limit is the compensation depth, that depth at which light-supported photosynthesis just balances the metabolic losses due to respiration. Usually the compensation depth is taken to be about 1% of the surface radiation. The column of water above the compensation point is the photic zone.

The depth of the photic zone varies greatly, depending on the amounts of suspended particles and dissolved substances in the water. Mixing of nutrient-rich waters in spring often reduces light penetration to just a few meters because of resuspended bottom particles, presence of large amounts of dissolved organic matter, and blooms of phytoplankton. Of course, light intensity varies during the day and the depth of photic zone

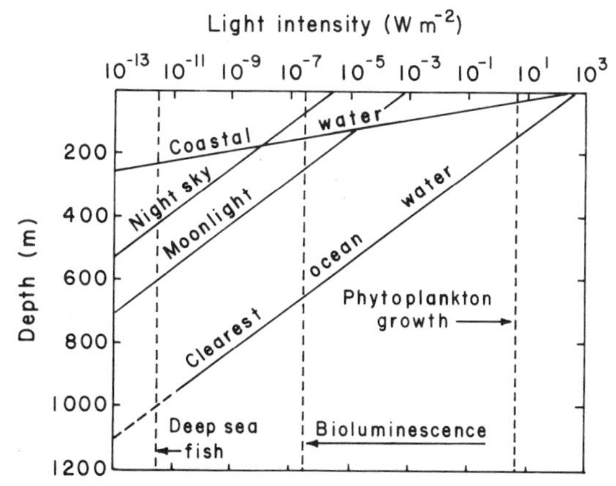

Figure 2-3. Logarithmic plot of $I_z = I_o\ d^{-kz}$ plotted versus depth in the water column (diagonal lines) for daylight in coastal water, and for daylight, moonlight, and night sky for oceanic water. The approximate limits at which certain processes are activated (algal growth, vision in deep-sea fish) and the intensity of ambient light below which luminescence can be seen are also indicated as dashed vertical lines. Adapted from Clarke and Denton (1962). © John Wiley & Sons, Inc., reprinted by permission.

changes accordingly. Even in the clearest oceans, this depth seldom reaches 200 m, so that all the primary production in the water column of the sea takes place in a very thin upper layer.

2.112 Light Intensity and the Rate of Photosynthesis

Vertical profiles of photosynthetic rates typically have a subsurface peak (Fig. 2-4, top). Although there are exceptions (cf. Fig. 1-10), this pattern is the most commonly found in nature and is primarily due to the effects of light intensity* on the photosynthetic process. Laboratory experiments where light intensity is experimentally manipulated have borne this out (Fig. 2-4, bottom).

The rate of the light reaction of photosynthesis is strictly dependent on light intensity. Increases in light intensity lead to greater photosynthetic rates until some maximum is reached, defined as P_{max}. At this point the producers cannot use any more light; the enzymes involved in photosyn-

* The measurement of light intensity is reported in widely varying units. Photosynthesis is driven by photons and so the preferred light unit quantifying photons is the einstein (E), where 1 E = 1 mole of photons. Full sunlight at the sea surface roughly equals 2000 μE m^{-2} s^{-1}. Rough conversion among other light units used in the literature are: 1 μE m^{-2} s^{-1} = 0.0187 ly (langley) hr^{-1} = 0.217 W (watt) m^{-2} = 51.2 lux = 4.78 ft candles, where 1 ly = 1 g cal cm^{-2}, 1 W = 10^7 ergs s^{-1}; 1 lux = 1 candle on 1 m^2; 1 ft candle = a candle at 1 ft distance; all conversions refer to light in 400–700 nm, the range of wavelength that spans photosynthetically active radiation (PAR).

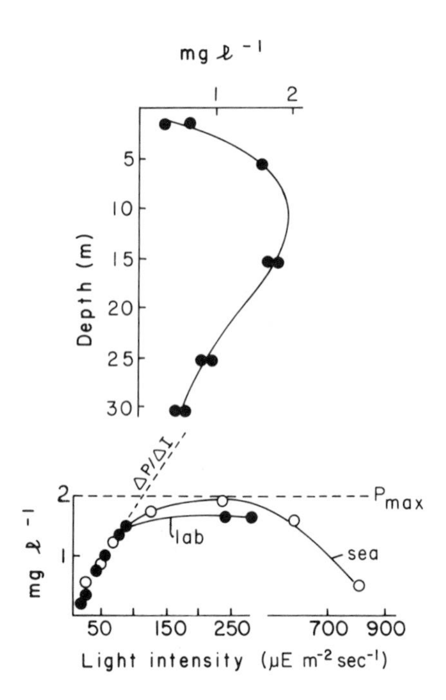

Figure 2-4. Top: Vertical profile of primary production in coastal water in milligrams of O_2 produced per liter. The experiment lasted three hours. Bottom: Comparisons of laboratory and field results of photosynthetic rate under varying light intensity. Adapted from Talling (1960).

thesis cannot act fast enough to process light quanta any faster, so rate of photosynthesis reaches an asymptote. The ratio of the change in photosynthesis with respect to the change in light ($\Delta P/\Delta I$) is a measure of how efficiently cells use changes in light. The ratio may be steeper for greens and diatoms than for dinoflagellates (Fig. 2-5), indicating that photosynthesis is saturated at lower light intensities for the former than the latter. Consequently greens and diatoms may be better suited for depths or latitudes where only dim light is available much of the year. Generalizations about the $\Delta P/\Delta I$ relationship are difficult, however, since the ratio varies substantially even within one species of algae and may be changed by nutrient deficiencies (Ketchum et al., 1958) and by cellular adaptations to low light in which the photosynthesis per cell may increase (Perry et al. 1981).

The actual light intensity at which P_{max} is reached depends on the algal taxon (Fig. 2-5). Dinoflagellates, at least in this data set, are light saturated at a more intense light level than diatoms or green algae. On this basis alone, we might therefore expect dinoflagellates to be more abundant nearer the surface, in midsummer, and in tropical latitudes relative to greens or diatoms. There are so many other factors that affect abundances and distribution that it is difficult to check this prediction.

Full sunlight usually inhibits photosynthesis in marine algae, as evidenced in the reduced photosynthetic rates at high light intensities (Figs. 2-4, 2-5). The reason for the inhibition of photosynthesis at high light

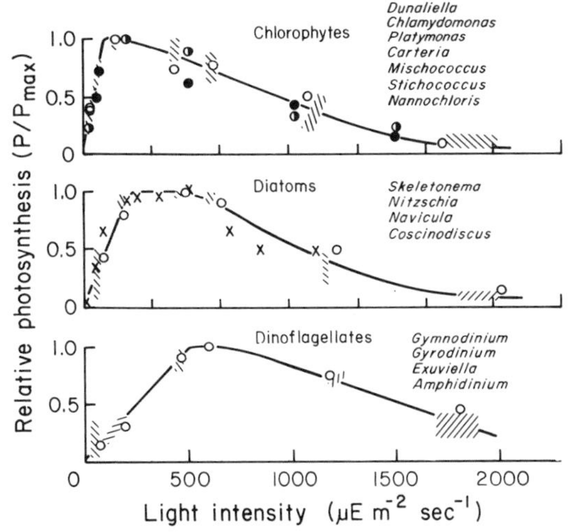

Figure 2-5. Curves of relative photosynthesis (P/P_{max}) versus light intensities for three algal groups. The chlorophytes include green and flagellated greens. Shaded rectangles represent the dispersion of points obtained experimentally using neutral filters in cultures grown with an irradiance of 1.3 μE m^{-2} sec^{-1}: open circles correspond to cultures in natural light: solid and half-solid circles are cultures at 0.48 and 1.98 μE m^{-2} sec^{-1} and measured in the harbor at Woods Hole. Crosses correspond to the data of Jenkin (1937). Adapted from Ryther (1956).

intensities is not well understood. Ultraviolet radiation is probably implicated (Strickland 1965), at least in some species, since in the laboratory under no UV radiation the $\Delta P/\Delta I$ curves do not show inhibition at intensities beyond the saturation point (McAllister et al., 1964; Harris, 1980).

The relation of photosynthetic rate to light intensity in macroalgae and seagrasses (Fig. 2-6) is similar to that found in single-celled algae. The species with thinnest fronds generally show higher values for the $\Delta P/\Delta I$ ratio. Inhibition by high light intensities is less marked than for single-celled algae, probably because of self-shading within frond or leaf tissues. This is especially evident in the green alga *Codium fragile*, which has thick, optically dense fronds (Ramus, 1978).

The photosynthesis to light intensity relation is not constant overtime. Producers may adapt physiologically to ambient light intensity, primarily by varying the amount of accessory pigments involved in harvesting photons. Photosynthesis in cells is carried out by photosynthetic units composed of a light-collecting antenna of accessory pigments and a reaction center made up of chlorophyll *a*. Under low light conditions, the size of the antenna increases. This shade adaptation improves photosynthetic efficiency or ability to collect light (Prezelin and Sweeney, 1979; Perry et al., 1981), and results in higher photosynthesis rates per unit chlorophyll (Fig. 2-7).

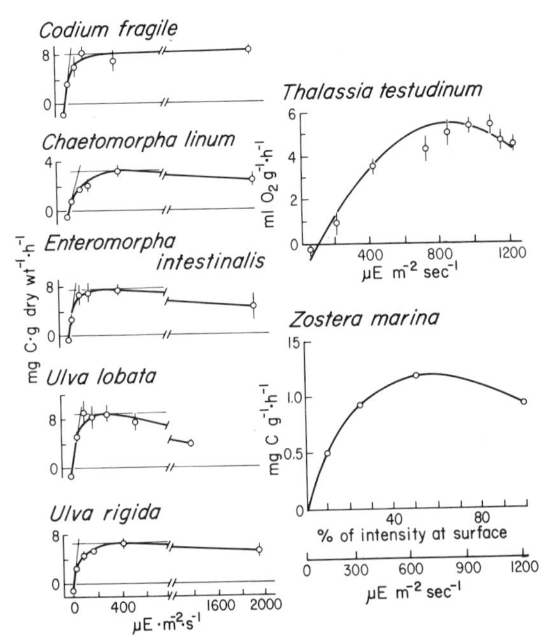

Figure 2-6. Net photosynthesis versus light intensities for five green macroalgae and two seagrasses. The $\Delta P/\Delta I$ and $P_{\max}$ are shown in the macroalgal graphs as straight lines. Points are mean and standard deviations, shown as vertical lines. Adapted from Arnold and Murray (1980), Buesa (1975), and McRoy (1974).

The induction of shade adaptation in the photosynthetic unit may take 10-12 hr. Phytoplankton, however, may be repeatedly advected to depths where light is low during the course of a day. In such circumstances, a 10–12-hr period of adaptation is too long, and it may be advantageous to maintain photosynthetic units furnished with large antennae, so as to be able to photosynthesize effectively at least during the repeated periods of

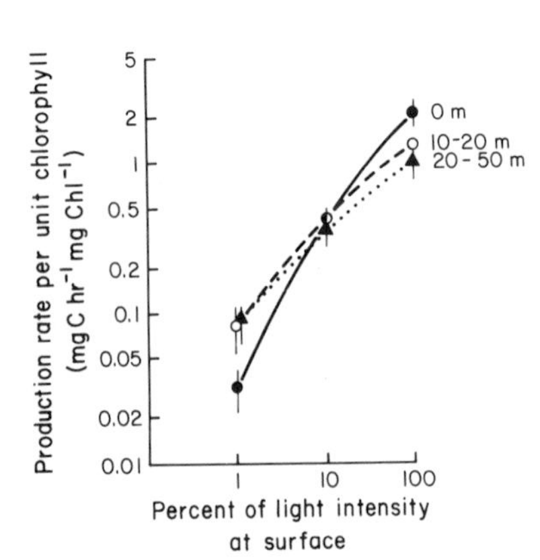

Figure 2-7. Light and shade adaptation in phytoplankton from the upwelling off northwest Africa. Samples were taken at the surface and at two deeper layers (10–20 m and 20–50 m) and incubated at 1, 10, and 100% of light at the surface of the sea. Adapted from Estrada (1974).

low light. Diatoms, a major component of phytoplankton, have such large photosynthetic units (Perry et al., 1981).

2.12 Quality of Light

2.121 Available Wavelengths in Water and Sediments

As light penetrates the water column, water molecules, dissolved substances, and suspended particles* not only attenuate the amount of light but also alter the spectral quality of light (Fig. 2-8, left and center). As light enters the water column, the longer (red and infrared) waves are absorbed quite near the surface and only blue light (short wavelengths) penetrates to some depth. Photosynthetic organisms in the water column therefore have to deal not only with a diminution of light quantity but also with a greatly altered spectral quality of light. These absorptional changes in quality occur much nearer the surface in coastal water than in oceanic water.

In sediments where the photic zone is at most a few millimeters in depth (Fig. 2-8, right), photosynthetic organisms experience a similar effect of depth on the light regime. In contrast to what happens in the water column, in sediments blue light is absorbed first, and the longer red and infrared wavelengths penetrate the farthest. This is because light penetration in sand is governed by refraction and scattering of light, and the refractive index of quartz, a typical component of sand, increases with shorter wavelengths (Fenchel and Straarup, 1971).

2.122 Absorption Spectrum of Marine Producers

The energy that drives photosynthesis is absorbed from sunlight by pigments found within internal membranes in chloroplasts in algae and within membranes located in the cytoplasm of bacterial cells. Each photosynthetic pigment has its own peculiar absorption pattern in the spectrum (Fig. 2-9). Absorption by ozone and water limits available radiation on the short (violet) and long (red) wavelengths of the visible spectrum (Figs. 2-1, 2-9), but the absorption peaks of the pigments commonly found in photosynthetic organisms cover virtually the entire remaining window of the energy spectrum of sunlight (Fig. 2-9). The range of photosynthetically active radiation for most algae and plants spans wavelengths of 400–700 mm, a range clearly adapted to available radiation.

Different groups of producers have somewhat different arrays of photosynthetic pigments, and show characteristic absorption spectra (Fig. 2-

* Phytoplankton absorb light primarily at wavelengths about 400 and 700 nm (Fig. 2-11). Water absorbs mainly near 700 nm, while the dissolved organic matter absorbs at wavelengths nearer 400 nm (Yentsch, 1980).

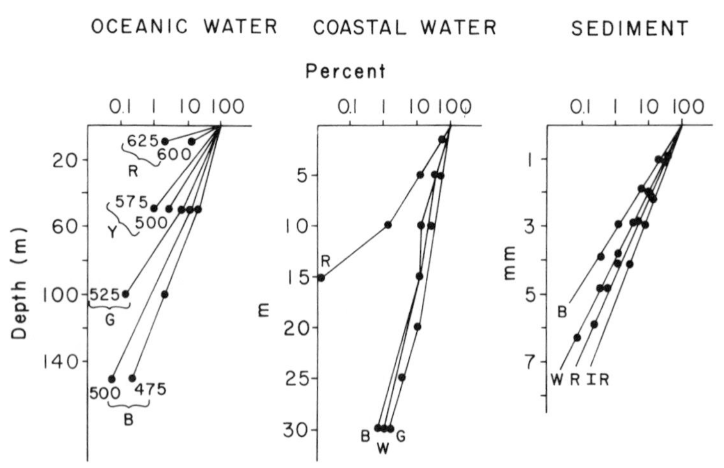

Figure 2-8. Percentage of incident light of differing wavelength penetrating to different depths in seawater and in a sediment. Note that depth is in meters for water and millimeters for sediment. Y, yellow; R, red; G, green; B, blue; and IR, infrared. W, white light is the sum of all wavelengths. Numbers on left-most graph are wavelengths in nanometers; adapted from Jerlov (1951, 1968) and Fenchel and Straarup (1971).

10). The green algae absorb light principally in the red and blue parts of the spectrum, due to the absorption spectrum of chlorophylls *a* and *b* (Fig. 2-9). Both green algae and higher plants have these pigments. Other marine producers do not have chlorophyll *b*.

Brown algae contain chlorophylls *a* and *c* and use the green and yellow wavelengths more efficiently than the green algae. The shoulder in the absorption spectrum of browns around 500–540 mm (Fig. 2-10) is due to the activity of fucoxanthin, a photosynthetically active carotenoid pigment. Peridinin, a similar carotenoid, is responsible for absorption of green-yellow light in dinoflagellates.

Red algae and cyanobacteria contain only chlorophyll *a* and water-soluble pigments that allow absorption of light in the blue and green wavelengths (Fig. 2-10). They also contain phycobilins, of which phycoerythrin—red in color—is often the most common. Red algae may be found at deeper sites than other algal taxa, and show remarkable absorption of precisely those wavelengths available in the deeper water (Fig. 2-10), but their pigment compositions are largely adaptations to low light intensity rather than to light quality (Dring, 1981).

The cyanobacteria contain only chlorophyll *a*, and their absorption spectrum resembles that of green algae, except for the marked absorption around 600–640 nm due to phycocyanin, a phycobiliprotein that is blue in color (Fig. 2-9). Some cyanobacteria show a much enhanced absorption in the green part of the spectrum due to phycoerythrin; *Oscillatoria*

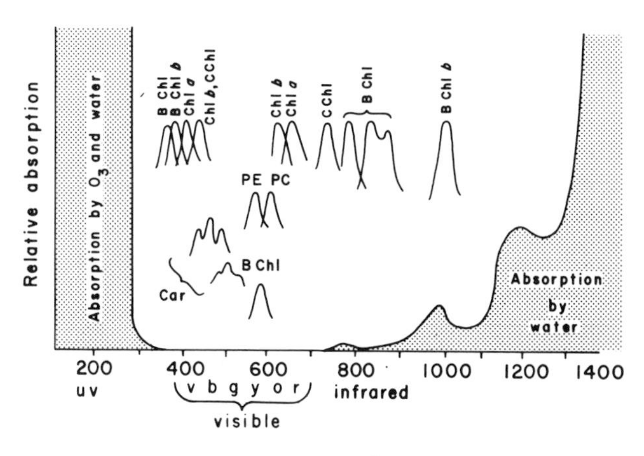

Figure 2-9. Absorption of radiation by photosynthetic pigments. For clarity only the principal absorption peaks are shown. Chl, chlorophyll; B Chl, purple photosynthetic bacterial chlorophyll; C Chl, green photosynthetic bacterial chlorophyll; Car, carotenes; PE, phycoerythrin; PC, phycocyanin. Adapted from Clayton (1971), Yentsch (1967), and Morel (1974).

thiebautii, for example, absorbs weakly at 600–640 nm but has a phycoerythrin peak at 495 nm (McCarthy and Carpenter, 1979).

The absorption spectrum of macro- or benthic algae seems well adapted to the light quality available (Fig. 2-11, top). The absorption spectrum of phytoplankton seems less well suited to make use of available wavelengths, especially in coastal water (Fig. 2-11, bottom). The large absorption peak in the red wavelengths might better suit phytoplankton for oceanic conditions, where red light penetrates somewhat deeper in the water column. Even in clear oceanic water, however, the red peak of chlorophyll absorption is not likely to absorb much light below 10 m (cf. Fig. 2-8, left) while the absorption of the blue band of chlorophyll will be functional deeper in the water column. There is also more energy in the blue range, since energy is negatively related to wavelength. Below 50 m absorption is mainly due to carotenoids that absorb near the blue wavelengths (Fig. 2-9). The presence of such accessory pigments is an important physiological adaptation, since turbulent mixing and sinking can readily transport phytoplankton cells to depths below 50 m.

Carotenoids are present in substantial amounts in virtually all major taxa of marine producers. In addition to absorbing light, they may also be involved in other functions. One such role may be to protect the photosynthetic apparatus from excess light. While in some green and brown algae three or four of the major carotenoids are photosynthetically active, it is not clear how photosynthetically active many of the remaining 60 identifiable carotenoids are, especially in the red algae (Halldal, 1974).

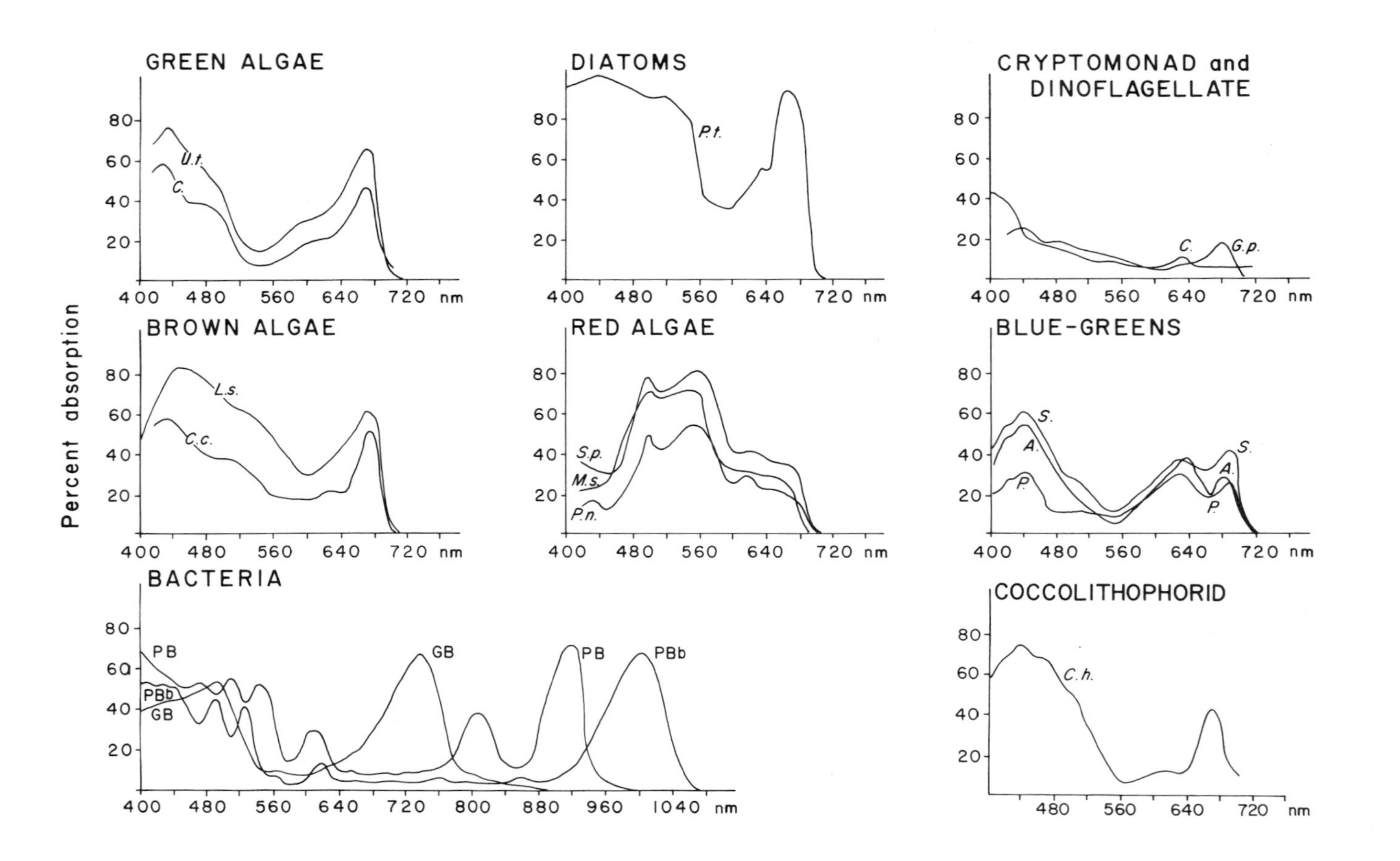
GREEN ALGAE
U.t.
C.
DIATOMS
P.t.
CRYPTOMONAD and
DINOFLAGELLATE
C.
G.p.
BROWN ALGAE
L.s.
C.c.
RED ALGAE
S.p.
M.s.
P.n.
BLUE-GREENS
S.
A.
P.
BACTERIA
PB
PBb
GB
COCCOLITHOPHORID
C.h.
Percent absorption
80
60
40
20
400
480
560
640
720
800
880
960
1040
nm

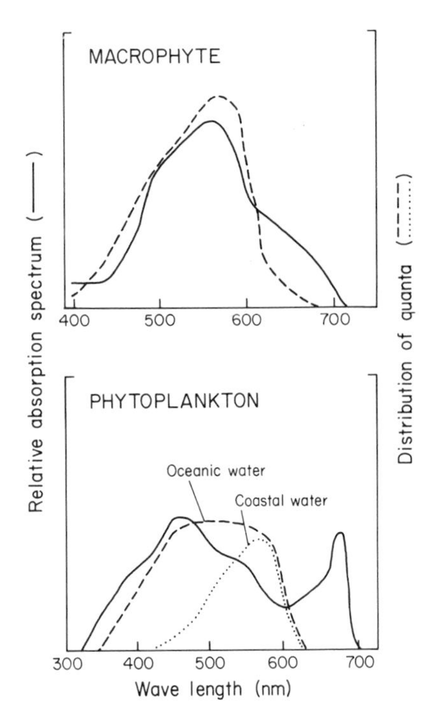

Figure 2-11. Top: Absorption of different wavelengths in relation to the quanta available to a red macroalgae in Kings Bay, Spitzbergen at 10 m. Bottom: Same comparison for a hypothetical mixed phytoplankton population at 10 m in oceanic and coastal waters. The dashed and dotted lines show the actual abundance of quanta at each wavelength in near surface waters. Adapted from "Light and photosynthesis of different marine algal groups" by P. Halldal in Optical Aspects of Oceanography by N. G. Jerlov and E. Steeman Nielson. © Academic Press Inc. (London) Ltd., reprinted by permission.

Details of algal pigments and their properties are reviewed by Jeffrey (1980).

The photosynthetic bacteria (Fig. 2-10) absorb energy far to the right of the algal absorption spectra; bacteriochlorophylls absorb in the red and infrared (Fig. 2-9). As we have seen in Sec. 1.21, photosynthetic bacteria require reduced compounds for their photosynthesis, so that they exist only where oxygen is absent, such as in sediments. They are often found with a layer of algae above them, and the algae absorb in the visible range of wavelengths. The bacteria have adapted to use the long-waved parts of the spectrum that penetrate the deepest into the sediment (Fig. 2-8) and that are not absorbed by the algae on the surface.

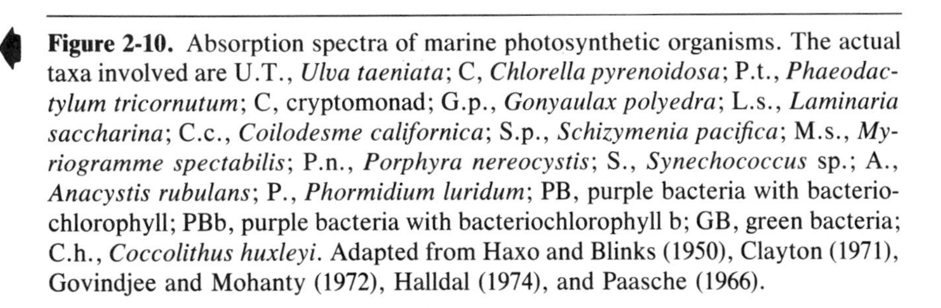

Figure 2-10. Absorption spectra of marine photosynthetic organisms. The actual taxa involved are U.T., *Ulva taeniata*; C, *Chlorella pyrenoidosa*; P.t., *Phaeodactylum tricornutum*; C, cryptomonad; G.p., *Gonyaulax polyedra*; L.s., *Laminaria saccharina*; C.c., *Coilodesme californica*; S.p., *Schizymenia pacifica*; M.s., *Myriogramme spectabilis*; P.n., *Porphyra nereocystis*; S., *Synechococcus* sp.; A., *Anacystis rubulans*; P., *Phormidium luridum*; PB, purple bacteria with bacteriochlorophyll; PBb, purple bacteria with bacteriochlorophyll b; GB, green bacteria; C.h., *Coccolithus huxleyi*. Adapted from Haxo and Blinks (1950), Clayton (1971), Govindjee and Mohanty (1972), Halldal (1974), and Paasche (1966).

2.2 The Uptake and Availability of Nutrients

2.21 Kinetics of Nutrient Uptake

If algae are placed in a nutrient medium, concentration of nutrients decrease over time (Fig. 2-12) in the solution as the nutrients are incorporated into the plant cells. The velocity at which algal uptake removes nutrients depends on the nutrient concentrations in the medium (Fig. 2-13). Such measurements are usually obtained using chemostats, devices by which specific nutrient concentrations can be provided continuously to populations of algae. The uptake rates measured in chemostat cultures increase as the concentrations of the limiting nutrient increase, until an asymptote is reached where the specific nutrient is no longer limiting. This pattern is arithmetically similar to that observed in the saturation of inducible enzymes and has been described by the well-known Michaelis–Menten equation

$$V = V_{max} S/(K_s + S). \tag{2-2}$$

This equation relates the velocity of nutrient uptake (V) to the maximum rate of uptake (V_{max}), the concentration of limiting nutrient (S), and a constant (K_s) that represents the substrate concentration at which $V = V_{max}/2$.

The estimates of V_{max} and K_s can be obtained by fitting a hyperbola to the data using a least-squares method, or more simply, by plotting the reciprocal of the hyperbola ($1/V$ vs $1/S$) to obtain a straight line. The y intercept in this Lineweaver–Burk plot gives $1/V_{max}$ and the x intercept furnishes $1/K_s$. Another procedure is the Woolf plot (Dowd and Riggs, 1964) (Fig. 2-10), where S/V is plotted versus S and the values of V_{max} and V_s are obtained from $S/V = K_s/V_{max} + S\ V_{max}$.

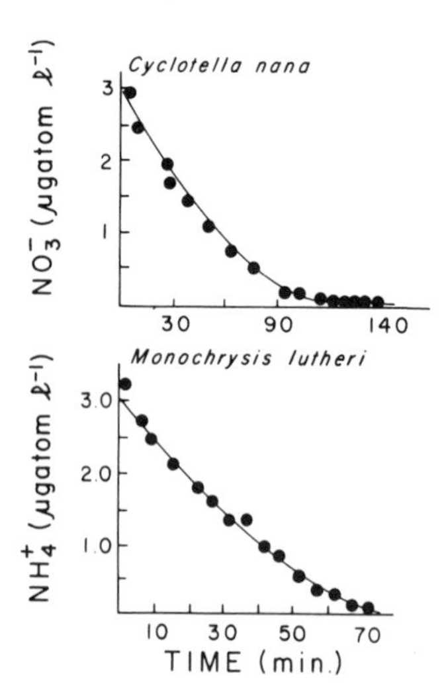

Figure 2-12. Consumption of nitrate and ammonium by two species of algae, a diatom *Cyclotella nana* and a flagellate *Monochrysis lutheri*. Adapted from Caperon and Meyer (1972).

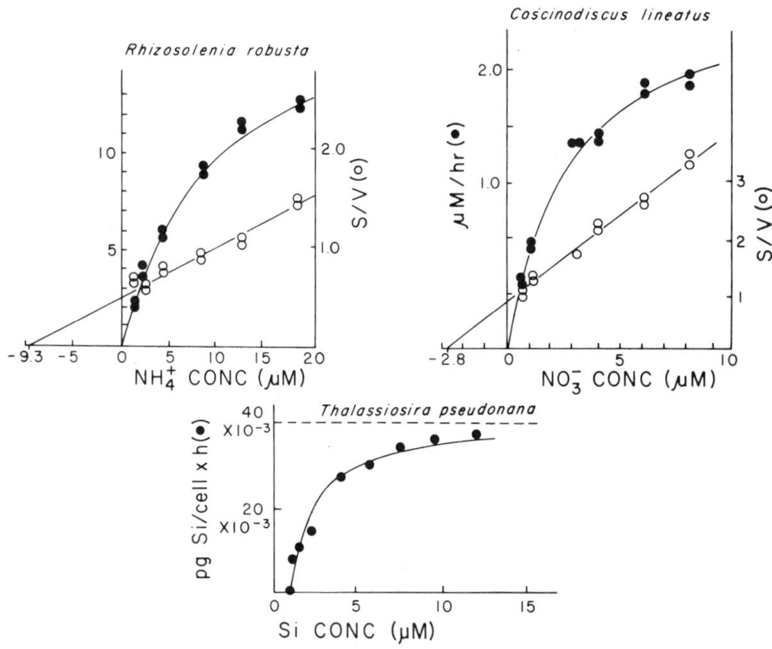

Figure 2-13. Michaelis-Menten curves (filled circles) and Lineweaver-Burk plots (open circles) fitted to data on uptake (μmole/hr) of ammonium, nitrate, and silica by three diatoms at different nutrient concentrations (μmole/liter). S is the concentration of nutrient being taken up, V is the uptake velocity. The x-intercept at the top two graphs provide the estimate of K_s. Adapted from Eppley et al. (1969) and Paasche (1973).

The Michaelis–Menten equation describes fairly well the dynamics of uptake of a variety of nutrients by many species of producers (Fig. 2-13). In certain diatoms, uptake of silica apparently does not start until a threshold concentration is reached, so a modified Michaelis–Menten equation is needed to describe these data (Paasche, 1973). The Michaelis–Menten equation can be used to estimate K_s and S when it is known that the nutrient in question is in fact limiting. If the specific nutrient studied is not the limiting nutrient, spurious values of V_{max} may be obtained, determined by the low supply of the unidentified limiting nutrient. V_{max} value may also be affected by the presence of a competing substrate. If nitrate uptake rates are measured in the presence of ammonium, the uptake of nitrate may be severely underestimated because of the preference for ammonium by many algae (McCarthy, 1980). In view of such difficulties, it is not surprising to find instances where uptake of nutrients does not follow the simple Michaelis–Menten equation (Brown et al., 1978; DeManche et al., 1979).

Uptake of nutrients is coupled to photosynthetic rate, and follows the same pattern as photosynthesis at different light intensities. Nutrients,

Table 2-1. Half-Saturation Constants (K_s) for Uptake of Nitrate and Ammonium by Cultured Marine Phytoplankton at 18°C[a]

	K_s (μg-atoms liter)		Cell diameter (μm)
	Nitrate	Ammonium	
Oceanic coccolithophores and diatoms	0.1–0.7	0.1–0.5	5
Neritic diatoms	0.4–5.1	0.5–9.3	8–210
Neritic or littoral flagellates	0.1–10.3	0.1–5.7	5–47
Oligotrophic, Tropical Pacific	0.1–0.21	0.1–0.62	—
Eutrophic, Tropical Pacific	0.98	—	—
Eutrophic, Subarctic Pacific	4.21	1.3	—

[a] From MacIsaac and Dugdale (1969) and Eppley et al. (1969). The Pacific data from natural mixed phytoplankton.

especially ammonium (Dugdale and Goering, 1967), can be taken up at slow rates even in the dark. Increased temperatures may somewhat increase nitrate uptake (Eppley et al., 1969; Thomas and Dodson, 1974).

In most situations, however, the major factor affecting uptake rate is the concentration of the limiting nutrient in the environment (Fig. 2-13). The half-saturation constant (K_s) varies, depending on the nutrient concentrations present where specific phytoplankton species were collected* (Table 2-1). Since an average of about 0.5 μgat-liter NO_3-N liter^{-1} or less is present in oceanic waters, with lower or comparable amounts of ammonium, it is not surprising to find half-saturation coefficients of oceanic phytoplankton around 0.5–1 μgat liter^{-1}. Diatoms found in inshore (neritic) water have much larger K_s values, consonant with the greater nutrient concentrations in such waters. Flagellates generally have a wider range of K_s values than other kinds of phytoplankton. K_s values for nitrate in phytoplankton from oceanic areas with very different nutrient concentrations show the same trend as the coastal–oceanic comparison: K_s values are high in more eutrophic (nutrient-rich) water and low in oligotrophic (nutrient-poor) water.

* There is some doubt about the validity of estimates of K_s obtained by incubation periods as long as hours, such as those in Table 2-1. Ammonium uptake can be very rapid during the first few minutes of exposure to nutrients (Glibert and Goldman, 1981), and slows later. Estimates obtained in exposures longer than 1 hr or so may therefore underestimate K_s (Glibert et al., 1982).

Table 2-2. Half-Saturation Constants for Nitrate of Three Species of Algae Obtained from Coastal and Oceanic Environments in the Atlantic[a]

Species	Source	K_s (Mean ± 95% confidence interval)
Cyclotella nana	Moriches Bay	1.87 ± 0.48
	Edge of shelf	1.19 ± 0.44
	Sargasso Sea	0.38 ± 0.17
Fragilaria pinnata	Oyster Bay	1.64 ± 0.59
	Sargasso Sea	0.62 ± 0.17
Bellerochia spp.	Great South Bay	6.87 ± 1.38
	Off Surinam	0.12 ± 0.08
	Sargasso Sea	0.25 ± 0.18

[a] From Carpenter and Guillard (1971). © Ecological Society of America, reprinted by permission.

The importance of ambient nutrient concentration is demonstrated by the occurrence of physiological races within the same species of algae or clones adapted to local nutrient regimes. Each of the bays in Table 2-2 are rich coastal lagoons; the Sargasso Sea on the other hand is one of the most oligotrophic regions of the ocean. In each of the three species of algae the K_s values for the strain from eutrophic waters are higher than those of the oceanic strain. These strains maintain their physiological distinction even after being cultured for many generations in identical media in the laboratory.

Another factor that affects uptake rates is size of the phytoplankton cell (Fig. 2-14): larger cells have higher values of K_s. The higher surface-to-volume ratio of smaller cells may provide an advantage in nutrient uptake in oligotrophic water, since uptake rates are limited by the number of uptake sites on the surface of the cells. Thus, smaller cells may be favored in oligotrophic waters. This fits in with the observation that oce-

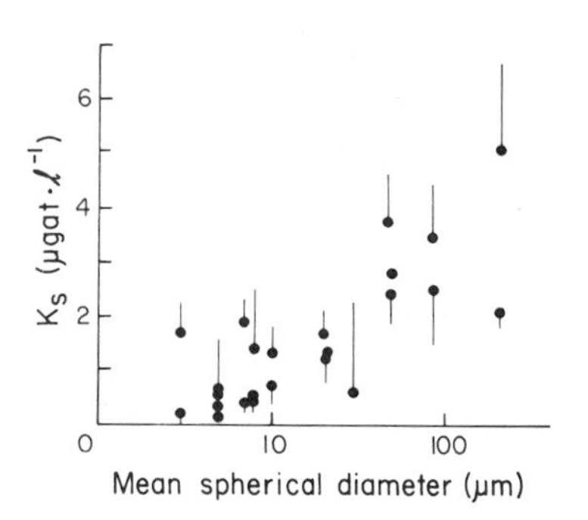

Figure 2-14. Half-saturation (K_s) values for nitrate uptake by phytoplankton of different size. The bars show the 95% confidence limits for the highest and lowest mean K_s reported. Adapted from Malone (1980).

anic phytoplankton tend, on the average, to be smaller than coastal algae, although there is great variability in the data [Table 2-1 and review by Malone (1980)]. The different average size may be a result of competitive exclusion of species of large algae with small surface-to-volume ratios by small algae with a high surface-to-volume ratio in oligotrophic waters. This may be one basis for the differences in species composition between coastal and oceanic floras.

2.22 Internal Nutrient Pools and Growth

The Michaelis–Menten equation can describe the growth rate (μ) of algal populations in chemostats

$$\mu = \mu_{\max} S/(K_s + S), \tag{2-3}$$

with $\mu_{\max}$, the maximum rate of growth, and the other terms being the same as in Eq. (2-2). By using Eq. 2-3, we assume that growth rate is constant regardless of the ambient nutrient concentration or the state of nutrition of the cells. This is not always so, since nutrient-starved algae placed in a fresh culture medium quickly deplete nutrients in the medium at rates exceeding the growth rate of the algae. This "luxury consumption" results in the buildup of an internal pool of nutrients that may be later shared by successive generations of cells. These internal pools of compounds such as polyphosphates or nitrogen storage compounds are responsible for the continued growth observed after the depletion of nitrate in the large bag experiment of Fig. 1-14.

The relation between ambient nutrient concentration and growth rate of algae can thus, in theory at least, be expressed in three steps. First, the rate of nutrient uptake increases hyperbolically as nutrients in the medium increase, as discussed in the previous section. Second, the rate of nutrient uptake is linearly related to the amount of nutrient stored in the plant cells. This may apply to nutrients such as nitrate, silicate, and phosphate, as well as to elements and compounds required only in trace amounts such as vitamin B_{12} (Fig. 2-15). Third, the rate of growth is hyperbolically related to the internal pool of nutrients (Droop, 1968), with the growth rate decreasing as the internal pool becomes saturated (Fig. 2-16). The Droop equation is best applied to describe growth where the limiting element is required in relatively small amounts (Goldman and McCarthy, 1978).

The description of the relations of available nutrients, uptake, internal pools, and growth, and other theoretical advances such as inclusion of the effects of internal pools on uptake rates (DeManche et al., 1979) suggests some of the couplings between nutrients in the medium and the actual growth rate of producers. Such advances have been eagerly received

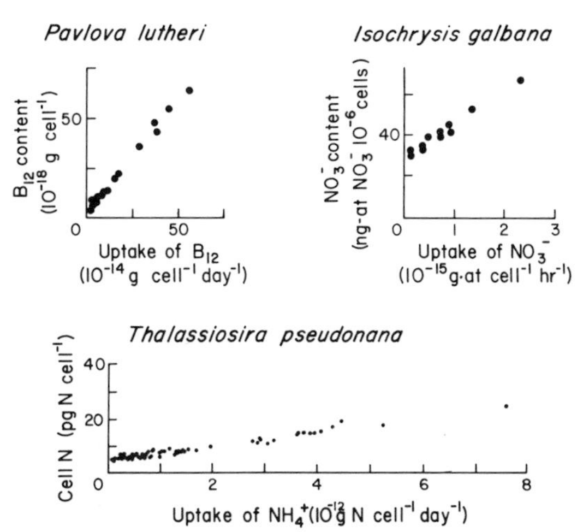

Figure 2-15. Relation of the internal pools and uptake of vitamin B_{12}, nitrate, and ammonium in three species of phytoplankton. Adapted from Droop (1973) and Goldman and McCarthy (1978).

because expressions such as Eqs. (2-2), (2-3), and (2-4) are needed for building predictive mathematical models that can be used as tools to gain further insight into the dynamics of marine ecosystems. There is as yet no consensus that the theoretical models discussed here are completely applicable to natural situations. The use of the equations requires many suppositions, including steady state growth, and growth limitation by a single identified nutrient. Moreover, the complexities of uptake in the presence of other organisms that may release or conserve nutrients have not been dealt with, nor have physical processes of fluid motion been considered. There have been some advances in introducing additional factors into the equations, for example, temperature (Goldman and Carpenter, 1974). Further work in this area may lead to useful, but necessarily complex predictive models.

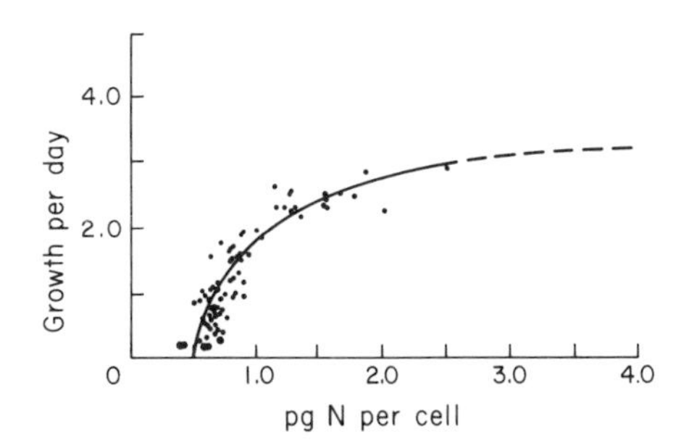

Figure 2-16. Specific growth rate and nitrogen cell content of the diatom *Thalassiosira pseudonana*. A picogram is 10^{-12} g. Adapted from Goldman and McCarthy (1978).

2.23 Availability of Nutrients

2.231 Availability of Nutrients in Sea and Freshwater

While seawater from different parts of the oceans may have variable concentrations* of different nutrients (cf. Chapter 11), representative vertical profiles for oceanic waters (Fig. 2-17) frequently have a marked decrease in nutrients near the surface. This reduction is due to uptake by phytoplankton in the photic zone. The profiles of Fig. 2-17 only show conditions for most of the year; in many regions of the sea there are times of year when physical processes result in vertical mixing of nutrient-rich water, so new nutrients reach the surface (cf. Chapter 14). Once in the photic zone, these nutrients are incorporated into particles and these particles in one fashion or another eventually sink from the photic zone into deeper waters where decay of the organic matter leads to the release of the combined nutrients (cf. Chapter 11).

Nutrients in coastal waters and estuaries are more abundant than those in the open ocean, as we will see below, but even in coastal water, nutrients are usually depleted in the photic zone. Thus, phytoplankton exert a major influence on the availability of nutrients and, by reducing the supply of their essential resources, limit their own growth.

It is not sufficient to speak merely of nutrients in general, since nutrient limitation of growth is invariably due to a specific deficit of an essential substance. We will therefore examine the availability of nitrate and phosphate. These two nutrients have received the most attention because analytic methods have long been available and, as will be seen, nitrogen and phosphorus are thought to be most often limiting to phytoplankton production. In addition, there is a sharp contrast in the role of nitrogen and phosphorus in regard to limitation of primary production in fresh and seawater, a comparison we will also examine.

Nitrate is generally far more abundant in freshwater than in seawater (Fig. 2-18, top), where nitrate concentrations are around 1 ugat NO_3-N liter^{-1} or less and rarely exceed 25 μgat liter^{-1}. The concentrations of phosphate in fresh and seawater cover approximately the same range, from 0 to about 3 μgat liter^{-1} (Fig. 2-18, bottom). Possible geochemical reasons for the differences in concentrations of these nutrients are discussed in Chapter 11.

* Oceanographers express concentrations in gram-atoms of an element per liter, since this makes it clear that, for example, a reported value refers to the nitrogen in nitrate (NO_3^-) and does not involve the oxygen. The actual expression is usually shown as μgat NO_3-N liter^{-1}. Concentrations are also often stated in molar (M) units. This is just as convenient a system of units, and μM NO_3^{-1} is equivalent to gat NO_3-N liter^{-1} in the case of compounds such as NO_3^- or NH_4^+, where one atom of the μ element in question is present.

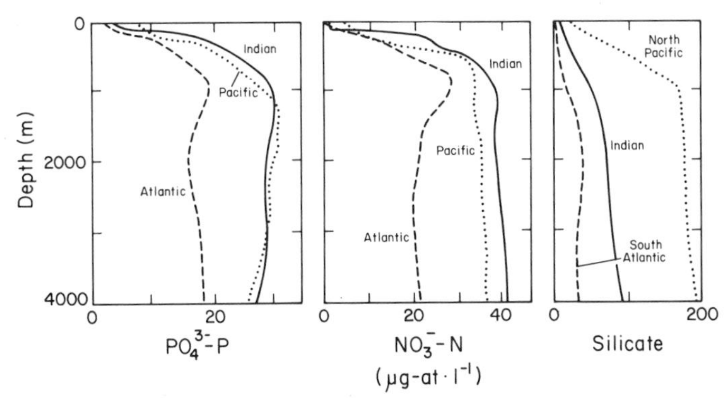

Figure 2-17. Vertical distribution of phosphate, nitrate, and silicate in the three major oceans. Adapted from Richards (1968).

Phosphate seldom achieves concentrations comparable to those of nitrate because it is strongly adsorbed to particles and forms insoluble salts (cf. Section 11.1), while nitrate is highly soluble (cf. Section 11.2). The concentration of other forms of nitrogen (reviewed by McCarthy, 1980)

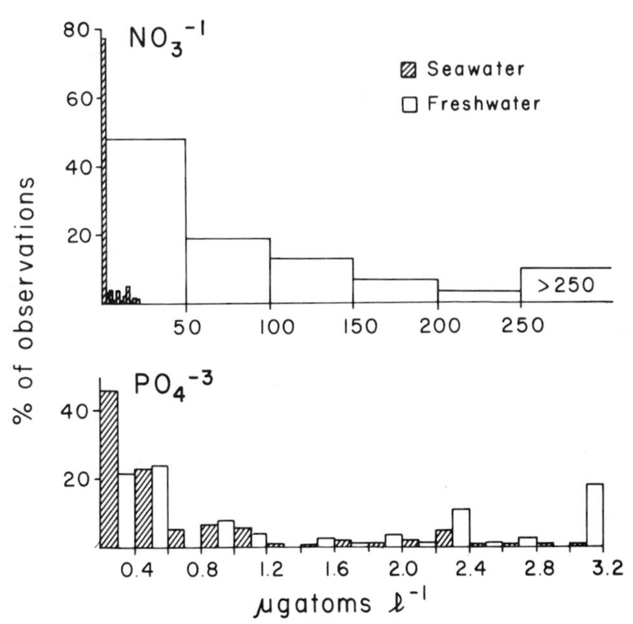

Figure 2-18. Concentrations of nitrate and phosphate in many samples of fresh and seawater. Values for seawater are for surface waters. Observations for this compilation were obtained by using values reported in *Limnology and Oceanography* since 1971 and reviews by Zenkevich (1963), Zeitzschel (1973), and Livingstone (1963). When many analyses of a specific site or station were available, representative values were chosen for each site.

used by producers—ammonium, urea, amino acids—are as variable and generally as low or lower than nitrate values. Different forms of nitrogen may be transported from surrounding watersheds into freshwater bodies. These inputs may have N:P ratios of 40:1, so that in freshwater nitrogen is considerably in excess of the 16:1 Redfield ratio* used by plants during growth. Further, nitrogen-fixing cyanobacteria are more plentiful in freshwater and their activity may also increase N:P ratios.

2.232 Nutrient Contents of Producers in Seawater and Freshwater Environments

Marine phytoplankton contain less nitrogen than freshwater phytoplankton relative to phosphorus (Fig. 2-19). The ratios of N to P span a very large range, indicative of varied supply and physiological state of cells, but marine phytoplankton most often have a N:P of about 10 or 20:1, comparable to the Redfield value of 16 : 1.

Marine macroalgae and vascular plants have ranges of nitrogen and phophorus content similar to those of freshwater species (Fig. 2-20). Note, however, that the modal nitrogen content of marine species is lower than that of freshwater species (Fig. 2-20, top). Marine species have a larger modal phosphorus content than do freshwater species (Fig. 2-20, bottom).†

The differences in nutrient contents of cells or tissues observed in Figs. 2-19 and 2-20 are likely to be due to differences in availability of nitrogen and phosphorus between fresh- and seawater (Fig. 2-18); this implies that in most marine waters nitrogen may be limiting to producers while in freshwater phosphorus is limiting. This contrast is corroborated by data in Fig. 2-19, where the dotted and dashed bars refer to N to P ratios measured in situations where the phytoplankton were clearly identified as being nitrogen or phosphorus limited. These data lead to the hypothesis that phosphorus limitation occurs more often in freshwater phytoplankton, while nitrogen limited growth is more common in marine phytoplankton. The relationships described above are only descriptive correlations; more critical tests of the hypothesis have been obtained by the

* Early studies in the chemical composition of plankton (Redfield, 1934) showed that particulate matter had carbon:nitrogen:phosphorus of 106 : 16 : 1, a relationship now referred to as the Redfield ratio. This was extended to claim that uptake of N and P from seawater would also follow a 16:1 ratio (Redfield et al., 1963; Corner and Davies, 1971). These ratios are convenient averages but may not always be matched under field conditions (Banse, 1979).

† The anomalous high frequency of very low (most are 0 or trace) phosphorus contents in Fig. 2-20 is due to coralline algal species whose cells are encrusted with calcareous deposits that increase the total weight. The data on these species are thus difficult to compare to other algal taxa. The very high values of phosphorus for freshwater plants are from sedges (*Cyperus*) in environments with high phosphorus. Other *Cyperus* species growing in low phosphorus environments lie in the class with lowest P content.

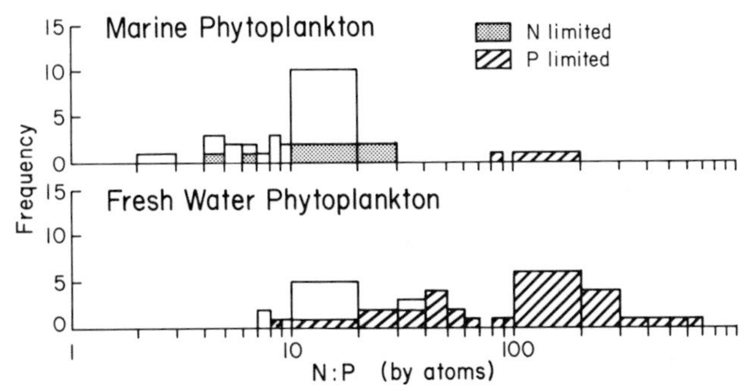

Figure 2-19. Nitrogen to phosphorus ratio in marine and freshwater phytoplankton. Where the ratio was obtained from cells growing in nitrogen or phosphorus limited cultures or come from low phosphorus environment (less than 0.1 mg P liter^{-1}), it is so indicated. Data from many sources, including Hutchinson (1975) and Jorgensen (1979).

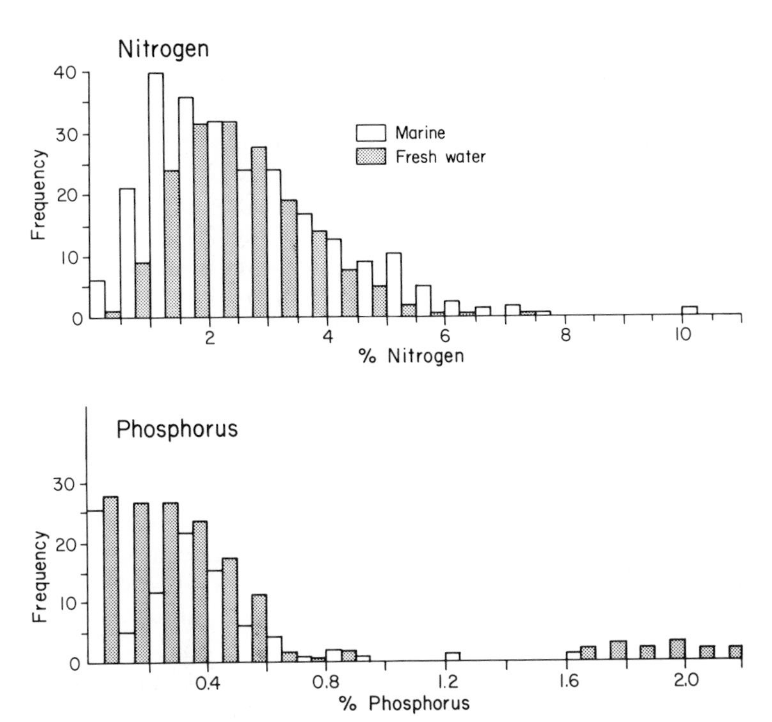

Figure 2-20. Nitrogen and phosphorus contents of marine and freshwater macroalgae and vascular plants. Compiled from many sources. The marine data from Vinogradov (1953) and articles in *Limnology and Oceanography, Journal of Phycology, Aquatic Botany, Botanica Marina, Marine Biology* of the last 10 years.

experimental work done in freshwater, coastal, and oceanic waters that is described in the next section.

2.233 Evidence for Nutrient Limitation in Phytoplankton

2.2331 Evidence from Freshwater Studies

For many years there was controversy among researchers working on freshwater systems about which element limited growth of freshwater producers. This became a specially important subject with the advent of phosphate-containing detergents, since the resulting phosphate-laden waste water entered water bodies and created eutrophic lakes and ponds, and obnoxious algal blooms. A series of laboratory and field experiments, especially the results of whole-lake fertilization experiments in Canadian lakes (Schindler, 1974, 1977), provided conclusive evidence that phosphorus is the limiting nutrient in freshwater bodies.

A typical response of phytoplankton in a freshwater-dominated coastal pond is seen in Fig. 2-21, bottom, where a growth response to phosphorus addition* is clear. There is a further stimulation of growth when both nitrogen and phosphorus are provided.

A nutrient is limiting only when in short supply; if concentrations of nitrogen, for example, are high, other elements become limiting to growth. Chemostat experiments where the ratio of N to P is varied show that there is a threshold below which algal cells are nitrogen limited and above which the culture is phosphorus limited (Fig. 2-22, top and middle). When one nutrient is limiting, the nutrient in excess accumulates in the cells (Fig. 2-22, bottom), as can also be seen in Fig. 2-19. The effect of the two nutrients is therefore not simultaneous, additive, or multiplicative, but rather follows Liebig's (1840) law that the resource in smallest supply is the limiting factor.

Different species of phytoplankton grow best at different concentrations and ratios of essential nutrients. In the Great Lakes, the ratio of Si to P may determine which of two species of phytoplankton predominate† (Fig. 2-23). Depending on the ambient ratios, therefore, competitive exclusion, or replacement of one species by another, can occur. Experiments where various concentrations of different kinds of nutrients are provided to natural phytoplankton do not, however, always provide easily interpretable results (Frey and Small, 1980). So far, it has not been possible to explain the species composition of phytoplankton only on the basis of the suite and concentration of nutrients available to the algae.

* This kind of laboratory experiment is artificial in that sedimentation and grazing losses from the water are curtailed. Further, the regeneration of nutrients provided by grazers is also absent. Nonetheless, such experiments are a convenient, simple description of the situation.

† There is little evidence that silica by itself limits phytoplankton in freshwater, even though it is a major component of diatom and other phytoplankton (Paasche, 1980).

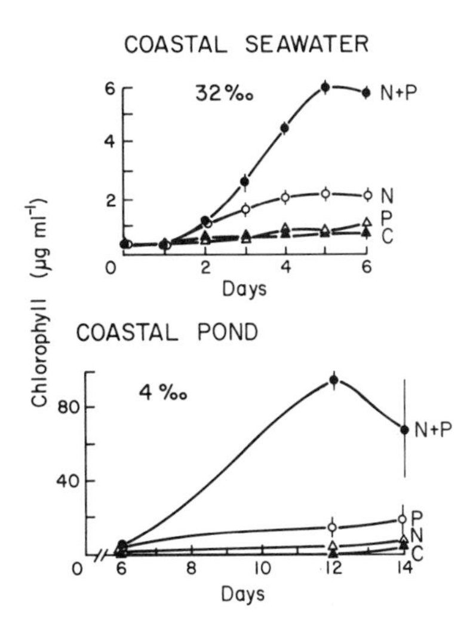

Figure 2-21. Enrichment experiments with coastal seawater of Vineyard Sound (salinity 32‰), Massachusetts, and a freshwater-dominated coastal pond (salinity 4‰) in Falmouth, Massachusetts. N+P, addition of nitrogen and phosphorus; P, addition of phosphorus; N, addition of nitrogen; C, control, no nutrient addition. Adapted from Vince and Valiela (1973) and unpublished data of Nina Caraco. Values are mean ± standard error of several replicates.

It seems odd that assemblages of multiple species of algae are the rule in the water column, an environment that appears to be very homogeneous. This is Hutchinson's (1961) "paradox of the plankton." In a homogeneous environment competition should have long ago led to dominance of the competitively superior species. Richerson et al. (1970) argue that, in fact, conditions in local, small-scale parcels of water change frequently enough so that the water column is quite patchy. Such local heterogeneities may consist of differences in nutrient content, or presence of absence of grazers. This local heterogeneity can potentially prevent one species

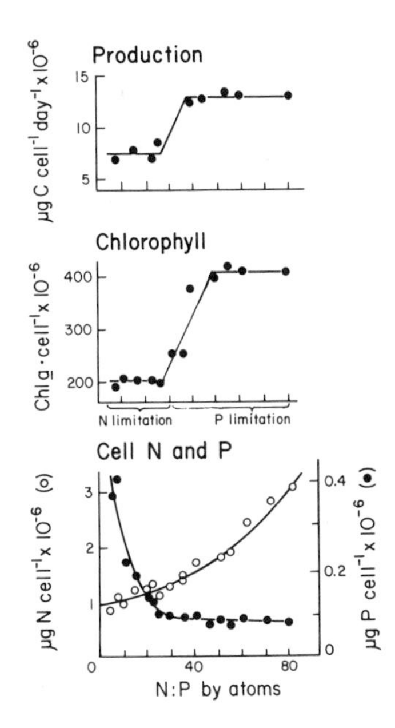

Figure 2-22. Production, chlorophyll, nitrogen, and phosphorus contents of *Scenedesmus* sp. in chemostats with varying N:P ratios. The N:P applies to both the cells and water in the system. Adapted from Rhee (1978).

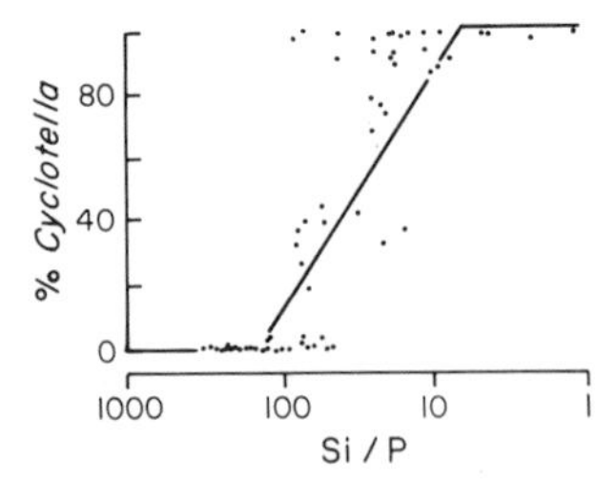

Figure 2-23. Abundance of *Cyclotella meneghiniana* compared to total of *C. meneghiniana* and *Asterionella formosa* in relation to the ratio of Si to P in water. The line is the expected proportion of *C. meneghiniana* based on chemostat experiments. Adapted from Tilman (1977). © Ecological Society of America, reprinted by permission.

from continually having a competitive advantage in and therefore monopolizing all patches. This "contemporaneous disequilibrium" may therefore be behind the presence of multispecies phytoplankton floras.

2.2332 Evidence from Studies in Coastal Waters

In coastal marine systems, nitrogen is the principal limiting element, as demonstrated by field observations and experiments. One of the early demonstrations of the role of nitrogen in coastal waters is the study by Ryther and Dunstan (1971) off New York Harbor, where organic particles and nutrients are high and decrease with distance from land (Fig. 2-24, top and middle). River and other freshwater inputs are often the major source of nutrients for some coastal waters. The decrease of dissolved inorganic nitrogen away from land is especially marked and suggests that this is the limiting nutrient. Stronger evidence is provided by experiments where seawater from a series of stations in the New York Bight area received additions of ammonium and phosphate (Fig. 2.24, bottom). Enrichment with nitrogen increased growth of the algae in 14 out of 15 stations, while additions of phosphorus did not increase growth. The stations nearest the nutrient source (a combination of eutrophic river water and urban sewage) were so nutrient rich that no response to further nutrient addition was seen. The phytoplankton from deeper waters responded to nitrogen additions, but the blooms were much smaller than those of harbor or shelf phytoplankton. This may be a secondary limitation due to a lack of trace nutrients or chelators of toxic substances, a topic discussed below.

The results of Ryther and Dunstan (1971) corroborated earlier conclusions by G. Riley and J. Conover based on interpretation of seasonal cycles of nutrients and plant growth, and also agreed with the results of the large bag experiments (Antia et al., 1963). Many other subsequent enrichment experiments (Fig. 2-21, top) (Laws and Redalje, 1979, for example) in coastal water also support the conclusion that nitrogen is the

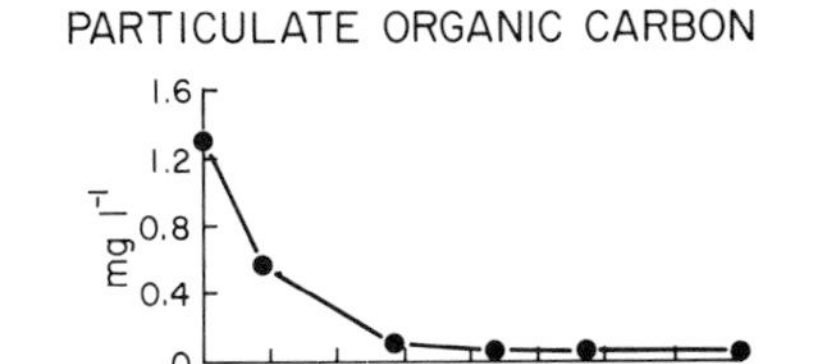

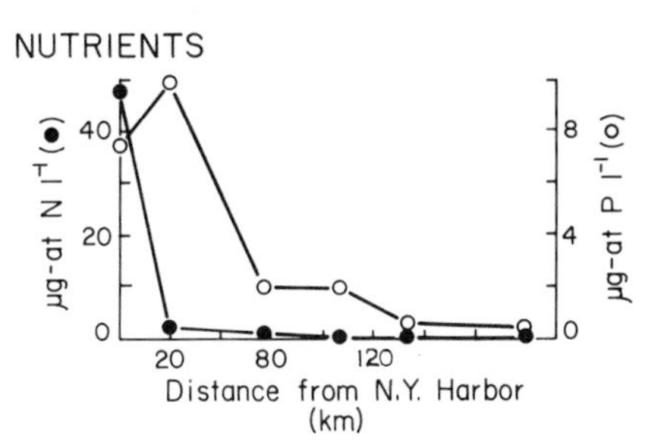

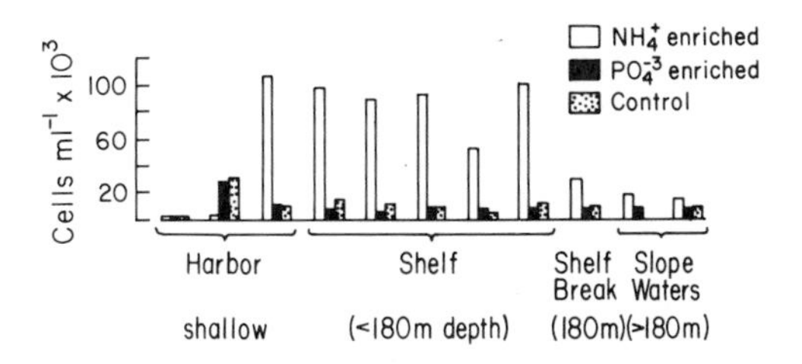

Figure 2-24. Top and middle: Concentrations of particulate organic carbon and nutrients in surface water in a transect from New York Harbor to offshore. Bottom: Growth of *Skeletonema costatum* in water samples that were enriched with ammonium or phosphate and in unenriched samples. The sequence of stations is in relation to their distance from the source of nutrients in New York Harbor. The inoculum with which the experiments were started was of the same size as the left-most station in the graph. Adapted from Ryther and Dunstan (1971). © AAAS, reprinted by permission.

primary limiting nutrient for coastal phytoplankton. This basic finding has fundamental applied importance: the nitrogen content of any effluent entering coastal waters is of considerable importance to management of the coastal zone, since it is closely linked to the level of primary production.

Other nutrients can be secondarily limiting. Once a nitrogen source is provided, there is a secondary limitation due to phosphorus (Fig. 2-21, top) and so on with other essential nutrients. Thus, if nitrogen is in excess of Redfield ratios in wastewaters entering a body of water, the added phosphorus may also be important.

Other essential or toxic elements may also be important. In recently upwelled coastal waters rich in nitrate the availability of dissolved metals may limit phytoplankton production. Iron and manganese, essential components in many biochemical reactions, may be limiting. Cupric ion at certain concentrations may reduce phytoplankton growth. On the other hand, appropriate concentrations of organic chelators may reverse the toxic effect of cupric ions (Sunda et al., 1981). It is not at all clear whether the role of such elements is widespread, or if they are only of local importance.

2.2333 Evidence from Oceanic Regions

Nitrogen may also be the primary limiting element in nutrient poor oceanic waters (Thomas, 1969). Nitrate and ammonium may be undetectable near the sea surface (McCarthy, 1980), undoubtedly because of uptake by phytoplankton. Sharp et al. (1980) find that phytoplankton grow slowly in the oligotrophic waters of the North Pacific, with doubling times of 5–9 days. The nitrogen-to-carbon ratios in the particulate matter are considerably lower than the Redfield ratios, suggesting a lack of nitrogen, a conclusion corroborated by Perry and Eppley (1981) based on uptake of radioactive isotopes.

Others claim that despite the low concentrations of nutrients in oceanic water, populations of oceanic phytoplankton grow at high rates. In the oceanic waters of the Sargasso Sea, specific growth rates of phytoplankton cells range between 0.05 to about 1 doubling per day, with modal values between 0.3 and 0.7 division day^{-1} (Goldman et al., 1979). This is comparable to the range of specific growth rates measured in coastal water, even though nutrient concentrations in the latter are higher. If these estimates are true, oceanic phytoplankton may therefore double every 2–4 days or so in spite of low nutrient concentrations.

Further, phytoplankton in the open Atlantic contain nitrogen and phosphorus in the 16:1 Redfield ratio, which in the laboratory is associated with maximum division rates (Goldman et al., 1979). This also suggests fast division rates in oceanic water.

Two hypotheses may reconcile the co-occurrence of fast growth and low ambient nutrients. The first hypothesis is that oceanic phytoplankton may saturate uptake rates and growth rates at very low nutrient concentrations, as in examples given by Goldman and McCarthy (1978) and McCarthy (1980). It is not known how common this is or if artifacts of measurement of nutrient uptake at low S are important. Oceanic assemblages may be made up of relatively rare phytoplankton species that are physiologically adapted to efficient and fast uptake of nutrients at low concentrations.

The second hypothesis is based on the idea that the average bulk concentration of a nutrient in a typical sample of several liters may not be relevant to an alga a few to tens of micrometers in diameter. An algal cell extracts nutrients from a few microliters of surrounding seawater, not from larger volumes of seawater. It may be that small parcels of water exist where the concentration of nutrients is considerably higher than in the rest of the water volume. McCarthy and Goldman (1979) and Turpin and Harrison (1979) hypothesize that excretion by zooplankton could produce tiny, nutrient-rich plumes that may be used by oceanic algae.

There are at least three difficulties with this second hypothesis. First, with current techniques it is difficult to measure nitrogen compounds in samples of a few microliters; new methods need to be developed to test directly the existence of such tiny parcels of high nutrient water. Second, recall that nonlimiting nutrients accumulate in internal pools (cf. Fig. 2-21). If some other essential element rather than N or P is limiting, the concentrations of N and P, and the N:P ratio could be high simply because the cells accumulate nitrogen and phosphorus while their growth or division are arrested by lack of something else, perhaps a trace element. Third, diffusion of nitrogen from tiny plumes may be faster than the rate of nitrogen uptake by cells, so that the plumes may be too short-lived to be useful to phytoplankton (Jackson, 1980). Lehman and Scavia (1982), however, demonstrated that freshwater phytoplankton in experimental vessels in which the water was not stirred could take up significantly higher amounts of PO_4 excreted by zooplankton than in vessels that were stirred. Stirring presumably destroyed the small-scale heterogeneities in nutrient concentration due to zooplankton excretion, and resulted in a homogenous but low nutrient concentration less useful to the algae. Lack of stirring seems to have allowed the high phosphate patches to last long enough for algae to make use of them.

Although the issue of the use of small patches by algae is not settled, excretion from animals must be a major pathway by which dissolved nutrients are regenerated in the oceanic water column (McCarthy, 1980; Eppley, 1980). These regenerated nutrients must be distributed in patches of large enough in extent and concentration so that phytoplankton can take up the nutrients, even though we lack adequate evidence.

The hypotheses offered by Goldman et al. (1979) and McCarthy and Goldman (1979) have stimulated new questions. Critical examination and testing of such insights in the lab and field will further understanding of the processes controlling primary production in the sea. These hypotheses deal with specific growth rates or the doubling time of cells. We still need to make the jump from this physiological measure of growth to a more ecological measure of production. Consider that oceanic phytoplankton occurs at very low cell densities. The production of phytoplankton in oceanic water, even with a division rate of four to five times a day, and average density of, say, 5 cells ml^{-1}, will not compare favorably with that of coastal water with a cell density of, say, 10,000 cells ml^{-1}, even if the coastal phytoplankton divide only once every 2 days. We need to understand not only the control of cell doubling times but also the control of abundances and seasonal cycles in oceanic water. This is discussed further in Chapters 8 and 14.

In oceanic waters it is likely that trace elements play more important roles limiting production rates that in coastal waters. There are some unusual oceanic areas that are relatively nutrient-rich (more than 2–6 μgat NO_3-N $liter^{-1}$). In one such area in the Equatorial Pacific, experimental addition of nutrients to the water column did not result in enhanced growth, much like some of the results obtained in enrichment experiments off New York. In these specific situations phytoplankton do not seem to be limited by nitrogen or phosphorus. We have so far emphasized nutrients—N and P—used in large amounts by producers,* but in oligotrophic waters limitation of growth by trace elements may be important. In very oligotrophic lake water, the short supply of molybdenum restricts algal growth (Goldman, 1972). In oligotrophic marine situations there have been a series of observations implicating iron as the primary controlling factor in the Sargasso Sea (Menzel et al., 1963) and in the Indian Ocean (Tranter and Newell, 1963). Iron additions, however, did not consistently stimulate growth in experiments in the Central Pacific (Thomas, 1969).† The response of mixed oceanic phytoplankton to additions of trace elements does not conclusively show limitation by such elements. Much more work is needed in this field, especially in regard to nutrient-metal interactions.

Trace substances may not only limit growth but also inhibit growth (Steemann Nielsen and Wium-Anderson, 1970). Cupric ion activity as low as 4×10^{-11} M can reduce cell mobility in red tide organisms (Anderson and Morel, 1978). Other substances, including some organic compounds, can also inhibit growth of microalgae and bacteria (Harrison and Chan, 1980; Conover and Sieburth, 1968; Sieburth and Conover, 1965). Trace metals can be chelated by natural organic matter. Humic substances are products of decomposition of plant materials that bind copper and other toxic ions so as to reduce the activity of metals in solution (Sunda and Guillard, 1976; Sunda and Lewis, 1978; Prakash et al., 1973; Barber et al., 1971). Such organic compounds are often transported by river water into the sea, where they enhance growth of diatoms and dinoflagellates (Fig. 2-25). Perhaps this is why red tide blooms are often recorded after periods of high discharge by rivers.

2.234 Evidence for Nutrient Limitation in Macroalgae and Vascular Plants

The relationships between ambient nutrients, uptake, storage, and growth by macrophytes are similar to those for one-celled algae. Growth of marine macroalgae and vascular plants is primarily limited by nitrogen supply. Experimental additions of nitrogen to salt marsh vegetation (Fig. 2-26) or to the growth medium of macroalgae (Fig. 2-27, top) result in increases in growth rate. As ambient nitrogen increases, growth rate increases up to a point, as in the case of phytoplankton. Beyond such a threshold, nitrogen is stored with no further increase in growth rate (Fig.

* There are low but detectable concentrations of Si over much of the oceans. In upwelling regions intense bloom of diatoms may deplete silicon, but there are few instances of direct evidence of Si limitations for the sea (Paasche, 1980).

† F. Morel (personal communication) proposes that this is because most iron exists as Fe^{3+} in seawater. Phytoplankton take up Fe^{2+} more readily than Fe^{3+} and have to rely on photoreduction of Fe^{3+} to Fe^{2+} to be able to take up iron. The iron added in the experiments may have been—or may have been quickly converted to—oxidized iron.

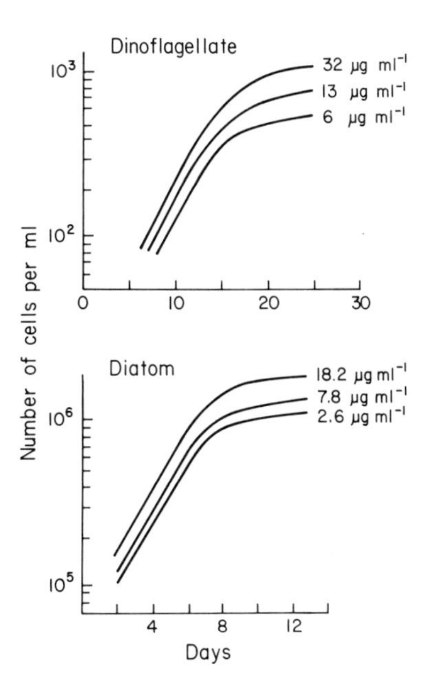

Figure 2-25. Top: Growth of the dinoflagellate *Gonyaulax tamarensis* in synthetic marine medium enriched with different amounts of humic acid extrated from marine sediments. Bottom: Growth of *Skeletonema costatum* in enriched seawaater medium under additions of different amounts of humic acid extracted from decomposed seaweed (*Laminaria*). Adapted from Prakash et al. (1975).

2-27, bottom). The lack of further increase is due to a secondary limitation by some other factor, again as in phytoplankton. In salt marsh plants, phosphate is the secondarily limiting factor (Fig. 2-26). If the major nutrients occur at high concentrations, trace metals may limit growth in coastal macrophytes (Prince, 1974). This is another example of Liebig's law: there is always one limiting factor.

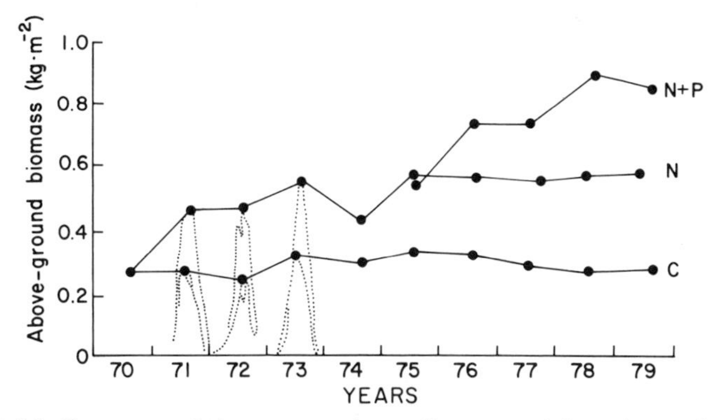

Figure 2-26. Response of above-ground standing crop of the salt marsh cordgrass (*Spartina alterniflora*) to experimental enrichment with nitrogen (1970–1979) and with a mixed nitrogen plus phosphorus fertilizer (1975–1979). Addition of phosphorus alone did not change the amount of biomass and is not shown. The dotted lines in 1971–1973 shows the seasonal pattern of growth of the plant. Adapted from Valiela et al. (1982).

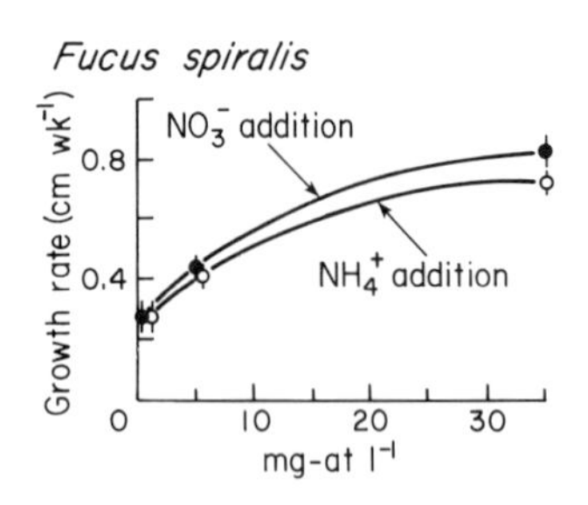

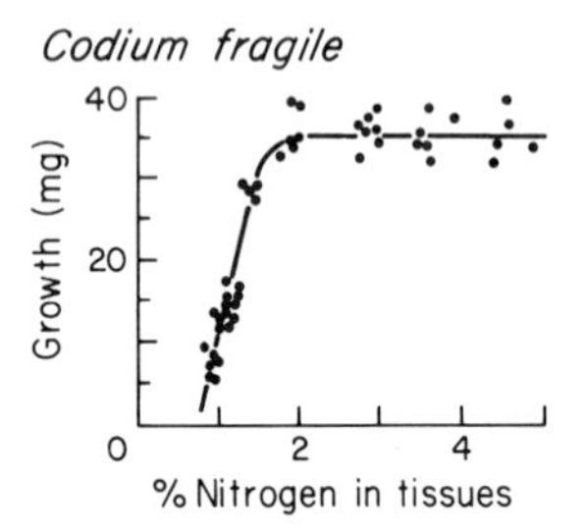

Figure 2-27. Top: Growth rate (elongation of fronds) of the brown alga *Fucus spiralis* in ambient concentrations (1.2 μg atom NO_3^- liter^{-1} and 1.7 μg atom NH_4^- liter^{-1} and in cultures where additional NH_4^- and NO_3^- were furnished. Adapted from Topinka and Robbins (1976). Bottom: Growth in weight of the green alga *Codium fragile* in relation to the percentage nitrogen in the tissues. Adapted from Hanisak (1979).

For rooted coastal plants where the sediment is anoxic, nutrient uptake may be inhibited by the reduced environment. Although nutrient concentrations may be high in interstitial water, anoxia inhibits the mechanisms by which plants actively take up nutrients, so that in such plants nutrient limitation may be mediated by anaerobic conditions (cf. Section 14.33).

2.3 Temperature and Interactions with Other Factors

We will see in Chapter 14 that temperature does have some influence in seasonal production cycles, but there are few observations that document important effects of temperature on rates of primary production in the sea. In certain situations under clear ice in the Antarctic Ocean, the rate of photosynthesis and growth may be limited by low temperatures (Bunt and Lee, 1970). The effect of cold is not general, since in most temperate waters, the spring bloom occurs at about the coldest part of the year.

There are measurable effects of temperature on growth rates of phytoplankton in laboratory cultures (Fig. 2-28), and local populations of many species of algae have peak growth rates in certain specific ranges of temperature (Hulburt and Guillard, 1968; Goldman and Carpenter, 1974). The doubling rates measured under optimal temperatures in the laboratory (up to 9 doublings day^{-1}) are much higher, however, than those observed in

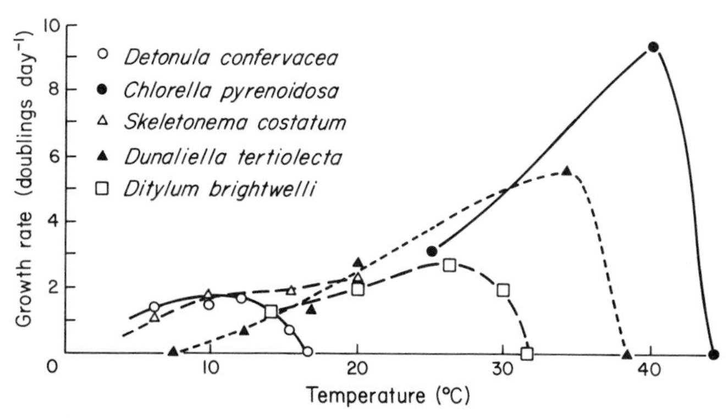

Figure 2-28. Growth rate of several phytoplankton species under different temperatures. Adapted from Eppley (1972).

the sea (0.25–1 doubling day^{-1}) (Eppley, 1972) so that it is hard to extrapolate the laboratory effects of temperature to the sea.

The small effect of temperature is further evidenced by the frequent occurrence of a poor correlation between ambient temperatures of the water where a particular species is found and its physiological temperature response. For example, *Thalassiosira nordenskioldii*, described as an Arctic species on the basis of where it occurs (Durbin, 1974), grows very well at temperatures warmer than 10°C. Such observations suggest that temperature does not normally determine the abundance and growth of algae in marine environments.

Temperature may be more important as a covariate with other factors than as an independent factor. For example, cells at low temperatures maintain greater concentrations of photosynthetic pigments, enzymes, and carbon (Steemann Nielsen and Jorgensen, 1968). This may result in more efficient use of light at low temperatures and higher photosynthetic rates of acclimated cells than would be predicted on the basis of temperature responses of unacclimated cells. Photosynthetic rates and growth rates of algae can be altered by light, as seen in Section 2.112, but increases in the ambient temperature can also increase maximal photosynthetic rates at any level of light intensity (Fig. 2-29, top left). In certain species, the day length may also significantly change rates of doublings, but does so only at high temperatures (Fig. 2-29, right). Higher temperatures also facilitate the uptake of nutrients (Fig. 2-29, top left), so that at any one concentration of phosphate in the medium, the maximum photosynthetic rate increases at warmer temperatures. In the case of algae under sea ice in the Antarctic Ocean, cited above, light was available, although low, while nutrients were high. In Tokyo Bay (Fig. 2-29, bottom left) nutrients were high, but the turbidity typical of shallow coastal wa-

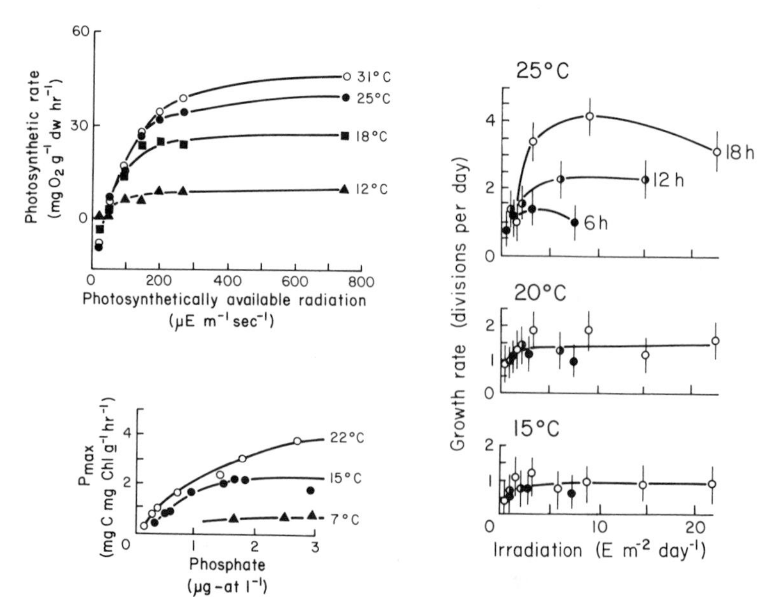

Figure 2-29. Interaction of temperature with light and nutrients. Top left: Photosyntetic rate of *Cladophora albida* under different levels of light intensity and temperatures in estuarine water. Adapted from Gordon et al. (1980). Right: Mean (± standard deviation) division rates during exponential phase of growth in *Talassiosira fluviatilis* at three temperatures and daylengths (18, 21, and 6 hr). Adapted from Hobson (1974). © Canadian Journal of Aquatic and Fisheries Sciences, reprinted by permission. Bottom left: Maximum photosynthetic rate (P_{max}) of natural phytoplankton of Tokyo Bay under varying phosphate concentrations and temperatures. Adapted from Ichimura (1967).

ters created low light. From the data available it is difficult to say whether light or temperature was most important.

Though temperature may not be a primary limiting factor in the sea it may have other consequences. There are increasing numbers of nuclear power plants over many of the world's coasts; these plants use seawater for cooling purposes and release heated water into the coastal zone. When the discharge results in increases in temperatures of the receiving water, changes in species composition may result. In freshwater, there is a transition from diatoms to greens to cyanobacteria in the immediate vicinity of the discharge canals. The abundance of certain dinoflagellates may increase in heated estuarine waters (Goldman and Carpenter, 1974). Disposal of heated water may therefore lead at least to changes in the species making up the phytoplankton community within the affected area. The species favored by higher temperatures, cyanobacteria and flagellates, moreover, may be less desirable as food for consumers than other kinds of phytoplankton.

2.4 Distribution of Phytoplankton Production Over the World Ocean

Knowledge of the concentration of nutrients, amount of light, and temperature are not sufficient to explain the distribution of primary production over the world's oceans (Fig. 2-30). A reductionist explanation including only these variables fails to explain the observed complexity. These variables certainly provide the basis for potential production, but actual production is the result of the movement of water masses coupled to nutrient and light conditions. Hydrography largely determines what the nutrient supply is within the photic zone, and how deep phytoplankton are carried by vertical mixing of water and hence how much light is available.

There are texts that describe water mass movements in the oceans (Pickard, 1964; Dietrich, 1968; Gross, 1977). Here we will only briefly discuss a few examples, enough to convey the idea of how major the effects of hydrography can be.

The most dramatic evidence of how hydrography controls production comes from regions of upwelling. Where upwelling of deeper water or other sorts of advection delivers nutrients in substantial amounts, primary production may be high. There are several types of hydrographic features where vertical advection of nutrients takes place, including coastal upwellings and divergences.

Coastal upwellings are common off Peru, northwest Africa, eastern India, southwest Africa, and the western coast of the United States. Their occurrence is marked by areas of high production located mainly along the western margins of continents (Fig. 2-30). Prevailing winds blowing somewhat parallel to a coastline move water in the direction of the wind; forces resulting from the rotation of the earth deflect the actual path of the water. In the western margin of continents, this Coriolis effect results in surface water being moved offshore. Colder, nutrient-laden deeper water moves up to the surface to replace displaced surface water (Fig. 2-31). This is far too simple a description of the dynamics of upwelling; recent work shows that the actual movement of water and nutrients in coastal upwellings is very variable and that the spatial and temporal patterns are complex (Boje and Tomczak, 1978).

The effects of the rotation of the earth on wind-driven currents are reversed in the two hemispheres, so in the western margin of continents northerly winds result in upwellings in the northern hemisphere while southerly winds lead to upwellings in the southern hemisphere. The Coriolis effect does not lead to high productivity in eastern margins of continents because the deflection moves nutrient-poor oceanic water toward the coast. This may also result from northerly winds in the southern hemisphere and southerly winds in the northern hemisphere. An example

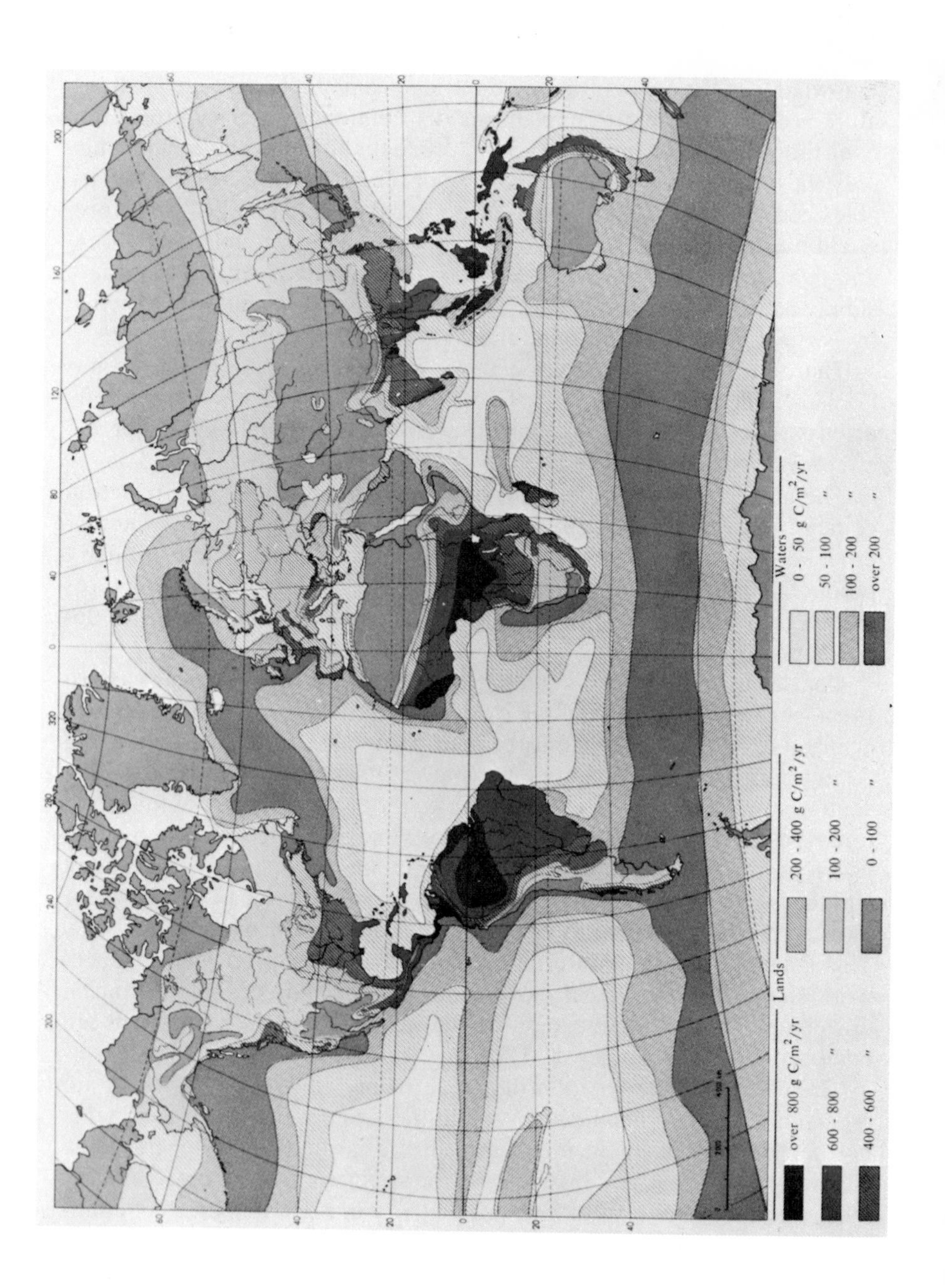
Lands
over 800 g C/m²/yr
600 - 800 "
400 - 600 "
200 - 400 g C/m²/yr
100 - 200 "
0 - 100 "
Waters
0 - 50 g C/m²/yr
50 - 100 "
100 - 200 "
over 200 "

Figure 2-30. Estimated world distribution of annual production. Notice that land patterns are markedly different from oceanic distribution of productivity. From Lieth (1975).

of the latter is the "downwelling" in the Washington-Oregon coast during winter (Duxbury et al., 1966).

Upwelling of nutrient-rich water may occur through a variety of hydrological mechanisms and is not limited to the coastal zone. In the open ocean divergences bring nutrient-rich water to the surface. A prominent feature of the Pacific is the fast subsurface Equatorial Undercurrent, leading from west to east in a fairly narrow band around the Equator. Near the Equator surface currents are slow. The contact of the moving equatorial surface water and the fast Equatorial Undercurrent results in shear stress that leads to a large amount of turbulent mixing and a renewal of nutrients in the photic zone. This produces a band of relatively high production (Table 2-3) in the mid-Pacific near the Equator (Fig. 2-30).

Another example of offshore upwelling of nutrient-rich water is the zone of high production found in the Southern Ocean (Fig. 2-30)—the waters surrounding Antarctica associated with the Antarctic Divergence (Fig. 2-32). We should note parenthetically that the high values of production reported in our map may reflect the concentration of studies in Antarctic waters during the spring and summer, times when phytoplankton production rates reach their highest levels (Glibert et al., 1982). The water near the surface is exposed to westerlies that drive the eastward-moving Antarctic Circumpolar Current. A northward component of flow of this current moves deeper water to the surface. This newly upwelled water,

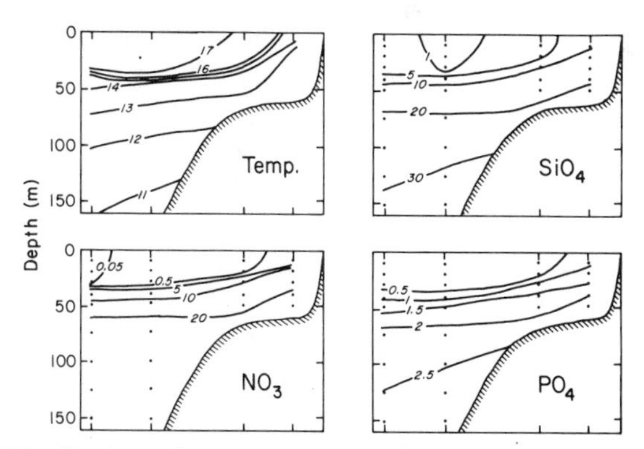

Figure 2-31. Distribution of temperature gradients, silicate, nitrate, and phosphate in the upwelling zone off Punta San Hipolito, Baja California. Temperature is in °C and nutrients in microgram-atoms per liter. The dots show the stations where water samples were taken. The farthest station was about 40 km from shore. Note the upward trend of the isoclines as the water approaches the coast. Adapted from Walsh et al. (1974).

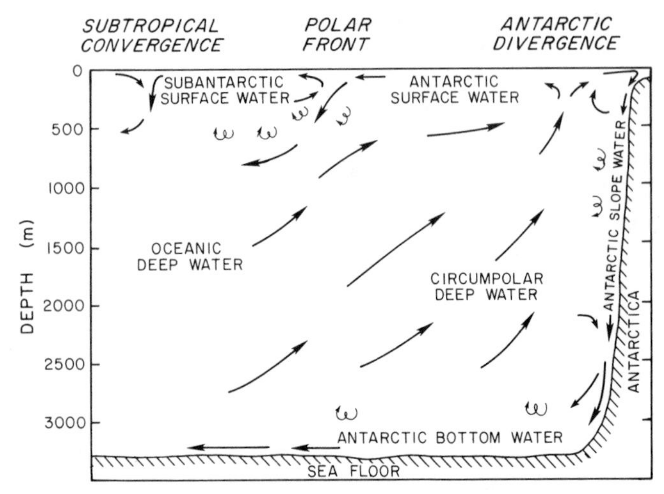

Figure 2-32. Schematic representation of flow of water masses in Southern Ocean, simplified from Gordon (1971). The arrows show direction of flow while the curlycues show areas of mixing.

now called the Antarctic Surface Water (Fig. 2-32) provides a very rich supply of nutrients that supports the very productive assemblage of phytoplankton, krill, penguins and other sea birds, seals, and whales of the Southern Ocean (cf. Section 9.42). The upwelled water extends far from Antarctica; the northern limit of this zone of high production is marked by the Polar Front or Antarctic Convergence, near 40–50°S, where the still-rich Antarctic surface water sinks below the warmer subantarctic surface water (Fig. 2-32). After sinking, the Antarctic surface water is renamed the Antarctic Intermediate Water and goes on to influence most of the oceans of the southern hemisphere. The northward-flowing Antarctic water masses are eventually warmed and their oxygen is consumed by biological activity deep in the North Atlantic. Eventually the deep water flows southward again. The conversion of deep water to surface water and back again with the consequent reacquisition of oxygen has been termed the "breathing" of the ocean (Gordon, 1971) and a key role in this process is played by the Southern Ocean.

Another prominent band of high annual production is at intermediate latitudes in the northern hemisphere. Its latitudinal location is more variable than that of its counterpart in the southern hemisphere due to the influence of land masses on circulation. Its geography is associated with the large-scale, clockwise gyres that circulate in the North Atlantic and North Pacific, and the western boundary currents (the Gulf Stream and the Kuroshio). Such major hydrographic phenomena lead to active turbulent mixing and larger eddies that result in renewal of nutrients in surface waters and hence in high production.

Other major areas of high production are the coastal zones (Fig. 2-30), where water- and wind-borne nutrients and particles originating from land are introduced into the sea. The shallowness of coastal water often allows surface disturbances to mix the entire water column vertically, a phenomenon that enhances transport of nutrients regenerated from the sediments into the photic zone. This frequent mixing, as we will see in Chapter 11, is one reason for high coastal primary production.

The low annual production rate evident in the Arctic Ocean is related to reduced amounts of light due to atmospheric filtering and the persistence of ice cover most of the year. Everywhere else in the ocean it is very hard to see broad regional effects of light limitation. For example, in the large subtropical gyres on either side of the Equator in the Atlantic and Pacific, where light is plentiful year-round, we find some of the lowest rates of primary production (Tables 1-2 and 2-3). In subtropical gyres such as the Sargasso Sea and in other open ocean environments, stable thermoclines prevent or reduce vertical mixing. The consequence of this is that nutrients, especially nitrogen, are very dilute or undetectable; production is thus characteristically low in much of the world ocean.

There are many other features of Fig. 2-30 that merit attention, but the few examples mentioned emphasize the major points of this section. We

Table 2-3. Estimated Nutrient Input, Plant Production and Other Properties of Three Different Subtropical Planktonic Ecosystems[a]

	Coastal upwelling	Equatorial divergence	Central gyre
Vertical velocity[b] (cm sec^{-1})	10^2–10^3	10^{-3}–10^{-4}	10^{-4}–10^{-5}
NO_3-N (μg atom l^{-1}) at 100 m	20–30	20–30	0–5
Primary production (g m^{-2} day^{-1})	1–10	1–5	0.1–0.5
Phytoplankton standing crop (mg C mg $chl.^{-1}$ hr^{-1})	60–180	45–90	15–30
Size of cells	Large (variable)		Small
Herbivore mean weight (μg C $animal^{-1}$)	40–100	4–40	4–40
Herbivore mobility	High	Lower	Low
Temporal variability of herbivore populations	High	Lower	Low
Steps in the food chain	1–2	3–4	5–6
Yield at top of food chain	High	Intermediate	Low

[a] Adapted from Walsh (1976).

[b] These are estimates of flow when the upwelling is active. At other times, there may be no vertical movement of water and nutrients.

have seen some evidence suggesting that low light may affect annual production; we will see more of this in Chapter 14, but if light were the dominant variable, one would not expect to find the distribution of production of Fig. 2-30. The same comments apply to temperature. In general, nutrients are the major factor limiting primary production in the world's oceans. The delivery and availability of nutrients, however, are mediated in a major way by the hydrography of the oceans, and this establishes the broad zones of primary production.

The same mix of effects of nutrients, light, and hydrography controls production by macrophytes in the margins of the sea, but local patterns of light and nutrients may be relatively more important to macroalgae and vascular plants than the regional patterns so obvious in phytoplankton. For example, salt marsh plants, sea grasses, or brown algae growing only meters apart can show differences in biomass and production as large as between their distributional extremes.

Part II

Consumers in Marine Environments

Combjellies (*mnemiopsis*), photo courtesy of Gray Museum, Marine Biological Laboratory, Woods Hole Oceanographic Institution.

Organisms that depend on organic matter produced by photosynthetic organisms range from bacteria to whales and are astonishingly diverse. Such consumers are heterotrophic and engage in secondary production.

Marine bacteria are hard to identify and many species have probably never been cultured. Some of the kinds found in the sea are discussed in texts such as Kriss (1963) or Fenchel and Blackburn (1979).

The small invertebrates are better known but still need work. Macroinvertebrates have been studied more thoroughly (see reviews by Barnes, 1980; Kaestner, 1970; Andersen, 1969). There are, for example, about 9,000 species of cnidarians, 20,000 molluscs (mainly bivalves), 8,000 annelids and perhaps 30,000 crustacean species. These are underestimates since, for example, even though about 1,300 species of copepods parasitic on fish and easily twice that many on invertebrates have been described, every new expedition to tropical reefs or deep sea collection finds many new species (A. Humes, personal communication).

Knowledge of natural history and systematics of marine organisms is a prerequisite for understanding their role in nature. The diversity of types of consumers precludes our presentation here; rather we focus on the ecological relations of representative populations. For many ecological purposes it is not necessary for ecologists to have an exhaustive taxonomic knowledge of the environment under study but the present trend of few young scientists specializing in systematics portends future problems. When the currently active generation of systematic specialists retires, we will have increasing difficulties in having specimens reliably identified.

In Chapter 3 we discuss the growth potential of consumer populations and the resulting changes in abundance over time, how such changes can be measured, and then introduce some theories on the reproductive strategies of consumers. Having established that populations can outgrow their resources, we show in Chapter 4 that competition for resources exists and that it has significant consequences for populations of marine consumers and for secondary production. In Chapter 5 we discuss consequences of changes in abundance of food, and outline the mechanisms by which consumers respond to changed resource abundance. Since food is usually in low abundance or is of suboptimal quality, selection of food of high quality must be important, so in Chapter 6 we examine the process of discrimination among food items. In Chapter 7 we describe how food consumed is used by consumers in metabolism, secondary production, and excretion, and discuss some important consequences of the various fates of energy. Chapter 7 concludes with an examination of energetics of populations and ecosystems, building on the information of previous chapters.

Chapter 3

Dynamics of Populations of Consumers

3.1 Elements of the Mathematical Description of Growth of Populations

The growth or decline of a population over a period of time can be expressed in terms of the changes in biomass (ΔB) or of numbers (ΔN). The changes that occur are the result of the sum of the individuals born (R) minus those dying (M) during the interval, the growth (G) of the individuals in the population, and of the net difference between emigration (E) and immigration (I). Thus we can write for biomass

$$\Delta B = R - M + G + I - E, \tag{3-1}$$

all expressed in units of biomass. The change in the number of individuals (ΔN) is expressed by

$$\Delta N = R - M + I - E. \tag{3-2}$$

In this case the growth of each individual is disregarded. The two equations are interchangeable by simply accounting for the biomass of the individuals in the population. We will return to biomass expressions in Chapter 7, where we deal with secondary production. In this chapter we concentrate on numbers of individuals, since most of the work on growth of populations is done on this basis.

The simplest case of population growth is where reproduction is by fission. After $0,1,2,3 \ldots n$ generations there will be x, $2x$, $4x$, $8x$, $16x, \ldots, nx$ individuals. This can be represented by the series 2^0x, 2^1x, 2^2x, $2^3x, \ldots, 2^nx$, so that after n generations the number of individuals (N_n) will be

$$N_n = 2^n N_0. \tag{3-3}$$

Under circumstances with ample resources, the rate of increase of number of individuals in the population accelerates exponentially while the time to double the population (t_d) is constant.

If we wish to measure the size (N_t) of the population after an interval of time (t) during which the population grows, we calculate

$$N_t = N_0 2^{t/t} d. \tag{3-4}$$

When the time elapsed equals the doubling time of the population, $t = t_d$, $N_t = 2N_0$. Equation (3-4) is usually linearized by taking logs of both sides of the equation,

$$\ln N_t = \ln N_0 \ln 2 \cdot t/t_d. \tag{3-5}$$

A plot of $\ln N_t$ versus t [or versus the number of generations in Eq. (3-3)] shows the slope or doubling time of the population. This doubling time is equal to $\ln 2/t_d$, with units of doublings/unit time.

A more general description of population growth, applicable to other forms of reproduction, is the differential form of the terms of Eq. (3-2) which describes instantaneous rates of population growth

$$\underset{t \to 0}{\text{limit}} \frac{\Delta N}{\Delta t} = \frac{dN}{dt} \tag{3-6}$$

One difficulty here is that individuals are discrete, but if populations are large the steps in abundance due to additions of discrete individuals will be small and will approximate a continuous variable.

If we standardize Eqs. (3-2) and (3-6) by dividing by the number of individuals, we obtain an instantaneous, specific growth rate

$$\frac{1}{N}\frac{dN}{dt} = b - d + i - e, \tag{3-7}$$

where b, d, i, e are the instantaneous rates of birth, death, immigration, and emigration. If we assume for now that emigration and immigration are approximately equal, and solve for N_t by integrating, we get

$$N_t = N_0 e^{(b-d)t}. \tag{3-8}$$

This is the basic equation for exponential increase (e is the base of natural logs) where N_t and N_0 are the population size at the start and end of the time interval. Taking logs and shuffling terms, we get

$$r = b - d = \frac{\ln N_t - \ln N_0}{t}, \tag{3-9}$$

where r is the instantaneous rate of growth of the population.

3.2 Survival Life Tables

3.21 Life Table Definitions and Calculations

The rate of growth of a population [Eq. (3-9)] is a function of the number of births and deaths, so to measure r we need to assess reproductive performance and survival rates of individuals throughout their life span. These properties of populations can be quantified by using schedules of births and deaths called life tables.

Survivorship tables consist of several columns. Table 3-1 illustrates a survivorship table for a barnacle population. The first column is x, the age classes into which we have divided the barnacle population. The next column, n_x, records the number of barnacles that were alive at the start of each age interval. The values of n_x are converted to a new column, l_x, by calculating the proportion of individuals initially present that survive to the start of x, $l_x = n_x/n_0$. In our example, for barnacles of 1 year of age, $l_x = 62/142 = 0.437$.

During any age interval deaths may occur, and an account of these is kept in the next column, d_x, the number of deaths during interval x to $x+1$. In the barnacle data, 80 individuals died between x_0 and x_1, leaving 62 individuals at the start of x_1. The age specific mortality rate during any

Table 3-1. Survivorship Table for the Barnacle *Balanus glandula* Settled at the Upper Shore on Pile Point, San Juan Island, Washington, U.S.[a]

Age in years (x)	Observed no. barnacles alive each year (n_x)	Proportion surviving at start of age interval x (l_x)	No. dying within age interval to$x$ + 1_n (d_x)	Mortality rate (q_x)	Age structure (L_x)	Mean expectation of further life for animals alive at start of age x (e_x)
0	142	1.000	80	0.563	102	1.58
1	62	0.437	28	0.453	48	1.97
2	34	0.239	14	0.412	27	2.18
3	20	0.141	(4.5)	0.225	17.75	2.35
4	(15.5)	0.109	(4.5)	0.290	13.25	1.89
5	11	0.077	(4.5)	0.409	8.75	1.45
6	(6.5)	0.046	(4.5)	0.692	4.25	1.12
7	2	0.014	0	0.000	2	1.50
8	2	0.014	2	1.000	1	0.50
9	0	—	—	—	0	—

[a] Data obtained 1–2 months after the 1959-year class settled and each year until 1968, by which time all of the 1959-year class had died. Entries in parentheses were estimated rather than counted directly. Data of Connell (1970), table adapted from Krebs (1978).

interval x to $x+1$ is the next column, $q_x = d_x/n_x$. During x_0 to x_1, $q_x = 80/142 = 0.563$.

One other important life table statistic is e_x, shown in the right-most column in Table 3-1. This is the mean expectation of future life for individuals that are alive at the start of x. The entries of the e_x column are calculated from $e_x = T_x/n_x$. L_x is the survivorship table age structure (the average number of live individuals during each age interval), and $L_x = (n_x + n_{x+1})/2$. T_x is the time units (years in our example) still to be lived by individuals alive at the start of x, $T_x = \Sigma_x^n L_x$, for n age intervals. For the 5-year-old barnacles, for example, $T_5 = L_5 + L_6 + L_7 + L_8 + L_9 = 16$. Then we can calculate e_x by dividing T_x, the years still to be lived, by the number of individuals (n_x) left alive at the start of x, and for the 5-year-old barnacles, $e_5 = T_5/n_5 = 16/11 = 1.45$ years. This is the average expectation of life for a 5-year-old barnacle.

3.22 Methods of Obtaining Numbers for Life Table Data

There are two different approaches that can be used in obtaining data for life tables. These two alternatives can be distinguished by considering the hypothetical population in Fig. 3-1. The occurrence of a series of births and deaths over time are indicated across the bottom of Fig. 3-1. This population has a more or less seasonal pattern of breeding, and individuals live for a maximum of 8 years. The "life line" of each individual is marked as a diagonal line that follows an individual from birth (top left) to death (bottom right) over time (indicated along the x axis), and through the various age classes (indicated along the y axis). We can follow the survival of a cohort of individuals born more or less synchronously, and from these data plot the survival curve of the cohort, shown in the small graph on the top left of the diagram. This approach to obtain survivorship data is often referred to as the cohort, dynamic, horizontal, or direct method.

Another way to obtain a survivorship curve is to record the age of individuals that die. In Fig. 3-1 the age of carcasses was recorded for a collection of carcasses accumulated during the expanse of time comprised by the x axis. The survivorship curve obtained by this approach—in this case plotted as percentage survival—are shown on the small graph at the right of the Fig. 3-1.* This second way to collect survivorship data is usually called the stationary, static, vertical, or indirect method (Caughley, 1966).

* A collection of all the individuals present at any one time, indicated by the vertical dotted line, would include all the live individuals present. This is L_x, the survivorship table age structure, and is shown in the histogram at top right of Fig. 3-1.

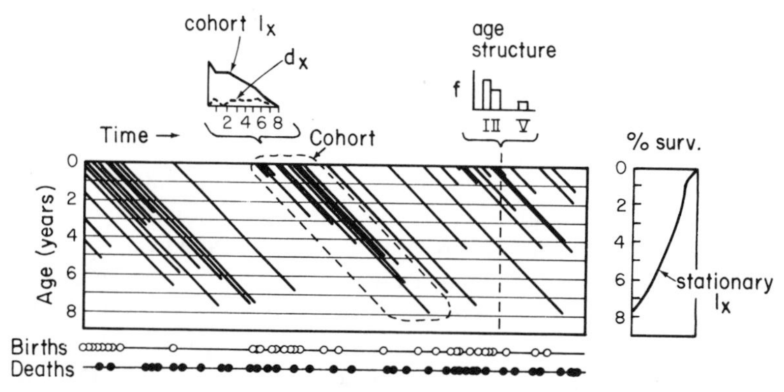

Figure 3-1. Diagram of the life and death of individuals of a hypothetical population over age classes and over time. The history of each individual is given as the diagonal lines. In the small graph of the top left of the figure, the vertical axis is the numbers of individuals, the horizontal axis is age in years. In the graph to the right, the vertical axis is age in years. Adapted from Margalef (1978). © John Wiley & Sons, Inc., reprinted by permission.

Specifically, the two approaches can be carried out by collecting data in several ways:

Direct methods. The direct following of a cohort over its lifetime can be done by either (1) recording the number of deaths occurring at each age, which provides entries for the d_x column, or (2) recording numbers of animals in a cohort still alive at various times, which provides data for the n_x column.

Indirect methods. In the more common situation where it is not possible to follow individuals in a cohort, there are four ways to obtain life table data: (1) record age at death of individuals marked at birth but not born at the same time, obtaining data for d_x; (2) age carcasses dead over a period of time, obtaining d_x data; (3) age carcasses from a population killed unbiasedly by a catastrophe, a method that provides n_x data; (4) census a living population, thus obtaining n_x data.

It does not really matter that data for different columns are obtained by different methods, since the columns of life tables are related: $n_{x+1} = n_x - d_x$, $q_x = d_x/n_x$, $l_x = n_x/n_0$.

The results of indirect methods will resemble those of direct methods for populations whose age distribution is stationary. In such stationary age distributions neither the abundance of the population nor the age structure changes over the period over which the censuses were taken. Since environmental conditions are seldom constant, there are usually changes in populations over time, so there will be differences among the two kinds of life tables. If there are large changes in the population during the period of study, the indirect methods will provide data in which some age classes will be over- or underrepresented in the sampling.* Caughley (1966) discusses ways to test whether stationary age distributions are involved.

In most life table studies, age is not directly known, so that some way of estimating age is often required. The best technique is to mark individuals and follow their survival, if the marking method does not affect survival. Many other indirect criteria have been used as ways to estimate age. These include size of the organism, growth rings on shells of molluscs, genital plates of echinoderms, growth rings on scales or otoliths of fish, teeth in marine mammals, and many others. Since young animals are the most fragile, have few hard parts, and have the highest growth and mortality rates, the young stages are almost invariably the weakest part of life table data.

3.3 Fecundity Life Tables

The survival life tables discussed above summarize the mortality schedule of a population. To determine the rate of population growth or decline we need to add reproduction schedules. Fecundity life tables (Table 3-2) contain an additional column, m_x, the number of female offspring per female of x age per unit time. The age-specific fecundity (m_x) is measured by examining the production of eggs or young for the various age classes of females. Histological examination of gonads of clams or fish, for example, can provide a measure of production of eggs.

To calculate the rate at which the population grows we have to estimate the number of young produced by each age class. This can be calculated by multiplying survivorship (l_x) by reproductive output (m_x) of each age class up to n. The sum of these products is the multiplication rate per generation (R_0),

$$R_0 = \sum_{x=0}^{n} l_x m_x. \tag{3-10}$$

In Table 3-2 we have survival and fecundity data for sardines. The n_x data were obtained and are expressed as l_x, and the initial value of l_{x0} is set at 1 for convenience in calculations. The right-most column is the product of age specific survival and fecundity; R_0 is the sum of the terms in the $l_x m_x$ column.

* This is the same problem presented by life insurance rates calculated on the basis of outdated human survivorship data. Since current human survival rates are nearly always higher than earlier rates, the fairer (and cheaper) rates will be those calculated for the very latest survival data. Incidentally, most of the development of life tables for ecological work was done based on actuarial techniques.

Table 3.2 Survivorship and Fertility Table for the Sardine *Sardinops caerulea* Off California[a]

x (years)	l_x	m_x	$l_x m_x$	v_x
0	1.000000000	0	0	1
2	0.000014084	36,543	0.5147	139,380
3	0.000009410	96,687	0.9098	215,814
4	0.000006328	119,414	0.7557	248,384
5	0.000004242	133,593	0.5667	269,760
6	0.000002843	143,824	0.4090	284,877
7	0.000001905	151,641	0.2890	296,288
8	0.000001277	158,743	0.2028	302,555
9	0.000000856	161,070	0.1379	300,818
10	0.000000574	161,070	0.0926	292,212
11	0.000000384	161,070	0.0619	274,862
12	0.000000257	161,070	0.0415	238,398
13	0.000000173	161,070	0.0279	161,070

[a] From Murphy (1967). © Ecological Society of America, reprinted by permission.

The rate R_0 can also be called the net reproductive rate, since it equals the ratio between the numbers of individuals in a filial generation (N_T) relative to the number of parent females (N_0):

$$R_0 = N_T/N_0, \tag{3-11}$$

where T is the time interval between generations. From Eqs. (3-8) and (3-9)

$$N_T/N_0 = e^{rT} = R_0. \tag{3-12}$$

Taking logs and rearranging the terms, we get

$$r = \ln R_0/T. \tag{3-13}$$

This is an important result, because it allows us to calculate the instantaneous rate of population increase at any one time, if we know T. T can be estimated by calculating the average age at which a female reproduces. If at various age intervals x there are are m_x young born to a female,

$$T = \frac{\Sigma x l_x m_x}{\Sigma l_x m_x} = \frac{\Sigma x l_x m_x}{R_0}. \tag{3-14}$$

In the case of the California sardine (Table 3-2), $T = 4.111$ years, and $R_0 = 4.0125$. This sardine population is quadrupling its numbers per generation, a rather rapid rate of increase. The sardine is a species that often suffers wide variations in density. The rapid net reproductive rate calculated here reveals that sardines are an opportunistic species able to increase quickly under suitable conditions (Murphy, 1968).

T is only approximately estimated by Eq. (3-14), since offspring in most populations are born over a period of time rather than all at once and not all females give birth simultaneously. Equation (3-14) may be a good estimate of mean generation length in situations where a cohort colonizes a new habitat or in populations with strongly seasonal breeding patterns. Leslie (1966) discusses other ways to obtain T based on the average age of reproductive and nonreproductive females.

In populations where generations do overlap, Eq. (3-13) is not an accurate way to calculate r. In such cases r is best calculated by substituting values of r in Eq. (3-15) until we find the value such that

$$\sum_{x=0}^{\infty} e^{-rx} l_x m_x = 1, \tag{3-15}$$

or a reasonable approximation to 1. In the case of the sardine (Table 3-2), $r = 0.338$. The above equations are mainly the work of A. J. Lotka and associates and are derived in Lotka (1956) and Mertz (1970), where their applications are discussed.

A convenient way to carry out the calculations of population growth in the case of complex life histories is by means of matrix algebra. The techniques are reviewed in Poole (1974) and Pielou (1977), among other books.

3.4 Some Properties of Life Table Variables

3.41 Survival and Mortality Curves

Most marine animals, especially invertebrates and fish, suffer very large mortalities during the early stages of their life history (Fig. 3-2). Such survival curves show a very steep initial slope, with a two to three order of magnitude change in percentage survival. The intense mortality of young is not due to intrinsically low survival, since in a protected laboratory environment many species show very low mortalities during the earlier stages of life (cf., for example, the two copepods on Fig. 3-2, top, whose survival was measured in the laboratory). Mortality rates taper off during later stages and life expectancies are high once the hazardous early stages are past. The mortalities of young plaice in their first few months of life can be 80% per month; by the time plaice are 5–15 years old the mortality may be down to 10% per year (Cushing, 1975).

In contrast, there are marine organisms, mainly mammals and birds, whose survival in the field is high during the early stages of their life history (Fig. 3-2, bottom). Marine mammals have especially low mortali-

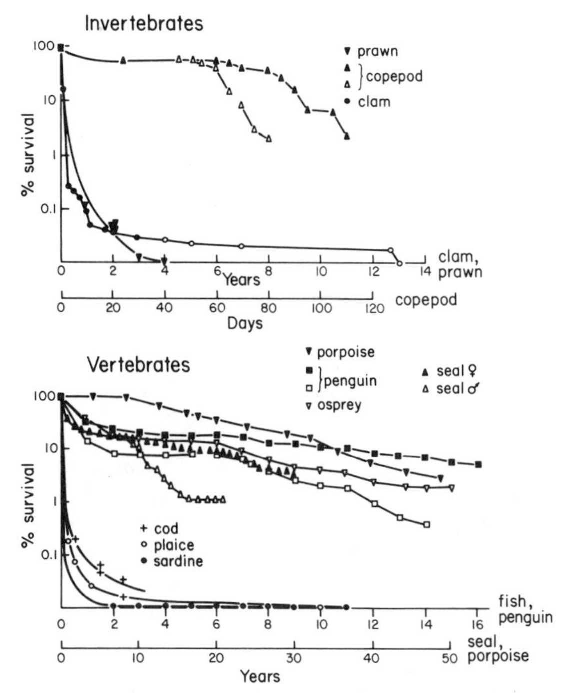

Figure 3-2. Survival curves for various animals. The copepods are *Tisbe reluctans* and *T. persimilis*; adapted from Volkmann-Rocco and Fava (1969). The prawn is *Leander squilla*; adapted from Kurten (1953). The clam is *Mya arenaria*; adapted from Brousseau (1978a). The porpoise is *Stenella attenuata*; adapted from Kasuya (1972). The osprey is *Pandion haliaetus*; data from Henny and Wight (1969), graphed on same scale as fish and penguin. The seal is *Halichoerus grypus*. Note that males and females of the same species can have very different survival curves. Adapted from Hewer (1964).

ties, clearly related to parental care and the birth of relatively large young in very small clutch sizes.

3.42 Age or Size Structure of Populations

The age structure of a population is often the easiest demographic information to obtain. If age or size structures are followed through time (Fig. 3-3), the effects of mortality (reduction of cohort size) and growth (shift to larger size) of the individuals making up the cohorts is evident. Quite often there is a broad overlap among cohorts but there are methods to separate the cohorts (Cassie, 1954; MacDonald and Pitcher, 1979).

Sometimes it is not practical to follow populations through time, nor to identify cohorts. In these cases the decrease in the abundance of age classes to the right of the mode of the entire population may be used as an estimate of mortality rates. Ebert (1977) and others review such calculations, all of which assume a stationary age distribution, constant mortality

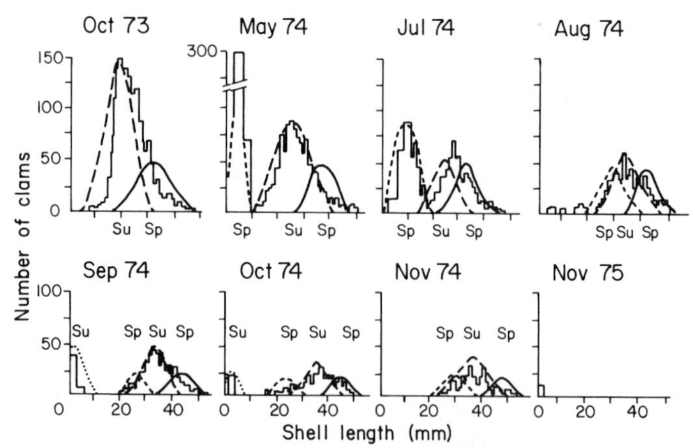

Figure 3-3. Age structure of the soft-shell clam, *Mya arenaria*, over time. The actual data are shown as histograms, while the curves are the cohorts separated by the use of probability paper (Cassie 1954). The pelagic larvae of this species settle onto sediments in spring (Sp) and summer (Su) cohorts. The pattern of growth and mortality can be seen by following, for example, the summer 1973 cohort (large dashes) through 1974. Adapted from Brousseau (1978).

over all sizes, and exponential growth of individuals (Bertalanffy, 1957). These assumptions do not always hold (Frank, 1965; Yamaguchi, 1975; Van Sickle, 1977), but such calculations can give reasonable approximations (Brousseau, 1979). If data on size specific growth and size frequency are available, the changes in numbers (u_j) of a population of size class j can be calculated (Van Sickle, 1977) as

$$u_j = [g_j N_j - g_{j+1} N_{j+1}]/N_j, \qquad (3\text{-}20)$$

where g_j and g_{j+1} are the growth rates of two adjoining size classes, N_j and N_{j+1} are the numbers of individuals in adjoining size classes. When frequency of size classes declines the growth rate will be negative, hence Eq. (3-20) also can be used to calculate mortality rates. Estimates of mortality rates for bivalves computed by Eq. (3-15) for size classes to the right of the mode can approximate estimates obtained by more complicated demographic means (Brousseau, 1979). In many cases there are few other feasible alternatives; for instance, most of the available estimates of mortality for whales use this general approach. It should be clear that there are limitations to this calculation, an obvious one being that only mortalities of older individuals are estimated.

A population subject to a constant schedule of births and deaths will increase geometrically [Eq. (3-8)] and will gradually approach a stable age distribution, in which the relative proportion of different age classes is fixed. This stable age distribution can be calculated as described by Mertz (1970). Seasonally breeding populations may not achieve a stable age distribution. We have earlier mentioned the stationary age distribution. The latter describes the age composition of a population exposed to given mortality rates (q_x), if birth rate were equal to death rate. Neither stable nor stationary age distributions are likely to exist for very long in nature, since populations seldom increase at the same rate for any length of time, and because the numbers of individuals in a population seldom remain constant.

The distribution of age classes can sometimes provide information on status of the population. Populations that are increasing tend to have more individuals in the youngest age classes; as each age class becomes older, the total numbers of the populations expand. Stable populations characteristically have about the same proportion of the population in all size categories; populations that are declining may have fewer individuals in younger age classes.

3.43 Fecundity and Reproductive Value

Fecundity, even if intrinsically high, does not guarantee reproductive success. A population must have sufficient individuals in the various age classes that contribute significantly to r. Age-specific fecundity can peak soon after the first reproduction, as in some copepods (Volkmann-Rocco and Fava, 1969), or later in reproductive life, as in clams (Brousseau, 1978) and sardines (Table 3-2).

The impact of individuals in different age classes on the reproductive rate of a population differs. Such a reproductive value is a result of the survival and reproductive performance of individuals in each age class. In an exponentially growing population the potential importance of each age class (x) for population growth is given by the reproductive value of that age class (v_x) relative to the reproductive value at birth (v_0)

$$\frac{v_x}{v_0} = \frac{e^{rx}}{l_x} \sum_{y=x}^{\infty} e^{-ry} l_y m_y . \tag{3-21}$$

What is summed in Eq. (3-21) is the number of offspring that will be produced by a female for age x as she passes through all the remaining age classes (y). Usually v_0 is made equal to 1 so v_x is expressed as multiples of v_0. The expression was first derived in its integral form by R.A. Fisher and its derivation is explained in Mertz (1970) and in Wilson and Bossert (1971).*

In our example of the Pacific sardine (Table 3-2), the reproductive value at recruitment (v_2) is much greater than at v_0, due to the very high mortality of larval sardines. The reproductive value increases in subsequent age classes when mortalities are less marked and m_x reaches its

* Equation (3-21) is not completely correct, even though it appears very frequently in the ecological literature. See Goodman (1982) for a discussion of the correct version, $v_x = e^{rx+1}/l_x \sum_{y=x}^{\infty} e^{-ry} l_y m_y$. The values calculated using Eq. (3-21) are in error by e^r. Since we are generally interested in the patterns of the v_x over x, the error is not critical, since the pattern will still be evident.

peak. There is a slight decrease in v_x in the last few years due principally to mortality but in this species v_x is significant for most age classes.

In a rapidly growing sardine population several age classes contribute significantly to reproductive rate (Table 3-2). During the 1940s fishing pressure on the larger classes was heavy enough so that only two age classes contributed to reproduction, compared to eight or so age classes before that. Perhaps this situation was implicated in the collapse of certain stocks in 1949–1950, after two consecutive spawning failures. Fishing had removed the mechanism—reproduction by several age classes—that the population had evolved to compensate for reproductive failures due to fluctuations in environmental conditions (Murphy, 1968).

The reproductive value can be useful to predict consequences of harvest of specific age classes, as for example, in prediction of consequences of size-selective fishing. Reproductive value could thus be an important tool in management of exploited stocks since, for example, it would be desirable to carry out harvests of stocks with the least effect on reproduction. Harvest of age or size classes with low v_x may be the most "prudent" fishing or predation strategy. In actuality, the optimization of harvest strategy is seldom so simple, and careful study of specific harvestable populations is required.

Knowledge of reproductive value may also be useful in studies of populations that colonize new habitats. The v_x of the propagule that has arrived at the new site may vary: if the colonizers are very young, reproduction and further expansion into the new habitat may be delayed until maturity; if very old, few offspring may be produced. Since colonization is often carried out by a few individuals, the reproductive ability of these few assumes a great importance (MacArthur and Wilson, 1967).

3.5 Reproductive Tactics

The various demographic characteristics of a population are interrelated traits, but some features affect population growth more significantly than others. For example, Cole (1954) found in simulation studies that age at first reproduction has a great effect on r (Fig. 3-4), so that delays in sexual maturity markedly lowered the attainable r.* Delays in age at first repro-

---

* In human populations there has been too much attention given to total family size and too little to age at first reproduction. Cole (1965) shows that human populations that started reproducing at 12 years of age could attain over twice the growth rate of a population whose age at first birth was 30 years of age. The effect of total number of children was small compared to the effect of age at first reproduction. The countries where growth of population has been most successfully curtailed are those where economic and social restrictions have led to delayed reproduction until the parents were in their 30s, as well as to reductions in the number of children per family.

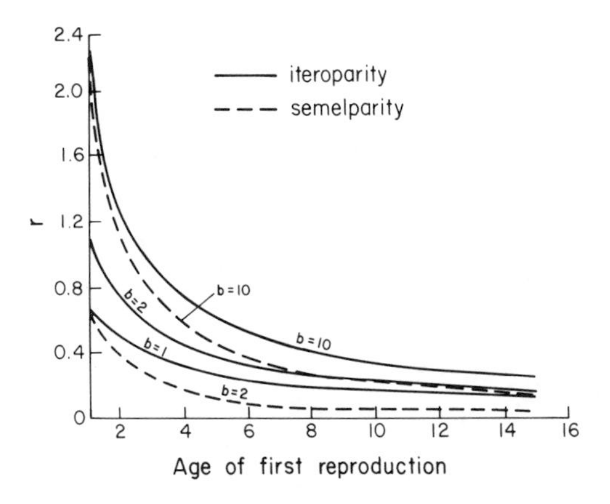

Figure 3-4. Simulation study of the effect of age at first reproduction and clutch size (b) on the rate of population increase (r). Age at first reproduction is shown for intervals representing the 1,2,3... intervals at which reproduction is possible for iteroparous species and for reproduction at the 1,2,3... interval of age for semelparous species. Adapted from Cole (1954).

duction have the greatest impact on r in the left side of the graph both in species that reproduce once during their lifetime (semelparity) or in cases when reproduction takes place at repeated intervals (iteroparity). The effect of age at first reproduction is lessened at low clutch sizes (compare clutch size $b = 10$ to $b = 1$ or 2). The importance of early reproduction is such that if reproduction took place at the earliest possible age (Fig. 3-4), there were minimal differences in r between iteroparous and semelparous species, and only a small difference in later reproduction.

The rich variety of demographic possibilities provided by semelparity, iteroparity, reproduction at any of a variety of ages, and variable clutch sizes allows species with different demographic repertoires to adapt to environments with different properties. Several theoretical constructs have been forwarded to codify the kinds of demographic features that would be most appropriate for environments of different properties. These include "the *r-K* dichotomy," "bet-hedging," and the "abundance/intermittency" idea.

3.51 *r-K* Dichotomy

Very different strategies of reproduction would be best suited for life in either sparsely inhabited but resource-rich environments where rapid and opportunistic population growth would be favored, or in resource-poor environments already saturated by organisms, where very slow population growth would be more adaptive. MacArthur and Wilson (1967) coined the terms "*r* selection" and "*K* selection" for these two alternatives.* Stearns (1976) reviews these theoretical constructs in some detail and summarizes the traits that are hypothetically characteristic of each:

* "*r*" refers to the intrinsic rate of increase of a population, and "*K*" to the carrying capacity of the environment, as used in the logistic equation of population growth (Chapter 4).

(1) *r* selection: early age of first reproduction, large clutch size, semelparity, no parental care, large reproductive effort, short generation time, small and numerous offspring, and low assimilation efficiency; (2) *K* selection: delayed reproduction, iteroparity, small clutches, parental care, smaller reproductive effort, longer generation time, a few large offspring, and high assimilation efficiency. Assimilation efficiency is taken up in Chapter 7. All the other characteristics are consistent with either prodigal or prudent reproductive tactics, to use Hutchinson's (1978) picturesque metaphor.

We should add a note about the "few large" and "many small" features, because we will come back to this in Chapter 9. Invertebrates and bony fish produce very numerous clutches, where each individual may have a 10^{-5}–10^{-6} chance of survival, as evident in the l_x curves of Fig. 3-2. Necessarily, the young are very small in relation to the size of the adult, and there is little or no parental care. Birds and mammals, on the other hand, produce far fewer young of a larger size in relation to the parents, and the chance of survival of each young can be 10^{-1}–10^{-2}. Birds and mammals carry out considerable parental care of the young.

Although the *r–K* dichotomy is an appealing idea and has stimulated much research, it is too much of a simplification. Not all species in an environment will be *r* or *K* selected; at the same time, any one environment may be *r* selective for a copepod but *K* selective for a whale. Further, there are many examples where a particular species shows combinations of traits that the *r–K* typology puts in the two separate categories. Finally, there are many species that are intermediate in demographic properties, so that the concept is probably best referred to as the *r–K* continuum. There has been, in general, insufficient independent verification of the *r–K* idea in part because we use many of the same properties to define environments that *r* or *K* select as we do to define whether a species is *r* or *K* selected. This circularity has made it difficult to test the idea.

3.52 Bet-hedging

Another theoretical scheme regarding adaptive reproductive strategies (Murphy, 1968; Schaeffer, 1974), is called "bet-hedging" by Stearns (1976), and is based on the idea that iteroparity may be a response to a changeable environment. The bet-hedging concept is derived from observations on reproduction in fish. In herring-like fish the species that suffer the most erratic breeding success are also the ones with the longest reproductive spans of life. Iteroparous breeding over a lengthy period of time may thus be a bet-hedging tactic that compensates for poor years in a variable environment. Results from terrestrial studies suggest that in variable environments, age at first reproduction is lower, size at sexual maturity smaller, reproductive effort higher, clutch size larger, and size of young smaller than in constant environments (Stearns, 1976). "Variable" and "constant" environments may not correspond exactly to r- and K-selective environments, so the predictions of the bet-hedging and r- and K-selection hypotheses may not be mutually exclusive. This makes it hard to devise hypotheses with which to assess the two schemes.

3.53 Abundance/Intermittency Scheme

Whittaker and Goodman (1979) and Grime (1977) propose a third hypothetical scheme. The scheme identifies three kinds of species, considering both the abundance and the intermittent pattern of availability of resources:

Adversity-selected species. Some environments are depauperate and can sustain only a small number of organisms: species living there are selected for survival under adverse conditions. Perhaps reproduction can only occur in the very minimal periods of more favorable conditions, and the reduced resources allow only a small amount of reproduction. The total abundance over the habitat seldom increases very much.

Exploitatively selected species. In environments where conditions for life unpredictably change and may suddenly but repeatedly provide bountiful resources, species capable of rapid exploitative activity will be present, and these species will have opportunistic characteristics. There will be strong selection for sharp increases in birth rate (b) on the frequent occasions during which resources increase; the rest of the time, selection will favor those able to reduce death rates (d). For example, a species can evolve some mechanism to survive long periods of bad conditions, if there is a frequent and reliable return to favorable conditions. Thus, to explain reproductive strategy in exploitation-selected species we need to consider both b and d, rather than just the rate of increase, $r = b - d$. These populations will be generalists in use of resources (taking whatever resources are available) and flexible in demography (both b and d can change).

Saturation-selected species. In environments capable of continually sustaining an array of species, and where there are few intermittent lapses in available resources, there will be selection for low rates of birth and death, demographies that lead to near-complete use of the available resources, and the species can be thought of as "saturation-selected." In such situations it is likely that, due to the high density, competition rather than extrinsic vagaries of the environment determines the actual birth and death rates of the populations.

This third scheme of adaptive strategies is as hypothetical as the two prior ones. Its three kinds of environments are not exactly parallel to those of the r–K or the bet-hedging alternatives, so comparisons are difficult. There is plenty of theory in this area; there is also a need for critical experiments to discriminate among competing ideas. Unfortunately, it is very difficult to make sure of just what is being tested. Reproductive strategies are an intrinsically interesting but perhaps intractable subject, one in which there is enough tautology to hinder the testing of hypotheses.

The reproductive strategy of specific species can, however, in some cases be clear, with important consequences for human exploitation of natural populations. There are many examples of this (Ricker 1975), but we can consider a representative example: the relative ability of bony and or cartilaginous fish to sustain fishing pressure and its relation to reproductive strategy.

Most marine bony fish may be "exploitatively selected" species that use what can be called an "*r* strategy," and produce very numerous small eggs, most about 1 mm or so in diameter; sometimes, as in salmon and halibut, eggs may be several millimeter in diameter. As many as 10^4–10^7 eggs may be released at each spawning. The small larvae (3–10 mm in length) hatch and then undergo a very hazardous period of drifting in the water column. Larval mortality rates can be very high, reaching 10% per day in haddock (Jones, 1973), 80% per month in plaice (Cushing, 1975) (cf. Fig. 4-9, right). A female Arctic cod may release about 4 million eggs, and on the average only 2 of those eggs may survive to reproduce.

Many cartilaginous fish (elasmobranchs such as sharks, skates, and rays) may be "saturation-selected" species, in contrast to bony fish. Cartilaginous fish may use a "*K* strategy," in which the young are well developed when born and resemble the adult of the species. The number of young per clutch is far smaller than other fish. For example, the ray *Raja erinacea* only produces about 33 eggs per year (Richards et al., 1963), while *R. cavata* releases about 100 eggs, and *R. brachyura* and *R. montaghi* 50 eggs per year (Holden, 1978). The weight of the young decreases as litter size increases, and the smaller young may be more susceptible to predators.

A consequence of the low rate of reproduction and growth of most elasmobranch populations is that they generally do not sustain exploitation very well. A number of fisheries, including those exploiting the California fin shark (*Galeothinus zyopterus*), Australian school shark (*G. australis*), the spiny dogfish (*Squalus acanthias*), plus skates and rays (Holden, 1978) have collapsed or produced reduced catches. Young are produced in low numbers and the recruitment is probably a function of parent density (and probably parent size). When fishing lowers density of the adult population, recruitment is affected.

In contrast, in most species of pelagic marine bony fish, recruitment can take place even with a minimal production of eggs (cf. Section 4.4). Thus, reduction in parent stocks is less likely to affect recruitment, since the removal of a substantial proportion of the reproductive population does not necessarily reduce the rate of population increase.

Chapter 4

Competition for Resources Among Consumers

4.1 Population Growth in Environments with Finite Resources

In the previous chapter we examined growth of populations whose reproductive rate was constant. In nature, the rate of population growth very often varies. If r varies randomly, a population may become extinct or, alternatively, suffer enormous increases in abundance, since the reproductive potential of any species is capable of producing very large numbers of new individuals. Since few environments are overwhelmed by living organisms, there must be mechanisms that inhibit increase in abundance as density increases. This, in brief, is the theory behind the logistic equation,* first suggested by P. F. Verhulst in 1838 and later derived independently by R. Pearl and L. J. Reed in 1920 (Hutchinson, 1978). The logistic equation is

$$dN/dt = rN\,(K - N)/K, \tag{4-1}$$

which is the exponential growth equation modified by a factor $(K - N)/K$ that reduces dN/dt as the number of individuals (N) increases and approaches the carrying capacity of the particular environment (K). This results in a sigmoid pattern of population growth, in which N has a smooth, decelerating approach to K.

---

* The logistic is just one form of sigmoid growth. There are many other ways to describe such growth, including applications of Michaelis–Menten curves (Slater, 1979), Gompertz curves ($dN/dt = rN \log (K/N)$) (Margalef, 1974) or exponential curves (Gallopin, 1971). Further alternative expressions are discussed by Hutchinson (1978).

The integrated form of the logistic equation can be fitted to actual data by first rearranging the integrated form of the equation ($N = K/[1 + e^{a-rt}]$) into ln $[(K - N)/N] = a - rt$, where a is a constant. Plotting ln $[(K - N)/N]$ on the y axis, and time in the x axis, the slope of the line is r, and a is the y-intercept.

The logistic model is one of the cornerstones of much ecological theory due to its broad generality. Precision and realism are additional desirable attributes of models (Levins, 1966), but it is not possible to achieve all three attributes in one model: the logistic has broad generality since it can apply to any population, but has only limited precision in predicting the abundance of a particular species (Krebs, 1978), since populations often overshoot K and subsequently crash. The logistic also has limited realism, since it assumes that all individuals contribute identically to r, that K is a constant, and that there are no time lags. None of these assumptions is very realistic, since there are very significant differences among individuals of different sexes and ages, most environments and resources vary markedly over time, and consumer populations and their food resources seldom if ever react to changes instantly. Nonetheless, the logistic model of population growth is heuristically attractive because it pinpoints, among other things, one very important concept: competition for resources must take place. The logistic has also been useful because it clearly identifies the topics—especially the assumptions—that need further examination. This clear identification of issues is one of the principal benefits of the construction of models, and has stimulated much work devoted to addressing the unrealistic aspects of the logistic and other models of population growth (Krebs, 1978; May et al., 1974).

4.2 The Nature of Competition

Competition takes place where resources are in short supply, and can take two forms (Nicholson, 1954; Elton and Miller, 1954). First, there is interference or contest competition, where access to a resource is denied to competitors by the dominant individual or species. Examples of interference competition are the release of antibiotics by microorganisms, territorial behavior, and social hierarchies. Interference competition results in unequal access to resources, and subordinate individuals may not be able to participate in reproduction. Interference competition tends to couple numbers to available resources and often reduces population fluctuations. Many proximate mechanisms involving behavior and physiology, such as responses to photoperiod, dormancy, and behavioral displays, are commonly involved in this kind of competition.

Most of the biological world, however, is more involved in exploitation or scramble competition, that is, the direct use of a resource, reducing its availability to a competing individual or species simply because of consumption. Exploitation competition tends, at least in theory, to involve large fluctuations in density since populations may be built up based on a temporarily abundant resource, and crash down when the resource is exhausted. Use of resources is not particularly efficient where allocation of resources is exploitative, since many organisms that will not reach maturity still consume resources.

Both interference and exploitation competition may occur simultaneously. *Patella cochlear*, for example, is territorial as are other limpets (Stimson, 1970; Branch, 1976) and actively prevents the encroachment of other animals in its feeding territory. *P. cochlear* preferentially eats the alga *Ralfsia* within its feeding territory, leaving only encrusting algae within it. *P. cochlear*, however, can continue to use the feeding territory by feeding on the encrusting algae, but other limpet species cannot eat encrusting algae. Thus, this is a case of both interference and exploitation competition mechanisms (Branch, 1976).

Competition pervades many dimensions of the life of any species, and what may appear as separate resources may in fact be inextricably linked. There is often no distinction, for example, between competition for food and space in rocky shore species. Experimental removal of the film of filamentous blue-green bacteria from territories of the owl limpet (*Lottia gigantea*) leads to a 25% increase in size of the territory 2 weeks later, while untreated territories changed by only −0.06% (Stimson, 1973). Another example involves encrusting bryozoans, organisms where overgrowth is a very common phenomenon. In shallow waters of the Pacific coast of Panama there are two bryozoans, *Onychosella alula* and *Antropora tincta*, that frequently live in close proximity. Buss (1979) fed radioactively labeled food pellets to colonies of these two bryozoans growing on separate surfaces and growing on the same surface. The feeding rate, indicated by the radioactive counts, did not change from the margin of a colony inward in monospecific colonies (Fig. 4-1, left). When the two species grow together on one substrate (Fig. 4-1, right) there was a nota-

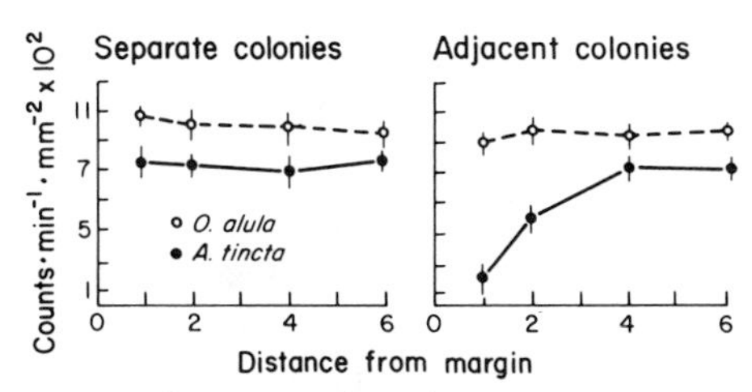

Figure 4-1. Feeding rates of two species of encrusting bryozoans (*Onychosella alula* and *Antropora tincta*) in colonies growing separately and with the two species growing adjacent to each other. Feeding rates are expressed as radioactivity due to ingestion of labeled food particles and were measured at various distances (in centimeters) away from the margin of the colonies. In the case of the adjacent colonies, the measurement was from the margin where the two species were in contact. Adapted from Buss (1979). © Macmillan Journals Limited, reprinted by permission.

ble decrease in feeding rate of *A. tincta* toward the margin of the colony in contact with *O. alula*. If growth rate depends on feeding rate, individuals on the edge of the *A. tincta* colony will grow slower, and the *O. alula* colony would overgrow the colony of *A. tincta*. This occurs despite the fact that growth of *A. tincta* in isolation is faster than that of *O. alula*. The specific proximate mechanism for this is that *O. alula* bears feeding tentacles that are considerably longer than those of *A. tincta*, and presumably can divert feeding currents away from *A. tincta*. Since in the thin layer of seawater above the rock surfaces where bryozoans live there are no other forces that move water effectively, *A. tincta* suffers a considerable reduction in food supply. Evidently, space and food are not independent resources.

Competition reflects the relative abundance of resources. The major determinant of relative abundance—since exploitation competition is so common—is the number of individuals in a population. The nature of the limitation of these numbers was heatedly debated in the 1950s and early 1960s. The crux of the argument was whether the rate of increase or decrease of a population could be controlled by factors that depended on the density of the population. These arguments are described in ecology textbooks (Krebs, 1978; Ricklefs, 1979) and elsewhere (Orians, 1962; Clark et al., 1967; Caswell, 1978) and need not be repeated here. Our interest below is to see if demographic properties are affected by changes in the density of a given population of marine organisms—intraspecific competition—and by changes in the density of other competing species—interspecific competition.

4.21 Intraspecific Competition

A very clear example of the effect of density within a species is provided by the studies of Branch (1975) on rocky shores of South Africa, where densities of the limpet *Patella cochlear* vary from 90 to over 1,700 limpets m^{-2}. This range of densities has a significant effect on survival and on fecundity.

High densities increase mortality of young limpets (Fig. 4-2, top), but once limpets grow older than about 20% of maximum age, survival is not affected by density. The increased mortality of the younger individuals at higher densities is due to the reduction of space available for settlement of small limpets. Limpets are territorial and aggressively defend their feeding territory so that at high densities higher proportions of larvae settle on the shell of older limpets (Fig. 4-2, bottom), presumably because these are the only surfaces that are not defended. Once the young limpets reach a certain age the shell surface of older individuals is not large enough to provide a sufficient grazing area, and the small limpets leave the shells of the older individuals and seek territories elsewhere. This is a hazardous stage in the life history that leads to the density-dependent early mortality

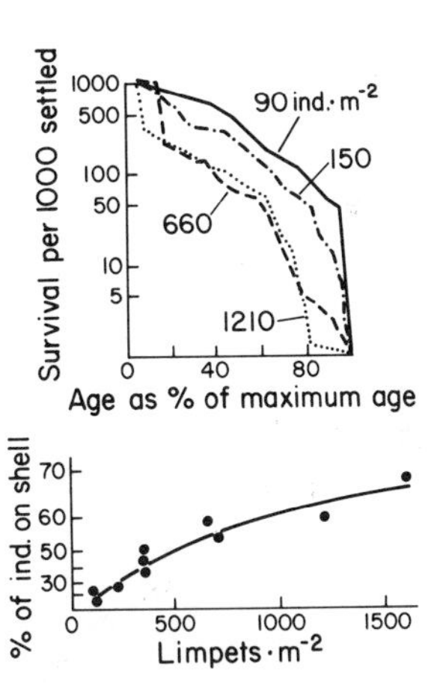

Figure 4-2. Effect of density on survival rate. Survivorship of the limpet *Patella cochlear* in areas of different density (top) and percentage of the population on the shells of adults (bottom). Adapted from Branch (1975).

seen in Fig. 4-2. Survival rate in the critical young stages is therefore density dependent.

As density increases, biomass of individual adult limpets increases, but the total biomass of the population eventually approaches an asymptote (Fig. 4-3, top). The more limpets per unit area, the lower the growth rate (Fig. 4-3, middle). Over time, reduced growth rates result in considerably smaller limpets at higher densities (Fig. 4-3, top and bottom). Smaller limpets release fewer larvae (Fig. 4-4, top). As density increases, gonad output per unit area of rock first increases due to the larger numbers of limpets present, but once competition becomes severe, there is a decrease

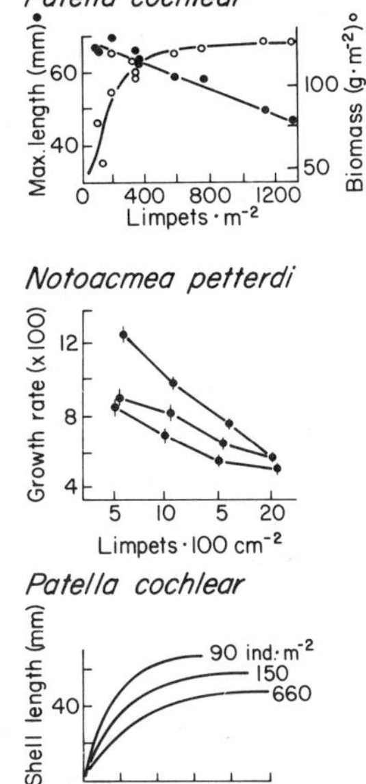

Figure 4-3. The effect of density on growth. Top: Length and biomass of *Patella cochlear* in areas of different density [adapted from Branch (1975)]. Middle: Growth rate ± standard error (= $[\ln(L_t—L_0)]/t$, L_t = final and L_0 = initial length of shell, and t = 4 months) of the limpet *Notoacmea petterdi* in relation to density. Measurements from Barrenjoey, Australia; the three sets of data come from areas at three elevations in the intertidal range, top line being the lowest location and bottom line being the highest [adapted from Creese (1980)]. Bottom: Growth over time of *P. cochlear* in areas of different density. Adapted from Branch (1975).

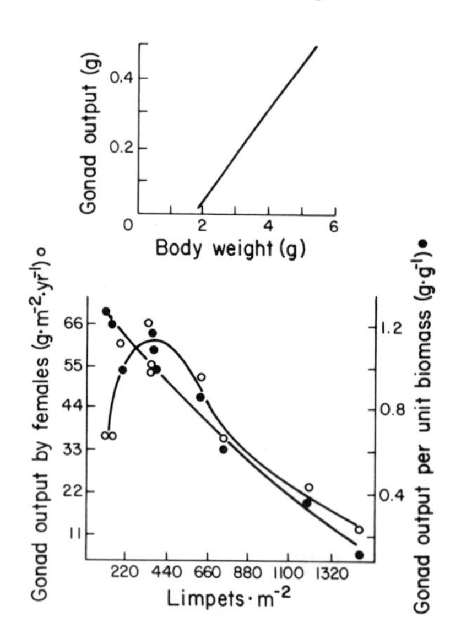

Figure 4-4. Effect of density and size on reproductive output. Release of reproductive materials by *P. cochlear* versus body weight (top) and versus density (bottom). Adapted from Branch (1976).

in reproductive output (Fig. 4-4, bottom). Gonad output per individual decreases rapidly as limpet density increases. Fecundity is therefore density dependent.

In limpets space and food supply are one variable and determine growth of individuals and fecundity. The feeding territories containing the algae that limpets use as food are smaller at high limpet densities and do not sustain high growth rates. Consequently the animals are smaller and low fecundities ensue. These relations of food supply, size, and fecundity are common, judging by the many cases where there are clear correlations between size of individual and fecundity (Fig. 4-5), for both invertebrates and vertebrates.

4.22 Interspecific Competition

4.221 Models of Interspecific Competition

The logistic model [Eq. (4-1)] describes intraspecific competition. Lotka (1925) and Volterra (1926) modified the logistic to incorporate the effects of interspecific competition by introducing a coefficient of competition, a_{12} or a_{21} that measures the effect of species 1 on species 2, or the effect of species 2 on 1. The Lotka–Volterra modification of the logistic is therefore, for species 1:

$$\frac{dN_1}{dt} = r_1 N_1 \frac{K_1 - [N_1 + a_{12}N_2]}{K_1} = r_1 N_1 \frac{K_1 - N_1 - a_{12}N_2}{K_1}, \quad (4\text{-}2)$$

and for species 2,

$$\frac{dN_2}{dt} = r_2 N_2 \frac{K_2 - N_2 - a_{21}N_1}{K_2}. \quad (4\text{-}3)$$

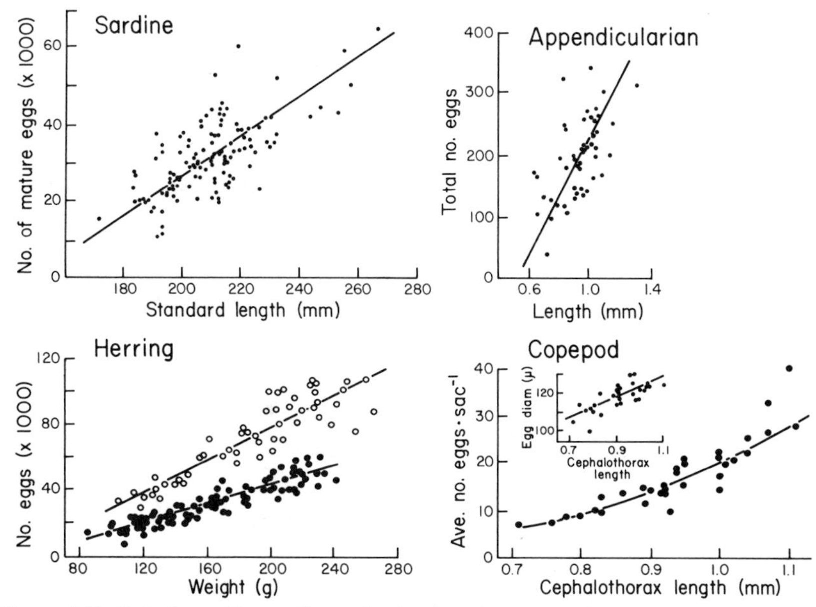

Figure 4-5. Relation of fecundity to body size. Sardine (*Sardinops caerulea*) data from McGregor (1957); appendicularian (*Oikopleura dioica*) data from Wyatt (1973); herring (*Clupea harengus*) data from Baxter (1959); copepod (*Pseudocalanus minutus*) data from Corkett and McLaren (1969).

These models have all the difficulties of the logistic and others. For example, it is seldom the case that there are only two competitors present in one environment. Nonetheless, it is instructive to use the equations to gain some perspective as to the possible consequences of competitive interactions. Details of how the Lotka–Volterra equation can be manipulated for this purpose are available in texts such as Begon and Mortimer (1981) or Emlen (1973). The relevant aspect here is that the Lotka–Volterra model can identify situations where, depending on the abundance of N_1 and N_2, and the values of a and K, the outcome of interspecific competition varies. Species 1 will outcompete and exclude species 2 if the effect of interspecific competition of 1 on 2 is greater than the effect of intraspecific competition of 2 is on itself. If both species are stronger interspecific than intraspecific competitors, the outcome is indeterminate and depends on the initial densities of the two species. Where there is clear partition of resources—that is, when intraspecific competition is more important than interspecific competition—it is possible to have stable coexistence of the two species. The model cannot predict exactly how much partitioning is needed to achieve coexistence.

Clearly then, the Lotka–Volterra models predict that interspecific competition must lead to the exclusion of one of the competing species. In fact, the competitive exclusion principle (cf. Section 4.5) was inspired by these conclusions. In spite of its limitations, the Lotka–Volterra model has made meaningful, testable predictions and has prompted much research, primarily on terrestrial populations.

4.222 Evidence for Interspecific Competition

Evidence supporting the idea that different species may compete for a resource has been collected, such as many well-documented cases of partitioning of resources or of displacement in morphological characters in sympatric species. These phenomena are suggestive but only circumstantial evidence that competition may have taken place in the past, and are discussed in Section 4.5 below.

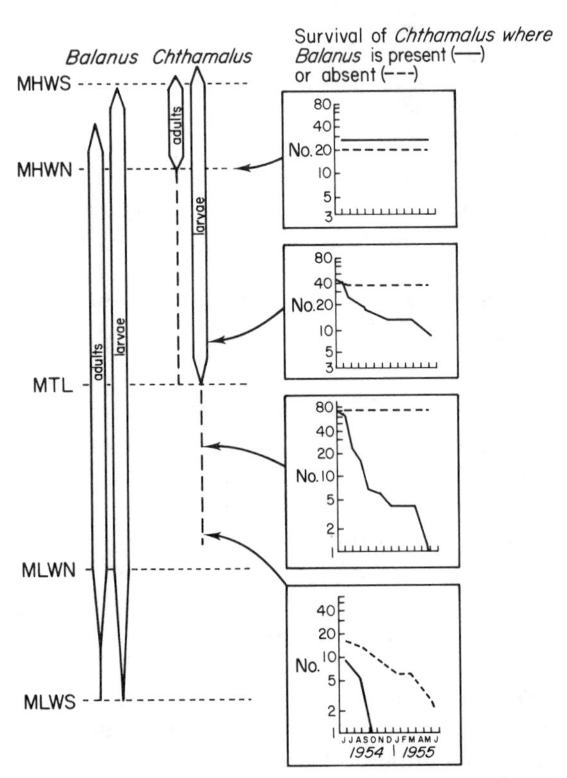

Figure 4-6. Vertical distribution of the barnacles *Balanus balanoides* and *Chthamalus stellatus* at Isle of Cumbrae, Scotland (left) and survival of *Chthamalus* in control situations and where all *Balanus* were experimentally removed (panels on right). The experiments on top two panels were done on the rocks and heights where *Chthamalus* had settled. The bottom two panels contain data from stones that were transplanted to the appropriate depths. The level of mean high and low water during spring and neap tides are indicated as MHWS, MLWS, MHWN, MLWS, MHWN, MLWN, respectively. Adapted from Connell (1961). © Ecological Society of America, reprinted by permission.

The actual occurrence of interspecific competition in the field, at least for animals, was first demonstrated in a study of two barnacles, *Chthamalus stellatus* and *Balanus balanoides*, on a Scottish rocky shore by Connell (1961). Adults of the two species coexist in the upper reaches of the tidal range (Fig. 4-6, left), where *Balanus* does not grow very vigorously or densely. Larvae of both species settle throughout a considerable portion of the tidal range, so a marked mortality of *Chthamalus* takes place throughout the lower to middle intertidal range. Connell set up pairs of experimental quadrats where he removed *Balanus* from one of the quadrats and left the other untouched. Then he followed the survival of the *Chthamalus* that settled on the quadrats. In the upper intertidal, above the level of mean high water during neap tides (MHWN), there was no effect of the removal of *Balanus* on survival of *Chthamalus*. At that elevation, these two populations do not interact, and the upper extent of the distribution of these barnacles is probably determined by desiccation. *Chthamalus* seems more tolerant of drying than *Balanus*. On the other hand, below MHWN, there is significantly better survivorship of *Chthamalus* in the absence of *Balanus*. Mortality rates of *Chthamalus* increase where rates of growth of *Balanus* are highest. *Balanus* grows faster and outcompetes *Chthamalus* for space by covering, lifting, or crushing *Chthamalus*. As in the case of intraspecific competition, interspecific competition reduces the growth of individuals and reproductive output (Table 4-1).

Intraspecific and interspecific competition are seldom separable in real situations. The growth of *Patella granularis* is negatively correlated to density of its own species and to the density of barnacles on the rock

Table 4-1. Size and Reproduction in the Barnacle *Chthamalus stellatus* in Experimental Situations Where Its Competitor, *Balanus balanoides,* Is Present or Absent[a]

Elevation in intertidal range above MLW (ft)	Average diameter (mm)		Percentage of individuals with larvae in mantle cavity	
	Balanus absent	*Balanus* present	*Balanus* absent	*Balanus* present
2.2	4.1	3.5	65	61
1.4	3.7	2.3	100	81
1.4	4	3.3	100	70
1.0	4.1	2.8	100	100
0.7	4.3	3.5	81	70

[a] From Connell (1961). © Ecological Society of America, reprinted by permission.

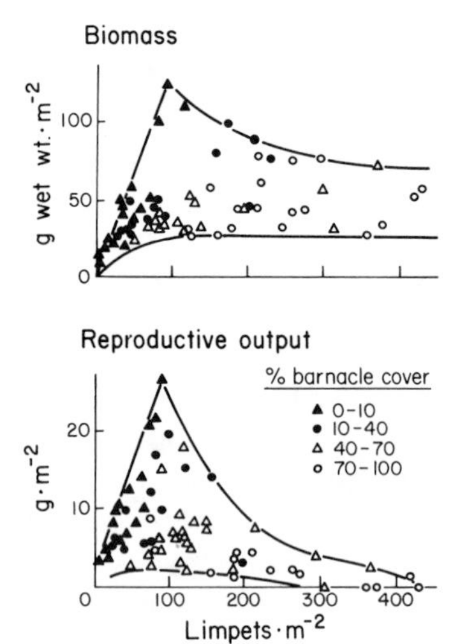

Figure 4-7. Effects of intraspecific and interspecific competition in *Patella granularis*. Adapted from Branch (1976).

surfaces (Branch, 1976). Where barnacles are scarce, biomass and reproductive output of limpets are directly related to limpet density and climbs steeply (Fig. 4-7) to about 100 limpets m^{-2}, beyond which biomass and reproductive output are reduced. This is the result of intraspecific competition. At any one density of limpets, however, the more barnacles that are present, the lower the limpet density and reproductive performance, so that interspecific competition is also important.

4.3 Density-Dependent Control of Abundance

Other examples of active competition besides those provided by Branch and Connell are now available. The collective results of these studies suggest that even though density-independent factors often greatly *affect* the numbers of organisms present, abundance is ultimately *controlled* by density-dependent mechanisms such as the ones reviewed in the previous section. *Control* implies that recruitment may increase at low density while it decreases at high densities.

Density-dependent mechanisms, such as proposed by Branch (1976) for the limpet *P. cochlear*, can hypothetically provide for control of abundance. High limpet densities lead to low growth rates and hence small size. At high densities more of the juveniles settle on adult shells, and this causes increases in mortality. Both small size and higher mortality depress reproduction, so the number of larvae released into the water is lower. Low densities of *P. cochlear* lead to low mortalities of juveniles, since most larvae are on the rock surfaces, and rapid growth, because large enough feeding territories are available, so large sizes are achieved. All this produces a large release of larvae into the water. These larvae then mix in the water column with others, regardless of the density of the parent population. Eventually, the larvae settle on some surface. The larvae apparently prefer to settle on sites with strong wave action, and such sites become high density areas, while areas of lower wave action become low density sites. The cycle of control is thus completed: even though different local sites within a coastal area may differ very markedly in density of *P. cochlear*, the regional population remains controlled within certain bounds by the processes just described. The scheme leaves many questions unanswered—for example, why would the preference for larval settlement in sites of strong wave action have evolved?—but nonetheless it is one of the most complete examples of density-dependent population control available.

The coupling of mortality, growth, and fecundity to density and food supply has also been demonstrated in other situations. S. B. Weiberg and V. A. Lotrich (unpublished data) and Kneib (1981) manipulated densities of the salt marsh killifish *Fundulus heteroclitus* within enclosed parcels of natural salt marsh. Mortality (Kneib, 1981), growth rate (Fig. 4-8, top), and the reproductive output of the fish population (Fig. 4-8, bottom) depend on the density of the extant population. The reduction in growth rate associated with greater densities is mediated by food supply, since much higher growth rates were observed where extra food was added to the experimental marsh parcels. In addition, fed fish showed no mortality and fecundity was high. In these fish, then, there is density-dependent control of abundance.

In our discussion so far we have considered that high densities lead to reduced growth of populations. This is probably true for many populations, but there must be some low densities where population growth increases as density increases. Low densities might, for example, lower reproduction due to difficulty of males and females finding each other; as density increases, reproduction rates might increase. Some other benefits of high numbers might also be important; in the guillemot (*Uria aalge*), a sea bird that nests colonially on rocky promontories, breeding success increases with density. When colonies are numerous and dense, gulls are less able to prey on eggs and chicks (Birkhead, 1977). A complete description of the effect of population size on population increase thus needs to include both a positively density-dependent range and a negatively dependent range. The humped curve that results has been known as the "Allee (1931) effect." One important consequence of positively density-dependent reproduction is that (assuming death rates remain constant) a species can decline to extinction if its density has become lower than a critical threshold, even though there are actively reproducing individuals still alive.

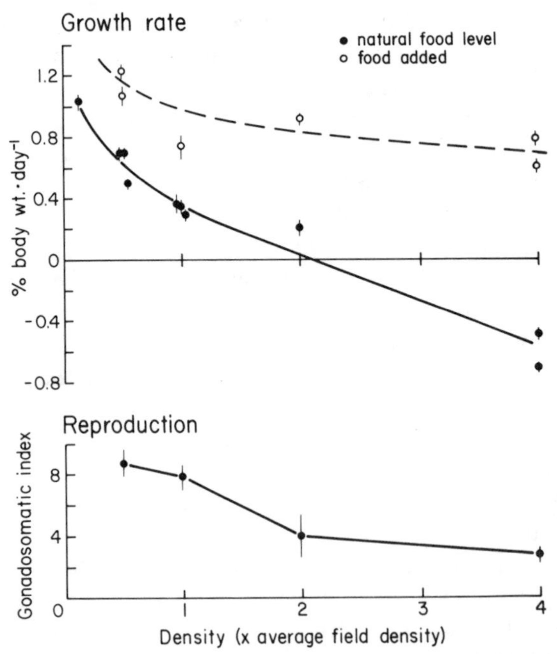

Figure 4-8. Changes in growth rate and reproductive potential in the killifish *Fundulus heteroclitus* in response to changes in density. Each point is the mean ±95% confidence interval. The average field density of fish was 180 fish per pen. Each pen consisted of an area of salt marsh artificially diked in a Delaware salt marsh. Gonadosomatic index = weight of the gonad/weight of fish × 100. Food additions were done within replicated diked areas by distributing fish flesh every 2-4 days into tidal pools within the areas. Data courtesy of Steven B. Weisberg and Victor A. Lotrich.

4.4 Density-Dependent Versus Density-Independent Effects on Abundance

There are examples of marine animals where there is evidence that density-dependent factors are not important. Experiments involving removal of a dominant species of damsel fish (*Pomacentrus wardi*) in the Great Barrier Reef of Australia show that rates of recruitment and survival of new recruits are independent of density of adult fish on the reef (Doherty, 1983). Further, the adult fish do not occupy all potential territories in the reef. The density of populations of reef fishes may be reduced by meterological events such as storms, cold temperatures, predators, or lack of enough larvae that can settle in the reef (Williams, 1980; Victor, 1983).

The lack of enough recruits—even though the reef damsel fishes are among the most fecund of vertebrates (Sale, 1978)—is especially likely in case of fishes that have pelagic larvae, since these larvae are exposed to many mortality factors during the period of larval drift in the water column and mortality would be very high. Manipulation experiments by Doherty (1983) show that the densities of damsel fishes do not saturate the possible habitats of the reef. Perhaps these fishes could saturate the reef after several periods of unusually successful settlement, but this seems to be unusual. Here we have a situation where populations appear not to exhaust resources; competition among similar species may thus be rare, since the obvious resources seem not to be in short supply.

It is not clear if such a density-independent situation is common among other marine fish populations. All the previous examples involve coastal, shallow water fish. It has not been feasible to carry out experiments in which density or food is manipulated in the open sea, but the economic importance of fish and concern with fishery yields have stimulated many analyses of density-dependent features of the population dynamics of pelagic marine fish.

The abundance of pelagic fish populations varies markedly with time, sometimes due to natural causes, sometimes due to changes in fishing pressure. Such changes in fish abundance offer the best glimpses we have of competitive interactions as possible controls of fish populations in the open sea. If the food supply of pelagic fish is limited, we would, as discussed in Chapter 3, expect that survival, growth, and recruitment depend on the abundance of adult fish. In the following sections we examine the available evidence for such density dependence in fish stocks, focusing primarily on intraspecific competition.

4.41 Effect of Abundance on Survival

The larger the density of larval and young fish, the lower their survival rates (Fig. 4-9). The survival is therefore uniformly negatively density dependent, in the species included in Fig. 4-9 and in others (Table 4-2).

The density dependence of survival has been specifically examined in salmonid fish. Salmonids on the northwest coast of North America hatch from eggs in freshwater and migrate as young fish through estuaries to the sea, where they grow into maturity. Cohorts from a particular river basin tend to migrate together to the same area of the sea when mature. The salmon eventually return to their native stream or lake to spawn and then die.* Salmon are regularly censused as they depart and as they return to various estuaries in North America, and these data enable comparison of the original numbers of young fish that left the estuary to the number of adults returning (Peterman, 1980). Although the data are very variable,

* Such a life cycle where fish return to freshwater to breed is referred to as anadromous. Fishes such as eels that live mainly in freshwater and return to the ocean to breed are referred to as catadromous. In the case of eels, breeding takes place in deep waters of the Sargasso Sea.

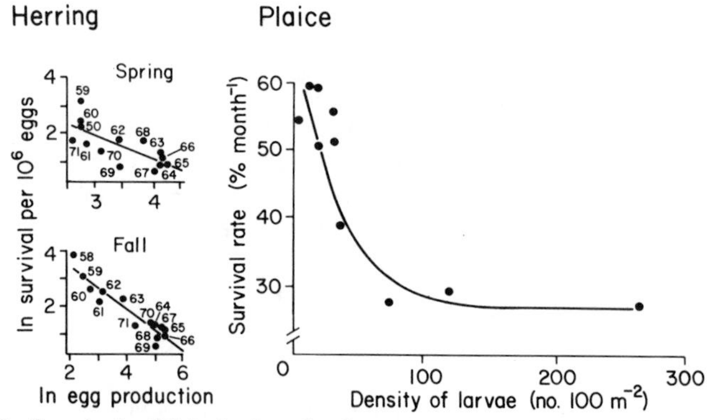

Figure 4-9. Survival of fish during the first year of life (age 0) at different densities. Data for spring and fall herring cohorts from Winters (1976). Plaice data from Lockwood (1978).

out of 18 data sets from various breeding grounds, 12 show that the more young fish that leave an estuary, the greater the mortality. There is of course no assurance that density in the sea is predicted by the abundance of stock leaving the estuaries. Members of specific stocks migrate over very large areas of the Pacific and mix with other stocks. Thus, the salmon data are only suggestive. The correlations of survival and abundance of young salmon have a great deal of variability, but suggest that young salmon may be food limited in the sea. This conclusion has consequences for management and stocking efforts (Simenstad et al., 1982) in

Table 4-2. Density Dependence of Survival, Growth, and Recruitment in Young and Adult Fish Stocks[a]

	Survival	Growth	Recruitment	Sources
Young Fish				
Herring (Gulf of St. Lawrence)	–	–		Winters (1976)
Herring (North Sea)	–	–		Iles (1967, 1968)
Plaice (Waddensea)		Weakly–		Rauck and Zijlstra (1978)
Plaice (North Sea)	–	–		Cushing (1975) Hubold (1978) Harding et al. (1978) Lockwood (1979)
Sole (Waddensea)		Weakly–		Cushing (1975)
Sole (North Sea)		–		de Veen (1978)
Winter flounder (Mystic River Bay)	–			Pearcy (1962)

Table 4-2. (*continued*)

	Survival	Growth	Recruitment	Sources
Halibut (N.E. Pacific)		–		Southward (1967)
Sardine (California Current)	–	–		McGregor (1957) Marr (1960) McCall (1979)
Gadoids (North Sea)		–		Raitt (1968)
Salmon (North Pacific)[b]	–	–	0	Ellis (1977) Peterman (1980) Rogers (1980)
Haddock (North Sea)[c]		–		Gulland (1970)
Adult Fish				
Cod (North Sea)	0	0	Weakly+	Daan (1978)
Cod (Arcto-Norwegian)		+	0	Garrod and Clayden (1972)
Plaice (North Sea)	0	Weakly–	0	Bannister (1978) Cushing (1975)
Sole (North Sea)		0		de Veen (1978)
Herring (Gulf of St. Lawrence)		–	Very weakly–	Winters (1976)
Herring (North Sea)	0	0	0	Iles (1967, 1968) Saville (1978) Cushing (1981)
Sprat (Gullmarsfjörd)		0		Lindquist (1978)
Mackerel (North Sea)		Weakly–	0	Hamre (1978)
Mackerel (Gulf of St. Lawrence)			0[d]	Winters (1976)
Pacific halibut		0	0	Southward (1967) Cushing (1981)
Sardine (California)	0	0		Marr (1960)
Haddock			0	Cushing (1981)
Whiting (North Sea)		Weakly–		Jones and Hislop (1978)

[a] 0, no significant relationship to abundance; + or – indicates positive or negative relationship to abundance.

[b] Several species of salmon involved. These data include individuals up to 5 years of age so are not only young fish; salmon spend lengthy periods in freshwater before entering the ocean, and density-dependent growth of salmon is less in the marine environment than in freshwater (Cushing 1981). West Coast salmon mature when they are ready to return to freshwater and reproduce just once.

[c] Two-year old fish show the relationship most clearly; the higher growth rates occurred in the stronger cohorts, in which years the food supply may have been considerably larger than average.

[d] Significant peak in recruitment during year with intermediate abundance of stock.

the freshwater and estuarine breeding areas: the stocking of individual estuaries by adding artificially reared young salmon may not result in a significant increase in the abundance of the stock, since there may not be enough food at sea for them. When a particular stock is almost depleted, of course, the stocking is useful.

The density-dependent survival of larval and young pelagic fish contrasts with the complete lack of density dependence of the survival of adult fish (Table 4-2). This general lack of density dependence of survival in adult fish could mean that adult fish, at least the ones in Table 4-2, are not abundant enough to tax their food resources.

4.42 Effect of Abundance on Growth

The rate of growth of young marine fish is lower when the density of the cohort is higher (Table 4-2). This is not true for populations of adult fish (Table 4-3). If growth rate depends on food supply, larvae and young fish must exist in food-limited conditions, while food may not be in short supply for adults.

Additional evidence on the relatively small effect of abundance on growth of adult fish can be garnered from fishing statistics obtained after the Second World War, when the intense fishing in the North Sea was halted. During the war years the total fish biomass in the North Sea about doubled, but most fish species showed minor or no responses in catch per unit of fishing effort or in growth (Holden, 1978; Richards et al., 1978).

One of the major commercial species in the North Sea is the plaice, which can be an example of the lack of density-dependent control of growth in adult fishes (Bannister, 1978). Before World War II, fishermen harvested perhaps 23% of the plaice in the North Sea each year. When the

Table 4-3. Average (± Standard Deviation) Body Length of the Isopods *Sphaeroma rugicauda* and *S. hookeri* in Four Areas of Shallow Water on the Danish Coast[a]

Site	Species	In body length (mm)	
		Sympatric	Allopatric
Sønderjylland	*S.r.*	1.64 ± 0.17	1.61 ± 0.17
	S.h.	1.38 ± 0.18	1.49 ± 0.16
Slien	*S.r.*	1.67 ± 0.17	1.61 ± 0.17
	S.h.	1.44 ± 0.20	1.49 ± 0.16
Storstrømmen	*S.r.*	1.68 ± 0.16	1.53 ± 0.21
	S.h.	1.43 ± 0.17	1.50 ± 0.16
Mariager	*S.r.*	1.88 ± 0.18	1.73 ± 0.20
	S.h.	1.32 ± 0.15	

[a] Adapted from Frier (1979).

fishing resumed at the end of the war, the stock had about doubled in number—yet the mean weight of postwar plaice had increased by 50%. Recruitment of plaice after the war was almost double that of prewar years. Different analyses of the pre- and postwar catch data yield different results, but the general conclusion is that the plaice did not show a very marked response to the increase in abundance, and neither did other species such as haddock (Gulland, 1970; Jones and Hislop, 1978). The implication is that at the time these species existed at an abundance considerably below the carrying capacity of the North Sea, and hence competition for food must not have been important for stocks of the adult fish.

There is an instance where more marked effects of density on growth of adult fish have been measured. Growth of herring in the Gulf of St. Lawrence (Table 4-2) is lower at higher densities. When the stocks of pelagic fish decreased by about 83% in the 1970's, the size of herring increased by about 53%. This area of the northwest Atlantic, along with the Grand Banks and Georges Bank, has supported a sizable fishery for a long time, known to Breton, Irish, and Scandinavian fisherman for centuries. These places of high fish density, therefore, are not representative of the sea in general; the high abundance of fish was remarkable enough to justify long, hazardous voyages in small vessels.

The herring in the St. Lawrence demonstrate that adult marine fish can be numerous enough to tax their food resources and hence produce some degree of density-dependent growth. The effects of density are less marked in the vast majority of other reported cases of stocks of adult fish . Perhaps more and better data will show otherwise, but for now we have to conclude that although marine fish *can* tax resources, such instances seem infrequent. Because of this, growth of adult pelagic fish seems not to be density dependent.

4.43 Effect of Abundance on Recruitment

The recruitment* of young fish into an adult population is not clearly related to the density of the adult fish (Table 4-2). The scatter of the data is too large to detect trends (Fig. 4-10) even though there are numerous published instances of models fearlessly fitted to such data.

The lack of density dependence is not due to lack of variation in the recruitment from year to year. This variation is quite large; in Norwegian herring, for example, recruitment may vary over two orders of magnitude over just a few years (Cushing, 1981). This may be an extreme example,

* For demographic purposes recruitment refers to the maturing of individuals into the adult age classes. In fishery publications, recruitment is defined as the appearance of a cohort into the catch due to their having grown large enough to be caught given the mesh size of the fishing gear.

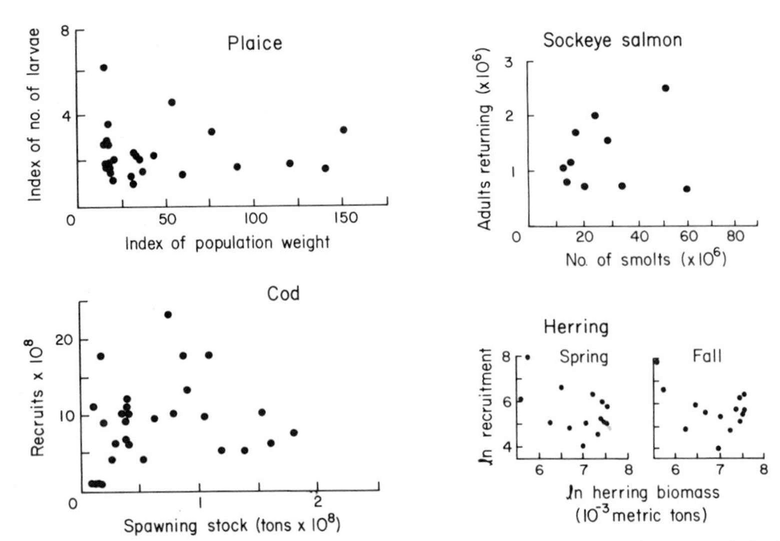

Figure 4-10. Recruitment in fish populations in relation to density. Top left: Recruitment of plaice in to the North Sea fishery in relation to stock density. The values are dimensionless indices obtained from catch statistics [adapted from Cushing (1975)]. Bottom left: Recruitment of cod in Arcto-Norwegian waters (1940–1969) in relation to the abundance of spawners [adapted from Garrod and Clayden (1972)]. Top right: Recruitment of sockeye salmon in Skeena estuary, British Columbia [adapted from Ellis (1977)]. Smolts refer to young fish leaving rivers for the sea. Bottom right: recruitment of herring in southern Gulf of St. Lawrence, Canada. Adapted from Winters (1976). © Canadian Journal of Aquatic and Fisheries Sciences, reprinted by permission.

but variability in recruitment is the rule although the causes of the variability are largely unknown.

The evidence on mortality, growth, and recruitment obtained from commercial fisheries leads to the conclusion that the densities of larval young in pelagic fish may be high enough for competition for food to be important. Adult fish exist at densities low enough that the carrying capacities of their environment are seldom taxed. Abundance of these fishes—and that of the reef fishes discussed earlier—is thus not completely controlled by factors that depend on density of the population. Rather, the control, if it does exist, is exerted by density-dependent mortality and growth during early life. Recruitment is not dependent on adult density, so there is little evidence that adult abundances are controlled by density-dependent mechanisms. In spite of large adult fecundities, mortality suffered by the early stages may be so intense that very few recruits survive and are available to enter the breeding population. If the mortality of early stages depends on many factors—as seems likely—the sum of these factors may impose what may in the aggregate appear to be random variation on mortality rates. Thus, for many fishes the number of recruits would vary randomly, and may not be higher when adult numbers are low or lower when adults are abundant.

The above considerations are a caution that even though in many species the density is probably determined by density-dependent factors, density-independent processes may be very significant in population dynamics. There is still, however, as discussed in Chapters 8 and 9, no agreement on the extent to which competition establishes the abundance of species in nature (Schoener, 1983).

4.5 Resource Partitioning

There is a long history of examples, primarily in birds, in which species that live together are described as partitioning resources. Such sympatric species divide up limited resources by behavioral, morphological, or ecological mechanisms. Lack's (1947) study of Darwin's finches in the Galapagos Islands is just one prominent example. Many of these examples are ambiguous, due to faulty data and interpretation, but the remarkable degree of partitioning of resources that exists in all environments suggests (although it cannot demonstrate, since the appropriate controls are lacking) that competition has been a pervasive factor during the evolution of present species.

Different aspects of habitat or food resources may be apportioned differently in some way by the species present in one place. The ubiquity of this partitioning gave rise to what has been said to be one of the few "laws" of ecology, the "competitive exclusion" or "Gause's" principle: complete competitors cannot coexist. This idea has a venerable history (cf. Krebs, 1978) but a contentious development, with opinions ranging from "one of the chief foundations of modern ecology" to "a trite maxim." The point to remember is that if enough effort is spent, it is possible to find differences in habitat selection or food partitioning between any two species. Partitioning of resources can occur either at one time and site, or may be accomplished at different times of day or year, and competing species may diverge morphologically enough to exploit different portions of resources.

Slightly different parts of a habitat can be used either by different age classes within a species or by several species. An example of the former is the distribution of the limpet *Patella compressa* on the surface of the kelp *Ecklonia* (Fig. 4-11). *Ecklonia* is almost the only place where this limpet is found. Juvenile *P. compressa* aggregate on folds on the bases of the fronds; older individuals move onto the broad flat "hand," while the oldest individuals establish themselves in shallow cavities in the stipe. The base of the shell of large individuals even grows concavely to adapt to the cylindrical shape of the stipe. The separation among the age groups

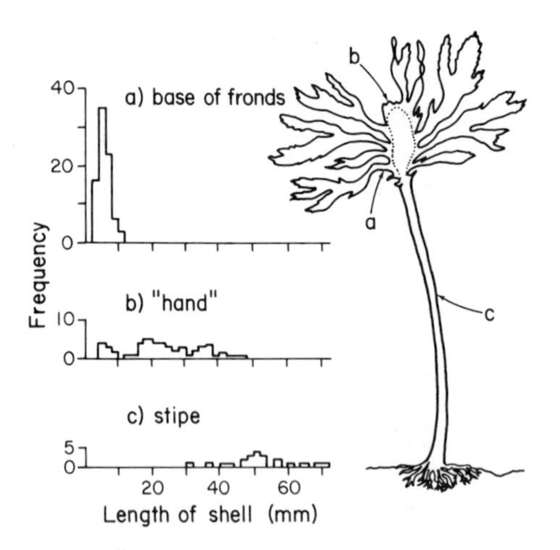

Figure 4-11. Location of size classes of the limpet *Patella compressa* on the kelp *Ecklonia*. The histograms show the frequency of individuals of various sizes. This kelp may reach 10 m in height. Adapted from Branch (1975a).

may partition space and food, presumably minimizing intraspecific competition.

Interspecific partitioning of a habitat by several species occurs in all environments. The assemblage of fish species found in New England salt marshes (Fig. 4-12, left histograms) partition the use of habitats within the marsh. Silversides, alewives, and menhaden are found primarily in the water column. The two sticklebacks are less stereotyped in use of habitat, and are found in the water column as well as on the bottoms of creeks. The two killifish are the least specialized in habitat choice. The last four species are found mainly on creek bottoms, with tautog, sea bass, and winter flounder using sandy bottom while eels were mainly found on muddy bottoms.

The partitioning of food among size classes of one species and other aspects of food selection are more fully discussed in Chapter 6. Here we can emphasize the extent of food partitioning by looking at food habits of salt marsh fish (Fig. 4-12, right histogram). Of the three species found in the water column, the menhaden eat mainly algae and silversides feed on larger zooplankton than the alewives (Werme, 1981). The three-spined sticklebacks eat much more plankton than the four-spined sticklebacks. The killifish are the most generalized feeders of all the fish species, but even so show differences in food selection. The sheepsheads feed principally on algae and detritus, the striped killifish on benthic invertebrates found on creek bottom, and the common killifish on algae and invertebrates found on the surface of the marsh, obtainable only during the time when the marsh is flooded. The remaining species of fish feed on benthos, and partition their food by using different combinations of size classes of prey.

Species often partition resources by using prey populations at different times. The three-spined stickleback is found primarily in the spring (April–June) while the four-spined is in the marsh for longer periods

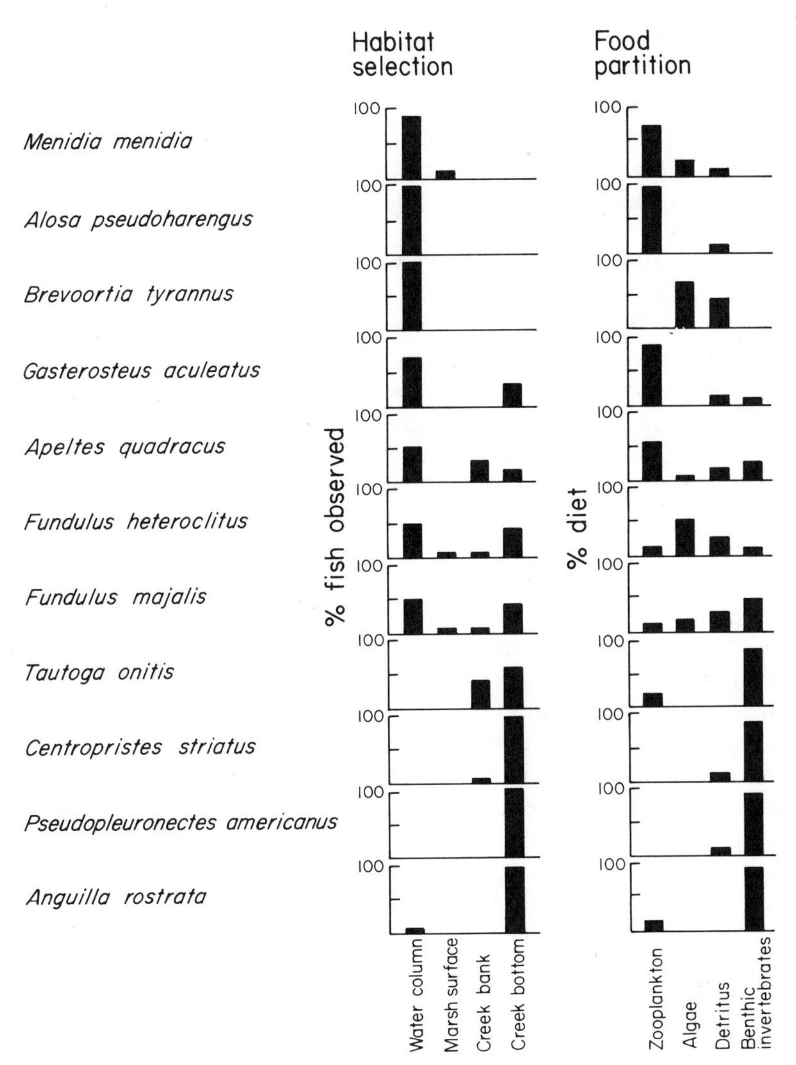

Figure 4-12. Partitioning of habitat and food by species of salt marsh fish. These results were averaged over an entire year and through many tidal cycles. The common names of the fishes, in order, are silversides, alewife, menhaden, three-spined stickleback, four-spined stickleback, common mummichog or killifish, striped killifish, tautog, sea bass, winter flounder, and American eel. Adapted from Werme (1981).

(April–October). The common killifish (and the related sheepshead minnow, *Cyprinodon variegatus*, not included in Fig. 4-12) feed primarily during high tide, while the striped killifish have no tidal feeding rhythm.

Partitioning of food and habitat may take place by changes in morphological characters in competing species (Brown and Wilson, 1956), although some of the original evidence for such character displacement has been interpreted in other ways (Grant, 1972). One example of character displacement has been claimed for two species of shallow-water marine isopods whose body sizes diverge when the two species are sympatric (Table 4-3). Size in isopods is important in various ways, one of which is that individuals of a given size occupy protective crevices of a given size. Perhaps displacement of size avoids competition among the two species for crevices.

Fenchel (1975) found another possible example of character displacement while surveying the size of two snails, *Hydrobia ulvae* and *H. ventrosa*, in estuarine sites off the Danish peninsula of Jutland. Shell lengths of the two species are similar in sites occupied by only one species, while in the 16 sites of sympatry there is a consistent divergence in shell length (Fig. 4-13, top left). Changes in size may have many ecological implications, one of which is that snails of different size feed on different size particles (Fig. 4-13, left bottom). This results in differences in the distribution of particle sizes ingested by sympatric snails compared to allopatric snails. The *Hydrobia* data suggest that character displacement is a real phenomenon and may result in reduced competition (Kofoed and Fenchel, 1976). Pacala and Roughgarden (1982) showed that in lizards the level of competition was inversely related to the degree of resource partitioning, as we would expect from the *Hydrobia* data. Schoener (1983) reviewed available results and concluded that overlap is inversely related to competition. Nonetheless, it is hard to prove that character displacement results from competition. Connell (1980) has pointed out that the localities studied by Fenchel varied in salinity, water turbulence, and

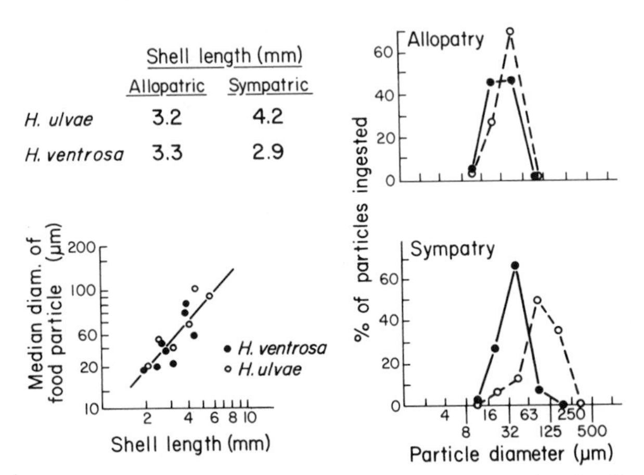

Figure 4-13. Character displacement and its consequences for feeding in *Hydrobia*. Top left: length of shells in two species of *Hydrobia* where they occur together and separately. Bottom left: relation of length of shell and median size of particles eaten by *Hydrobia*. Right: distribution of particles sizes ingested by the two species of *Hydrobia* where they occur together and separately. Adapted from Fenchel (1975).

particle size of sediment, so that it is not possible to prove that competition alone led to character displacement. Furthermore, in any one site the species may occur together 1 month and alone the next, so there may not be enough time for displacement to occur in species that disperse so actively. Character displacement remains a suggestive but unproven consequence of competition.

Static documentation of food and habitat partitioning and character displacement only constitutes circumstantial evidence of competition; it is not possible to attribute cause and effect. Competition could cause partitioning, but also a low degree of partitioning or displacement could reduce competition. In contrast, where experimental manipulation such as in Connell's work has been done, we were in a position to say that competition was responsible for the phenomena observed. Of course, it is difficult to carry out experiments on processes that take place over relatively long periods of evolutionary time, such as character displacement, but other aspects of competitive interactions are subject to an experimental approach.

4.6 Niche Breadth and Species Packing

Resource partitioning due to interspecific competition implies that very specialized species should be found in habitats containing many species. Much space has been devoted to this topic in the ecological literature. If we accept that the maxim "jack-of-all-trades, master of none" is applicable to how organisms use resources, we can argue that more specialized species will make more effective use of whatever resources are available. Hence more species should be able to be packed into an environment containing specialists. This tendency to increased specialization can go on until constrained by two factors. First, the variability in the resource being partitioned sets a limit. If a specialist becomes dependent on a specific resource, but the resource is not available in sufficient abundance or during a certain time, that specialist species is more liable to disappear. Second, intraspecific competition will favor some degree of nonspecialization so that not all individuals or age classes within a species exploit the same resource. Recall, for example, habitat partitioning due to intraspecific competition in limpets (Fig. 4-11).

The above arguments are completely hypothetical, with precious little evidence to support them. There are several reasons for the lack of evidence. First, it is extremely difficult to define a niche. Hutchinson (1957) first developed the concept of the niche as a multidimensional hyperspace, comprising all the variables important to a species. While interesting, this idea is not testable. The best that can be done is to isolate a few potentially important variables and see how species align themselves along these few dimensions of the niche. The range of values of the chosen variables that is tolerated by the species is then called the niche width and is measured in various simple ways (Colwell and Futuyma, 1971; Pianka, 1973) or by using multivariate statistics (Green, 1971). While the variables chosen may appear very important to the ecologist, they may be less critical to the organisms under study. Second, there is always the problem of scale of measurement; we may measure the temperature range in terms of a few degrees while the organisms may respond simply to hot or cold, while being far more sensitive to minute changes in another variable. Third, some dimensions of niche breadth are more relevant to some taxonomic groups than to others, and comparisons are difficult even within similar ecosystems. For example, Kohn (1971) finds remarkably high degrees of specialization in food habits of *Conus* snails in Pacific coral reefs. On the other hand, Sale (1977) concludes that the diet of coral reef fishes is rather generalized. The diets of species of *Conus* tends not to overlap, while those of fishes do. As Anderson et al. (1981) point out, however, this does not mean that the fishes are complete generalists, since reef fishes have quite specialized requirements in terms of habitat, for example. Further, Roughgarden (1974) shows that some reef fishes do show partitioning with regard to prey size (cf. Chapter 6). Such considerations have made it difficult to generalize about how niche breadth should vary in different kinds of environments.

The study of niches and how species are "packed" into an environment requires a more dynamic understanding of what a species does than is provided by an account of items in the diet. The morphology and behavior of an assemblage of sympatric predators, for example, determine the kinds and sizes of their prey, and how closely apportioned the prey are (Schoener, 1974 and others). To see what is involved in the acquisition of prey, it is necessary to learn a considerable amount about the feeding process and feeding selectivity involved, the subject of the next two chapters.

Chapter 5

Feeding and Responses to Food Abundance

5.1 Introduction

5.11 Modes of Feeding in Marine Consumers

Marine heterotrophs have evolved a very large variety of feeding mechanisms to obtain their rations, and any given species usually makes use of several feeding mechanisms (Pandian, 1975). Heterotrophs may use organic matter dissolved in water or on larger particles. Dissolved food may be taken up through the body or cell surface, as occurs in microorganisms, many invertebrate parasites, and pogonophorans. Other consumers take up fluids through their mouth, as in the case of some nematodes, trematodes, leeches, parasitic copepods, and young mammals. Dissolved food is also obtained by some heterotrophs from their endosymbionts, as in the case of zooxanthellae in corals and other cnidarians, and sulfur bacteria in some pogonophorans and bivalves.

5.111 Uptake of Dissolved Organic Matter

Heterotrophic uptake* of dissolved organic matter from water is primarily carried out by bacteria, fungi, and other obligate heterotrophs. Facultative heterotrophic uptake may also be carried out by algae and some invertebrates.

---

* The kinetics of uptake of dissolved organic compounds can often be described by Michaelis–Menten kinetics (Parsons and Strickland, 1962; Wright and Hobbie, 1965). This implies, as in the case of nutrient uptake reviewed in Chapter 2, that there is a maximum uptake rate at some concentration of substrate, and that perhaps the number of transport sites in each cell sets this rate.

There are many measurements of heterotrophic uptake of dissolved organic carbon (DOC) by algae, usually done by experimental addition of high concentrations of organic substrates. If uptake of DOC by bacteria can take place at low (nano- or micromolar) concentrations of DOC, one would suppose that in nature bacteria do nearly all the heterotrophic uptake, since they would easily maintain the DOC concentrations below the level where algal uptake can take place. This supposition has not been conclusively verified. The best evidence to support the idea that bacterial heterotrophy is much more important than algal heterotrophy comes from studies in which uptake of organic substrates by populations of living microorganisms is fractionated using filters. About 90% of the heterotrophic assimilation of tritium-labeled organic compounds is due to organisms smaller than 5 μm (Azam and Hodson, 1977), most of them smaller than 1 μm and largely bacteria.

In sediments, where the concentration of specific organic substrates tends to be in the micromolar to millimolar range rather than the nanomolar to micromolar range, there is some evidence that some benthic diatoms can take up certain low-molecular-weight organic compounds in competition with bacteria (Saks and Kahn, 1979). The presence of diatoms with no photosynthetic pigments in sediments certainly supports this idea (J. Hobbie, personal communication). Conceivably, in very rich interstitial water and in detritus, heterotrophic uptake by algae could be significant.

There is a growing literature on the uptake of organic substances by invertebrates—reports by Costopulos et al. (1979), Stewart and Dean (1980), and Slichter (1980) are but a few of the current examples. Uptake takes place against very considerable gradients, and seems to be carried out by saturable processes describable by Michaelis–Menten kinetics. Measurements of uptake, however, have usually been carried out in the presence of bacteria, which has confounded the measurements, since bacteria can take up organic compounds very rapidly (Siebers, 1979). The measurements have also been done at extremely high substrate concentrations relative to ambient values, and so have not shown convincingly that invertebrate heterotrophy was important in nature. Recent developments of analytical techniques have made it feasible to demonstrate that uptake does take place very quickly at ambient concentrations as low as 38 nmoles amino acid liter^{-1} in the blue mussel *Mytilus edulis* and in other invertebrates (Manahan et al., 1982). Concentrations of DOC are higher in interstitial water of sediments (5–16 μM of amino acids, for example). Meiofauna (Meyer-Reil and Fauvel, 1980) and polychaetes may take up amino acids at these concentrations (Jørgensen and Kristensen, 1980, 1980a). The actual extent to which invertebrates can use heterotrophic uptake to complete their ration is not known. It seems likely that this is a small source of dietary requirements.

5.112 Symbiosis

There are many instances of heterotrophs making use of DOC released by closely associated autotrophs (Trench, 1979). Such cases of symbiosis range from very loose to very close associations. An example of a loose symbiotic arrangement is that of a species of *Flavobacterium* found associated with cells of diatoms (Jolley and Jones, 1977). Presumably these symbionts exchange organic substances which only one of the partners manufactures. In anaerobic protozoans, fermentation may yield end products of use to the bacteria found within them; the protozoans in turn may depend on the bacteria for cellulases (Fenchel et al., 1977). Symbiotic associations are common in corals and other cnidarians that contain dinoflagellate phytosynthetic inclusions—zooxanthellae—that fix and are thought to release DOC used by the heterotrophic coral host. Some host species, such as corals and some bivalves with reduced digestive systems (Cavanaugh et al., 1981) may be able to survive without symbionts, but there are obligate symbiotic relationships where the host has no independent source of energy. Examples of this are gutless pogonophorans (Southward and Southward, 1980; Cavanaugh et al., 1981) and some oligochaetes (Giere, 1981).

Some deposit or suspension feeders that live in environments with abundant sulfur resources have facultative symbionts that use the sulfur to drive the reactions of carbon fixation. Examples of the latter have been found in the spreading centers in the deep ocean, where geothermal sulfur is released, and in coastal environments where anoxic muds containing sulfide are common (Cavanaugh et al., 1981).

The best-studied cases of symbiotic exchange of DOC involve corals and other cnidarians. Lab experiments demonstrate that zooxanthellae provide significant amounts of carbohydrates for the metabolism of sea anemones. When particulate food and light are available to anemones with symbionts, the respiratory quotients (CO_2 evolved/O_2 consumed) of the anemones are between 0.9 and 1, indicating carbohydrate metabolism (Fig. 5-1, left). Anemones without symbionts that have no particulate

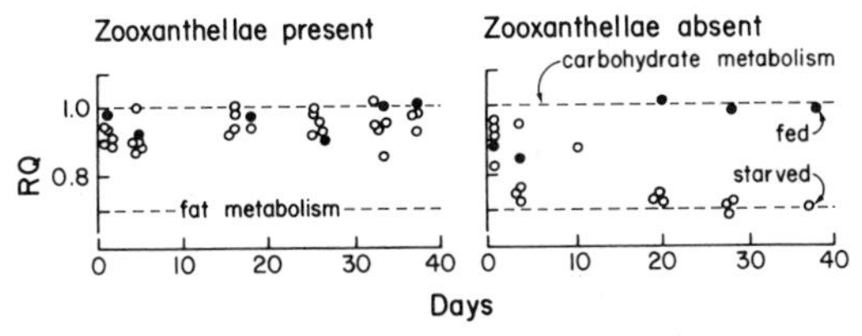

Figure 5-1. Respiratory quotient (RQ = CO_2 evolved/O_2 consumed) in the anemone *Anthopleura elegantissima* where zooxanthellae were present or absent. Anemones were either fed shrimp or starved. Adapted from Fitt and Pardy (1981).

food available switch to metabolism of fat in a short time, evidencing that their carbohydrate reserves are depleted (Fig. 5-1, right).

The inclusion of organisms in the tissues of other organisms has an extremely long evolutionary history, and Margulis (1981) believes that many of the radical steps in the early evolution of organisms are based on symbiotic associations. Many of the organelles of cells, for example, are the result of incorporation of previously free-living species into a host species and subsequent modification of the symbionts. The evolution of symbiotic relationships is widespread and the ensuing adaptations are extremely varied. The acquisition of such symbionts has made possible behavioral and morphological adaptations that demonstrate that for such species symbiotic procurement of energy is as or more important than predation.

The importance of maintaining the symbiotic relationship can be seen in the evolution of behavior and morphology designed to facilitate symbiosis. Certain free-swimming medusae with zooxanthellae have daily vertical migrations that take them to strata of water where there is enough ammonium to satisfy the needs of the zooxanthellae (Muscatine and Marian, 1982). Certain sessile sea anemones in coral reefs may have their zooxanthellae and nematocysts located in different organs. The anemones expand the organs bearing zooxanthellae during the day (Fig. 5-2, left diagrams), when light is available; the organs bearing nematocysts are expanded at night (Fig. 5-2, right diagrams) when the zooplankton prey are most active.

In such anemones and in corals, the zooxanthellae provide carbon compounds to the host, while the nematocysts, through the prey they provide, furnish the nitrogen needed by the symbiotic algae (Trench, 1974). Seawater around coral reefs tends to be very nutrient poor, and the assurance of a ready supply of nitrogen is an important asset for producers. It is therefore not surprising that most of the primary production in reefs is carried out within corals.

Corals are only partially heterotrophic. Among Caribbean reef-building corals, species more adapted to live autotrophically seem to have the smallest polyp sizes (Porter, 1976; Wellington, 1982a,b), since large polyps seem best suited for capture of zooplankton. Even the most heterotrophically adapted coral species, however, only obtain up to 25% of their ration by predation (Porter, 1978). Johannes et al. (1970) calculated that the energy requirement of corals near Bermuda was more than an order of magnitude greater than the energy yield from capture of zooplankton. Further evidence of the dependence of corals on photosynthetic symbionts is that the ratios of stable carbon isotopes of the tissues of the host coral match those of their zooxanthellae, instead of the isotope ratio of the zooplankton (Land et al., 1975). The coral-zooxanthellae unit therefore functions mainly as a primary producer, with the internal exchange of DOC being an essential feature for the success of the arrangement.

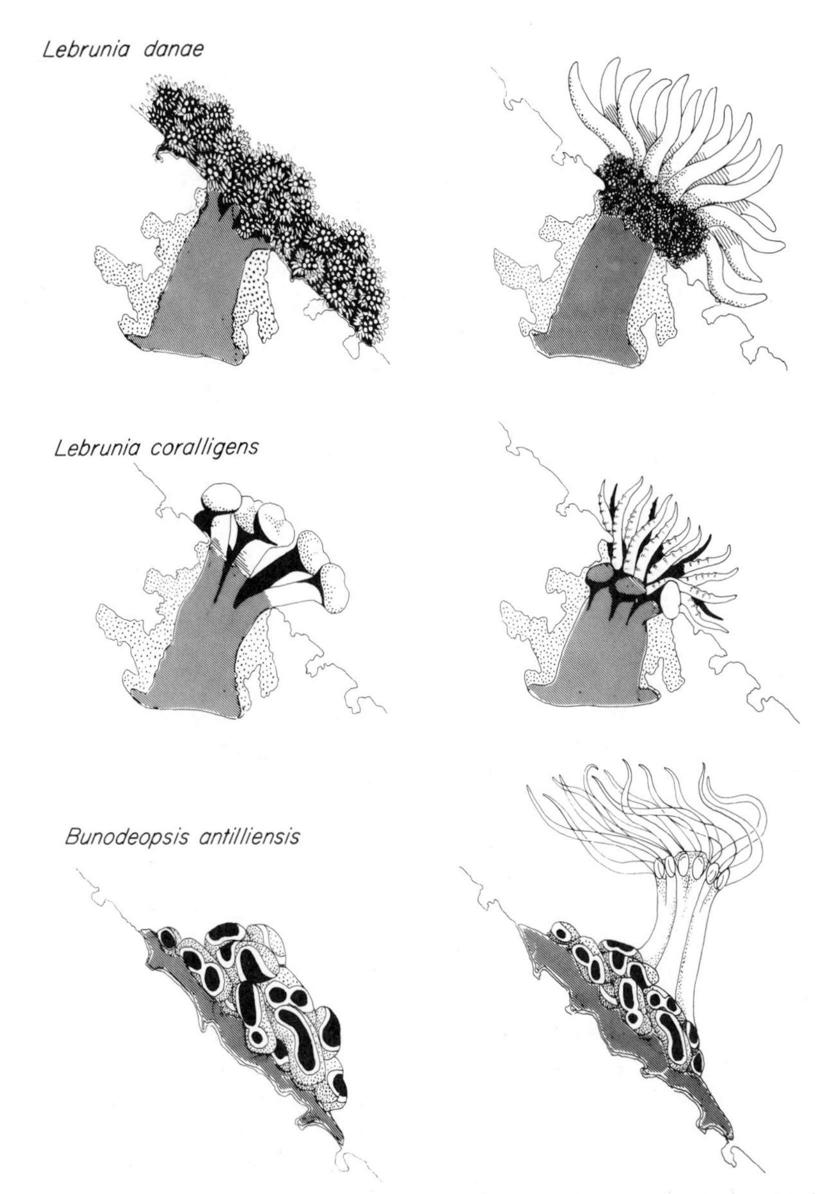

Figure 5-2. Appearance of sea anemones of a coral reef in Jamaica during the day (left diagram) and night (right diagram). Adapted from Sebens and De Riemer (1977).

5.113 Consumption of Food Particles

The majority of consumer species eat particulate foods. Food particles may be consumed by engulfing prey with pseudopods, as in, for example, radiolarians and foraminiferans. Suspension feeders collect food particles in various ways. Ciliates, sponges, bivalves, many annelids, and brachiopods create feeding currents by the beating of cilia or appendages to bring food to the animal. Tunicates and some gastropods produce mucus sheets that are used as sieves or sticky surfaces to capture food. Copepods and barnacles use setae- and setule-bearing appendages to capture particles much smaller than the consumer. Some fish strain particles by passing water through gill rakers, and whales use baleen for the same purpose. Deposit feeders such as many sea cucumbers, bivalves, polychaetes, and shipworms collect and ingest sediment that contains food particles, while raptorial feeders (gastropods, cephalopods, crustaceans, cnidarians, turbellarians, polychaetes, fish, birds, turtles, snakes, and some mammals) capture food particles that are relatively large compared to the size of the consumers. Feeding studies have been done mainly on species that feed on particles, so the remainder of this chapter focuses on particulate feeding.

5.12 The Study of Feeding Rates by Consumers

The mechanics of finding, capturing, and ingesting particles of food by consumers are similar whether they are carnivores, herbivores, or detritivores. The time and effort allotted to each of the components of feeding—search, pursuit, capture, and handling of food items—differs in a snail feeding on benthic microalgal mats compared to tuna pursuing mackerel. In both cases the food procurement process is made up of the same *components*, even though, for example, the time spent pursuing food differs. One approach to the study of feeding by consumers is to find out how each component is affected by the abundance of consumers and food. Once the effects of consumer and food abundance are understood, a synthesis of the process can be provided by models that incorporate these variables. The models, ideally at least, can then be used to assess the role of feeding by consumers in nature.

Models of the feeding process based on an analysis of components of feeding and their dependence on density of food and consumers have been constructed by Holling (1965, 1966) and others (Beddington, 1975; Beddington et al., 1976; Griffiths and Holling, 1969; Hassell et al., 1976; among many others). Since the work on these models is stated in terms of predator and prey, we will use these terms in this and the following chapter, but with the understanding that they may apply to herbivores, carnivores, or detritivores, and their food particles.

The rate at which consumers acquire food depends largely on their abundance and on the abundance of food and consumers. The following sections of this chapter examine the three major ways in which predators respond to abundance of their prey. These responses include (1) the functional response (Solomon, 1949), in which predators increase prey consumption as prey abundance increases; (2) the numerical response (Solomon, 1949), in which the number of predators increases as prey abundance increases; and (3) the developmental response (Murdoch, 1971), in which predators grow to a larger size faster when more prey are available.

5.2 Functional Response to Prey Density

5.21 Types of Functional Response

As prey numbers increase, predators eat more prey. This so-called functional response to prey density may take three different shapes (Fig. 5-3). In the type I response, the consumer eats an increasing number of prey items as prey density increases, in linear proportion to prey abundance, until a satiation point is reached. The satiation point occurs because the predator cannot handle prey any faster, so that the ingestion rate then remains approximately constant even if prey density increases further. A combination of two such linear relationships has been used to describe the functional response of both herbivorous and carnivorous copepods (Frost, 1977).

Predators exhibiting a type II response increase consumption of prey at a decelerating rate as prey density increases. This results in a curve that rises to an asymptote, reflecting increased costs or constraints associated with the consumption of more prey per unit time. At increased prey densities either the mechanism of prey capture may become increasingly inefficient or the motivation for increasing feeding rate may be reduced.

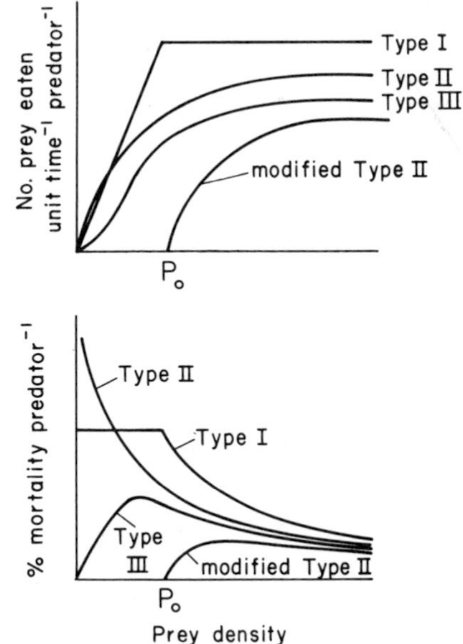

Figure 5-3. Shape of hypothetical functional response curves (top) and the percent mortality expected per predator by each of the types of functional responses (bottom). The height of the asymptotes in the top figure is not an important consideration. Adapted from Holling (1959) and Peterman and Gatto (1978).

The type II response has been described by many authors (Ivlev, 1961; Parsons and LeBrasseur, 1970; McAllister, 1970; Frost, 1972; Corner et al., 1972; Robertson and Frost, 1977; among others). Holling derived it in the equation

$$I = aP/(1 + abP), \tag{5-1}$$

where I is the ingestion rate at prey density P, a is the rate at which prey are encountered, and b is the time required to capture and eat a prey item. Ivlev (1961) adapted a curve derived earlier by Gause:

$$I = I_m(1 - e^{-dP}), \tag{5-2}$$

where I_m is the maximal rate of ingestion and d is a constant specifying the rate of change in I with respect to P. Cushing (1978) and Crowley (1973), arguing by analogy to Michaelis–Menten uptake (cf. Section 1.421), predicted ingestion rate using

$$I = I_mP/(K_d + P), \tag{5-3}$$

where K_d is the prey density at which ingestion equals $I_m/2$. These three expressions of a type II functional response are conceptually and arithmetically related (Parker, 1975) and the major point is that the acquisition of food items by the type II response can be represented by a variety of equations that describe a process that becomes increasingly saturated at higher densities of prey.

Parsons et al. (1967, 1969) claimed that in zooplankton no feeding took place below some low concentration of particles. They therefore modified Ivlev's equation to allow for such a threshold (Fig. 5-3):

$$I = I_m[1 - e^{d(P_0 - P)}], \tag{5-4}$$

where P is the prey density and P_0 is the threshold prey density below which no feeding takes place. Others reached similar conclusions (McAllister, 1970; Adams and Steele, 1966), and Steele (1974a) used equations incorporating this feeding threshold as a cornerstone of a model of the dynamics of grazing by copepods on phytoplankton in the North Sea (cf. Section 8.31). The attraction of the threshold idea in the type II response is that if grazers cease exploiting prey below a threshold density, this provides a sanctuary in which a few phytoplankton can survive to regenerate the population when grazers are no longer present near the depleted patch of phytoplankton. Inclusion of the hypothetical feeding threshold in the model thus furnishes a mechanism for recovery of a depleted patch of phytoplankton. The modeled populations of phytoplankton are thus not exterminated by grazers.

It is not yet clear whether this threshold is a general phenomenon in the sea. *Calanus* can remove an average of 92 cells day^{-1} even at the relatively low cell density of 270 cells liter^{-1} (Corner et al., 1972). In these experiments the copepods still fed at 50 cells liter^{-1} but counts were too variable to allow conclusions. Frost (1975) found that the rate of filtration by *Calanus* was depressed at about 200 cells ml^{-1} (type I response, Fig. 5-3) and some filtration still took place at about 100 cells ml^{-1}. In fact, when copepods were fed on larger cells,* filtration went on even at lower densities, and ingestion did not decline, let alone stop, until cell densities dipped below 4–11 cells ml^{-1}. These cell densities, while low, are in the range of concentrations of oceanic phytoplankton (Fig. 1-2). Thus, zooplankton, while showing lowered ingestion rates at low food densities, can still obtain food even at very low densities of phytoplankton.

Type III responses to increasing prey density initially show an accelerated rate of ingestion (Fig. 5-3). It is not clear how this acceleration is brought about; one thought is that the accelerating attack rate is perhaps related to the buildup of a search image (Tinbergen, 1960). As encounters with a given kind of prey are more frequent, the predator perhaps 'learns' that that particular type of prey is adequate and concentrates feeding on that kind of prey. Several other mechanisms that could accelerate the density-dependent portion of the type III functional response have been proposed (Hassell and May, 1974; Crawley, 1975; Hildrew and Townsend, 1977; Hassell et al., 1977; Cook and Cockrell, 1978) but are not well established. Explanations of how these hypothetical alternative mechanisms might work are lengthy and can be found in the references cited above.

After the accelerating portion of the type III functional response there is an inflection point beyond which the rate of change of prey eaten per predator slows down, much as in the type II functional response.

Behavioral adjustments in feeding rate by predators can bring about all three types of functional responses, with no significant lag between the time when there is a change in number of prey consumed.

Of the three types of functional response, type III would seem to have the greatest potential to regulate prey numbers, since it generates a negative feedback on prey density (Fig. 5-3, bottom): the percentage of the prey eaten by the predator *increases* as density of prey increases, at least for the lower range of prey densities.

Work on functional responses of predators has emphasized theory, sometimes verified in laboratory situations, and dealt mainly with terrestrial species. How generally applicable are functional response curves to marine organisms? In Fig. 5-4 we have the results of selected experiments with a variety of marine invertebrate species where the number or amount of food eaten was recorded in situations where density of food varied. In all cases, there is some sort of functional response: all the invertebrate species consumed more prey when prey density increased. There is, how-

* Note that cell size and cell density are not independent; this is the first of many examples we will see that although we discuss components separately as an expository device, the components may be inextricably related.

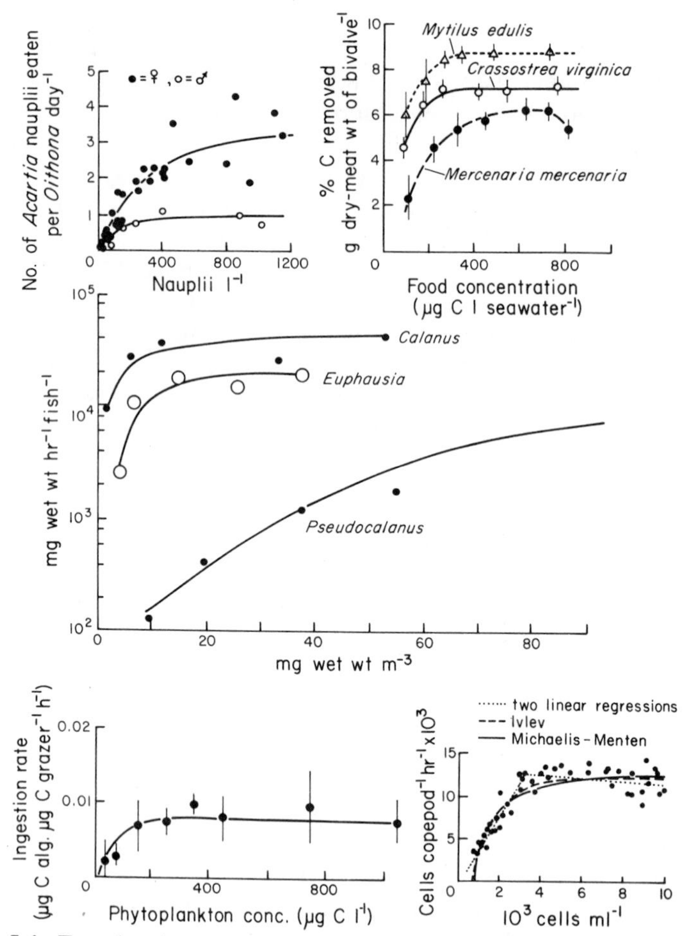

Figure 5-4. Functional responses of various marine invertebrates to increases in density of food. Top left: Predation by *Oithona nana* on nauplii of *Acartia* (Lampitt, 1978). Top right: Amount of particulate carbon removed from suspension by three species of bivalves (Tenore and Dunstan 1973). Middle: Ingestion of three prey zooplankton species of varying size by juvenile pink salmon (Parsons and LeBrasseur 1970). Bottom left: Ingestion rate of phytoplankton carbon by copepods (McAllister, 1970). Bottom right: Ingestion of the alga *Thalassiosira* by the copepod *Calanus*, adapted from Frost (1972) and Mullin et al. (1975), with three alternative descriptions of the functional response.

ever, considerable scatter in the results, so that for most cases it is not possible to decide if curves of type I or type II responses best fit the data. In fact, where (Fig. 5-4, lower right) the fit of three models (two straight lines, Ivlev's and the Michaelis–Menten equations) has been explicitly tested, no one model fitted better than any of the others. Given the variability in even the best available results, it is not possible to decide which model fits best, let alone infer the mechanisms of feeding by the fit to a specific curve.

Examples of type III functional response curves from the field are harder to find in the literature, but predation by shore birds on bivalves may follow this pattern. Feeding by certain fish and starfish in the laboratory also may show type III functional responses (Fig. 5-5). In the experiments with starfish, after exposure to the single prey species for 10–12 weeks, the entire predator population had "learned" that these were adequate prey and spent less time in investigative behavior and fed more readily on the prey offered. The number of prey eaten at any one density increased, and the curve changed to a type II response whose shape was largely due to the relative efficiency with which prey were processed. In cases where there is one or just a few prey species, the learning phase may thus be a temporary phenomenon. The term "learning" is used here in a very broad sense, including not only the strictly behavioral notion of changed behavior due to experience but also physiological adaptation such as induction of enzymes and changes in assimilative efficiency.

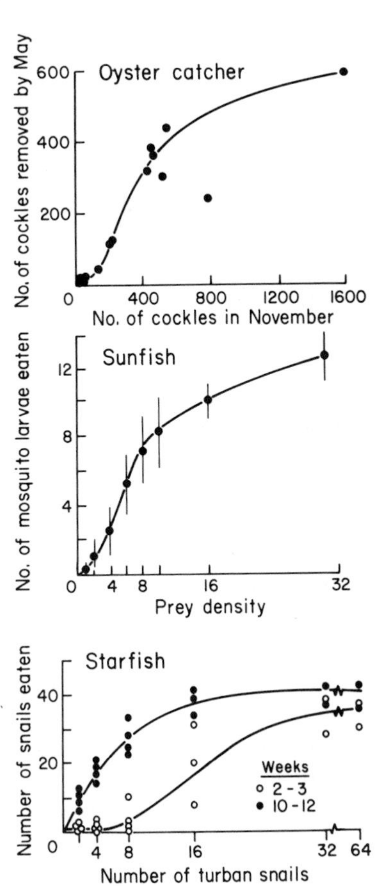

Figure 5-5. Type III functional responses. Top: Oystercatchers (*Haematopus ostralegus*) feeding on cockles (*Cerastoderma edule*) in Llanridian Sands, Burry Inlet, South Wales. The densities of cockles were measured in November and again in May, from 1958 to 1970. Missing cockles are inferred to have been eaten by the overwintering population of oyster catchers. Curve is fitted by eye, ignoring the one outlying point. Adapted from Horwood and Goss-Custard (1977). Middle: Feeding by bluegill sunfish (*Lepomis macrochirus*) on mosquito larvae in the laboratory. Data of R. C. Reed in Murdoch and Oaten (1975). Bottom: Feeding by starfish on turban snails (*Tegula*) in the laboratory. The data shown are those from weeks 2–3 and 10–12 in a 3-month-long experiment. Data of D. E. Landenberger in Murdoch and Oaten (1975).

Feeding by some ctenophores on copepods (Reeve et al., 1978) also shows sigmoid (type III) functional response curves. Because of the mechanisms involved in ctenophore feeding (passive capture of particles in mucus-covered tentacles), it is difficult to imagine a way that learning can produce the sigmoid curve. This once again points out that using theoretically derived curves to infer mechanisms is an uncertain business. Feeding mechanisms are better studied directly, such as by *in situ* observations using SCUBA (Hamner et al., 1975), filming methods (Alcaraz et al., 1980), and radioactive tracer experiments (Roman and Rublee, 1980).

There are several aspects of feeding responses to increased food abundance that have not as yet been considered but may be important. First, none of the responses discussed above predicts that ingestion may be lower at very high prey densities, even though some of the data suggest lowered ingestion rates at the highest prey densities (Fig. 5-4). It is likely that in many cases there are additional mechanisms that lead to lower feeding efficiency or clogging at very high densities of food particles (Harbison and Gilmer, 1976). Such mechanisms have not yet been included in models of the functional response.

Second, there are instances where there is no asymptote in the feeding rate of copepods as the density of food increases (Mayzaud and Poulet, 1978) so none of the three responses described above may apply. Feeding by adult *Mnemiopsis* (a ctenophore) is carried out by extension of mucus-covered lobes, and does not become saturated at even the highest densities found in the natural environment (Reeve et al., 1978). This contrasts with the evidence described above and makes it clear that short-term experiments may not provide the entire picture of the functional response of predators to prey density.

Third, since the density of prey varies through a season, species of zooplankton may show different functional responses at different times of year. For example, the rate at which *Calanus pacificus* filters cells out of the water varies over the season (Runge, 1980), with a fivefold increase between early summer and the colder months. Most of the experiments discussed above involve collecting predators and placing them in a container with a range of food densities for short periods of time. Such experiments generally show that the functional response is saturated at higher densities. Given long enough periods of time, however, zooplankton may adjust their functional responses to the ambient prey density and composition. Several species of copepods in Canadian coastal waters acclimate their feeding rates to seasonally varying food densities, and feeding increases as ambient concentration of food particles in the water column increases (Mayzaud and Poulet, 1978). Both behavioral and physiological acclimation took place, since not only were more prey taken at higher prey densities but also there was higher activity of various digestive enzymes in the zooplankton. Short-term experiments carried out at different times during the year yielded type II saturated feeding curves, but the density at which the asymptote was reached varied with ambient food density. Although such acclimation may not occur in all predators (Fox and Murdoch, 1978), it is important to consider such changes before interpreting field data or developing theories of grazing in the sea.

5.22 Components of the Functional Response

5.221 Components Related to Prey Density

The components of the functional response that are related to prey density were listed by Holling (1965) as the rate of successful search for prey, the time that the predator spends searching, the time required for the predator to handle prey, the degree of hunger in the predator, and the manner in which the prey manages to inhibit the predator. We will examine each of these in turn.

5.2211 Rate of Successful Search

The rate of successful search is determined by three subcomponents, including the relative mobility of predator and prey, the size of the perceptual field of the predator relative to the density and size of prey, and the proportion of attacks that result in a successful capture of a prey.

The relative mobility of predator and prey defines in part what prey are available to the predators. It also may reflect whether the predator uses stalk or chase tactics. Predatory fish attack prey using bursts of speed (about 10 lengths s^{-1}) that may be over three times the cruising speeds. If the predator is much larger (say 15 times) than the prey, bursts of speed are not needed: herring of 25–30 cm do not need to accelerate when feeding on *Calanus* (0.2–0.3 cm in length). The herring may need to find over 1,000 *Calanus* day^{-1} to provide daily food requirements, so a very large proportion of each day must be spent seeking copepods. On the other hand, a cod (1 m in length) may only need to feed on three whiting about 20 cm in length per day, but to capture the whiting the cod needs high chase speeds (Cushing, 1978). Thus, there are tradeoffs that species have made in how they have adapted to find food. Predators that use easily overtaken prey, necessarily small sized, must spend long periods feeding to satisfy daily requirements. Chase predators may use large prey, spend only a fraction of their time in actual attack, but must have physiological and behavioral adaptations for acceleration and pursuit. Many species may alternate the use of both strategies depending on available food supply.

Predators may use complex behaviors to modify the rate of successful search. Humpback whales round up zooplankton prey either by herding (Howell, 1930) or by the use of "bubble curtains" (Jurasz and Jurasz, 1979). Bubbles are released by a whale while swimming in a tight horizontal circle. The curtain of bubbles then rises around the clusters of prey.

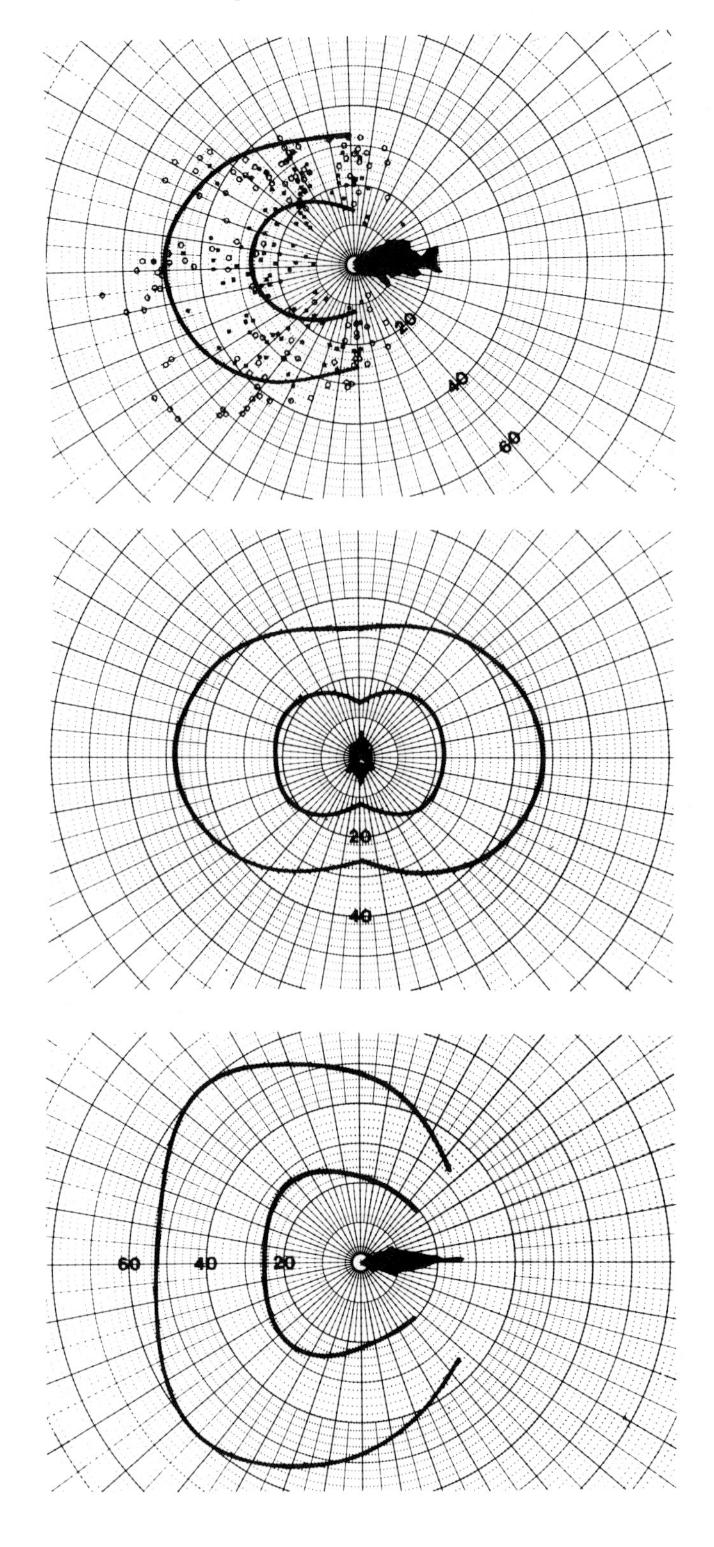
20
40
60
20
40
60
40
20

The zooplankton move away from the bubble walls, and as a result concentrate toward the center of the circle. The whale then harvests this temporarily increased density of zooplankton by swimming up the cylinder of seawater contained within the bubbles. Such complex behavioral adaptations may be the exception rather than the rule for most predators.

Another example of a predator that reduces search time is provided by siphonophores that bear tentacles that resemble copepods, mimicking even the presence of long antennae (Purcell, 1980). The "copepods" are even contracted at variable intervals that resemble the darting swimming of copepods. Examination of the gut contents of these siphonophores shows crab larvae, large copepods, and euphausiids, all of which eat copepods. Other siphonophores have specialized organs resembling fish larvae.

The size of the perceptual field of the predator relative to the density and size of the prey (Kerr, 1971a) also influences whether a predator can find a prey. Visual predators are more likely to detect and eat prey nearby than further away (Fig. 5-6), and for any given visual acuity, larger prey will be detected at further distances (O'Brien et al., 1976). This may result in the consumption of proportionately more large than small prey (cf. Section 8.21). Prey that are sparse will require considerable searching, and if they are small, may not compensate for energy expended. The shape and size of perceptual fields (Fig. 5-6) depend on the anatomy of the predator, and may be quite variable. We have discussed only visual means of prey detection, but mechanoreceptor, chemosensory, or other means may be used to detect prey (cf. Section 8.1).

The proportion of attacks that result in a successful capture of a prey varies widely for different prey and can be affected by many variables. For example, the hunting success of blackbellied plovers hunting in mud flats for benthic invertebrates is lower in the cooler parts of the year. Presumably, warm temperatures encourage prey organisms to be active nearer to the surface of the sediment, where they may be more susceptible to predation (Baker, 1974). Further, whether or not a capture attempt is successful determines what happens in subsequent intervals of time. Once larval anchovies find a patch of food organisms, they remain feeding within it (Hunter and Thomas, 1975). Very similar behavior is shown by other predators. When a prey is successfully captured by a foraging black-

Figure 5-6. Prey detection field for a bluegill sunfish at about 5,900 lux. The bold lines indicate the zone within which 95% (closest to the fish) and 5% of the prey present were detected. Left: Lateral view, with black points showing sites where prey were detected and open points indicating where prey were placed but not located by the fish. Top right: Front view, points omitted. Bottom right: Top view, points omitted. The numbers show centimeters away from the fish. In all cases the prey were 2-mm-long zooplankton (*Daphnia magna*). Adapted from Luecke and O'Brien (1981). © Canadian Journal of Fisheries and Aquatic Sciences, reprinted by permission.

bellied plover, the bird mainly stays where it is and continues hunting. When a peck is unsuccessful, however, the plover on the average takes one to several steps and therefore moves to a new area (Baker, 1974). Thus capture success leads to apportionment of time to further feeding, while failure leads to nonfeeding behavior, designed to move the predator in search of patches with higher densities of prey. In situations such as mud flats where feeding may be restricted to certain hours of the days during favorable stages of the tide, it is essential for the predator to use the available time carefully. The result of the pattern of feeding is that the predators aggregate on patches where prey is more abundant.

5.2212 Search and Handling Time

The time spent searching and handling prey can be partitioned into subcomponents: (1) time spent pursuing and subduing a prey, (2) time spent eating the prey, and (3) time spent in a digestive pause during which the predator is not hungry enough to renew hunting activities. Differences in the nature of the predator and prey may require different apportionment of time spent in each of these three subcomponents. Different predators may use very different tactics in finding prey.

The time that predators spend handling each kind of available prey may vary substantially. The overall capture method used by a predator broadly sets limits on time spent handling prey. For predatory snails that bore through thick shells of bivalve prey by rasping and chemical secretions, handling time of a prey specimen may be on the order of days. Starfish feeding on mussels may take hours to process one prey. At the other end of the spectrum, handling time may be less important for intermittent suspension feeders such as copepods. These suspension feeders may have very short handling times, on the order of seconds or less (T. Cowles, personal communication).

Handling time may also vary with the type of prey, even though a predator may be using the same overall method of capture. For blackbellied plovers hunting in mud flats, some clams and polychaetes take more time to capture than smaller prey (Baker, 1974).

5.2213 Hunger Level

The level of hunger of the predator is determined by the rate at which food is digested and assimilated, and by the capacity of the gut of the predator. Availability of food is usually not continuous, so feeding, and hence the relative fullness of the gut, may vary over time. Fluctuations in feeding behavior may involve changes over months (Jones and Geen, 1977) or they may have much shorter time scales, days or hours (Elliot and Persson, 1978). Fish that feed early in the morning do not feed again until the meal is digested, so that the fullness of stomachs and the consumption of food have a bimodal pattern over the course of a day. In fish, mechanical stimuli provided by the relative gut fullness may be the determinant of hunger level. In many arthropods there may be no feeding until the peritrophic membrane lining the gut is regenerated (Reeve et al., 1975). This membrane is involved in processing food in the midgut and is excreted as the lining around fecal pellets. Whatever the specific mechanism involved, hunger level varies over time. The quality of food available and the digestive properties of the predator then determine how quickly the next meal is sought.

The size of the prey relative to the stomach volume of the predator also is important in determining frequency of attack. Larger prey take longer to digest (Jones and Geen, 1977), so that hunger levels rise more slowly when the prey are large and therefore the next attack is delayed. Gelatinous zooplankton fishing with mucus nets capture very small particles, so they need to feed almost continuously. On the other hand, many other species show very intermittent feeding. Of course, expenditures of energy are required for the acquisition of prey, so that effort involved in capturing either very small or very large prey have to be metabolically balanced relative to the ration obtained. This topic is discussed more fully in Chapter 7, and Lehman (1976) furnishes a model that incorporates feeding, ingestion, and digestion in filter feeders. There are other factors that affect rates, such as temperature. In cod, for instance, ingested food may disappear from the gut in 25 hr at 10°C, while at 2°C food lasts about 60 hr (Tyler, 1970). Thus ambient conditions also affect feeding rates.

5.2214 Inhibition of Predation by Prey

Prey can reduce their susceptibility to being eaten by behavioral or morphological adaptations. Such inhibition of predation by the prey is very well established in marine molluscs (see citations in Harrold, 1982), where there are many stereotyped responses to starfish predation. One example is provided by two kelp forest snails (*Tegula pulligo* and *Caliostoma ligatum*) that are eaten by the sea star *Pisaster giganteus*. Both snails twist their shells violently on contact with *Pisaster*, crawl away as fast as possible, and release their hold on the substrate and tumble away if contact persists. *Calliostoma* also covers its shell with the mucous secretions of its foot, and apparently this hinders efforts of *Pisaster* in grasping the shell. In addition, *Calliostoma* may aggressively bite *Pisaster* when the star has made contact with the snail. *Tegula* neither uses mucus nor biting for protection, and as a result, is eaten by *Pisaster* three times more frequently than *Calliostoma* even though it is encountered by *Pisaster* much less often. The difference is due to the effectiveness of the defensive behavior, not in preferences by *Pisaster*, as shown by a field experiment with narcotized snails (Harrold, 1982). Although many marine organisms have no way to actively defend themselves from predators, not all are just passive prey.

Another mechanism for inhibition of predators is through morphological adaptations,* a phenomenon known as mimicry. If a prey species is morphologically recognizable, and unpalatable, predators may learn to avoid it. This is related to density in that a certain frequency of encounter between predator and prey species is needed before learning by predator takes place. Simulation studies by Holling (1965) showed that Batesian mimicry—where a palatable prey species resembles an unpalatable species—could provide considerable protection to the palatable prey even though predators may feed on some individuals of the palatable prey species, or even if the palatable prey species was more numerous than the unpalatable model. In the case of Mullerian mimicry—where relatively unpalatable species resemble each other—the resemblance developed between the species helps both prey species because the predatory burden is shared. The model further predicts that this second kind of mimicry also helps the predator because it allows a more effective avoidance of distasteful species, since just one image has to be learned as unpalatable. Holling (1965) also identified a third kind of mimicry where a group of edible species could share the burden of predation by evolving to resemble each other, or at least not evolving obvious differences in appearance.

The proportions of any one species consumed by predators would be smaller than it would be in the absence of the mimicry (Van Someren and Jackson, 1959). This protection is provided particularly at high densities of prey species. High densities could be the result of high absolute numbers of the mimic species or of local high densities resulting from schooling behavior (Brock and Riffenburgh, 1960). Below these high densities, predation favors high diversification of prey genotypes (and phenotypes) in Holling's model, a feature actually demonstrated for a predaceous echinoderm and the butterfly clam by Moment (1962). Presumably the presence of many different phenotypes would reduce the effectiveness of the learning component of the functional response to prey density.

Throughout this discussion of predation, we have focused on biological mechanisms. There are also nonbiotic variables that are important in determining the overall rates of attack. For example, a larger number of prey may be eaten (Fig. 5-7, top) at higher temperatures, because warmer temperatures allow faster movement of cold-blooded invertebrates and therefore the time they spend handling their prey (Fig. 5-7, bottom) is much shorter. This in turn makes it possible for a faster rate of attack and a more marked functional response to the density of the prey. Thus, temperature is important at least in affecting handling time. Similar effects can be expected on many of the other components and subcomponents of predation.

---

* There is also the intriguing but largely unstudied possibility of chemotactic mimicry in the case of prey of predators that use chemotactic means of detecting prey.

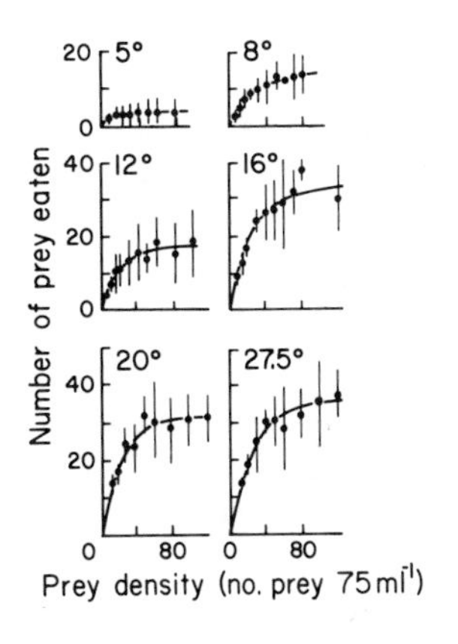

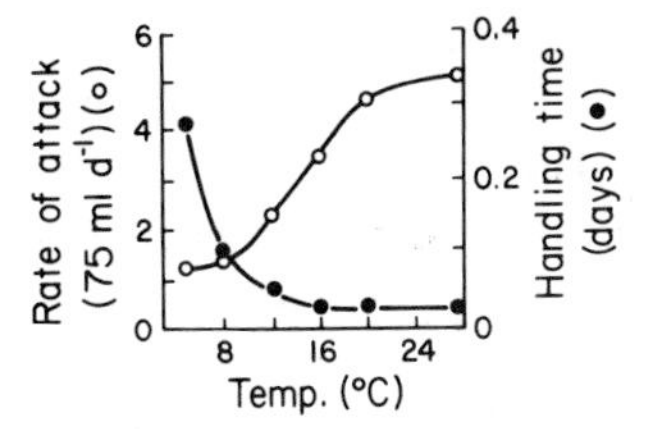

Figure 5-7. Top: The effect of temperature on the functional response (mean ± standard error of number of prey eaten) of a dragonfly naiad (*Ishnura elegans*) feeding on a cladoceran (*Daphnia magna*). Bottom: The effect of temperature on the time to handle prey and on the rate of attack by *Ishnura* on *Daphnia*. Adapted from Thompson (1978).

5.222 Components Related to Predator Density

There are additional components of predation that are strongly influenced by the density of the predator. These include the impact of social facilitation and avoidance learning by prey, intensity of exploitation, and interference among predators.

5.2221 Social Facilitation and Avoidance Learning by Prey

As the number of predators increases, the competition for a limited number of prey may increase, and the probability of any one predator finding an unattacked prey decreases. In some cases, there may be an increase in the number of prey eaten as density increases if social facilitation of attack exists. Examples may be a group of predatory fish shoaling a school of smaller prey fish, or groups of swordfish coursing through fish schools and feeding on the frightened or distracted prey. Escape from one predator may make the prey temporarily more susceptible to another predator (Rand, 1954; Charnov et al., 1976). In such instances, increased contact with conspecifics leads a predator into eating more, searching more effectively or capturing prey more readily. On the other hand, as predator numbers increase, some species of prey may, through associative learning, avoid a specific predator encountered (and survived) before. The greater the density of the predator, the greater the probability that each prey will have acquired a successful way to avoid attack (Holling, 1965).

5.2222 Intensity of Exploitation

In general, the intensity of exploitation of prey by a predator is complicated by the fact that predators and prey are usually patchily distributed over space (cf. Chapter 13). Thus, it is difficult to define the relevant densities of prey and predators, and therefore the distribution of predator attacks. Although Griffiths and Holling (1959), Hassell and Varley (1969), Free et al. (1977) treat such complications in models and certain laboratory situations, it is not clear how these features apply in the field, nor are there field studies where such difficulties are dealt with.

The extensive work on the effect of fishing effort on fish stocks does not usually consider spatial distributions but is the only body of data available on the effect of various levels of exploitation on prey. Much theoretical fisheries work has centered on the idea that as the intensity of exploitation or fishing effort increases, the predators (fishermen) obtain an increased harvest until a peak yield is reached. This peak may be sustainable for some time if fishing effort is maintained, but any increase in predation pressure beyond this maximum yield decreases the amount of food harvestable by the predator. This sort of response of yield to exploitation pressure can be observed in laboratory studies (Silliman, 1968) and in commercial catches in the sea (Fig. 5-8). The mechanism behind such an optimal yield curve* may vary from prey to prey, but it is thought to involve compensatory adjustments by the exploited populations through changes in growth, survival or fecundity. At some point, the removal rate is greater than the rate at which the prey can compensate demographically, and both the yield to predators and the stock of prey are reduced.

5.2223 Interference Among Predators

As density of the predator increases, predators may mutually interfere with each others' activity: as the number of nudibranch predators increases, the number of polyp prey eaten per nudibranch is lowered, regardless of the density of the prey (Fig. 5-9). The nudibranchs are typically solitary and at high densities the increased encounters with conspecifics somehow disrupt normal feeding behavior.

There is a second kind of interference due to the functional response: patches of dense prey are exploited more intensely than low density

* The maximum yield curve has been a key concept for management strategies and models of commercial fisheries (Cushing, 1975; Ricker, 1975). There are many reservations as to its usefulness in actual management of fish stocks (Larkin, 1977), since (1) the concept is not helpful in dealing with the not infrequent catastrophic declines in stocks or in allocating fishing to different geopolitical areas so that no one local stock is depleted, thus maintaining genetic variability of the species; (2) maximum yield ideas have not been developed to include multispecies stocks, so do not include interactions among various species in one area; and (3) maximum yield may not be practically sustainable and may be economically undesirable, since perhaps only a lower supply of certain species may be marketable.

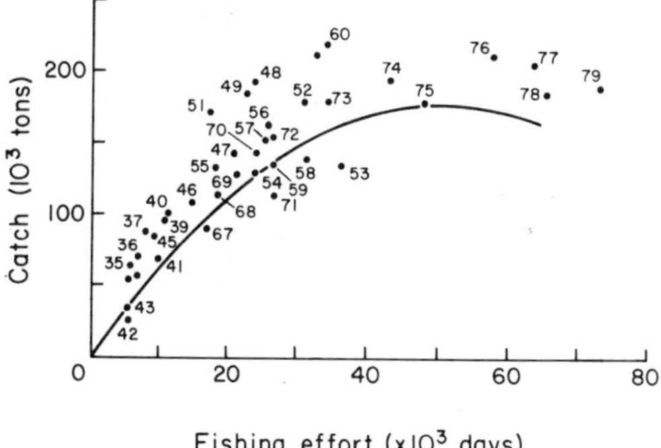

Figure 5-8. Relation between catch of yellowfin tuna (*Thunnus albacores*) in the tropical Pacific between 1967 and 1979 and fishing effort in days. The line represents the theoretical equilibrium yield curve for the species. Adapted from Inter-American Tropical Tuna Commission (1980). Values from 1935-1960 from Schaefer (1957).

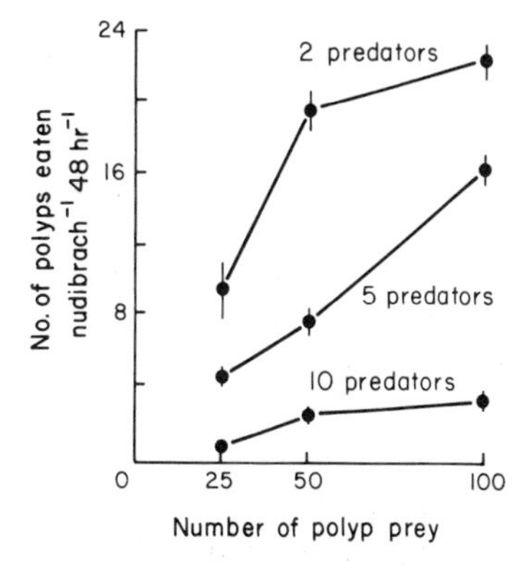

Figure 5-9. Predation rate (mean ± standard error) by the nudibranch *Coryphella rufibranchalis* on polyps of the hydroid *Tubularia larynx*. The experiments were run with sets of 2, 5, and 10 nudibranchs present to compare the effect of increasing density of the predator. Adapted from MacLeod and Valiela (1975).

patches (Free et al., 1977). The result of this is that at high densities of predators it is harder to find patches of prey that are not already exploited (Rogers and Hassell, 1974; Beddington, 1975). This indirect effect of predator density reduces the number of prey available over time to the average predator. The relative importance of the two types of interference is not well known.

There is a further complication in that schooling may protect individuals of a predator species from their own predators (Brock and Riffenburgh, 1960). Schooling by predators, however, may reduce foraging efficiency if area of discovery of prey among members of the school overlaps as a result of the schooling. This would reduce the prey available per predator, and it may be advantageous for schooling predators to disperse to an extent while feeding (Eggers, 1976). School of hungry sticklebacks (*Gasterosteus aculeatus*) (Keenleyside, 1955) and jack mackerel (*Trachyrus symmetricus*) (Hunter, 1966) are less compact than after feeding or when well fed.* The cost of schooling may, theoretically at least, be less

* This response of predators to scarce prey is quite general and applies to a variety of quite different predators. The Athapaskan Indians in the Pacific Northwest of North America lived inland and lacked the abundant food supply of the coastal tribes. When hunting the Athapaskans dispersed into small groups to maximize chances of encountering game.

when prey of schooling fish occur in very dense patches, since searches would then be rewarded by groups of many prey. Perhaps if the prey are very patchily distributed the tradeoffs of schooling by predators may become favorable (Eggers, 1976).

The information available is insufficient to evaluate how widespread or important mutual interference by predators is in nature. One serious problem is that the operational densities of predators in the actual environment may be much different than in the very simplified laboratory settings where interference has been demonstrated. This is a basic problem needing further research.

5.3 Numerical Responses by the Predator to Density of Prey

5.31 Means by Which Predators Respond Numerically to Prey Density

When density of food particles increases, predators can increase their numbers per unit area by aggregation, increasing fecundity, and lowering mortality.

5.311 Aggregation

The first way in which abundance of predators may increase where prey are more numerous is by means of aggregation behavior, where the predators simply seek and move to arcas of high density of prey. Aggregations of predators are usually found in sites with high densities of prey. As discussed earlier in relation to feeding by larval sardines and blackbellied plovers, predators search for areas with abundant prey. In mudflats, predatory wading birds are generally most abundant where benthic organisms are denser (Fig. 5-10). Overwintering sanderlings (*Calidris alba*), a common shore bird, defend feeding territories in beaches, and the area of the territories is inversely proportional to density of the food organisms. In some cases the reduced territories may simply be the result of sufficient food being available even in the smaller territory. In the specific case of sanderlings, the relationship of territory size and density is not due directly to the effect of the food density itself but occurs because it is so costly to defend a larger territory from the increased number of intruder sanderlings that are attracted to places of high prey density (Myers et al., 1979). Additional nonmarine examples of aggregation of predators in areas of high prey density are reviewed by Hassell et al. (1976). All such aggregations are the result of interaction of the feeding behavior of predators with densities of prey and yet again make evident the artificiality of our attempts to discuss the effects of various components of predation separately.

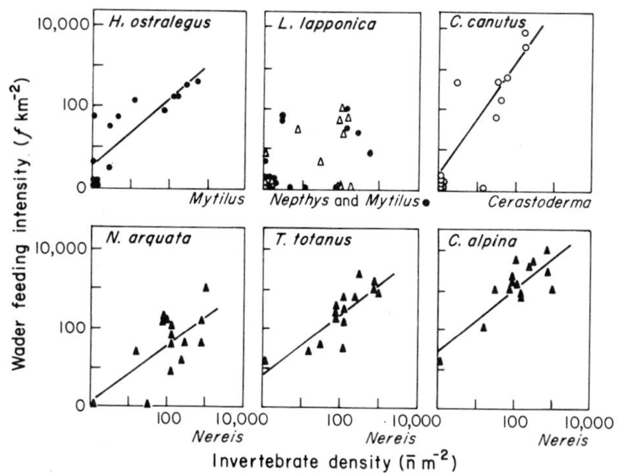

Figure 5-10. Relationship of hours of feeding by shorebirds km^{-2} (f km^{-2}) and density (average number prey m^{-2}) of invertebrate prey on intertidal flats. The names of the shorebirds are inside the panels, and, in order, are oystercatcher, curlew, bar-tailed godwit, redshank, knot, and dunlin. The invertebrate prey are named below the panels. A regression line is included where the relationship was significant. Adapted from Bryant (1979). © Academic Press Inc. (London) Ltd., reprinted by permission.

The behavior that results in aggregation of predators requires only a small time lag before the population of predators adjusts its local abundance to the abundance of prey. As long as predators are mobile and available, aggregative numerical responses are well suited to exploit, and may limit the number of prey organisms in some sites; of course, for the aggregative response to be effective there have to be enough predators at large to aggregate. In many instances there are no such dispersed populations, and there the aggregative response is less important than the fecundity response discussed below.

5.312 Increased Fecundity at High Prey Density

The second mechanism that may bring about a numerical response to prey density is the production of more young when food particles are more abundant. When the copepod *Calanus* is fed on cultures of *Skeletonema* of increasing cell density, the result is a nearly linear increase in fecundity (Fig. 5-11, top). Beddington et al. (1976) review other aquatic and terrestrial examples and provide analysis and a model of the coupling between fecundity and prey supply.

This fecundity response is of interest because more predators are produced even if the initial population of predators dispersed at large was not

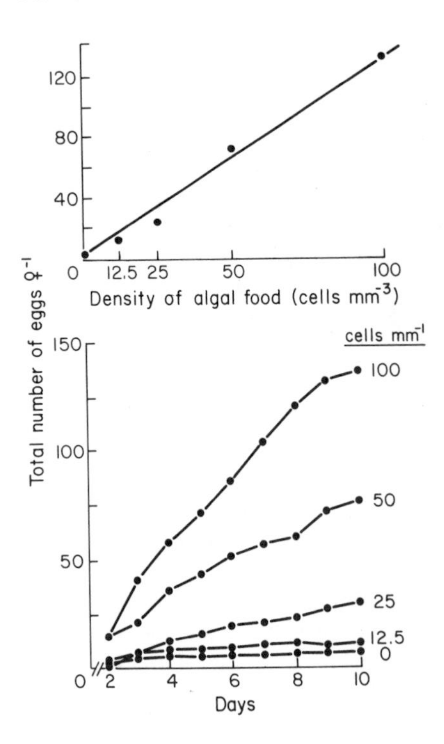

Figure 5-11. Numerical response of *Calanus* fed in the laboratory on *Skeletonema*. Top: The total production of eggs as influenced by the density of *Skeletonema* after 10 days. Bottom: The development over time of the numerical response to density of *Skeletonema*. Adapted from Marshall and Orr (1964).

very numerous—as for example, at the start of the growing season in temperate water columns. Note (Fig. 5-11, bottom) that there is a time lag involved in this response: in the copepod, it takes several days for the full response of fecundity to increased availability of algae. This lag is important in that it may restrict the ability of grazers to limit populations of phytoplankton. Some predator species may have slower fecundity responses, perhaps requiring months, while others require less time. In the laboratory *Paracalanus*, a small copepod, can increase production of eggs after only 12 hr of feeding on high concentrations of algae. The plasticity of the fecundity response evidences once again that the repertoire of predators can be extremely varied, and that there are many slightly different tactics that predators employ to obtain and make use of available food.

5.313 Increased Survival at High Prey Densities

Well-fed predators show better survivorship (Beddington et al., 1976; Cushing, 1975), and since population size is the balance between birth and death rates (emigration and immigration are part of the aggregative response), any decrease in death rate may lead to a relative increase in abundance. The implications of this specific mechanism have not been explored.

5.32 Numerical Responses in the Sea

Most of the data discussed in the previous paragraphs are laboratory results, but numerical responses to changes in food abundance have actually been documented in the sea. The phytoplankton abundance in an area of the North Sea varied over time and was measured during a series of cruises (Table 5-1). The copepod *Calanus helgolandicus* fed on the phytoplankton, and the daily ration available to the copepod varied from almost 400% to less than 1% of the copepod's body weight. Production of considerable numbers of eggs by *Calanus* only occurred while food supply was very high. Consumption of large quantities of phytoplankton may be needed to enable the copepod to achieve a reasonable portion of its potential fecundity. Perhaps some trace element or essential vitamin is required for maintenance, growth, and egg production, and the amounts needed are only obtainable by eating many algae.

5.33 The Numerical Response and Control of Prey

We have referred earlier to the fact that the type II functional response, the most common of the three types, did not seem to offer possibilities of controlling prey populations, because rate of attack did not accelerate as density of prey increased (Fig. 5-3, bottom). A combination of the type II functional response and the numerical response (specifically the aggregation of predators plus the increase in fecundity) do provide at least a theoretical mechanism for such a negative feedback (Hassell et al., 1976; Beddington et al., 1976), although the arguments supporting this statement are too detailed to elaborate here. Presumably, a numerical response plus either type I or type III functional responses may also enhance the chances of control of prey population. The role of the numerical response in nature has not been studied enough, and there is a very large

Table 5-1. Food Availability and Egg Production of *Calanus helgolandicus* in a Large Patch in the North Sea[a]

Cruises	Algal stock (mm^3 liter^{-1})	Daily ration/body wt. × 100	Egg production (Eggs female^{-1} day^{-1})
5–7	0.5–0.6	370	18.5
7–10	0.01–0.2	26	2.0
10–13	0.003–0.004	0.8	—

[a] The data were obtained in a series of 13 cruises over several months. Algal stock was assessed by volume of cells in a standard sample. Adapted from Cushing (1964). Paffenhöffer (1971) believes that the high food consumption rates (third column) are overestimates; corrected values show about 195% of body weight of *Calanus* consumed per day with dense phytoplankton, but the point still remains that food is consumed at remarkably high rates.

gap in studies about the numerical response in general and in its integration with other aspects of predation.

In the field, the numerical response must be very important, for instance, in the case of overwintering populations of zooplankton in temperate or cold seas. These populations are usually in low abundance and must undergo a numerical response during spring blooms of phytoplankton. Whether or not the zooplankton control phytoplankton depends on whether the zooplankton densities achieved by aggregation or reproduction are high enough. Where zooplankton population density is low on a large spatial scale, aggregation probably is less important than the fecundity response. The lag inherent in the fecundity aspect of the numerical response is critical, as will be discussed in Chapters 8 and 14.

5.4 Developmental Response to Prey Density

When food density is high, individual predators may, because of their functional response, consume more prey than at low prey densities. When animals can eat more, they become larger (Fig. 5-12). In turn, the resulting larger predators may eat more prey than smaller predators (Murdoch, 1971; Beddington et al., 1976). This phenomenon is the developmental response.

The growth rate of consumers is related to density of food [Fig. 5-13; cf. Chapter 7 and Brockson et al. (1970); Fenchel (1982)]. Since at least some of the food that the predator eats needs to be expended for maintenance of metabolism, there is a threshold of prey density below which the predator will not grow, and in fact growth can be negative in such circumstances (Fig. 5-13, top). Increased food supply may also accelerate the completion of developmental and shorten generation time stages, thus hastening the numerical response to prey density. This is yet another example of the interconnections among components of predation, one

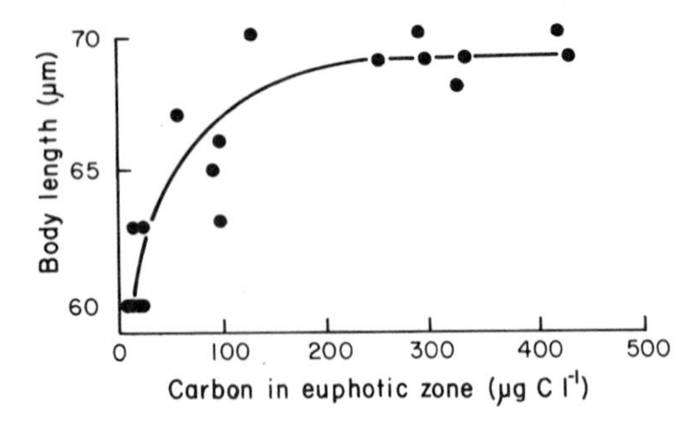

Figure 5-12. Prosome length of *Calanus pacificus* in Puget Sound versus the phytoplankton carbon concentration in the euphotic zone during 1965. The carbon was estimated using a time-varying carbon to chlorophyll ratio. Adapted from Frost (1974).

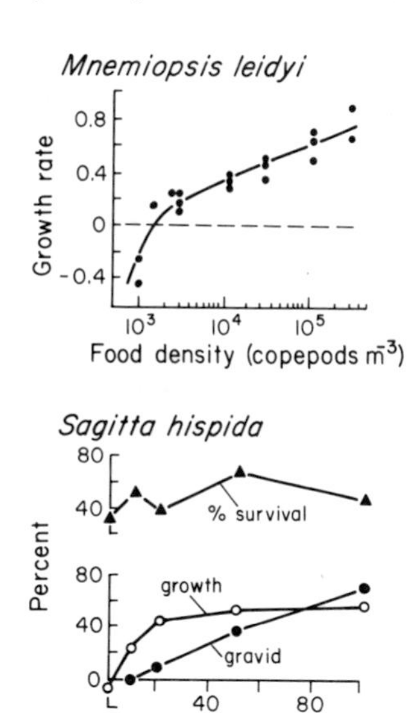

Figure 5-13. Laboratory studies of developmental responses to increased density of food particles in two marine invertebrates. Top: Growth rate of the ctenophore *M. leidyi* under increasing abundance of copepods. The growth rate is expressed as a growth coefficient k obtained by solving the equation $W_T = W_0e^{kt}$. A growth coefficient value of 0.8 implies that the organism is more than doubling (actually 2.2 times) in size each day. Adapted from Reeve et al. (1978). Bottom: Growth rate (expressed as increase in length) for the chaetognath *S. hispida*, including for comparison the percentage survival and percentage of the individuals that were gravid, two components of the numerical response to prey density. Adapted from Reeve (1970).

that brings us to the relationship of food supply and growth, a topic that is considered further in Chapter 7.

The relative importance of the developmental versus the other responses to prey density has not been assessed for populations in the field. Laboratory studies suggest that the increase in survival component of the numerical response due to larger size may not be as marked as that of the fecundity component, as least for chaetognaths (Fig. 5-13, bottom).

The presence of a developmental response provides another potential way by which predation can exert density-dependent control of prey populations. In computer simulation studies, Murdoch (1971) concluded that the type II functional response by a predatory marine snail could not by itself control prey numbers. When the developmental response was added, however, there was a range of prey densities (toward the lower end of the range studied) where the percentage of prey attacked increased as density of prey increased. Beyond this density, prey numbers escaped control by the predator.

As prey are more readily available, the predators achieve larger sizes, and, as we will see in Section 6.41, larger predators are proportionately more effective at feeding. Larger size also often leads to greater reproductive output (Section 7.52), so there are very close couplings among the functional, numerical, and developmental responses to prey density. The way in which these three major components articulate with each other varies with each instance of predator and prey species. An enormous variety of combinations of density, availability, and quality of prey exist. Each species of predator can make use of the many components and subcomponents of predation in somewhat differing combinations to exploit available prey in the best manner.

The foregoing analysis of feeding on particles painstakingly went over many components and subcomponents of the process. Many further details could have been added. Holling's analysis of components ultimately intended to combine mathematical representations of the components in a model that synthesized how predation took place. The analysis has identified many mechanisms, but unfortunately, the details needed to specify the model exceed current knowledge about most predator-prey relations. Efforts at synthesis of predation have therefore focused on partial aspects of the predation process. Some of these models are discussed in Chapters 8, 9, and 14. In spite of the difficulties of synthesis, component analysis has made evident the mechanics involved in the consumption of particles in a systematic way.

Chapter 6

Food Selection by Consumers

6.1 Introduction

In the last chapter we focused on the importance of food quantity. It turns out, however, that not only abundance but quality of food items must be considered. Moreover, even though there may be sufficient abundance of food items of adequate quality, they may not be easily vulnerable to predators. In this chapter we take up the quality and availability of food particles. First, we need to consider the cues used by consumers in finding and selecting food of appropriate quality and the role of properties of the food and consumer, particularly size and chemical makeup, in food choice. We then consider the relative vulnerability of food items, mainly in reference to the morphology of prey and the physical structure of the habitat. Last, we consider consequences of food selection, especially in regard to stability of prey populations.

6.2 Behavioral Mechanisms Involved in Finding and Choosing Food

The behavioral mechanisms by which consumers find food vary widely. Marine microbes are attracted to a wide variety of stimuli, but in most cases it is not clear whether such trophisms result in finding food. Larger vertebrate and invertebrate consumers have better-known mechanisms with which to seek food. Many marine consumers are visual, so that size, color, and shape of food items are important.

Chemosensory organs are commonly found in many marine organisms. Cod, for example, can detect chemical compounds with organs on their pelvic fins and the barbel (Brawn, 1969), although they often use both chemoreception and sight for locating invertebrate prey. Shrimp can also use chemical cues to track and find prey (Hamner and Hamner, 1977). Experiments in which copepods were fed microcapsules containing phytoplankton homogenates show that chemosensory cues can be used by these organisms since the ingestion rates on such particles are considerably higher than ingestion of untreated particles. The specific compounds involved are not well known, but there is considerable specificity in feeding cues; for example, in *Acartia clausi* L-leucine stimulated feeding while L-methionine or glycolic acid prompt no response (Poulet and Marsot, 1980). Chemoreceptors in lobsters (*Homarus americanus*) are sensitive to 35 individual compounds tested, many of which—particularly the nitrogenous compounds—are in high concentrations in prey of lobsters (Derby and Atema, 1982). The lobsters were able to detect remarkably low concentrations of several compounds, in the range of 10^{-6} to 10^{-14} M.

In copepods, chemosensory detection of algae may be the primary mechanism used in feeding by herbivorous species (Friedman and Strickler, 1975), while mechanoreception may be used principally by raptorial feeders. Copepods may use mechanoreceptors located on their first antenna (Strickler and Skal, 1973) to sense hydrodynamic disturbances caused by moving prey (Strickler, 1975). Removal of the first antenna does not affect the rate of feeding on phytoplankton but sharply reduces the rate of capture of moving prey (Landry, 1980). Chaetognaths also respond to mechanical vibrations with pursuit and capture behavior (Feigenbaum and Reeve, 1977). Certain sharks can detect electromagnetic fields generated by life activities of their flatfish prey and can use this sensory ability to capture prey buried in sediments (Kalmijn, 1978).

6.3 Factors Affecting Food Selection by Consumers

The selection of food items by consumers is a complex phenomenon whose details are only beginning to be understood. The size and chemical composition of the item, however, are probably the two major cues that tell a consumer that it has found adequate food. Other properties of prey, such as specific morphological features and foraging methods, also are important in prey selection.

6.31 The Influence of the Size of Prey and Consumer on Selection of Food Particles

6.311 Sizes of Prey Particles Used by Consumers

Some predators attack prey larger than the predator itself, as in the case of the predaceous ctenophore *Beroe nata* feeding on the larger ctenophore *Bolinopsis* (Hamner et al., 1975).* The other extreme is the feeding by consumers on particles measuring less than 1 to a few μm, usually free-living bacteria. Particles of this size may constitute 25 to 50% of the daily ration of the larvacean *Oikopleura dioica* (King et al., 1980); bivalves and some crustaceans can remove particles of bacterial size from suspension (Peterson et al., 1978). Pelagic tunicates, tens of centimeters in length, secrete mucus webs that may functon as passive filters and can retain particles as small as 0.7 μm in diameter (Harbison and Gilmer, 1976).

Other consumers fall between the extremes illustrated above, but the size of the food particle relative to the size of the predator is usually an important criterion. Although predators use many kinds of cues to choose prey, much prey selection is accomplished using the size of the prey as the discriminating variable.

6.312 Size Selectivity in Consumers and Some Consequences

Work on freshwater fish (Ivlev, 1961; Brooks and Dodson, 1965) first highlighted the fundamental importance of size of prey for predators. Ivlev made use of an index of electivity† to assess the degree of selection:

$$E = (r_i - p_i)/(r_i + p_i), \tag{6-1}$$

where r_i is the proportion of item i in the diet and p_i is the proportion of the same item available in the environment. Positive values of E show selection, while negative values show relative rejection by the predators. Using E, Ivlev could demonstrate that fish show clear preferences in the size of prey they elect to feed on (Fig. 6-1). Prey selection patterns such as found by Ivlev imply that predators exploit certain size classes of prey population more intensely than others and that at least one way by which predators can partition a food supply is to have different preferred size classes. A further implication is that the sizes below and above the size

* A curious and unusual feeding pattern is for a predator to attack prey of similar size to itself but to eat only a portion of the prey. Examples of this are typhloscolecid polychaetes feeding on heads of chetognaths (Feigenbaum, 1979) and young flatfish feeding on the siphons of bivalves (Trevallion et al., 1970). In both cases, the prey survive and regenerate the lost parts.

† Jacobs (1974) and Cock (1978) assess Ivlev's electivity index and suggest modifications.

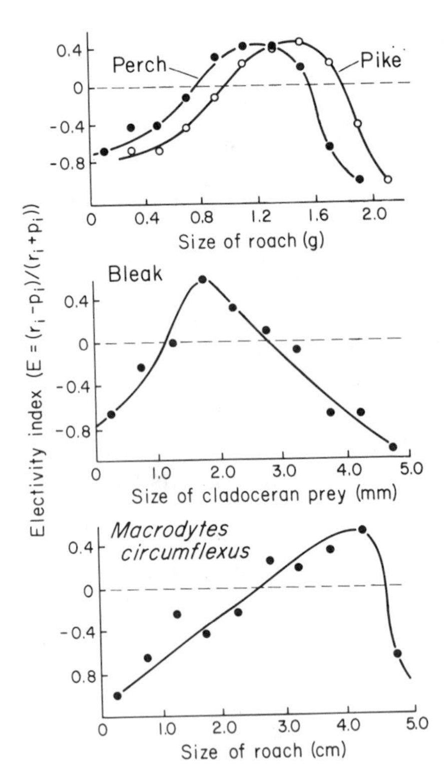

Figure 6-1. Selectivity of prey by predaceous fish in the laboratory. Top: Perch (*Perca fluviatilis*) and pike (*Esox lucius*) feeding on roach (*Rutilus rutilus*). Middle: Bleak (*Alburnus alburnus*) feeding on several species of cladocera of varying size. Bottom: Larvae of *Macrodytes circumflexus* feeding on small roach. Adapted from Ivlev (1961).

classes preferred by predators are refuges from predation pressure, and that a prey may thus outgrow its vulnerability to predation.

Although the phenomenon of size selection was first studied in fish, it is widespread among various taxa and quite often selection for larger items is evident. Predatory birds such as the redshank, a common Eurasian shorebird, preferentially eat prey at the larger end of the size distributions available on mudflats. This can be seen by comparing the histogram of sizes of amphipods eaten by the birds relative to those found in the mud (Fig. 6-2, top), and by the regression of average size of amphipods eaten and available in the mud (Fig. 6-2, bottom).

The preference for larger particles may also result in faster feeding rates when food of larger size is found. Filter-feeding fish such as menhaden filter more actively (or, in other terms, show increased functional responses) when large prey are available (Fig. 6-2). Unlike Ivlev's fish (Fig. 6-1), menhaden do not show a lowered feeding at the largest sizes offered (Fig. 6-3), bottom). This is due to the relative size of gape of the mouth to size of prey given in the experiment; if increasingly larger prey were offered to menhaden there would be some point where feeding would decrease. It is therefore important in studies such as these to offer particles that span size spectra usually found by the consumer in the field.

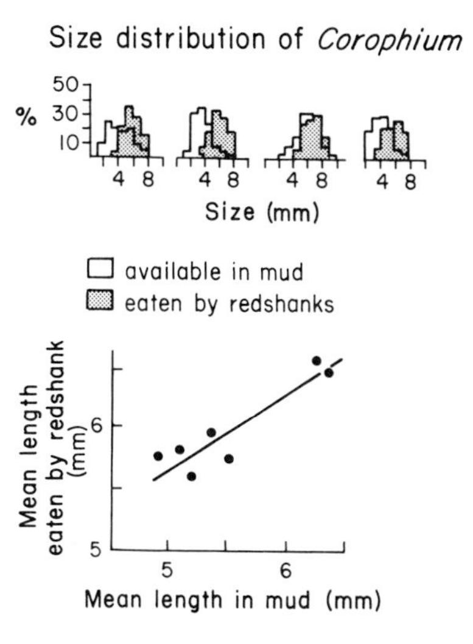

Figure 6-2. The sizes of the amphipod *Corophium volutator* taken by redshank (*Tringa totanus*) in comparison to the sizes available in intertidal flats, Ythan estuary, Scotland. The top histogram shows size classes for four sites, while the lower graph relates mean lengths of amphipods in redshank diets and in the mud. Adapted from Goss-Custard (1969).

As consumers grow and increase in size, they feed on relatively larger food items. Larvae of the plaice (a flatfish of great commercial importance) feed on various species of small zooplankton, but in certain years their diet in the North Sea consists almost entirely of appendicularians. As the larvae of the flatfish grow (Fig. 6-4), the modal size of prey in the guts of the larval fish increases in size: the diet of the larval fish changes and eventually matches the available food supply.

Thus, increased size of predator is followed by increases in the preferred size of prey. Starfish (*Pisaster giganteus* and *P. ochraceus*) are com-

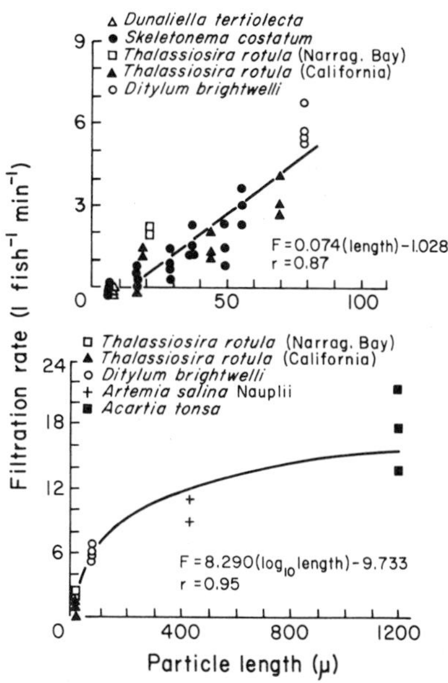

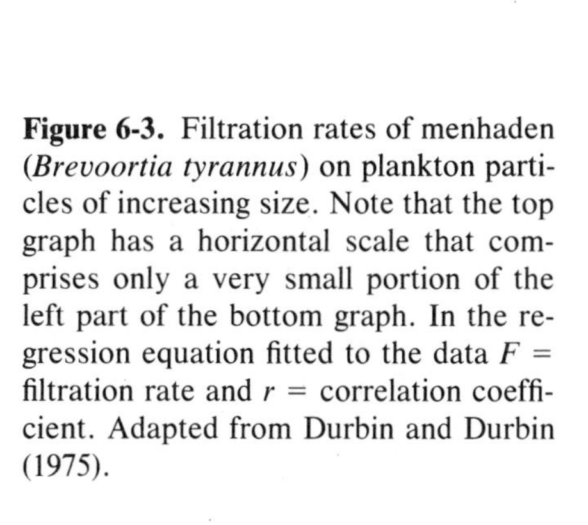

Figure 6-3. Filtration rates of menhaden (*Brevoortia tyrannus*) on plankton particles of increasing size. Note that the top graph has a horizontal scale that comprises only a very small portion of the left part of the bottom graph. In the regression equation fitted to the data F = filtration rate and r = correlation coefficient. Adapted from Durbin and Durbin (1975).

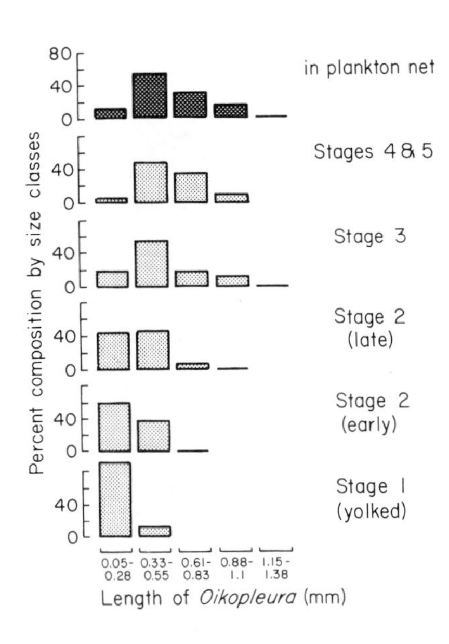

Figure 6-4. Size composition of prey (the appendicularian *Oikopleura dioica*) eaten by larvae of plaice (*Pleuronectes platessa*) in different stages of development (lower five histograms) and captured in plankton nets (top histogram). Adapted from Shelbourn (1962).

mon predators of the rocky shore in California. The starfish increase in radius as they age and feed on larger specimens of mussels (*Mytilus californianus* and *M. edulis*) (Fig. 6-5, left). The predators do not discriminate so much between species of mussels, but rather among the different sizes of the mussels. Fishes typically also feed on larger prey as they grow (Fig. 6-5, right). Again size of prey rather than species of prey is the more

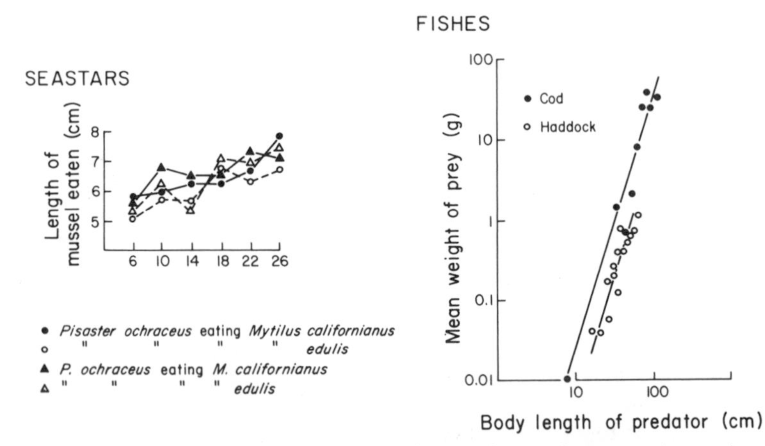

Figure 6-5. Relation of predator and prey sizes in invertebrates and vertebrates. Left: Size of mussels eaten in relation to the size of the starfish predator. Mussels were offered in the laboratory in a size range of 3–10 cm in shell length. The starfish only fed on mussels of 5–8 cm. Open circles also refer to *P. giganteus*. Adapted from Landenberger (1968). Right: Length of two fish species in relation to mean weight of prey found in their stomachs. Adapted from Jones (1978).

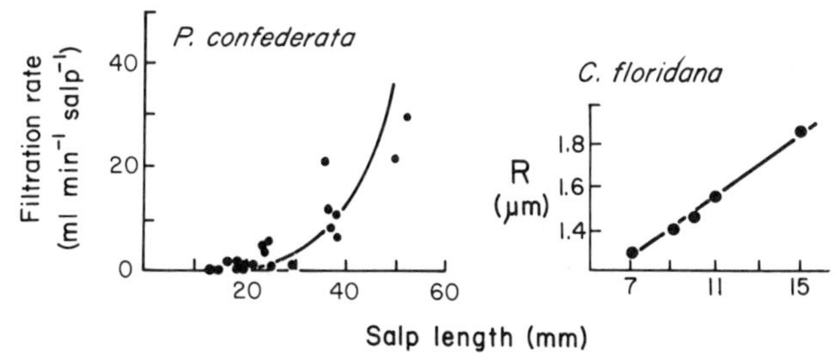

Figure 6-6. Left: Filtration rate as a function of body length for the salp *Pegea confederata* [adapted from Harbison and Gilmer (1976)]. Right: Retention index (R = particle size retained with 50% efficiency) as a function of body length of the salp *Cyclosalpa floridana* [adapted from Harbison and McAlister (1980)].

important variable. Many species of bottom-feeding (demersal) fish have pelagic larvae; as the young fish grow, they require larger prey; at some point they are unable to find large enough prey in the plankton and become demersal.

Larger predators do not merely eat larger prey, they also become more effective predators. The filtration rate of some pelagic tunicates, for example, increases exponentially with size of the tunicate (Fig. 6-6, left). A larger tunicate, therefore, is capable of capturing more particles per unit time. In fact, the retention efficiency of the gelatinous sieves used by the tunicates depends on the size of the tunicate, with larger specimens being able to retain larger particle sizes more effectively (Fig. 6-6, right).

The performance of consumers of different size can be compared in Table 6-1. *Pseudocalanus* sp. and *Calanus pacificus* are both primarily herbivorous suspension feeders, but *C. pacificus* is about 17 times heavier. *Pseudocalanus* require only about 1,500 cells ml^{-1} to reach 90% of its maximum feeding rate, while *Calanus* needs to feed on phyto-

Table 6-1. Some Properties of Feeding by Three Species of Copepods on *Thalassiosira fluviatilis* in the Laboratory[a]

	Mean dry body weight (μg)	Maximum ingestion rate (cells cop^{-1} hr^{-1})	Cell conc. at which 90% of max. ingestion rate is achieved (cells ml^{-1})	Max. ration as fraction of body carbon
Pseudocalanus sp.	10	1356	1515	1.13
Calanus pacificus	170	12652	4489	0.62
Aetideus divergens	47	1201	6598	0.21

[a] From Robertson and Frost (1977).

plankton about three times as dense to do so. *Calanus*, however, can filter at a maximum rate which is over nine times that of *Pseudocalanus*. The end result is that *Calanus* can obtain about twice as much carbon per unit time as *Pseudocalanus*.

Data on *Aetideus divergens* are included (Fig. 6-7) to show that even though larger than *Pseudocalanus*, a species with feeding appendages ill-adapted to handle small particles will perform poorly. *A. divergens* shows a type II functional response to prey density, but the saturation level is not reached when small particles make up the available food. The maximum feeding rate achieved by *A. divergens* feeding on small particles is only about a fifth of what it can obtain when feeding on larger particles and does not seem to reach saturation, even at very high prey densities (Fig.

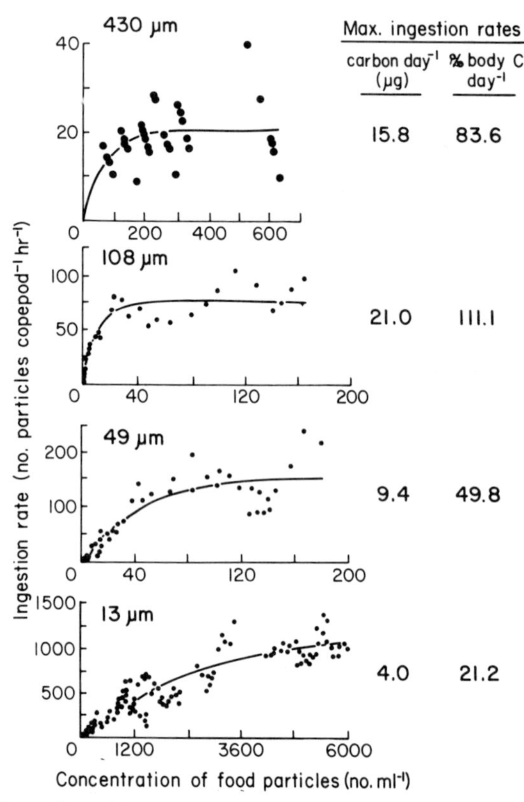

Figure 6-7. Effect of particle size (largest on top graph, smallest on bottom graph) and density (horizontal axis) on the feeding rate of the copepod *Aetideus divergens*. The foods are 430 μm: nauplii of *Artemia salina*; 108 μm and 49 μm: cells of *Coscinodiscus angstii*; 13 μm: cells of *Thalassiosira fluviatilis*. The numbers on the right are the calculated maximum rates of ingestion of carbon for each particle size and the carbon that this maximum rate would yield in terms of the percent of body carbon. Adapted from Robertson and Frost (1977).

6-7). Feeding on the smallest cells yields a little over 20% of body weight per day, only enough for sustenance and perhaps a small reproductive effort.

6.32 Selection of Food on the Basis of Chemical Cues

The palatability of food items depends on chemical composition. There are two major aspects to this notion. The first is the chemical defenses used by a potential food or prey to deter feeding by a consumer, and the second is the presence of cues that convey the nutritional adequacy of the food.

6.321 Chemical Deterrents

The use of chemical signals to deter consumers is widespread among organisms. The presence of chemical deterrents to feeding by herbivores is especially well established in terrestrial plants (Rosenthal and Janzen, 1979, for example), and it is becoming evident that marine producers also have such defenses.

Dinoflagellates have well known toxins, but it is hard to know what the target for these compounds is, since many invertebrate grazers are not affected. Vertebrates can be severely disabled by the toxin, so perhaps dinoflagellates originally evolved such compounds to keep from being consumed by grazing fish (White, 1981). Some brown and red algae contain deterrents that prevent grazing by herbivorous snails (Geiselman and McConnell, 1981). Certain phenolic compounds in salt marsh grasses make tissues less palatable to geese (Buchsbaum et al., 1982). Extracts of eelgrass leaves containing phenolic acids inhibit both grazing by amphipods and microbial decay (Harrison, 1982).

The deterrent substances are collectively known as secondary plant compounds to suggest that for the most part they are not involved directly in the primary metabolic pathways of producers. In marine producers these secondary deterrent compounds may include (a) phenolics: cinnamic acids in vascular plants, tannin-like compounds in brown algae, and simple, often halogenated compounds in red algae; (b) nitrogen compounds: carotenoids in diatoms, pyrrole alkaloids in red algae, alkaloids in monocots; (c) terpenoids: halogenated terpenoids in red algae, saponins in monocots; and (d) acetogenins typical of red tide dinoflagellates and toxic lipids in some green algae (Glombitza, 1977; Hoppe et al., 1979; Swain, 1979; Geiselman and McConnell, 1981; Fenical, 1982).

Organisms with sessile growth habits, regardless of whether they are a macroalgae, vascular plant, or animal, often acquire chemical defenses against consumers. In the sea, where many taxa have a sessile, plant-like growth habit, there are many examples of chemical defenses against browsers. Marine sponges (Green, 1977) and ascidians (Stoecker, 1980) are just two examples of taxa that have such defenses, and chemical deterrents are widely distributed in many taxonomic groups (Jackson and Buss, 1975). In the Great Barrier Reef of Australia, 75% of common coral reef invertebrates (including four phyla and 42 species) are toxic to fish (Bakus, 1981). The importance of chemical deterrents is demonstrated by the presence of some structural adaptation against predation by fish in most of the species of invertebrates not protected by chemicals. In addition, of the invertebrates that usually are hidden in some protective microenvironment, and therefore are less exposed to being eaten, only 25% are toxic to fish.

The production of chemical deterrents by marine primary producers not only affects herbivores but also detritus feeders. Many of the substances that deter herbivores remain after the plant dies and their presence in detritus inhibits feeding by detritus feeders. For example, the consumption of salt marsh plant detritus by amphipods and snails is inhibited by high concentrations of ferulic acid, a phenolic compound that reduces palatability of marsh grasses (Valiela et al., 1979). Thus, a device used by plants to reduce herbivore has consequences for decay and the regeneration of organically bound nutrients, since the inhibition of consumption by detritivores slows decay of detritus.

6.322 Cues of Nutritional Quality

The nutritional quality of food is of obvious importance (cf. Chapter 7), and evolution must have provided for a correlation between the palatability of a particular food item and its nutritive value to a consumer. The need to maintain a diet of sufficient quality and containing specific dietary requirements while feeding on materials of uneven or low-quality forces grazers to display remarkable feeding selectivity. Selectivity may result from a preference hierarchy among food types based mainly on chemical cues. The cues that determine the hierarchy must include both cues of quality as well as deterrent cues. In some instances the nutritional quality of a food may override the effect of deterrent substances. Feeding on detritus by salt marsh snails (*Melampus bidentatus*) is inhibited by enhanced concentrations of ferulic acid in the food. When the nitrogen content of the detritus is experimentally increased, however, the inhibition of feeding by ferulic acid disappears (I. Valiela and C. Rietsma, unpublished data). The cues that determine palatability are hierarchically arranged; in the salt marsh snail, nitrogen content can overcome chemical deterrents; other variables such as pH and salinity have places lower in the hierarchy. In other species of consumers the hierarchy of cues may differ, or other cues may be more important.

The importance of food quality is made evident by the many unusual adaptations in feeding behavior that lead to improvements in the quality of food eaten. For instance, the green turtle (*Chelonia mydas*) is a herbivore that feeds on turtle grass (*Thallassia testudinum*) in tropical shallow coastal waters. The tissues of this plant are low-quality food, especially mature leaves that are rather high in fiber and low in protein. Turtles partially improve their diet by maintaining grazing plots where they repeatedly crop young leaves richer in nitrogen (Bjorndal, 1980).

6.33 Other Properties of Food Items

For organisms that ingest sediment particles the abundance of the microbial flora growing on the particles is important. In the archiannelids *Protodrilus symbioticus* and *P. rubriopharyngens* the more bacteria per gram of sediment the more attractive the sediment (Gray, 1966, 1967).

Other less obvious properties of particles may also be important, at least for benthic animals feeding on soft deposits (Self and Jumars, 1978). Many worms collect particles by means of long, sticky tentacles. The transport of these particles down the tentacles into the mouth by ciliary movement is less effective for heavier particles, so that ingestion is therefore "selective" for particles of lower specific gravity. Other mechanisms of particle transport in polychaetes also result in sorting of particles within the gut, since particles of low specific gracity have shorter residence within the gut. Other characteristics, such as surface texture and shape, are also likely to be important in ingestion by consumers, and are beginning to receive some study.

6.34 Morphology and Foraging Tactics of Consumers

Copepods, fish, and other consumers may have several ways to find and eat prey. By slight variations in the combinations of components of predation (or consumption in general), there are possibilities for widely varying ways in which consumers can partition food resources. The anatomy of a consumer is of course adapted to the kind of food it eats: the morphology of the mouth of nematodes, for example, varies, and details of musculature and teeth betray the species that grasp prey from those that feed by bulk ingestion of tiny microbial cells (Tietjen and Lee, 1977).

The species of consumers in any one community usually differ morphologically and display different foraging tactics. One example of just such an assemblage of predators is the group of terns that use Christmas Island in the Central Pacific as a nesting ground (Table 6-2) and forage around the island and offshore for food to feed their young. All the species of terns feed on prey captured very near the surface of the sea. Other predators that forage in the same area, for example, the shearwater *Puffinus nativitatis*, differs from the terns by feeding deeper below the surface. Except in the case of the blue-gray noddy, prey for the terns become

Table 6-2. Foraging Tactics of Species of Terns Nesting on Christmas Island, Pacific Ocean[a]

Species	Wgt (g)	Diet	Modal fish length (cm)	Modal squid mantle length (cm)	Approximate fishing range (km)	Web area in feet (cm)
Sooty tern (*Sterna fuscata*)	173	Fish, squid	2–4	4–6	>100	4.8
Brown noddy (*Anous stolidus*)	173	Fish, squid	2–4	4–6	80	13.8
White tern (*Gygis alba*)	101	Fish	2–4	2–4	—	3.4
Black noddy (*Anous tenuirostris*)	91	Fish	0–2	2–4	8	10.9
Blue-gray noddy (*Procelsterna cerulea*)	45.4	Fish, invertebrates, some squid	0–2	0–2	8	10.1

[a] Adapted from Ashmole (1968).

available principally when shoaled by tuna or other predatory fish. There are, incidentally, two other species of tern nesting on Christmas Island. *Thalasseus bergii* is ecologically isolated from the others since it feeds primarily on beaches, and *S. lunata*, a small relative of *S. furcata*, is not well studied.

The size of the terns varies (Table 6-2), with the sooty and the brown noddy being similar and larger than the others. The blue-gray noddy is the smallest while the remaining two are intermediate in weight. The taxonomic composition of the diet of the terns is similar, except that the blue-gray noddy eats a larger proportion of tiny crustaceans and marine insects (*Halobates*).*

The two species intermediate in size—the white tern and the black noddy—feed on fish and squid, with the noddy using less squid. The white tern feeds mainly at dawn, so it has access to a different group of species of fish than the noddy, since nocturnal and diurnal fish species are not the same. The noddy returns to its nest site at night, feeds on smaller fish than the white tern, and uses dipping tactics more frequently.

The blue-gray noddy is the smallest of the species and can use only the smallest fish and squid, since its small bill cannot hold large fish. It fishes near the island, with lots of surface dipping and pursues whatever small arthropods it can get.

* A survey of knowledge on marine insects is given in Cheng (1976).

The pair of large species is about the same weight and feed on similar kinds and sizes of foods. Although they can be found fishing together, the sooty has a far larger foraging range than the brown noddy. Its search time therefore must be considerably longer. The sooty, furthermore, largely relies on shallow dives from the air to capture prey, while the brown noddy also uses the better developed webbing on its feet to engage in dipping from a swimming position on the surface of the sea. The difference in attack probably provides different prey, since the speed, detection distance, and acceleration of prey and predator differ in diving and dipping.

Although these tern species resemble each other in morphology, feed largely on similar taxa, and use only a few foraging methods, there are very substantial differences in the details of how they obtain food. Thus the differences in species of prey in the diets of these predator species arise primarily because of slight differences in anatomy and foraging patterns. We can generalize that species of predators that coexist sympatrically have evolved enough differences in foraging tactics (and/or prey selectivity) so that the food particles used by the various species of predators are on the average different. The very diverse approaches to consumption of particles, made possible by the variety of components of the process of prdation, thus allows for the partitioning needed by groups of similar species.

The example of the Christmas Island terns should have reminded us of the competitive partitioning of resources in salt marsh fish (Chapter 4). The similarity should indicate that competition and predation often are intimately linked, and both potentially are important factors affecting what consumers are present and what they eat. The relative role of competition and predation in thus structuring communities are evaluated in Chapters 8 and 9.

6.4 Examples of Feeding Mechanisms at Work: Suspension Feeding

Feeding by plankton on suspended particles has received a great deal of attention. An examination of some of the details of suspension feeding highlights many of the specific mechanisms we have discussed above and how they interact.

The feeding of copepods had long been thought to involve the use of appendages to direct feeding currents through sieves. The sieves were thought to be the array of setae on certain mouthparts, particularly the second maxillae (Fig. 6-8). The setae have setules set at specific distances apart, and water was thought to be forced past the setal sieves. Particles above a certain size would be retained in the sieves and could then be ingested. This filtration process has been called passive filter-feeding.

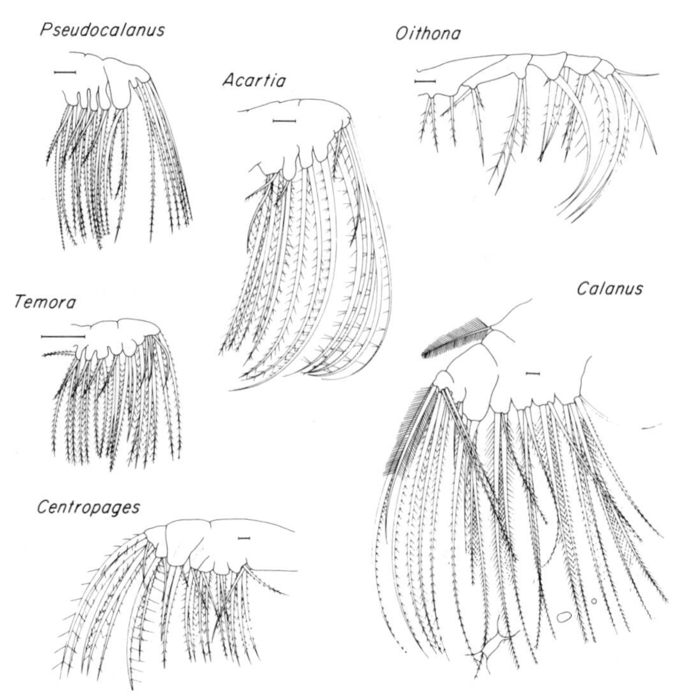

Figure 6-8. Feeding appendages (second maxillae) of some common copepods, showing setae and setules. Silhouettes of phytoplankton cells of differing size are superimposed on the second maxillae of *Calanus*, including, in order of increasing size, a microflagellate, a coccolithophorid and a diatom. The horizontal bars show 20 um. Redrawn by Masahiro Dojiri from Marshall and Orr (1955, 1966) and Steele and Frost (1977).

Recent work using slow-motion microcinematographic records of feeding by individual copepods shows that we need to revise our view of copepod feeding. Purely passive mechanical particle capture seldom occurs; copepods use their mouthparts to chemotactially select and capture algal cells out of the feeding currents (Paffenhöfer et al., 1982). Kohl and Strickler (in press) point out that the viscous nature of water at the small size scales of feeding appendages would make sieving of water very difficult, and they conclude that in most cases particles such as algal cells are detected and actively captured individually by feeding appendages. The postulated action of feeding appendages is supported by the abundant chemo- and mechanosensory organs found in mouthparts (Friedman and Strickler, 1975).

It seems, however, unlikely that copepods can actively capture particles smaller than 5 μm individually, since they would be difficult to detect. Yet, they do consume such particles, so perhaps some bulk processing of water resembling passive filtering may be used by copepods when only small food particles are available (Paffenhöfer et al., 1982). In any case, it is clear that the term filter-feeders is inadequate; we therefore will refer to suspension feeding, a broader term that includes the possibility of other feeding behavior besides passive filtering.

Many suspension feeders also can pursue and capture relatively large individual prey by grasping. Any of a number of appendages are used, but certain species have mouthparts that are stronger and may be better suited for raptorial behavior (cf. *Oithona* in Fig. 6-1). Most zooplankton species appear to readily switch from one mode of feeding to another, depending on the abundance of small and larger prey. *Calanus pacificus*, for example, prefers to feed on copepod nauplii, but if phytoplankton are abundant enough it will feed preponderantly on algal cells (Landry, 1981). Euphausiids and very likely many other plankton consumers can also switch from feeding on phytoplankton to raptorial capture where necessary (Parsons and Seki, 1970).

Switching from one prey type to another is common in some species of copepods, who often feed on whatever particle size is available in the water. If the size distribution of particles changes over time (for example, compare the June to the December data, Fig. 6-9, top row), the particle distribution ingested by the copepods present (Fig. 6-9, bottom five rows) changes accordingly (Richman et al., 1977; Poulet, 1978; Cowles, 1979).

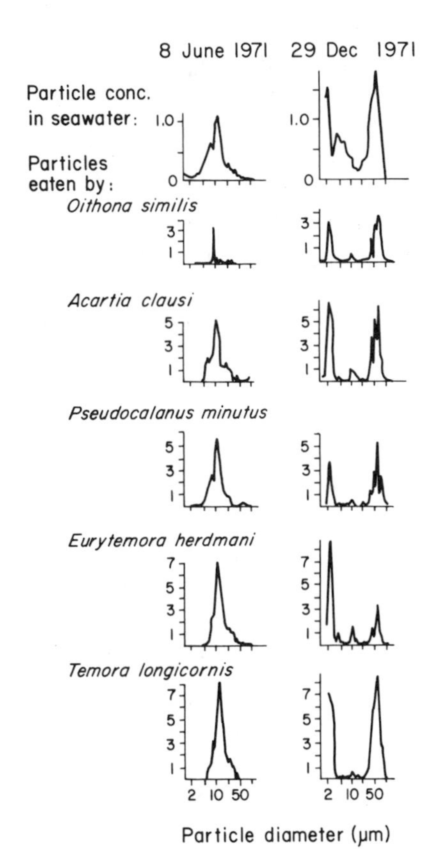

Figure 6-9. Particle size spectrum in sea water (ppm by volume) and consumed by several copepod species in Bedford Basin, Nova Scotia, Canada, on two dates. Particle consumption (the vertical axis of the bottom five rows of graphs) is in mg hr^{-1} $copepod^{-1} \times 10^{-4}$. Adapted from Poulet (1978).

In many other studies of suspension feeders there is, in contrast, evidence of a clear preference for larger particle sizes (Richman et al., 1977; Donaghay and Small, 1979). Some copepods, for example, retain and eat larger algal cells more efficiently than smaller cells (Fig. 6-10). When hungry, such copepods show more marked discrimination than fed copepods, and also favor large particles (Fig. 6-10). The latter is a counterintuitive result, since it might be supposed that a hungry individual would accept food in any size.

We thus have that size selective behavior is very variable from one suspension feeder to another. Some copepod species are more flexible with regard to the particle sizes on which they will feed than are other species (Fig. 6-11). *Centropages typicus*, for example, shifts its feeding in response to the abundance of suspended particles. On the other hand, *Undinula vulgaris* ingests particles toward the small end of the spectrum of available sizes in spite of the abundance of smaller and larger particles. Perhaps certain environments provide more constant supplies of particles than others and the feeding behavior of copepods may be adapted to such patterns of abundance. It is not yet clear whether the abundance and distribution of food particles in certain environments could favor a flexible or a sterotyped strategy in selection of food sizes. This topic could sustain further interesting research.

We have seen repeatedly that the specific feeding tactics of different predators can differ markedly and that feeding behavior can change depending on circumstances. Such changes take place by adjustments in the many components of predation and there are innumerable changes potentially available to even very similar predators. The resulting plasticity of hunting behavior of different species studied is no doubt responsible for the divergent results often obtained by different researchers.

There is a further ambiguity in regard to size selectivity which is caused by the way in which feeding selectivity is measured. Most work on

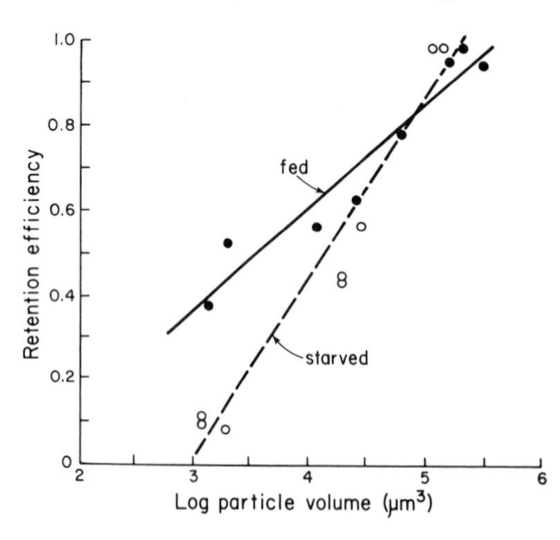

Figure 6-10. Retention efficiency of female *Calanus pacificus* feeding on the diatom *Thalassiosira*. Experiments run with previously fed and starved females to measure the effect of increased hunger level. Adapted from Runge (1980).

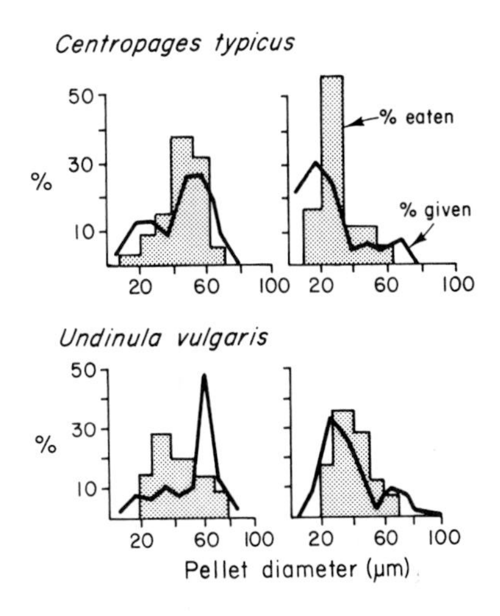

Figure 6-11. Shipboard experiments where the frequency of particles (plastic pellets) added to seawater were skewed to the large and small end of the size spectrum of suspended particles. The experiments with *C. typicus* were carried out with specimens from a station in the coastal shelf (200 m deep) off Cape Cod in February. The *U. vulgaris* data were obtained in the Sargasso Sea, at a 3,000-m deep station during September. Adapted from Adams (1978).

particle selectivity in zooplankton has been done using electronic particle counters that measure the distributions of volumes rather than linear dimensions. Sieves such as hypothesized for some suspension feeders, however, remove particles in proportion to linear dimensions rather than particle volumes. Copepod sieves and electronic counters may thus not sort particles in the same fashion (Malone et al., 1979). Harbison and McAlister (1980) passed suspensions of particles through man-made sieves and obtained results similar to earlier work interpreted by others as selective feeding involving behavior. The shape of the phytoplankton used as food, whether spherical or elongate cells, or occurrence of cells furnished with spines or elaborate architecture, seriously affects the retainment by passive sieves, perhaps independently of the volume of the cells. All this further emphasizes the need to make direct observations of feeding behavior (Alcaraz et al., 1980; Paffenhöfer et al., 1982).

The different results of studies of size selectivity in suspension feeders and the ensuing controversies in the literature thus are fueled by the use of species with different hunting tactics and by the lack of a method free of artifacts. We have come full circle with regard to technology: it was to a large extent because of the advent of the Coulter counter in the 1960s that so much attention could be focused on grazing on small particles by zooplankton. Recent direct observations have shown that, at least for some copepods, there is a complex feeding behavior whose study will modify the passive-sieve interpretation. It remains to be shown how widespread this is and whether or not it is more quantitatively important than filtering through sieves.

A better example of passive filter feeding is provided by the gill rakers of certain fish. These organs serve as retaining nets for particles carried into the mouth of the fish by swallowed water. The size of the pores in the sieves involved are considerably larger than those of copepods and viscosity of water is less of a problem. Sardines, anchovies, herring, and other fish can thus filter feed by swimming with an open mouth, but they also pursue and capture individual prey of a larger size. When offered plentiful food of a large size, anchovies—and many other marine animals, including copepods (Runge, 1980)—go into a brief feeding frenzy and consume prey at higher rates. The rate of food consumption by anchovies decreases as the maximum gut capacity is approached. Filter feeding by sardines on smaller particles is not an effective way to obtain a ration, so that even after 1 hr of feeding only two thirds or so of the maximum capacity of the gut is filled. Leong and O'Connell (1969) conclude that anchovies could not obtain sufficient rations by filter-feeding alone. These fish must supplement their diet with raptorial feeding on large particles. Perhaps the feeding frenzy is an adaptation to exploit as much as possible of a patch of large particles as quickly as possible.

The mechanisms for detection and capture of prey do not always detect and provide consumable prey. Some prey will be too big, some too small, some will be distasteful or have too many protective spines, and these will be rejected. There is, therefore, considerable selectivity even in suspension feeders.

6.5 Optimization in Food Selection by Consumers

Throughout its life, a predator has to decide repeatedly where to find prey and what item of food to eat. In fact, predators have a very well-developed ability to discriminate among prey items and adjust their feeding accordingly. Goss-Custard (1970) allowed captive redshanks (a common European shorebird) to feed on a series of prey placed out sequentially in a row of depressions bored on a wooden plank. The prey were cut sections of mealworm larvae, and each trial provided the redshanks with "small" (one segment) and "large" (eight segments) prey in a ratio of 1 large : 3 small prey (Fig. 6-12). The procedure was repeated, and as a bird began to feed, it took 30–90% of small prey during the early trials. Within five to six trials, however, having encountered and evidently preferring large prey, the redshank searched for and ate large prey, missing only 1.9% of the large prey. Note that during this stage, virtually all small prey were neglected. So far, all we have is a demonstration of size-selective predation, with rejection of less desirable prey. If the ratio of large to small prey in the trials was 1 : 43 rather than the 1 : 3 (inset, Fig. 6-12), the bird perceived the change in ratio and in just a few feeding trials adjusted feeding so that the less desirable but much more abundant small prey were taken. This may be an extreme example of a change in feeding to suit available food, but makes the point that predators are quite capable in this

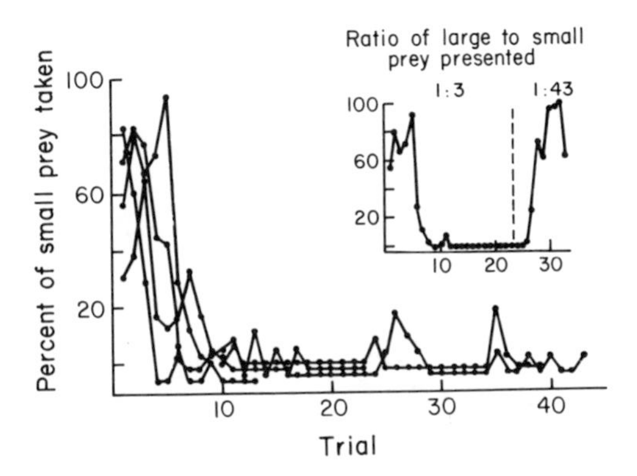

Figure 6-12. Proportion of small prey taken by redshanks (*Tringa totanus*) in a series of trials where they were allowed to feed sequentially on large and small prey offered in a ratio of 1:3. The inset shows the switching due to changing the ratio of large to small prey from 1:3 to 1:43. Adapted from Goss-Custard (1970).

respect. It is thus not farfetched to think that a predator may choose prey items so as to maximize the net energy intake per unit foraging time (Emlen, 1968; Shoener, 1971; Pulliam, 1974; Charnov, 1976; Comins and Hassell, 1979; and others). Experimental evidence for such optimal foraging behavior is available for a variety of predators (Charnov, 1976; Emlen and Emlen, 1975; Kislalioglu and Gibson, 1976; Werner, 1977; Werner and Hall, 1976).

Predators have to spend time searching, subduing, handling, and eating prey. As yet no one has put together a complete accounting of the cost in time of each of these components of predation relative to the benefits in terms of energy or biomass ingested, but in certain predators and prey one or two components are the most important and a cost/benefit analysis is easier.

Carcinus maenas is the common green or shore crab of the coasts of both sides of the Atlantic, and feeds on a variety of prey, including sessile mussels (Elner and Hughes, 1978). Mussels are detected by chemoreceptors in the antennae and mechanoreceptors in the walking legs. The crabs examine an individual mussel for 1–2 s and then either reject it or proceed to try to crush the shell, turning it around until a weak spot is found, and if they are successful they then feed on the flesh. These activities may be time-consuming, on the order of 10^2–10^3 s for breaking and 10^2–10^3 s for eating mussels.

Both of these component activities require more time as the size of the prey increases (Fig. 6-13), and in addition, the size of the predator affects their duration (Fig. 6-13, right). A rough estimate of value of the prey versus the cost of pursuing it can be obtained by a ratio of the grams of biomass eaten relative to time invested in handling (Fig. 6-14, top). Because of the effect of size of predator on handling time, larger prey are of more value to larger predators than to smaller predators. Clearly, there are many other factors that potentially may affect cost/benefit, including time to eat, different assimilation efficiency of crabs of different size, and nutritive quality of mussels of different size. The use of time to break as

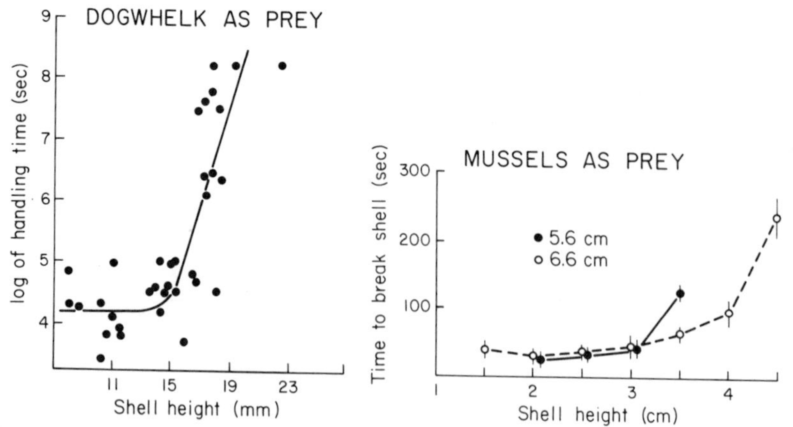

Figure 6-13. Left: Handling time for crabs (*Carcinus maenas*, 6–7 cm in carapace width) feeding on dogwhelks (*Nucella lapillus*) of varying size. From Hughes and Elner (1979). Right: Shell breaking time for *C. maenas* of two sizes (5.6 and 6.6 cm carapace width) feeding on the ribbed mussels, *Modiolus demissus*. Unpublished data of J. Wagner.

the cost, however, gives a good approximation to the actual consumption of mussels (Fig. 6-14, bottom). Green crabs do seem to feed on a size distribution that optimizes the weight eaten relative to the time it takes to handle the prey.

If crabs do not often encounter mussels of the optimal size, they may shift to a different, much more time-consuming method of attack on larger mussels, involving grinding away the edges of the valves until a chela can be inserted into the shell. Prey of larger than optimal size are taken if they are in high abundance relative to "optimal" mussels. The crabs at all times attack prey of all sizes; the size of the crab, the size of prey, and their density determine the model frequency with which a prey of given size is eaten.*

If prey are very rare, even very large mussels may be opened by crabs given enough time and hunger. There may be some size classes of prey that cannot be successfully attacked by crabs, no matter how long the crab persists. Starved crabs may persist in attacking invulnerable prey for up to 15 min, but for less hungry crabs with no other food, the handling time is usually cut short and the crab goes on to seek more suitable prey. For rare prey with invulnerable size classes, the optimal tactic may in fact be to cursorily examine the prey, eat the easily handled prey, and not persist too long on the more difficult items.

* Similar relationships have been proposed in models of zooplankton grazers (Lam and Frost, 1976; Lehman, 1976). Taghon et al. (1978) and Doyle (1979) extend these models to benthic animals feeding on particles of sediment, where the food is the microflora growing on the sediment particles.

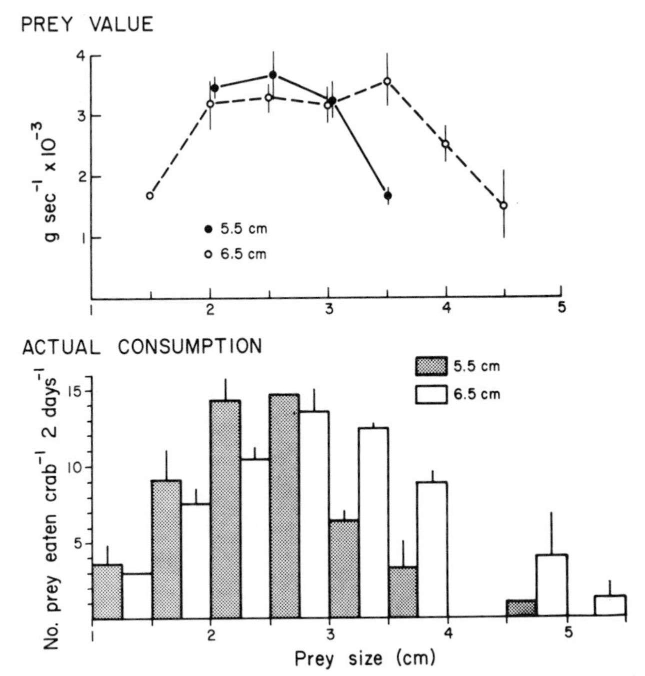

Figure 6-14. Top: Prey value, expressed as grams of mussels eaten by green crabs of two sizes (5.5 and 6.5 carapace width) per second. Bottom: Actual consumption when crabs were offered a choice of prey sizes. Unpublished data of J. Wagner from material collected in Great Sippewissett Marsh, Massachusetts.

In addition to the choice of specific prey size, foragers also have to choose where to feed. It would be most profitable to find patches where density of food is higher than the average density of food items in the entire habitat. Once feeding has reduced the density within a patch to near the average for the habitat, the foragers should move on. Mathematical expressions of this aspect of optimal foraging are available (Hughes, 1980; Krebs et al., 1981) and can be coupled to optimization of food obtained and time spent.

Optimal foraging theory is general enough to be applicable to many instances of food gathering by consumers. One difficulty with the theory is what common currency to use in optimizing food acquisition and expenditures during foraging. Most of the work on this area has been done on the basis of energy content, but it should be clear from our examination of properties of food items (Section 6.3) that other criteria could also be the subject of optimization. Redshanks may, for their own reasons, actually prefer energy-poorer amphipods to energy-richer worms (Goss-Custard, 1981); other consumers may optimize energy and nitrogen content while at the same time avoiding certain chemicals (Glander, 1981). Time is also a constraint of importance to many predators. It is therefore difficult to define just what should be optimized in an optimal foraging model. A second difficulty is that multiple factors may be optimized simultaneously by foragers. Food choice in the field, then, could differ substantially from that in the laboratory, where only one variable may have been considered in the experimental study of optimization. The available theory does not yet include multiple optimization, and in fact it may be very difficult if not impossible to test such a theory in the field because of the complex observations needed (Zach and Smith, 1981).

6.6 Vulnerability and Accessibility of Food Items

While some prey may be unavailable because of deterrents, low quality, or inappropriate size, there may be perfectly adequate prey that still may not be easily available to predators. We have seen earlier that one way a prey may escape predation is merely to hide until it has outgrown its predator. This is a common occurrence and is well demonstrated in a whelk, *Nucella lapillus*, that escapes predation by green crabs after it reaches a shell height of about 15 mm (Hughes and Elner, 1979). In the field the susceptible size classes hide under stones or in crevices and are thus less available to green crabs. There is thus an evolutionary race between the searching ability of predator and the hiding ability of prey.

A somewhat different kind of spatial refuge from predation protects *Balanus glandula*, a common barnacle found primarily high in the intertidal in the coast of Washington. Whelks (*Thais* (= *Nucella*) spp.) eat all the *Balanus glandula* that settle toward the low end of the intertidal range. *B. glandula* does not grow to a size where they are safe from predators, but high in the shore the period between low tides (during which *Thais* is active) is too short to allow the whelks to complete drilling into the barnacles, and so above a certain height in the intertidal range the barnacles are in a spatial refuge (Connell, 1970).

Another kind of sanctuary from predation is that afforded by the complexity of the arena in which predation has to take place, as demonstrated in a classic paper on mites by Huffaker (1958): increased complexity of the habitat decreases the relative vulnerability of prey. This idea has broad application, and can be demonstrated to occur in salt marshes on the east coast of North America, where there are two major vegetated intertidal habitats, low marsh (dominated by cordgrass, *Spartina alterniflora*) and high marsh (dominated by marsh hay, *S. patens*). Many types of benthic organisms are found in the sediments of these two habitats and are eaten by killifish during high tide. Laboratory experiments where killifish were allowed to feed on amphipods show that the func-

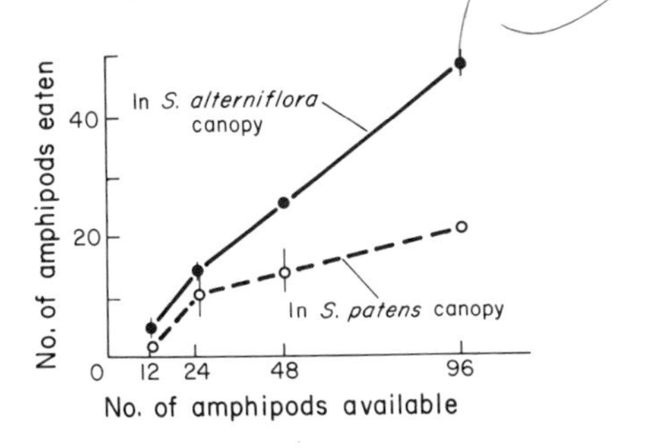

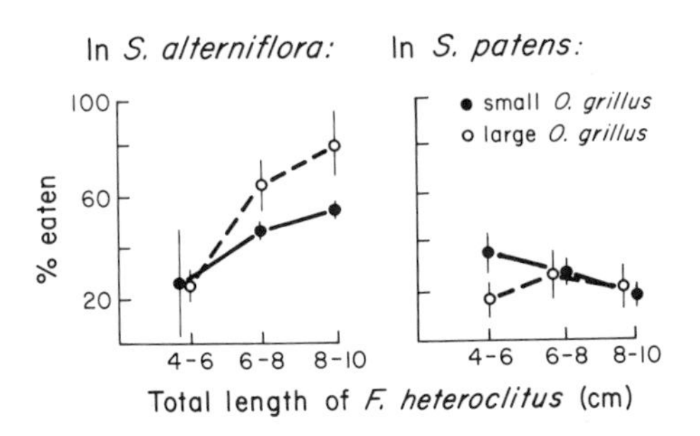

Figure 6-15. Functional response of *Fundulus heteroclitus* to densities of amphipod prey in two habitats with different physical structure. Bottom: Percent of amphipods prey eaten by *F. heteroclitus* of differing size in each of the two habitats. Adapted from Vince et al. (1976).

tional response to amphipod density differs in the two habitats (Fig. 6-15, top). In fact, only in the low marsh was it possible for fish of large size to evidence their superiority as predators (Fig. 6-15, bottom). The explanation for this has to do with the physical character of the two habitats. Low marsh plants are more sparsely distributed than high marsh plants. On average, the distance between nearest neighbor stems is 1 mm in high marsh, while stems in low marsh are spaced almost five times as far. The dense stems of high marsh make it difficult for the fish to enter the habitat, and at the same time provide hiding places for prey. The prey are therefore less vulnerable to attack by fish in high marsh. Since it is more difficult for larger fish than for smaller fish to find their way into a complex, dense habitat, this explains the lack of effect of predator size on the number of prey eaten (Fig. 6-15, bottom).

The above experiments are laboratory results but there is evidence of the importance of habitat complexity in the field. We can compare the distribution of prey sizes and abundance in salt marsh plots where fish had access to prey with results from areas where they were excluded by means of fences (Fig. 6-16). The number (top right within each graph) of available prey in samples from low marsh was much lower than those from high marsh. Where fish were present, fewer prey were found than where fish were excluded. From our discussion of prey size, we know that predation is usually size specific, and the snail can escape predation by growing to an invulnerable size (Vince et al., 1976). In low marsh the smaller size classes of snails are removed where predators are present. The size distribution of snails in high marsh where fish were excluded

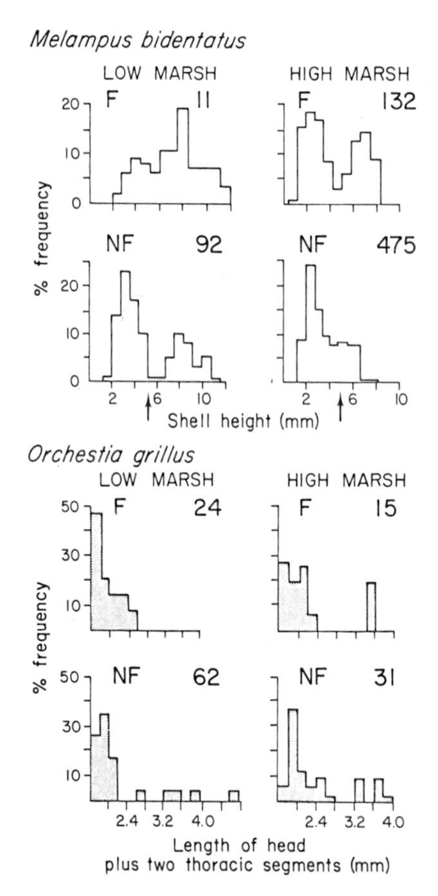

Figure 6-16. Abundance (numbers on top right of each histogram, in individuals per 2 cores of 9 cm radius) and size distribution of the prey species (histogram) in two salt marsh habitats. Area labeled "F" are exposed to natural predation by *F. heteroclitus* during high tide, while these fish were excluded from the "NF" areas by means of wire mesh fences. The arrows show the threshold above which *M. bidentatus* is not eaten by *F. heteroclitus*. Adapted from Vince et al. (1976).

resembles that of low marsh with no fish. The more remarkable result, however, is that the size distribution of prey in high marsh also has a large proportion of small individuals. This corroborates the laboratory results, and shows that predation by fish is so reduced in the more complex habitat that it cannot affect abundance or size distributions of prey.

Very similar conclusions emerge out of the data of field distributions of the amphipod (Fig. 6-16), the difference being that unlike the snail, the entire size distribution of amphipods is subject to predation and, in fact, fish prefer larger amphipods (Vince et al., 1976). The results of these experiments demonstrate that the physical architecture or structure of the habitat in which predation takes place has critical consequences for both predator and prey.

It is not clear that such structure is important to predation in planktonic communities. We tend to see the water column as relatively homogenous but there is an increasing amount of evidence that there are important heterogeneities in temperature, salinity, density, nutrient content, and also the motion of water. These heterogeneities pervade all spatial scales (cf. Chapter 13) and may actually provide a complex structure to the water column, with still to be explained consequences for predator–prey interactions.

6.7 The Importance of Alternate Prey

6.71 Prey Switching by Predators

In discussing optimal foraging we concentrated on the size of individuals within a species of prey. Prey choice, however, also involved choice among different species. If predators can feed on several species, they might consume the most abundant prey first, perhaps by the buildup of a search image, as mentioned earlier. Then when numbers of this species are lowered, they might concentrate on whatever other prey species is most numerous at the time. This feeding behavior, similar to the patch exploitation pattern hypothesized in optimal foraging, has been called "switching" (Murdoch, 1969). Switching has received attention because such behavior on the part of the predator could exert maximum predation pressure on the most abundant species, and thus prevent any one species from monopolizing an environment, an important feature that may explain how multispecies communities are maintained (Elton, 1927).

Predators with strong preference among possible prey, such as carnivorous snails, seldom switch regardless of relative abundance of the prey species (Murdoch, 1969).

Switching has, however, been found in consumers that have less marked preferences among the alternate prey available. Switching occurs in copepods (Landry, 1981), for instance, presumably because although suspension feeders are size selective they may not have strong preferences among prey of different species. Pelagic food webs, in fact, have been described as "unstructured" (Isaacs, 1973) precisely because any predator is thought, within constraints imposed by its size, to be able to eat most other species in the assemblage of species present.

In species where preference among alternate prey species is not too marked, the relative density of the alternate prey species may be important. The snail *Lepsiella vinosa* preys on two species of barnacles that settle on pneumatophores of Australian mangroves. Switching by the snail takes place provided that the density of the one slightly preferred prey species is not too high (Bayliss, 1982).

It is not clear exactly how switching takes place. It may not just be a reflection of relative abundance of alternate prey types, and learning and rejection may not be involved in every case. Possible mechanisms for switching might involve changes in foraging methods, predation on high density patches of different prey species, and the differential benefit of alternate prey to the predator.

Changes in foraging method: Switching may take place due to changes from one foraging method to another (Akre and Johnson, 1979). Damselfly naiads are a very common predator in freshwater ponds and lakes and themselves are subject to visual vertebrate predators. To avoid detection by fish and amphibians, naiads use the less risky ambush mode of hunting as long as it results in sufficiently frequent encounters with prey, in this case two cladoceran species. There seems to be no clear preference by the predator for one or another type of cladoceran. One of the prey species is more easily captured using ambush methods, but if it becomes too rare the hunger level of the naiad reaches a point where the naiads start hunting by searching. Since the second species of prey is more easily encountered while searching, this type of prey then becomes more abundant in the diet. In this instance, no learning or rejection of less preferred prey is involved. Rather the hunger level relative to prey density is the driving force that changes feeding tactics.

Predation of high-density patches of different species: Switching may take place in predators that concentrate foraging in patches of high density of prey (cf. Section 5.3). If different patches where prey are found at high density are dominated by different types of prey, as the predator moves from one patch to another the result is effectively switching, with no rejection or learning behavior (Murdoch and Oaten, 1975). One important consequence is that this sort of predation leads to a type III functional response, as does the shift in hunting behavior seen in the naiads. This sort of switching could be important in thus allowing a density-dependent acceleration of prey consumption.

Differential benefit of alternate prey to the predator: Sea otters select prey species that provide them the most energy per time spent foraging and handling (Ostfeld, 1982). As a given prey becomes rarer, search time increases, and thus the otters would probably switch to a more abundant prey where the ratio of energy obtained to time spent may be more favorable. The criterion used to switch would be a function of profitability rather than merely depend on abundance of alternate prey.

The diversity of ways in which feeding interactions can be carried out demonstrate the manifold ways in which animals can carry out similar activities, both by flexible behavior within one species and by different adaptations by several species. This is a major lesson to be learned from this section, and is one of the reasons why we examined the plethora of examples set out in this chapter. A second major point is that the flexibility in behavior seen in the above examples, plus a weak initial preference when alternate prey are about equally abundant are preconditions to switching when prey densities may become unequal. As such, one may imagine that if prey preference is very marked there are potentially few situations where the mechanism of switching may be important in affecting stability of prey numbers. The degree of prey preference—and the related idea of profitability—are key elements in how feeding on alternate prey may be important in governing abundance of prey populations.

6.72 The Impact of Preference in Multiple-Prey Situations

In the field, the usual condition is for a foraging organism to be faced with choices among a variety of very different food types. A situation where a predator had a choice among various prey was examined by Landenberger (1968), in experiments in which two species of carnivorous starfish fed in an arena where seven species of prey molluscs were available. Their prey included three species of mussel, two species of whelk, a turban snail, and a chiton. Mussels were the preferred food of the starfish. The data are reported as to the percentage of mussels in the first, second, third, etc., groups of 30 prey eaten (Fig. 6-17, left panels). The first point to be made [one made earlier by Holling (1959)] is that the presence of alternate prey reduces the predation pressure on all prey, even the most preferred: mussels made up rather less than 100% of the first group of prey.

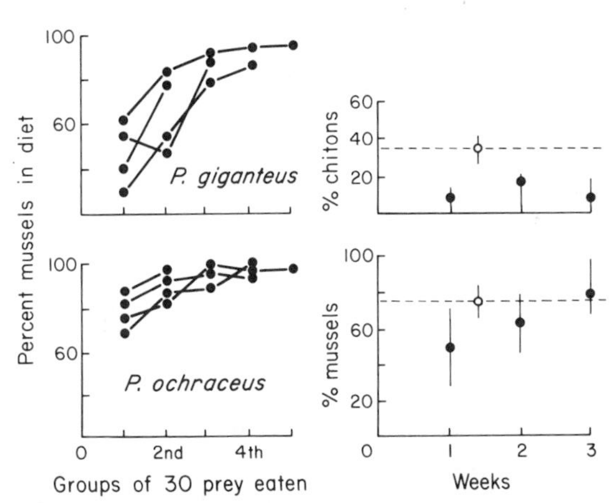

Figure 6-17. Effects of alternate prey on predation by two species of starfish *Pisaster giganteus* and *P. ochraceus*). Left panels: Percentage of the preferred prey (mussels) in the diet, expressed as the 1st, 2nd, etc., groups of 30 prey eaten. Right panels: Percentage of alternate prey (mean ± range) eaten by starfish trained for 3 months to eat turban snails. For comparison, the dashed line and open circle show preference of untrained control starfish are included. Adapted from Landenberger (1968). © Ecological Society of America, reprinted by permission.

The second point is that as the predators gained experience with the second, third, etc., groups of prey eaten, the proportion of preferred prey (mussels) in the diet increased, so that after five weeks, starfish consumed mussels almost exclusively. For these predators the effects of alternate prey are not permanent.

A third points (not obvious in Fig. 6-17) is that the predators were less discriminating when more kinds of prey were available. Only broad taxonomic categories (e.g., pelecypods, gastropods) were recognized by starfish in situations with many alternate prey species, while when an experiment allowed the starfish a choice of only two species of mussels, starfish were capable of distinguishing species, and showed preference for one of the two species offered.

Since past experience with given species of prey seemed to be an important feature, Landenberger then tested the impact of alternate prey on starfish fed for three months on turban snails (Fig. 6-17, right panels). When the alternate prey were chitons (a prey not liked by starfish), only 10% or so of the prey of the snail-conditioned starfish were chitons, the rest being turban snails. When mussels were the alternate prey, there was an initial consumption of 50% turban snails, still to an extent reflecting their long conditioning on turban snails. As time passed, the starfish switched to feeding on mussels: the predators overcame previous history and used the most preferred food items.

In Landenberger's study, as in Ivlev's (1961) work on feeding by fish, the rate of feeding increased with time. The predators were also most selective when feeding at high rates.* With experience, the starfish learned how to be more effective predators on the array of prey available.

6.8 Predation and Stability of Prey Populations

In our discussion of Chapters 5 and 6 we have so far ignored whatever coevolution has taken place between predator and prey. We continue to do so, but evolutionary processes are clearly involved. Recall the discussion of chemical feeding deterrents and mimicry. We have already noted, for example, that some polychaetes may secrete chemical deterrents active against some predators. There may be morphological adaptations. Zooplankton may grow spines or processes that prevent ingestion. The

* We have already seen this counterintuitive effect of hunger level in Section 6.4 when discussing copepods. It is not clear exactly how generalizable is the relation of greater selectivity to feeding rate. In blackbellied plovers search times are shorter when the birds feed on a more varied diet (Baker, 1974). Shorter search times would allow faster feeding rates, so that the less selective the predator, the faster the feeding rate. The relation of feeding rate, alternate prey, and selectivity needs further study.

shells of many marine snails show structures and sculpture that seem to reflect the presence of many molluscine predators through the evolutionary history of snails (Vermeij and Covich, 1978). The architecture of snail shells is much less ornate and more conservative in freshwater environments where typically there are fewer predators specializing in snails than in marine habitats.

Predators show adaptive responses of their own. In nature we find that, by and large, there is an interplay between prey and predator whereby both continue to be present, at least in ecological if not evolutionary time scales. Predators could potentially eat all prey, and prey could potentially fill the environment with their progeny, yet on average neither of these situations takes place.

What potential ecological mechanisms does the process of predation provide by which control of prey numbers can take place? In terms of preventing prey from reaching excessive numbers, the functional response does not seem like the most important contributor, except where a type III curve is found; we have seen that this may not be a very common or permanent feature. We cannot be sure that the rarity of type III functional response is not a laboratory artifact, but switching, patchiness of prey, and presence of alternate prey favor this response. The much more common type II response is by itself less likely to provide the means by which predators control prey. The functional response does not act alone, however. Predators also respond to changes in prey density or to prey patchiness by aggregation. This numerical, almost lagless response, combined with the increased recruitment of young predators, could make predators effective enough to control prey numbers, since both increase markedly as prey density increases. The developmental response, although involving a lag in time, as we have seen, could add to the effectiveness of predation in the control of prey.

On the other hand, complete extirpation of a prey species is prevented by the existence of refuges, of habitat complexity, and by the growth of prey to invulnerable size classes. These devices allow survival of certain proportions of the prey population, making extinction less likely. Another way to escape control by consumers is for prey to grow so fast as to overwhelm the ability of predators to respond to prey density. This aspect is discussed in Chapters 8 and 14 in regard to phytoplankton and their grazers.

In the complex phenomenon of "predation"—including animals feeding on plants, animals, microbes, and detritus—there are a variety of mechanisms that could lead to the control of prey numbers. A very rich array of predation tactics exists and similarly, a variety of prey defenses, and each case may have its own distinctive features. The generalizations that we can make are thus not about the specific ways in which consumers obtain food particles. Rather, the generalities concern the components of the predation process. Time is the most limited resource. Each predator must allot time to detect, pursue, capture, and handle prey. The apportionment of time among these various phases of predation, the deployment of the components of predation to carry out the processes, and the properties of the prey determine what a predator does and how important it may be in structuring the abundance of other species in the community in which it lives.

Chapter 7

Processing of Consumed Energy

7.1 Flow of Energy Through Consumers

In the previous two chapters we have discussed the various means by which consumers obtain sufficient rations. In this chapter we focus on the various fates of ingested material and how this apportionment affects consumers. In this section we will look at the overall scheme of flow of ingested material; in the following section we examine the processes that affect the various fates of matter ingested by consumers and some consequences of this partitioning. In many instances the rates of each process vary so widely that it is necessary for comparative purposes to use ratios between two of the processes, which we refer to as efficiencies. After completing the examination of partitioning of food in specific populations, we consider how the partitions are put together in energy budgets for populations and ecosystems.

It is customary to consider the partitioning of specific elements, usually carbon, nitrogen, or phosphorus (Butler et al., 1970; Corner and Davies, 1971; Kuenzler, 1961; Jordan and Valiela, 1982), or the energy* contained in ingested food. Our emphasis will be the fate of consumed energy in a consumer (Fig. 7-1). A certain portion of the energy of ingested material is not assimilable and is released back into the environment as unused feces (F); the remainder of the ingested food is assimilated (*A*). Some of the

* There has been much research on energy in consumers and many measurements of energy content in terms of calories of biologic materials (Paine, 1971; Cummins and Wuychek, 1971). In general, the ash-free caloric content of consumers lies between lower values typical of carbohydrates (3.7 kcal g^{-1} for glucose, 4.2 kcal g^{-1} for cellulose) and higher values of lipids and fatty acids (9.4 kcal g^{-1}). Most consumers have a far narrower range of caloric content, averaging about 5–6 kcal g^{-1} (Holme and McIntyre, 1971).

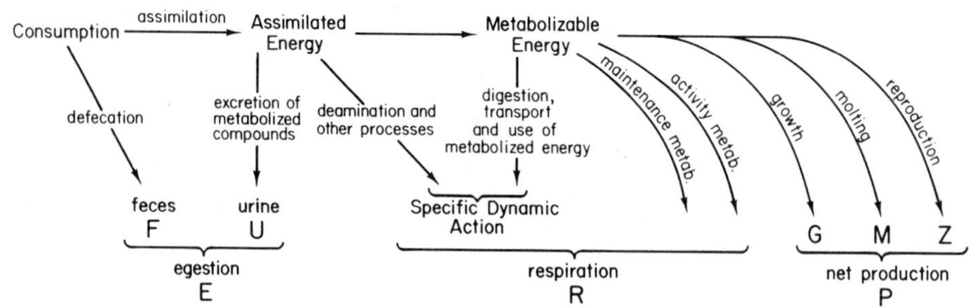

Figure 7-1. Scheme of the various fates of consumed energy and the terms used in discussing processing of energy. Many marine species release mucus in large quantities. This is not shown in this figure but could be a significant loss of metabolizable energy that should be included in net production for such species. Modified from Warren and Davis (1967).

products of metabolic reactions are energy-rich nitrogenous compounds such as urea and uric acid; these compounds are voided (U) and are usually measured together with fecal losses (F) in a measurement of egestion (E). Some of the energy in ingested food is lost in the deamination of amino acids not incorporated into consumer protein, and there are further losses of energy in the digestion, transport, and metabolism of food. The sum of these losses of assimilated and metabolizable energy is referred to as the specific dynamic action and is evidenced as an increase in oxygen consumption soon after feeding occurs. The major losses of ingested energy are due to expenditures required to maintain metabolic processes and to sustain levels of activity sufficient for life activities. The sum of specific dynamic action plus metabolic and activity losses is respiration (R). Respiration is the oxidation of organic matter and results in the release of energy and CO_2 [cf. Eq. (1-2)]. Respiration is thus the release of energy fixed by producers.

The energy that is assimilated but not respired is devoted to production (P). Production consists of the energy invested in reproductive products (Z), and in growth (G). If the species has the kind of development that involves shedding of outer layers, there is a further loss due to molting (M) that is also included in production. The sum of these various terms is the energy consumed (C),

$$C = P + R + E. \tag{7-1}$$

C for field populations has often been estimated by some version of Eq. (7-1), usually using the assimilation efficiency (AE),

$$AE = [(C - E)/C]\ 100, \tag{7-2}$$

which can be measured in the laboratory with known rations (cf. Section

7.21).* Then, by substitution into (7-1),

$$A \cdot C = C - E = P + R, \qquad C = (P + R)/A. \tag{7-3}$$

The scheme of Figure 7-1 was derived for consumers that feed on particles and are fairly large. With some modification it could be applicable to smaller consumers such as aerobic heterotrophic microorganisms. The modifications would include release of vacuolar waste products instead of feces, secretion of mucopolysaccharide sheaths instead of molting, reproduction by fission instead of release of reproductive products.

7.2 Assimilation

7.21 Measurement

The simplest measurement of assimilation would be to obtain the ash-free weight of food and feces, but it is often difficult to weigh feces, and reingestion of feces and losses of soluble materials will affect the measurement. Animals may also assimilate the ash portion of food (Cosper and Reeve, 1975); prawns, for example, can assimilate over 30% of the ash in their food (Forster and Gabbott, 1971). Feeding of radioactively labeled food, followed by measurement of labeled CO_2, feces, and soluble compounds, is a more acceptable approach. This procedure requires adequate controls to assess bacterial uptake of label and a long enough interval of time during which the consumer is exposed to the label so as to allow tracer content in the body of the organism to reach equilibrium.

There may be a substantial amount of organic matter lost during feeding by consumers. This means that ingestion is not equal to the removal of food organisms. The loss during such "messy" feeding by zooplankton can be up to 30% of the ingested food (Williams, 1981). ^{14}C studies of feeding by a marine isopod on *Calanus* spp. showed that perhaps 25% of the biomass of the prey was lost during feeding. Such losses, if not considered, may lead to overestimates of assimilation and assimilation efficiency. Inert radioactive tracers such as ^{51}Cr (Calow and Fletcher, 1976) can also be useful in measuring assimilation.

7.22 Factors Affecting Assimilation

It is usual to refer to the ratio of food assimilated to food ingested as the assimilation efficiency. Measurements of assimilation efficiencies for marine species can vary significantly even within one species. For example,

* There is some ambiguity in Eq. (7-2) in that $E = F + U$ (cf. Fig. 7-1), and most of F is material that has not been assimilated. It is generally not practical to measure this nonassimilated fraction separately.

the assimilation efficiency of *Calanus helgolandicus* may be anywhere from 12.4 to 61%; that of *Centropages typicus* 16 to 79%; *Temora stylifera* 0 to 91% (Gaudy, 1974). Such variation in assimilation efficiency may be due to a variety of factors including food quality, amount of food, and age of the consumer.

7.221 Food Quality

It is obvious that the higher the quality of a food, the higher the assimilation efficiency. Food quality has many dimensions; a relatively simple property is the percentage of inorganic ash: the higher the percentage ash in food, the lower the assimilation efficiency (Fig. 7-2).

Differences in the food habits of various species lead to significant differences in assimilation efficiency. Estimates of assimilation efficiency obtained in various ways for a very wide variety of marine, freshwater, and terrestial animals are compiled in Figure 7-3. The classification of species into detritivore, herbivore, and carnivore is based on the actual food types eaten by the consumers (dead plant material plus microbial flora, living producers, and living animals, respectively) during the measurement. Individual measurements of assimilation efficiency are subject to many experimental errors and are very variable, but considering the data set as a whole, there are clear trends. Carnivores characteristically show high assimilation efficiencies; their food is of high quality and readily digestible, since it lacks cellulose and other undigestible compounds. Herbivores span a wide range of assimilation efficiencies, but peak in intermediate values of assimilative ability. As discussed in Sections 6.32 and 8.22, producers may be inadequate food; for example, many plants often contain secondary chemical compounds that bind proteins, lowering the quality of food used by grazers, and this low food quality results in low assimilation efficiencies. Single-celled algae, yeasts, and bacteria tend to be assimilated with higher efficiency than macroalgae and higher plants, although not as efficiently as animal food. Several short-lived, opportunistic macroalgae, referred to as ephemerals in Section 8.211, are also consumed with high efficiency. This agrees with what we know

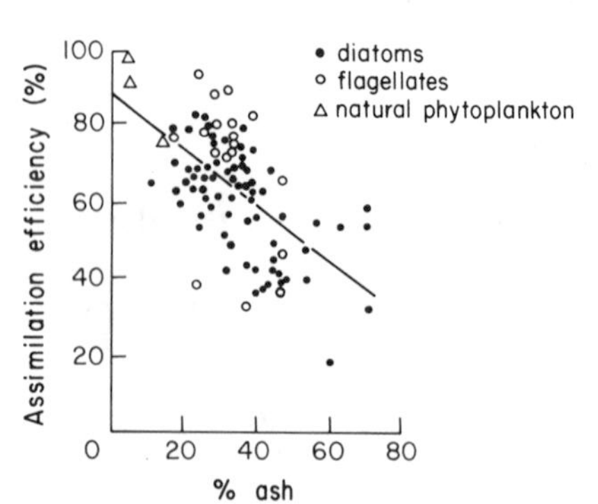

Figure 7-2. Assimilation efficiency in *Calanus hyperboreus* feeding on phytoplankton of different percent ash content. Adapted from Conover (1966).

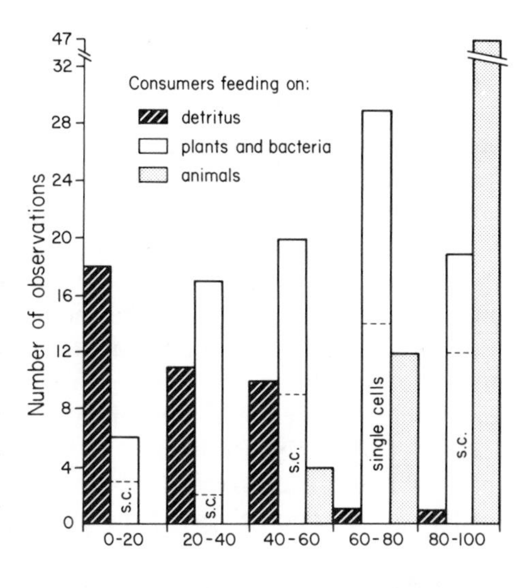

Figure 7-3. Frequency of assimilation efficiencies in a wide variety of animals that feed on detritus, plants, and animals. Herbivores that feed on single-celled algae, bacteria, or yeasts are grouped in the lower portion of the open bars. One species may have been included more than once if experiments with different food types were done. Values from many sources, including the compilations in Welch (1968), Conover (1978), Forster and Gabbott (1971), Dickinson and Pugh (1974, vol. 2), Krebs (1976), Winter (1978), Ricklefs (1979), Kiørboe et al. (1981), and many individual papers.

about the lack or presence of structural or chemical defenses in these kinds of producers. The result is that single-celled or ephemeral species are often preferred food of herbivores. Other measurements of high assimilation efficiency in Figure 7-3 were obtained with terrestrial herbivores that feed on seeds or sap; seeds have typically high protein contents and are high quality food, while feeding on sap avoids many of the secondary defensive compounds, since these are to a large extent associated with structural plant tissues. Herbivores that feed by ingesting tissues of vascular plants tend to have lower assimilation efficiencies, since these plants tend to have well-developed structural tissues and chemical defenses.

In most ecosystems grazing does not consume all the plant matter (Table 8-1), so that senescent algal and plant tissues largely become detritus. Even in oceanic water columns, where grazing can consume all that is produced by the phytoplankton, much of the organic matter is returned to the environment by the grazers in the form of fecal matter (cf. Chapter 10). Detrital matter is therefore plentiful and the majority of species in most environments consume detritus. Yet, detritus is poor food; detritivores are subject to both nitrogen and carbon deficiencies in their diet. Animals require that carbon be in roughly a 17/1 ratio to nitrogen (Russell-Hunter, 1970). Detritus is often lower in nitrogen; C/N values of 20–60/1 are not unusual in detritus from vascular plants in salt marshes and 15–30 in eelgrass (Godshalk and Wetzel, 1978). The C/N ratio of particulate matter in the oceanic water column in the Atlantic is in the order of 7–21 (Gordon, 1977; Knauer et al., 1979). Particles at 0–250 m in depth are richer in N (C/N = 7.5) while at depths of 3000–4000 m the nitrogen is relatively lower (C/N =15.1) (Gordon, 1977). In coastal areas off California, C/N vary between 8.8 and 15, depending on depth and intensity of production near the surface, while in the open Pacific the C/N values lie between 13 and 29, the high values associated with deeper water. The factors that govern the ratio of C/N in the water column are not well known, but differences in sampling and analytic methods may be responsible for part of the observed variability.

Sediments have C/N ratios of 10 or more (Degens, 1970). Even though bulk C/N ratios may not exceed the desirable ratio of 17/1, much of the nitrogen in particulate organic matter is unavailable to suspension or deposit feeders due to binding of nitrogen compounds with clays, lignins, and other phenolic compounds, including nitrogen-containing humic acids (Odum et al., 1979; Rice, 1982). This immobilization of nitrogen means that even where C/N ratios of organic matter are low, benthic deposit feeders or other detritivores generally do not get enough nitrogen (Tenore et al., 1979).

Fecal pellets are abundant in soft sediments, and they have C/N = 9–15 (Honjo and Roman, 1978), a more favorable ratio than the 17/1 presumably required by consumers. Fecal pellets are nevertheless a poor food. *Calanus helgolandicus* fed on fecal pellets develop more slowly from the third copepodid to adult stages (20–22 days) than specimens fed on phytoplankton (8–15 days). Survival is also lower in pellet-fed (57–71%) than in phytoplankton-fed copepods (66–99%) (Paffenhöfer and Knowles, 1979). Growth is reduced when detritus is included in diets, and copepods on detrital diets have higher respiration rates as more detritus is included in the ration (cf. Fig. 7-8).

The caloric content of detritus is a rough index of the potential energy available to detritivores. The polychaete *Capitella*, for example, incorporated more detritus of *Spartina* when it had a higher caloric content, while consumption and assimilation were not increased by increased nitrogen content (Tenore et al., 1982). Caloric content may be thus more important than nitrogen content (Tenore, 1983). Detritivores therefore also have a problem obtaining enough carbon and energy. Even though detritivores may have the carbohydrases needed to digest detritus (cf. Section 10.1343), little of the detritus may be assimilated (Table 7-1). Cammen et al. (1978) calculated that the polychaete *Nereis succinea* can only obtain 26–45% of its carbon requirement by feeding on detritus. Prinslow et al. (1974) found no growth for the killifish *Fundulus heteroclitus* on a diet of detritus alone, even though detritus was often found in their gut.

As we would expect, then, detritivores show lower assimilation efficiencies, lower for the most part than herbivores (Fig. 7-3). Detritivores can efficiently assimilate food of high quality, for example, microflora (Table 7-1), even though their assimilation of detritus may be low. Detritivores are therefore not physiologically restricted to low assimilation efficiency; it is the quality of the food that determines the low assimila-

Table 7-1. Assimilation Efficiencies in an Amphipod, *Hyalella azteca* (Hargrave, 1970), a Polychaete, *Nereis succinea* (Cammen et al., 1978), and a Holothurian, *Parastichopus parvimensis* (Yingst, 1976) Feeding on Various Foods

Species of consumer	Food	Assimilation efficiency (%)
Hyalella azteca	Bacteria	60–83
	Diatoms	75
	Blue-green bacteria	5–15
	Green algae	45–55
	Epiphytes	73
	Leaves of higher plants	8.5
	Organic sediment	7–15
Nereis succinea	Microbes	54–64
	Spartina detritus	10.5
Parastichopus parvimensis	Microorganisms	40
	Organic matter	22

tion. The low quality of detritus as food for consumers is probably due to two features of detritus. First, if the detritus is derived from a vascular plant, it bears some of the antiherbivore compounds produced by the living plants. Although the soluble fraction of secondary compounds may be leached out of detritus, a large enough portion remains bound to cell walls in detritus to affect palatability, digestibility, and assimilability of detritus by detritivores (Harrison, 1982). Second, there is a marked loss of soluble nitrogen and other compounds from detritus soon after senescence of the parent plant (cf. Chapter 10); this loss decreases the quality of detritus as food for consumers.

In view of the low quality of detritus as food, it seems likely that detritus feeders do not merely consume bulk detritus. They may be facultative predators, enhancing their diet with an occasional high-quality prey item ingested along with the detrital particles, or they may also selectively feed on nitrogen-richer fractions of detritus (Bowen, 1980; Odum et al., 1979).

Part of the difficulty in assessing the assimilation efficiency of detritus feeders may be due to methodological limitations, since techniques to estimate microbial biomass and microbial products are not yet completely developed. Bacterial biomass is not a straightforward measurement to make. In the past phase-contrast microscopical counts or plate counts have been used together with an average value of weight per cell. The phase-contrast microscopy, however, does not allow discrimination of small cells from detrital particles. Plate counts inevitably record only the growth of cells that can grow in the specific medium on which the samples are inoculated. Plate counts provide severe underestimates of bacterial abundance. Recently, direct methods of counting bacteria have been made available, where samples of water or sediment containing bacteria are stained with fluorescent dyes and observed with epifluorescent illumination (Hobbie et al., 1977). There are equally difficult problems with current methods of estimating fungal populations, so that fungal biomass may be frequently underestimated (Newell and Hicks, 1982). There may also be abundant but as yet unmeasured energy-rich exudates released by microbes that could contribute to the diet of detritus feeders (Hobbie and Lee, 1980). Advances in the measurement of these components of detritus may provide new perspectives in the nature of detritus as food for consumers.

The results compiled in Figure 7-3 were calculated on the basis of organic matter, energy, or carbon, but the general results also apply to other specific nutrients. For example, the predaceous pteropod *Clione limacina* assimilates 99% of the nitrogen it consumes (Conover and Lalli, 1974), while the mussel *Modiolus demissus*, feeding on a mixed diet of phytoplankton and detrital particles, assimilates only 50% of the nitrogen consumed (Jordan and Valiela, 1982). In general then, ingestion of high-quality food leads to high assimilation efficiencies of most components of the food eaten.

7.222 Amount of Food

Assimilation efficiency is reduced as the amount of food ingested increases (Fig. 7-4). Excess feeding may waste time, energy, and food, and most consumers probably reduce feeding rates rather than simply ingest in excess and defecate what they do not use. For example, the rate of filtration by bivalves and zooplankton increases sharply as food abundance increases, up to some relatively high level of filtration determined by the maximum feeding effort exerted by the animal (Lam and Frost, 1976; Winter, 1978). If abundance of food particles increases further, the filtration rate remains at the constant maximum rate and provides a linearly increasing amount of food that can be ingested. At some higher food density, ingestion rate reaches the maximum number of particles that can

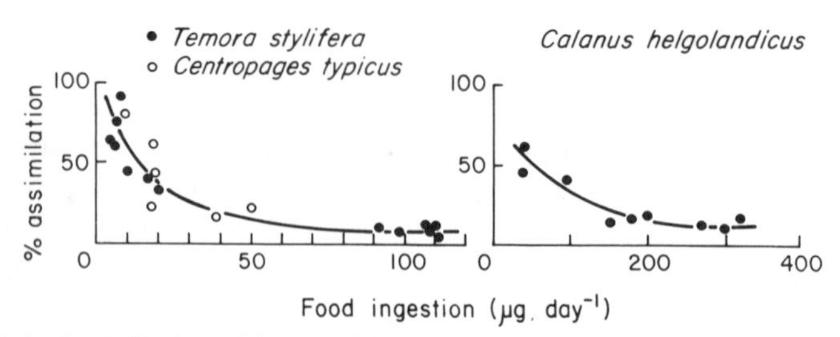

Figure 7-4. Assimilation of ingested food by three copepod species in relation to the amount of food ingested. Adapted from Gaudy (1974).

be ingested and can pass through the digestive tract; if food density increased beyond this threshold, the rate of filtration is lowered, and the amount ingested remains near the maximum rate of ingestion.

The assimilation efficiency depends on the abundance of food particles up to the point where the maximum ingestion rate is reached; at higher densities of food, assimilation rate may be independent of food density, as occurs in copepods (Conover, 1978) and bivalves (Winter, 1978; Navarro and Winter, 1982). In general, there is probably some upper limit to the rates at which food can be assimilated, and at higher ingestion rates

Table 7-2. Age-Specific Assimilation Efficiencies of Various Consumers

		Assimilation efficiency (%)
Fiddler crab (*Uca pugnax*)[a]	1 yr (juveniles)	44.3
	2 yr (juveniles)	40.2
	3 yr (peak reproduction)	42.5
	3 yr (past peak reproduction)	34.5
Amphipod (*Orchestia bottae*)[b]	9 mg	50.6
	80 mg	30.4
Scallop (*Patinopecten yessoenis*)[c]	1 yr (prereproductive)	79
	2 yr (reproductive)	66
	3 yr (reproductive)	68
Crab (*Menippe mercenaria*)[d]	Zoea stages 1	42
	2	40
	3	49
	4	50
	5	71
	Megalops larvae	85
Blue whale (*Balaenoptera musculus*)[e]	Suckling	86
	Pubertal	79–83
	Adult	80
Fin whale (*Balaenoptera physalis*)[e]	Suckling	93
	Pubertal	79–83
	Adult	80
Sea urchins (*Strongylocentrotus intermedius*)[f]	1 yr	70
	2 yr	69
	4 yr	68

[a] Krebs (1976).
[b] Sushchenya (1968).
[c] Fuji and Hashizume (1974).
[d] Mootz and Epifanio (1974).
[e] Gaskin (1978), data of Lockyer.
[f] Fuji (1967).

assimilation decreases and rate of defecation may increase. The hydroid *Clava multicornis*, for example, defecates 26% of food eaten when food is scarce and 39% when food is abundant (Paffenhöfer, 1968).

7.223 Age of the Consumer

Younger (or smaller) individuals generally show higher assimilation efficiencies than older specimens (Table 7-2). The exception to the trend (crab in Table 7-2) may be due to a switch from an algal to a carnivorous diet; choice of food thus confounds the age comparison. The need for high-quality food (and relatively greater quantities, cf. Sections 3.5, 7.4) may be an important factor in life histories and habitat use of marine animals. Many coastal fish, for example, use food-rich salt marshes and estuaries as nurseries (Werme, 1981), since this provides the juveniles with a better and more ample supply of food than is available in deeper waters.

7.3 Respiration

7.31 Measurement

Respiration of organisms is usually measured by the rate of oxygen consumed in either the water or the air that surrounds the organism. For species of very small size, a Warburg or Gilson respirometer, or oxygen microelectrodes can be used; in larger specimens the amount of oxygen consumed over a given time interval may be determined by an oxygen electrode or a Winkler titration (Edmondson and Winberg, 1971).

Respiration, food consumption, and energy use are roughly interconvertible using so-called oxycalorific coefficients. Very approximately, one liter O_2 consumed equals approximately 5 kcal of energy. With a respiratory quotient ($RQ = +\Delta CO_2/-\Delta O_2$) of about 1, 1 mg dry weight of food is equivalent to about 5.5 kcal of energy (Parsons et al., 1977). Elliot and Davison (1975) review oxycalorific coefficients of different substrates. Rates of respiration can be converted to carbon by using the relations

$$\text{mg } O_2 \text{ consumed per unit time} \times (12/34) \times RQ = \text{mg C used per unit time,} \quad (7\text{-}4)$$

$$\text{ml } O_2 \text{ consumed per unit time} \times (12/22.4) \times RQ = \text{mg C used per unit time.} \quad (7\text{-}5)$$

The values of *RQ*—determined mainly in mammals—vary from above 0.7 in organisms using fat for energy to 0.8 and 1 for consumers metabolizing proteins and carbohydrates, respectively.

7.32 Factors Affecting Respiration Rates

Respiration is an extremely variable process for two reasons: (a) it is markedly affected by many external factors and (b) its four components (Fig. 7-1) may all vary simultaneously and in different ways. Here we are concerned mainly with external factors such as temperature, size of the organism, level of behavioral activity, and biochemical composition of the source of energy.

7.321 Temperature

Transient increases in temperature prompt increases—often exponential—in respiratory rates (Fig. 7-5). Such increases are a very common effect of temperature on many biochemical and biological processes (Somero and Hochachka, 1976). The actual respiration rate at a given temperature varies with the species involved (Fig. 7-5, top left), but there are many examples of the general increase in respiration with temperature (Teal and Carey, 1967; Smith and Teal, 1973a,b, for example). There are temperatures beyond the range of acclimation of a species (Fig. 7-4, top left and top right), where temperatures are detrimental.

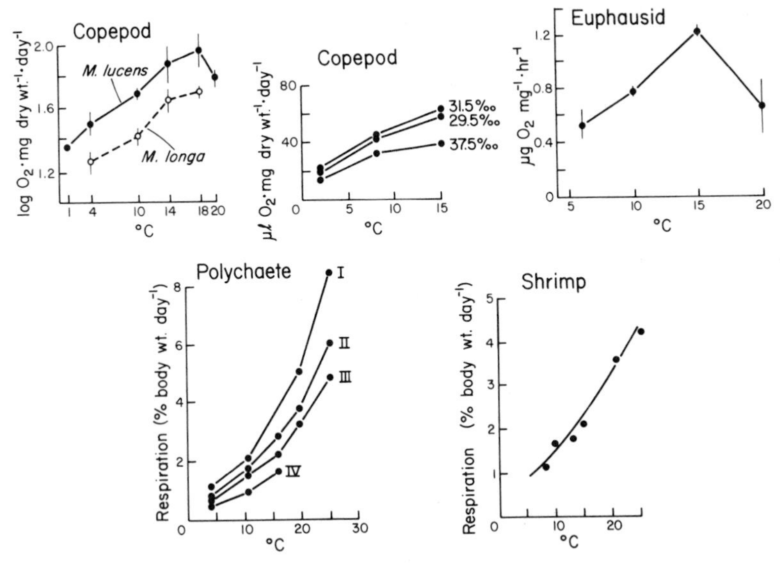

Figure 7-5. Respiration rate per unit biomass relative to temperature. Copepods: *Metridia lucens* and *M. longa*, from Haq (1967), and *Calanus finmarchicus* at three salinities, from Anraku (1964). Euphausiid: *Euphausia pacifica*, from Paranjape (1967). Reprinted by permission of the Canadian Journal of Fisheries and Aquatic Sciences. Polychaete: *Nereis diversicolor*, four size groups, I being the smallest, from Ivleva (1970). Shrimp: *Leander adspersus*, from Ivleva (1970).

The response of a process to temperatures (t) can be represented by

$$Q_{10} = [r_1/r_2]^{[10/(t_1 - t_2)]}, \qquad (7\text{-}6)$$

where r_1 and r_2 are O_2 consumption rates at temperatures t_1 and t_2. The Q_{10} is an indication of the dependence of the rate in question on temperature; it can be interpreted as the increase in reaction rate for a 10° change in temperature. Most biological rates have a Q_{10} between 2 and 3. In addition to the Q_{10} relationship, there are several other empirical formulas that nicely fit actual data (McLaren, 1963), among which is Belehradek's (1935) equation

$$\text{metabolic rate} = a(t + c)^b, \qquad (7\text{-}7)$$

where t is the temperature, and the constants, a, b, c, are obtained empirically. A scale correction (c) shows the temperature at which the rate is 0, b determines the curvature of the line (in other words, the degree of dependence at different temperatures) and a is determined by the units in which the rate is measured.

The marked dependence of respiration on temperature stimulated comparisons of metabolic rates in species from cold and warm environments. Early measurements of metabolic rates of polar and tropical fish (Scholander et al. 1953; Wohlschlag, 1964) led to the conclusion that species of cold latitudes had a 5- to 10-fold larger respiratory rate. This elevated metabolic expenditure was taken to be the cost of maintenance at cold temperatures, and since maintenance costs were thus high, growth rates had to be low. This fitted well with the low rates of growth often recorded in polar species. It turns out that polar fish are rather sensitive to stress, and the procedures involved in the measurements resulted in overestimation of standard respiratory rates* (Holeton, 1974). Subsequent careful experimental work has shown that standard metabolic rates of fish (Holeton, 1974) and invertebrates (Everson, 1977) of polar regions are very low. Other aspects of energy use—growth, activity, reproduction—are also low, so that the total food consumed is relatively low compared to that of similar species in warmer waters (Clarke, 1980).

Other factors besides temperature affect respiration. Since osmoregulation may be an important metabolic expenditure, salinity of the medium may also affect respiration (Fig. 7-5, top middle). Pressure may increase respiration rates if an animal is deeper than its normal depth range. Invertebrates within their normal depth range have respiration rates unaffected by pressure, even if they live at great depths (Smith and Teal, 1973b).

7.322 Body Size

Total respiration rates increase as size of the organism increases (Fig. 7-6, top three graphs), but if respiration is expressed on a weight-specific basis (Fig. 7-6, bottom two graphs), smaller organisms have higher weight-

* See Section 7.323 for definition.

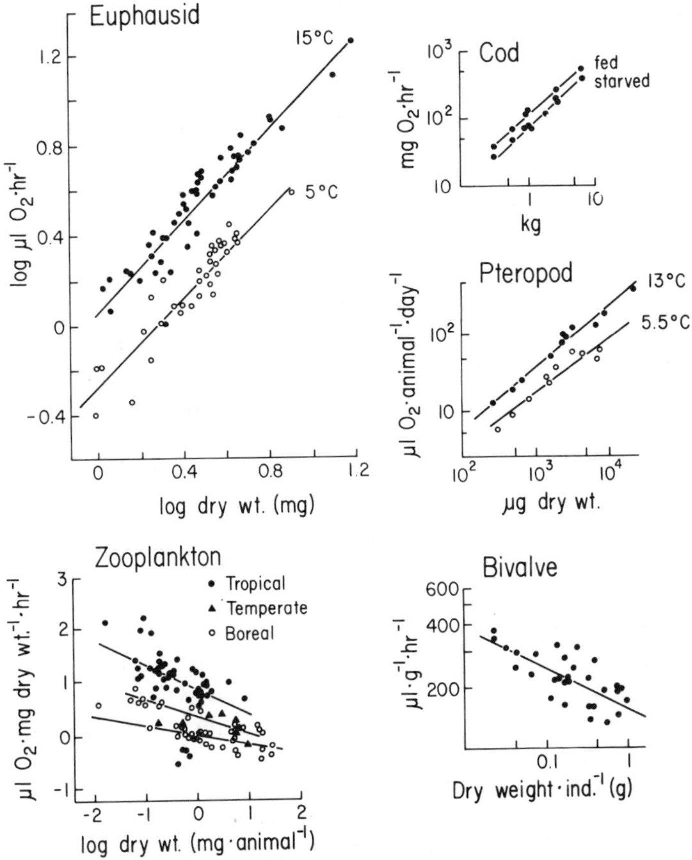

Figure 7-6. Respiration rates relative to size of organisms. The units in which respiration is expressed in the literature are perversely varied, as can be seen in these examples. The top three graphs show respiration rate on a per-animal basis; the bottom two graphs show respiration rate on a per unit weight basis. The euphausiid, *Euphausia pacifica,* at two temperatures, from Paranjape (1967). © Canadian Journal of Fisheries and Aquatic Sciences, reprinted by permission. The pteropod *Clione limacina,* at two temperatures, from Conover and Lalli (1974). Cod (*Gadus morhua*), in recently fed and in starved individuals, from Saunders (1963). Seventy-five species of zooplankton from the Pacific, sorted out by geographic area, from Ikeda (1970). Bivalve data (*Scrobicularia plana*) from Hughes (1970).

specific respiration rates. Although there are exceptions to this rule, it is true for most species. The effect of size on respiration rates applies interspecifically (Fig. 7-6, bottom left) and intraspecifically (Fig. 7-6, bottom right). Smaller organisms may also be more subject to the effect of temperature, as evidenced by the divergence of the regression lines for the three geographic regions of Figure 7-6, bottom left.

The size dependence of respiration (R) for most organisms (Banse 1976) can be described by

$$R = aW^{0.75}, \tag{7-8}$$

where a is a constant and W is the size or weight. The exponent may vary somewhat for different taxonomic groups, but generally the values fall between 0.7 to 0.8. In fish, for example, this relation of metabolism to weight is (Winberg, 1971)

$$\text{rate of } O_2 \text{ consumption} = 0.3\ W^{0.8} \tag{7-9}$$

in ml O_2 hr^{-1} at 20°C.

In animals where growth takes place by stages that differ morphologically and physiologically the usual respiration-size relation may not apply. For example, in lobsters (*Homarus americanus*), late larval stages undergo very rapid changes, and to accommodate this metamorphosis increased energy expenditures are needed (Capuzzo and Lancaster, 1979). The respiration per unit weight in these larvae peaks during the later, larger stages when changes are most demanding. It is not clear whether this is unique to lobster larvae or more general.

7.323 Level of Activity

In warm-blooded species there is a "basal metabolic rate" set by the homeothermic mechanisms. For cold-blooded species there is no such thing, since body temperature depends largely on external heating, and it is usual to refer to the metabolic rate of unfed animals as the "standard metabolic rate." Short-term increases in behavioral activity markedly increase respiration above either basal metabolism or standard rates (Fig. 7-7). In the Pacific sardine (*Sardinops caerulea*) bursts of swimming activity may increase oxygen consumption by two or three times the rate observed when the fish is cruising slowly. Activity generally exceeds standard metabolic levels; in young fish an activity level of at least 1.5 times the standard metabolic level is required to obtain a maintenance ration (cf. Fig. 7-14, left). In king penguins the metabolic activity needed

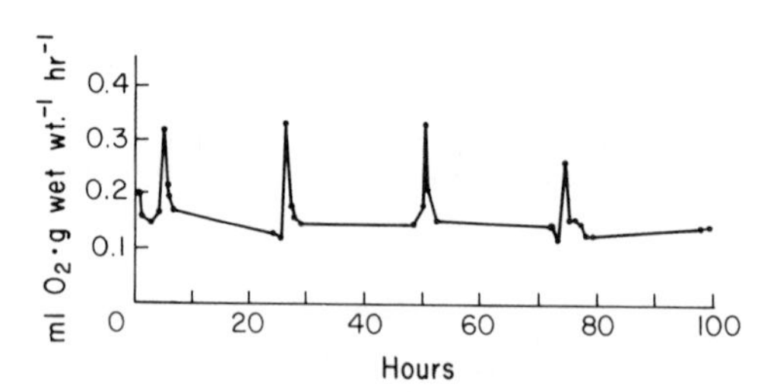

Figure 7-7. Rate of oxygen consumption by a single Pacific sardine (*Sardinops caerulea*) at normal cruising speed (baseline) during and after feeding (sharp peaks) on brine shrimp introduced into the tank. Prey introduced into the tank were quickly consumed, so that feeding lasted only a few minutes. Adapted from Lasker (1970).

to forage for food to feed chicks plus maintain the parent bird is about two to three times the basal metabolic rate (Kooyman et al., 1982).

Respiration rates also increase after feeding, but rates do not immediately return to basal metabolism after feeding is ended; there is lag of several hours, during which metabolic rates remain elevated, a phenomenon referred to as the specific dynamic action of food. Similar differences due to recent feeding can be found in many animals (Fig. 7-6, top right), and in fact, in certain fish that feed very intermittently the effects of a single feeding may last for 2 days (Conover, 1978). Calculations of the energy requirements of a species therefore requires some estimate of the level of activity and specific dynamic action over the time of interest.

Processes related to activity, such as the functional response to food abundance, take place at temperatures within the range of acclimation of the species. At low temperatures, for example, the activity of cold-blooded organisms or microbes is curtailed. Copepods in early spring may thus be unable to respond to algal densities, and do not consume many algae until the water warms up later in the season.

7.324 Source of Energy

The chemical makeup of the substrates used by animals to obtain energy can to some extent also affect the rates of respiration, since the respiratory quotients for different organic compounds vary. The oxidation of carbohydrates supplies the animal about 4 kcal g^{-1}, protein about 4–4.5 kcal g^{-1}, and fat about 9.5 kcal g^{-1}.

It is common to consider the rate at which O_2 is used relative to excretion of nitrogen (as ammonium) as a standardized way to compare metabolic rates of zooplankton (Conover, 1978). Use of protein should yield a O : N of about 8, while oxidation of average marine organic matter should produce a ratio of 17. Values of O:N for various marine invertebrates compiled by Conover (1978) range from 1.3 to 15.6. The low values in this range are probably due to the use of inadequate methods.

There are seasonal patterns of O : N ratios that convey a consistent picture: when zooplankton or bivalves (Bayne, 1973) are rich in lipids due to plentiful food during the spring bloom, respiration of protein is low and therefore O:N is highest. In winter, food is less available and lipid reserves are depleted, and the ratio decreases. Thus, within one species and one season, the relative respiration rate and O : N ratio can be higher or lower depending on the substrate within the organism used to supply energy. The ratio may also depend on the stage of development in invertebrates, if lipids become less important as a primary source of stored energy as the animal grows (Capuzzo and Lancaster, 1979).

Since internal lipid resources can be exhausted, the quality of the diet can effect respiration rate within relatively short periods of time. As we have seen (Section 7.221), detritus is a low-quality food, and the more

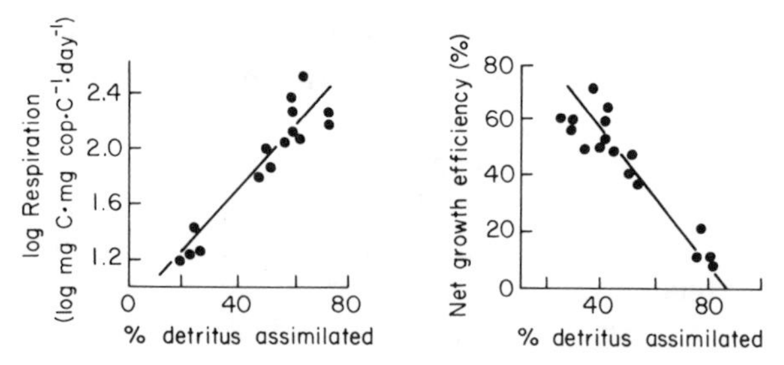

Figure 7-8. Effect of increased assimilation of detritus on the respiration and net growth efficiency of copepods. Adapted from Chervin (1978).

detritus in the diet of the copepod, presumably the lower the lipid and protein content of the ration and the more energy expended to collect a sufficient ration (Fig. 7-8 left). It is not surprising then that respiration increases as more detritus is included in the diet.

7.4 Growth

7.41 Measurement

The growth of individual microbial consumers in the field is not measurable currently with any accuracy (Brock, 1971) but that of larger consumers has been measured extensively. The relative ease of measurement of individual growth of animals is very taxon specific. Increases in dry weight over intervals of time are the most direct measurement, but linear measurements can often be more easily done—for example, the width of carapace in decapods. Growth layers in mollusc shells, polychaete mandibles, whale teeth and earplugs, fish scales, and otoliths have all been used to determine the age and growth of individuals (Rhoads and Lutz, 1980). Such criteria for size and age are related to weight by an equation such as $w = ql^b$, usually in the linearized form

$$\log w = \log q + b \log l, \tag{7-10}$$

where w = weight, l = length (or some other size criterion) and q and b are constants determined empirically; growth rate can be calculated from the slope of the line.

Many quantitative descriptions of growth have been based on equations of von Bertalanffy

$$L_t = L_\infty[1 - e^{-k(t-t_0)}] \tag{7-11}$$

where L_t = length at time t, L_∞ is an asymptotic or maximum length typical of the species, k is a constant that indicates the rate at which L_∞ is approached, and t_0 is the time at which growth starts. With slight modifi-

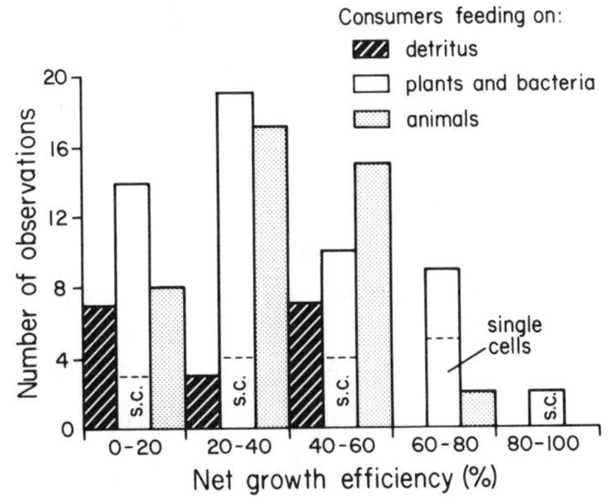

Figure 7-9. Net growth efficiencies for many species of consumers, including insects and other invertebrates, and vertebrates both in the field and in the laboratory. Most entries are for marine organisms, but some freshwater and terrestrial species are included. Single-celled foods are indicated in the lower part of bars in the plant category. Values from many individual papers and compilations by Paffenhöffer (1968), Trevallion (1971), Krebs (1976), Hughes (1970), Jones (1978), Shafir and Field (1980), Bougis (1976).

cation, weights can be used in Eq. (7-11) instead of lengths, for instance, in the Gompertz equation

$$w_t = w_0 \, e^{G(1-e^{-gt})}, \tag{7-12}$$

where w_t and w_0 are the final and initial weights within the period of interest, G is the instantaneous rate of growth at t_0 and when $w = w_0$, and g is the instantaneous rate of decrease of the instantaneous rate of growth. The Gompertz curve allows for sigmoid growth while the von Bertalanffy model is a decelerated monotonic curve. At the inflexion point of the Gompertz curve g equals the instantaneous rate of growth. The differences among the various growth models such as Eqs. (7-11) and (7-12) are slight when compared to the variability of field data. These expressions assume constant weight changes over time, an assumption that may be unrealistic; nevertheless, the fit of such models to actual growth data is quite good (Ricker, 1975). Some examples that show how well theoretical growth curves such as Eq. (7-11) can fit actual data are included in Figure 7-15 (top left).

7.42 Net Growth Efficiency

In comparisons within a species, growth rate is closely related to the net growth efficiency (K_2)* (Calow, 1977; Kiørboe et al., 1981), the proportion of assimilated food that is incorporated into growth,

* Net growth efficiency is often labeled "K_2" in the literature; "K_1" is the gross growth efficiency, equal to (G/I) 100.

$$K_2 = [G/(C - R)] \times 100. \tag{7-13}$$

Net growth efficiencies for many species of cold-blooded consumers range very broadly, but most consumers apportion 20–40% of assimilated energy to growth (Fig. 7-9).

Warm-blooded species are not included in Figure 7-9. Their expenditures for maintenance are larger than for cold-blooded species, so that the proportion of assimilated food that is devoted to growth is much less than in fish and invertebrates. Marine birds and mammals therefore put relatively less assimilated energy into growth than other marine taxa. Cold-blooded species are thus better candidates for mariculture, since the interest there is to grow as much animal protein as possible with as little food as possible.

7.43 Factors Affecting Growth

7.431 Temperature

Growth, like most other physiological processes in animals, is markedly influenced by temperature (Fig. 7-10). Growth rates are low at the lowest temperatures, peak at more intermediate temperatures, and decrease

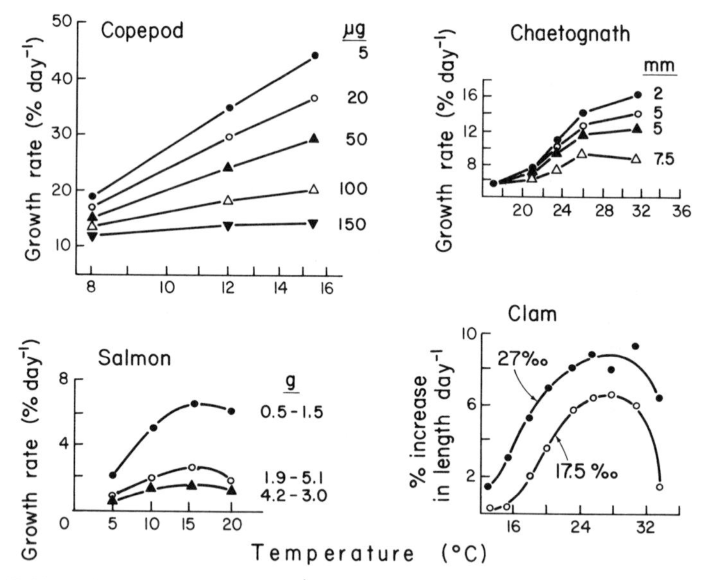

Figure 7-10. Influence of temperature on growth rate. Data for copepod (*Calanus pacificus*) from Vidal (1980), chaetognath (*Sagitta hispida*) from Reeve and Walter (1972), sockeye salmon (*Oncorhynchus nerka*) from Shelbourn et al. (1973), and quahog clam (*Mercenaria mercenaria*) at two salinities from Davis and Calabrese (1964). Data for salmon "include values for fish of different weight. Data from chaetognath includes values for individuals of various lengths.

again if temperatures are so high as to require large energy expenditures for maintenance (Fig. 7-10, salmon and clam examples). Differences in growth rate result in individuals of differing size.

Lower tempertures—within limits—may allow a reallocation of energy from respiration to growth. The colder water temperatures usually found at depth may be responsible for increases in size in taxa found deeper in the oceanic water column (Tseytlin, 1976, for example). There are exceptions to this trend, but the advantage of lower metabolic expenditures due to colder waters at depth may be one reason for the very common occurrence of daily vertical migration of many plankton species (cf. Chapter 14).

Higher temperatures abbreviate the duration of life history stages in cold-blooded organisms. In general, the warmer the water, the faster the development of eggs and juveniles, and the shorter the generation times (Fig. 7-11). The effects of temperature are quantitatively significant: for instance, an increase of 6°C can reduce the generation time of a chaetognath by nearly 2.5 times (Fig. 7-11, top right).

The result of all these temperature effects is that developmental and numerical responses by consumers such as zooplankton depend on tem-

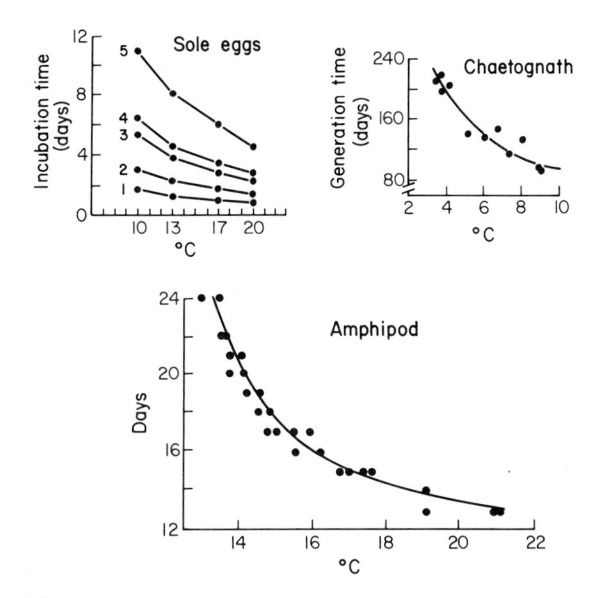

Figure 7-11. Effect of temperature on rate of development of life history stages. Incubation time (time elapsed before hatching) in five different stages in the development of eggs of sole (*Solea solea*), from Fonds (1979). Generation time of chaetognath (*Sagitta elegans*) from Sameoto (1971). © Canadian Journal of Fisheries and Aquatic Sciences, reprinted by permission. Intervals between successive molts in adult female amphipods (*Gammarus zaddachi*) from Kinne (1970).

perature at times of temperature changes such as during spring. There are of course limits; if temperatures exceed the range to which a species is adapted, other species better suited to warm temperatures take over from the more cold-adapted species.

7.432 Abundance of Appropriate Food

Food type seems to have a far less clear effect on net growth efficiency (K_2) than it does on assimilation efficiency. There are no obvious differences in K_2 values for species feeding on animals or plants (Fig. 7-9). If food provides enough of a ration, the biochemical transformations of assimilated compounds are the same regardless of the origin of the compounds. A detrital diet may not provide sufficient nutrition, as suggested by the somewhat lower K_2 values. Where the growth efficiency of copepods was measured in experiments where the diet contained increasing amounts of detritus, K_2 was reduced (Fig. 7-8, right). Other studies with zooplankton confirm the reduction of growth under detrital diets as well as slower development and reduced survival (Paffenhöfer and Knowles, 1979).

We have already discussed some aspects of the effect of ration size on growth under the guise of the developmental response to prey density (Section 5.4). Increases in rate of food consumption result in increased growth rates of consumers (Fig. 7-12,), with a deceleration at higher rations. This is a measure of the cost of maintenance of larger body sizes.

Growth efficiency is negatively related to ration in laboratory experiments. Within any one size class of young salmon, the growth efficiency is

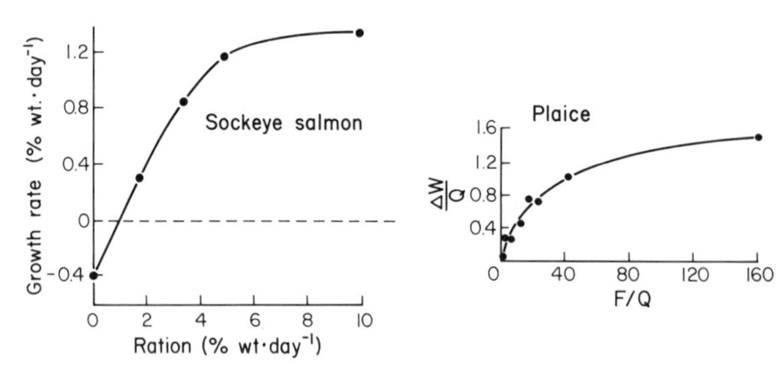

Figure 7-12. Growth rate in relation to food consumed. Sockeye salmon (*Oncorhynchus nerka*). Adapted from Brett et al. (1969). Reprinted by permission of the Canadian Journal of Fisheries and Aquatic Sciences. Young plaice (*Pleuronectes platessa*) feeding on siphons of the bivalve *Tellina tenuis*. Growth (ΔW) and food consumption are both normalized to Q, the standard metabolic rate. Adapted from Edwards et al. (1970).

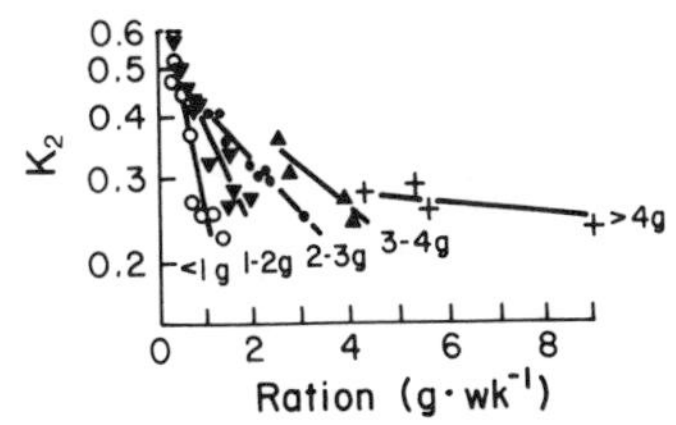

Figure 7-13. Net growth efficiency (K_2) of chum salmon (*Oncorhynchus keta*) in the laboratory at different weekly food rations. The straight lines refer to young salmon in 1-g size classes. These "K_2-lines" have been obtained only for the right side of the curves; there must be extensions to the left that have a positive slope, since K_2 is 0 when ration available reaches a certain minimum threshold. Adapted from Kerr (1971). © Canadian Journal of Fisheries and Aquatic Sciences, reprinted by permission.

lower at higher rations (Fig. 7-13). Fish that have more food items available tend to be more active (Kerr, 1971).* Since the energetic cost of swimming in fish varies approximately as the square of the velocity (Fry 1957), the respiratory requirements increase more than proportionately to level of activity. It is not clear, however, whether this phenomenon of lower growth efficiency at higher prey densities occurs in the field.

As consumers increase in size, the growth to consumption ratio is lower and the effect of ration on growth efficiency is considerably less (i.e., the slopes of the so-called "K_2-lines" flatten out, Fig. 7-13). This is probably due to the greater total metabolic costs of large individuals.

It should be evident that curves such as those on Figure 7-12, based on single food types in the laboratory, do not convey the complete picture of how growth is regulated. Size (Section 6.31) and spatial distribution (Section 13.4) as well as abundance of food particles play some role in setting growth rates.

No one has documented all the variables affecting growth efficiency, but computer simulation studies show that based on what is known of fish feeding, growth—expressed as K_2—is best when relatively large prey are available, even if they are rare (curve for 100 g prey, Fig. 7-14, left). Small prey can support growth of the predator if the smaller prey are relatively abundant (compare the curves for prey of 0.1 g, Fig. 7-14, left). A diet of small and rare prey demands increased metabolic expenditures because of the high level of activity needed to gather enough prey. In such circumstances energy is therefore devoted to maintaining activity rather than to growth, and growth of the predator is therefore curtailed.

* An extreme of this tendency is the "feeding frenzy" that takes over many predators when exposed to a very dense aggregation of prey.

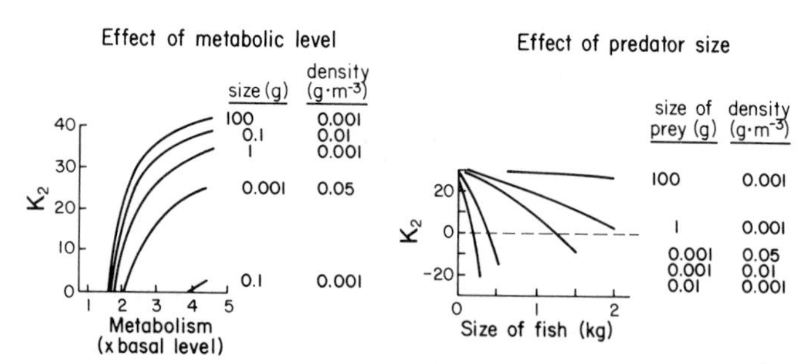

Figure 7-14. Results of computer simulation studies on the effects of metabolic level and size of brook trout (*Salvelinus fontinalis*) on net growth efficiency (K_2). Left: K_2 as function of metabolic level for a 1-kg trout for different sizes and abundance of prey. Right: Changes in K_2 as a predator grows, for different combinations of size and abundance of prey. Metabolic level is assumed to be 2.5 × basal metabolic rate. Adapted from Kerr (1971a). © Canadian Journal of Fisheries and Aquatic Sciences, reprinted by permission.

7.423 Size of Consumer

Growth of individuals decelerates as individuals age (Fig. 7-15). In both vertebrates and invertebrates there are physiological mechanisms that reduce growth per unit weight as time passes. Other physiological controls make for differences among sexes (Fig. 7-15, bottom). The effects of size have been thought to apply not just to stages within the life span of a species but interspecifically as well, since in general, small species grow faster than larger species (Fenchel, 1974). Growth efficiency, however, does not vary with size in comparisons among species (Banse, 1979), although it is not clear why this is so.

In many crustaceans, as animals reach adulthood, the yearly growth may be considerable but is converted to reproductive products and cast molts (Krebs 1978), so the net annual growth for the individual is close to nil. In many animals the reduction in growth rate with age within a species may come about by the increased food energy required to supply maintenance expenditures by larger individuals (Section 7.322).

7.424 Interactions Among Variables

In Chapter 2 we discussed the interacting variables that determine rates of primary production. For consumers there are also important interactions among the variables discussed in this chapter.

Temperature and size. Smaller specimens of a species generally grow proportionately faster than larger individuals as temperature increases. This can be demonstrated in serpulid polychaetes (Dixon,

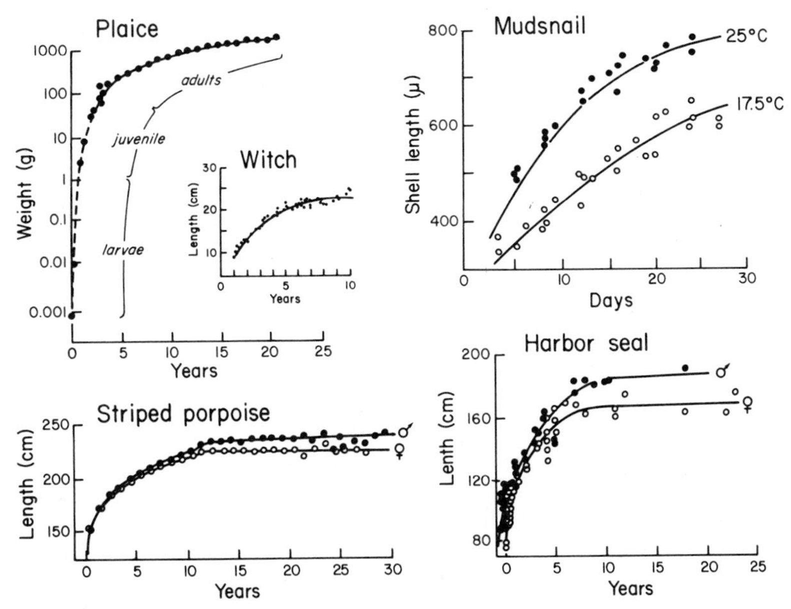

Figure 7-15. Growth of consumers over time. For plaice (*Pleuronectes platessa*) the curve fit by von Bertalanffy equation is shown as a complete line; smaller specimens do not fit equation. Adapted from Cushing (1975) after Beverton and Holt. Witch (*Glyptocephalus cynoglossus*) data for female fish only, fitted by a von Bertalanffy equation by A. B. Bowers. Mud snail (*Nassarius obsoletus*) from Scheltema (1967). Striped porpoise (*Stenella coeruleoalba*) from Miyazaki (1977). Harbor seal (*Phoca kurilensis*) from Naito and Nishiwaki (1972).

1976), and can be seen in the growth rates of mud snails of various sizes exposed to different temperatures (Fig. 7-15, top right).

Temperature and ration size. At low temperatures growth rates are low (Fig. 7-16). As temperatures increase so do growth rates, but the increase is greatest where food is abundant; loss of weight may occur where food is scarce (Fig. 7-16, left). The increase in growth at higher temperatures reflects the general increase in metabolic activity as, for example, the increased rate at which digestion takes place in warmer temperatures (Fig. 7-16, right, inset). The combined effect of ration size and temperature produces a peak growth rate which shifts to the right in peak growth at each ration. This shift is an expression of the balance between the increased growth due to higher physiological activity of higher temperatures and the greater amounts of food needed at higher temperatures. At the highest temperatures, the resulting increased respiration reduces growth.

Consumer size in relation to prey size and density. Kerr (1971a) followed changes in growth efficiency in growing fish in his simulation study (Fig. 7-14, right). For the smallest fish, growth is possible

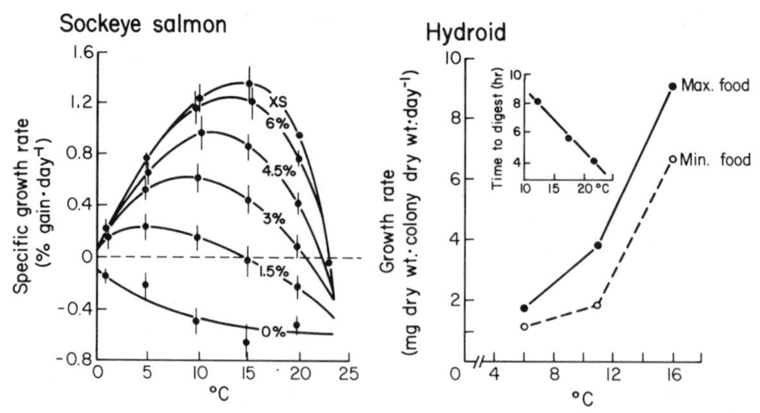

Figure 7-16. Combined effects of temperature and food supply on growth rate. Left: Young sockeye (*Oncorhynchus nerka*) grown in the laboratory at rations of 0–6% of body weight per day and at excess (xs) ration. Means ± standard deviation. Adapted from Brett et al. (1969). © Canadian Journal of Fisheries and Aquatic Sciences, reprinted by permission. Right: Growth rate of the colonial hydroid *Clava multicornis* in the laboratory at various temperatures and a maximum level of consumption and at the lowest level that allowed growth. Adapted from Paffenhöfer (1968) and Kinne and Paffenhöffer (1965).

(K_2 is positive) even on diets of small and rare prey. As growth takes place, the fish move toward the right, and the "K_2 lines" slope downward. There is a point where a fish needs to seek either larger or more abundant prey to maintain a positive K_2. The largest fish can only sustain growth on large prey. This is the metabolic basis behind the pervasive tendency for size selection in the feeding behavior by predators (Chapter 6).

Temperature, ration size, and size of consumer. Growth rates of *Calanus pacificus* increase with temperature and decrease as body size increases (Fig. 7-10, left). Note, however, that the slopes of the lines decrease with larger copepods, so that the growth response of large copepods to temperature is much reduced. This may be the result of the amount of food required to sustain maximum growth increases with temperature and size of copepods. Food densities capable of sustaining maximum growth for small copepods are therefore insufficient to support maximum growth of large copepods. It may be, therefore, that at low food concentrations the growth rate of small copepods is not determined by food but by temperature, while in large copepods food supply determines growth rate (Vidal, 1980). In fact, larger copepods may attain their largest size at low temperatures if enough food is available. This very different response of large and small copepods to temperature gradients may in part explain why there have been reports of reduced body sizes as temperature increased (McLaren, 1963; Sameoto, 1971), while other studies show no such relation (cf. review in Vidal, 1980).

Oligotrophic tropical waters have mainly small species of zooplankton while rich temperate waters hold communities of larger zooplankton. It may be that small species predominate in depauperate environments, where they can grow despite low food supplies, but only do so at high temperatures. Larger species may be excluded from oligotrophic warmer water and may do best where food supply is high, even if temperatures are low (Vidal, 1980).

7.5 Production

Production rate of a population is the number of individuals present times the net growth in these individuals. In microorganisms that divide by fission, production is just the increase in number of cells, since growth of individual cells is difficult to measure. More complicated modes of growth and reproduction add further components to production, as pointed out in Section 7.1. In crustaceans, for example, net production is the sum of growth, reproduction, and molting. The measurement of production must therefore include the relevant components, and the factors that affect production are those that apply to its component parts.

7.51 Measurement

There are many methods available to estimate production by animals, each devised for certain taxa and conditions. Reviews of procedures are available in Winberg (1971), Edmondson and Winberg (1971), Mann (1969), and Greze (1978), on which the discussion below is based.

Some of the methods to estimate production require estimates of biomass. This is a very important property of populations but the methods to measure biomass are varied. Reviews of methods to estimate biomass of microbes can be found in Hobbie et al. (1977) and in Paul and Johnson (1977), and for biomass of invertebrates and fish in Southwood (1978), Jones (1964), and Holme and McIntyre (1971).

Neither we nor any of the above references deal with production by microbial consumers. Rates of production by microbial heterotrophs have been hard to measure in nature, and methods are still controversial. Fuhrman et al. (1980, 1980a), Karl et al. (1981, 1981a,b), and Fuhrman and Azam (1982) provide some approaches being investigated. Ducklow (1983) cites other references on measurement of bacterial production. The approach is to measure uptake of a labeled organic compound, such as thymidine, and use this plus biomass estimates to calculate production. One problem with this approach is that the background concentration of thymidine in the water is usually not known. Another method is to calculate the growth rate of the population from the frequency of dividing cells (Hagstrom et al. 1979). Not enough comparisons among methods have been done.

7.512 Methods for Animal Production

7.5121 Cohort Methods

Where distinct cohorts are evident, in any given time interval $t_0 - t$, production (P_{t-t_0}) is the increase in biomass due to growth ($B_t - B_{t_0}$) plus the biomass lost due to predation and other mortality (B_m)

$$P_{t-t_0} = B_t - B_{t_0} + B_m. \tag{7-14}$$

This general approach was first used by P. Boysen-Jensen and involves measurement of biomass at different times. B_m is calculated as the sum of the products of the number of individuals at t_0 by the weight of the individuals at t. This provides the biomass of animals not present at the time of measurement.

K. R. Allen and V. N. Greze independently improved the Boysen-Jensen method, and calculated production of a cohort as

$$P_{t-t_0} = (n_t - n_{t_0})[(w_t - w_0)/2], \tag{7-15}$$

where n and w are the numbers and average weights at the beginning (t_0) and end (t) of the interval of measurement.

The time over which production is measured needs to be short relative to changes in the life history of the species, since both growth and mortality can be exponential and the use of linear interpolations such as ($n_t - n_0$) or ($w_t - w_0$) would not adequately assess rates over relatively long periods of time. Parsons et al. (1977) recommend that measurement intervals should be about 10% of the generation time of the species.

7.5122 Cumulative Growth Methods

Where there are no clear cohorts and reproduction is continuous or repeated, production can be calculated based on the growth curve of the species. The basic idea is to obtain the product of number of individuals and average growth increment over a certain interval, and then sum over the entire life history (up to the kth age or size classes):

$$P = \sum_{i=1}^{k} n_i g_i, \tag{7-16}$$

where g_i = the mean increase in weight during the interval for each age class. Values of g_i should be obtained by direct measurement. Values of g_i can be calculated from studies of net growth efficiency (K_2) and respiration (R) as

$$g_i = R[K_2/(C - K_2)], \quad (7\text{-}17)$$

where $R = aw^b$, but this is less desirable. This latter procedure based on efficiencies should be only used as a check on production rates determined by other methods, since R and K_2 are rarely constant over all the age groups. The calculation of growth is more realistic if the effects of temperature are included by using curves such as those in Figure 7-9. This method may overestimate production because individuals that die between sampling periods are included in the estimate of production. The cumulative growth method is more frequently applicable than cohort methods but can be quite complicated in certain species and generally only provides a rough estimate of production.

7.5123 Life Table Methods

Elster (1954), Edmondson (1960), Hall (1964), and others have used life table methods to calculate the growth of populations of freshwater zooplankton based on number of eggs and their development time relative to ambient temperature. The calculation of number of eggs produced by the population (cf. review by Paloheimo, 1974) and their rate of development allow the estimation of N_t, the potential increase in numbers of the population within a period of time $(t - t_0)$. The difference between the potential (N_t) and actual $(N = N_t - N_0)$ increase in the population is the mortality (N_m), which can be calculated as

$$N_m = N_t - N_a = [N_{eggs} \times 1/\text{development time}] - N_a. \quad (7\text{-}18)$$

Multiplication of N_t by mean weight per individual provides estimates of total production. Heinle (1969) applied this approach to measure production in copepods, including naupliar and copepodid stages.

7.5124 Energy Budget Approach

Another way to measure production is to measure consumption, respiration, egestion, and excretion, and from this calculate production using Eq. (7-1). This procedure is cumbersome, time-consuming, and subject to large errors. The inherent variability is large because the error of the estimate of production is the cumulative error associated with measurement of each of the components. There are additional ambiguities, such as whether to enter reproductive products separately or as part of growth. In most cases the growth term will have included biomass of reproductive products as the eggs and sperm developed within the adults. As with all the other methods, many assumptions need to be made as to conversion coefficients and the environmental conditions, for example, temperature, during the time interval of interest. Identification of the diets of consumers in the field are seldom well known, and many assumptions usually have to be made prior to the calculation of consumption.

A rough way of estimating production is to measure O_2 consumption or CO_2 production—in other words, to measure respiration—and then calculate production using an empirically derived relationship to production, such as that developed by Engelmann (1961)

$$\log R = 0.62 + 0.86 \log P, \tag{7-19}$$

or by McNeill and Lawton (1970)

$$\log P = 0.8262 \log R + 0.0948, \tag{7-20}$$

where R and P are respiration and production.

This was originally done for terrestrial species but fits marine animals tolerably well (Hughes, 1970). The application of log–log relationships such as Eqs. (7-19) and (7-20) needs to be interpreted with care, since such straight lines drawn with logarithmic scales may hide large absolute deviations at the upper end of the scales. The usefulness of respiration as a predictor of production is in part due to respiration consuming a very significant proportion (70–80%, cf. Section 7.6) of the consumed food. A relation between production and respiration thus seems reasonable and may be useful as a measure of total consumer activity. This approach has been applied to faunas in sediments of very deep oceans, where no other method to measure production can be applied (Smith and Teal, 1973a,b).

7.5125 Comparisons of Methods to Measure Secondary Production

Each of the various methods discussed above is subject to large errors, and it is necessary to make many assumptions when using all the methods. In fact, consumer production measurements are some of the most variable and error-prone of the variables in ecology. Other approaches to measure consumer production have been suggested, for example, use of ^{14}C-labeled food (Chmyr, 1967), but these also require many assumptions and ancillary measurements. Methods to measure primary production apply to the whole assemblage of species engaged in photosynthesis, in contrast to methods to estimate secondary production, which generally measure production of a specific population.* This requires that ancillary data on numbers, size specific growth rates, etc. be gathered. This adds to the difficulties, since the reliability of the production measurement, in addition to the problems inherent in each procedure, also depends to a large extent on the quality of the additional data sets needed. In spite of all these caveats, Greze (1978) after reviewing a few studies where production was measured on the same population by two or three different

* One exception to this is that measurement of O_2 consumption or CO_2 production can provide assessment of respiration by the assemblage of consumer populations present. This approach has been used to measure "community respiration" in marine benthic communities, as will be seen in Chapter 10.

methods, concluded that the results were roughly comparable. When choosing a method it is necessary to consider carefully all the methods suitable for the population to be studied, and then to carefully design a research program based on the idiosyncrasies of the specific case.

7.52 Biomass and Production of Consumers

7.521 Planktonic Consumers

The epifluorescence microscopy method (Hobbie et al., 1977) plus the use of electron microscopy have provided estimates of bacterial number and biomass for seawater and sediments (Meyer-Reil, 1982). The range of abundance is relatively narrow (10^4 to 10^6 cells ml^{-1} in seawater, equivalent to 1-200 μg C $liter^{-1}$). Most cells (80–90%) are probably free bacterioplankton, while the rest are attached to particles. Most of these bacteria are taken to be heterotrophic. Other methods to estimate bacterial biomass involve measuring the abundance of compounds specific to bacterial cells, such as certain lipopolysaccharides (Watson et al., 1977) or muramic acid (Moriarty, 1977).

As mentioned earlier, methods to measure bacterial production are not well established, and the few measurements available are still controversial. Bacterial production may be 0–500 μg C $liter^{-1}$ day^{-1} in coastal waters, and 0.5–05 μg C $liter^{-1}$ day^{-1} in oceanic waters (Ducklow, 1983). The abundance of bacteria in seawater is about $0–10 \times 10^9$ cells $liter^{-1}$ (Ducklow, 1983). Bacterial production may amount to about 5 to 50% (Es and Meyer-Reil, 1982; Fuhrman and Azam, 1982) or 0.75% (Ducklow, 1983) of primary production by phytoplankton. In other marine environments dominated by macroalgae, bacterial production may amount to 6–33% of macroalgal production (Newell et al., 1981; Stuart et al., 1982).

Production by the larger planktonic consumers varies tremendously. Some rough estimates of the biomass and production of microzooplankton are provided in Table 7-3. Note that production is higher in the nutrient-richer coastal waters than in more oligotrophic deeper waters. To make comparisons it is convenient to take the ratio of production to the average biomass; this is often not a reliable measure, since, for example, there are large seasonal variations in biomass. Nonetheless, comparisons of *P/B* are instructive, since this ratio can be thought of as the turnover rate of these populations (Fig. 7-17).

The *P/B* of the few available—and controversial—measurements for bacteria indicate a turnover of 100–400 times a year (Fig. 7-17, top). This is faster than most turnover rates for zooplankton, including copepods, cladocerans, and euphausiids. Zooplankton have a wide range of *P/B*, but the modal turnover rate is about 10-20 times a year. Unfortunately there are no measurements for the planktonic ciliates and flagellates; their *P/B* should lie closer to the bacterial values than to the larger zooplankton.

Table 7-3. Biomass and Production of Zooplankton (Excluding Salps) and Fish from Various Marine Regions[a]

	Depth[b] (m)	Biomass (g dry wt. m^{-3})	Production (g C m^{-2} yr^{-1})
Zooplankton			
Inshore waters	1–30	122	15.3
Continental shelf	30	25	6.4
Shelf break	200	108	5.5
Open Sea	200	20	5.7
Benthos		(g C m^{-2})	
Estuaries	0–17	5.3–17	5.3–17
Coastal seas	18–80	1.7–4.8	0.7–12
Continental shelf	0–180	23	2.6
Continental slope	180–730	18	2.4
Deep Sea	More than 3,000	0.02	—
Pelagic Fish			
Continental shelf	0–180	2.6	0.3
Continental slope	100–730	10.6	1.3
Demersal Fish			
Continental shelf	0–180	8.6	0.3
Continental shelf	180–730	4	0.2

[a] Adapted from Mann (1982), Mills (1980), and Whittle (1977). Note the difference in units in biomass of zooplankton and other entries. Fish production calculated roughly from catch statistics. Demersal fish live and feed near the sea floor.
[b] Depth to which sampling took place.

Measurements for pelagic fish are few, in spite of the extensive fishery work. Production by pelagic fish species is relatively low (Table 7-3) and *P*/*B* values are small (Fig. 7-17, top).

7.522 Benthic Consumers

Bacteria are more abundant in sediments (10^{10}–10^{11} cells ml^{-1}) than in the water column (Meyer-Reil, in press), but there are no direct measurements of total bacterial production in sediments, although it appears that the biomass of sediment bacteria turn over more slowly (Fig. 7-17 bottom) than that of planktonic bacteria. Sediment bacteria may be relatively inactive. This needs further study. Other microbes such as fungi and yeasts also need study.

Consumers larger than bacteria (3–4 to 45 μm) are called microbenthos; these are mainly ciliates and some flagellates. Organisms between 45 and 500 μm in diameter (nematodes, foraminifera, harpacticoid copepods, some ostracods, and polychaetes) are called meiofauna, while the macro-

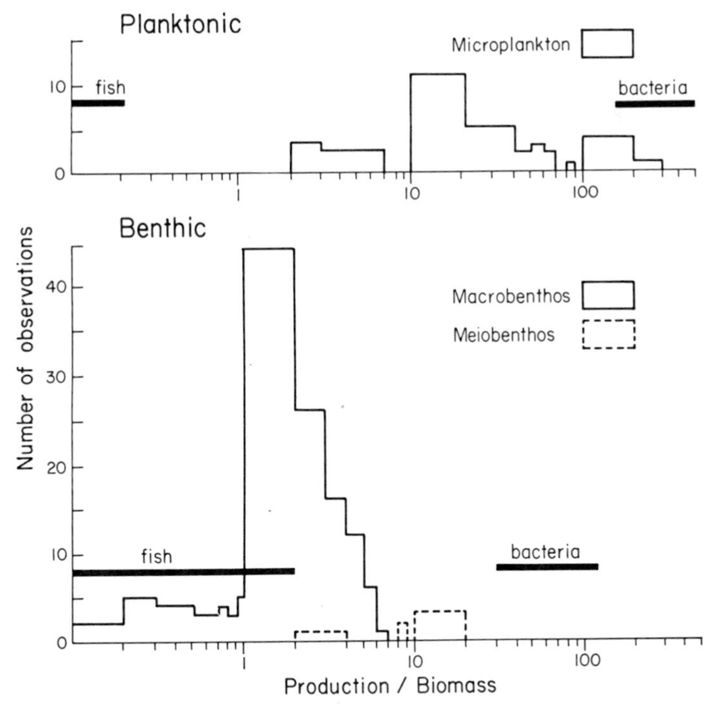

Figure 7-17. Production to biomass ratio for planktonic and benthic species. A few freshwater species included; their *P*/*B* did not differ from those of related marine species. The ranges of the few available estimates for bacteria and fish are included as bars to show how they compare with invertebrates. Data compiled from Allan et al. (1976), Ankar (1977), Banse and Mosher (1980), Berkes (1977), Burke and Mann (1974), Cederwell (1977), Conover (1974), Fleeger and Palmer (1982), Feller (1977), Gerlach (1978), Hibbert (1977), Josefson (1982), Peer (1970), Siegismund (1982), Mullin (1969), Lasker et al. (1970), Warwick and Price (1975, 1979), Sanders (1956), Mills (1980), and Warwick (1980).

fauna are larger than 500 μm—or 300 μm if we wish to include more of the immature stages. Macrofauna are very diverse, with amphipods, various annelids, and bivalves being typical representatives. These size categories are to an extent arbitrary, but Schwinghamer (1981) shows that the biomass of marine benthic organisms peaks at size ranges of 0.5–2 μm (bacteria), 32–250 μm (meiofauna), and 8–32 μm (macrofauna). Thus, the classification does have some real meaning.

As in the plankton, benthic biomass and production are very variable (Table 7-3). Biomass and production are higher in shallower waters and decrease at great depth. The few available *P* to *B* ratios of meiofauna are considerably larger than those for the larger macrofauna (Fig. 7-17 bottom); the meiofauna turn over about 10 times per year, while the modal macrobenthic turnover rate is one to two times a year. When data become available, the microbenthos and bacteria *P/B* will probably lie to the right of both macrobenthos and meiobenthos. Vernberg and Coull (1974), for example, estimate that the turnover rates of ciliates in shallow marine sediments is between one and four orders of magnitude greater than that of the meiofauna and macrofauna. Some *P/B* ratios of fish associated with the benthos are included in Figure 7-17, bottom; these species turnover rather slowly, perhaps once a year, and production is smaller (Table 7-3) than that of smaller-bodied consumers. The larger *P/B* ratios for fish included in Figure 7-17 are from rich estuaries; fishes of deeper water have very low *P* to *B* ratios.

The faster turnover—the result of growth and mortality—of smaller organisms (Fig. 7-19, left) means that although the biomass of small-sized species may be smaller than that of larger species, the higher specific production—what we referred to above as the *P/B*—of smaller species often makes them proportionately more important producers than larger species (Fenchel, 1969; Gerlach, 1978; Vernberg and Coull, 1974).

7.53 Factors Affecting Production

Anything that affects growth, reproduction, metabolism, or molting affects production. Temperature, size of individuals, and food quality and quantity are therefore important since they affect growth as we have seen earlier in this chapter. For example, weight-specific excretion increases with temperature (Mayzaud and Dallot, 1973; Ikeda, 1974), so that it is not surprising to find also that production depends on temperature (Fig. 7-18).

Respiration rates are size dependent, and therefore production, as mentioned above, is inversely related to size of the consumer [Fig. 7-19, left, and Warwick (1980)]. The effect of size on production holds both for comparisons of species of differing adult size and for comparisons of smaller and larger or younger and older specimens within one species (Fig. 7-19, right). There is a general pattern of higher specific production with smaller size*; there are also slightly different relations for taxonomic groups, whose slopes are somewhat offset from each other. Sheldon et al. (1972) review production rates of many species, using time to double numbers as an index of production. They find that microbes and algae of 1–100 μm in diameter double in less than 10 to about 100 hr. Invertebrates (primarily zooplankton 100–10,000 μm in average diameter) double numbers in less than 100 to 1,000 hr. Fish, 10^4 to 10^6 μm, take the longest time to double their number, about 10^3 to 10^4 hr.

---

* The production-size relationship is good enough to have encouraged Sheldon and Kerr (1976), based on the dimensions and food availability in Loch Ness, plus more speculative guesses as to size, to predict that there may be 10–20 monsters present in the Loch, assuming they are top predators,.

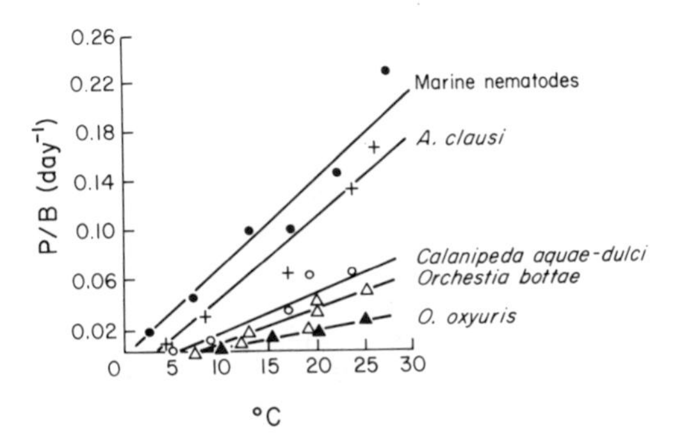

Figure 7-18. Relation of specific production (production/biomass) to temperature in various marine species, including nematodes, two copepods, and amphipods. The ratio *P*/*B* is used here to normalize production in the case of species with different biomass. Adapted from Zaika and Makarova (1979).

Both the quantity and quality of foods affect production. In terms of quality, we have seen earlier that nitrogen content of food is important to herbivores; in experiments with a detritus-feeding polychaete the production of biomass increased as the percentage of nitrogen in detritus increases (Fig. 7-20). In terms of quantity of food, in the same experiments, production was clearly higher where the amount of detritus is larger (Fig. 7-20). Notice, however, that when enough nitrogen was available, there was a threshold beyond which production was no longer stimulated. Some other essential constituent of detritus, perhaps energy content, limited production by the polychaetes (Tenore et al., 1982).

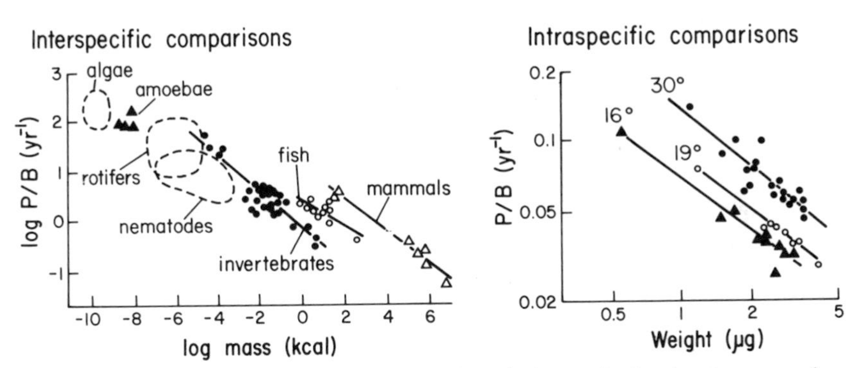

Figure 7-19. Specific production per year in relation to body size (expressed as energy content or weight—1 kcal is roughly 1 g wet weight). Left: Comparisons for various taxonomic groups: Fast-growing single-celled algae, soil amoebae, rotifers, nematodes, other invertebrates, fishes, and mammals. Dashed lines surround inferred ranges. Adapted from Banse and Mosher (1980). © Ecological Society of America, reprinted by permission. Right: Intraspecific comparison for the copepod *Acartia clausi*, Azov Sea, at three temperatures. This copepod strain is adapted to rather warm temperatures. After Kinne (1970), modified from V. E. Zaika and L. M. Malowitzkaja.

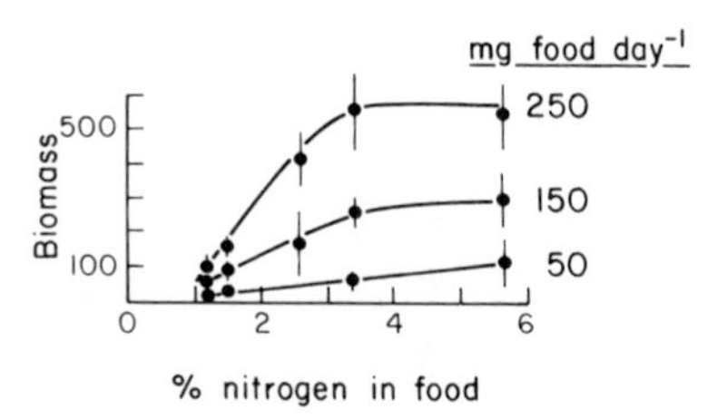

Figure 7-20. Biomass of the polychaete worm *Capitella capitata* (mg ash-free dry weight/0.1 m^2 tray) when fed on detritus from a series of foods of different percent nitrogen (*Spartina alterniflora, Zostera marina*, mixed cereal, *Fucus* sp., and *Gracilaria* sp., respectively, in the graph). The diets were offered at three dosages, shown in the right of the graph. Adapted from Tenore (1977).

7.54 Production Efficiency

The production efficiency of a population ($[P/A]100$) is the proportion of assimilated energy devoted to production. Humphreys (1979) surveyed many studies and concluded that birds and mammals had production efficiencies of 1 to 3%. In fish and invertebrates the production efficiencies were 10–25%. This difference is probably due to the higher metabolic costs of warm-bloodedness mentioned above, and is evident in a graph of production versus respiration (Fig. 7-21). For any one level of respiration, warm-blooded birds and mammals are less productive. The magnitude of production or respiration did not affect production efficiency in any of the taxa, since the slopes of all regressions were equal to one. There was also no effect on size of the organism and only meager evidence that food quality affected production efficiency.

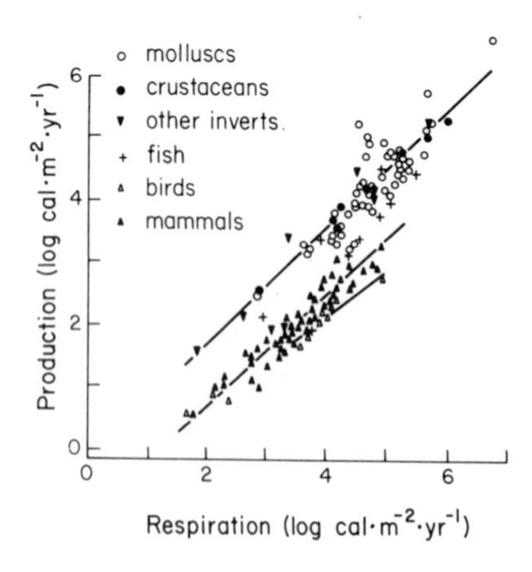

Figure 7-21. Relation between production and respiration in some major taxonomic groups. Species from marine, freshwater, and terrestrial environments are included. The regression lines for non-insect invertebrates, mammals and birds are included. Adapted from Humphreys (1979).

7.6 Energy Budgets for Populations

There are many studies where the various fates of matter consumed by a population have been identified and the fluxes measured. Specific values for each component vary enormously with local conditions, and comparisons among species are sometimes difficult because not all components are separated in each study. Rather than list individual measurements, it is instructive to look at the frequency distributions of measurements of the percentage of assimilated energy invested in production, respiration, reproduction, egestion and molting.

Producers use light energy, consumers use energy stored in organic matter; respiration produces energy from stored products. The common currency in all these transformations is energy, so that it has become customary to measure them in energy units.*

Different species allot their assimilated energy in widely different ways, but there are some recognizable patterns (Fig. 7-22). Most consumers convert about 0–30% of their assimilated energy into production and about 40–80% for respiration. Investment in reproduction is low, generally considerably lower than 10% of assimilation. In organisms with ecdysis, molting also generally consumes less than 10% of assimilation. Egestion is hard to interpret since both feces and urine are included and fecal material is not assimilated, while urine is the waste product of assimilated food; in any case, 10–20% of assimilation is the usual extent of this modified "excreted" loss. For respiration, the standard maintenance levels are usually comparable to the losses due to specific dynamic action, while active metabolism may account for another third of total respiration. These proportions, however, are extremely variable.

Of course, the processing of energy by populations changes over time, both through seasonal patterns and through aging. Consumption by most species has a strong seasonal signature (Fig. 7-23), increasing in the warmer months, as we expect from our prior discussions of respiration, assimilation, growth, and from seasonal changes in food abundance. Seasonal fluctuations of growth and standing crop may not necessarily be correlated over time (Fig. 7-23), since mortality and migration also play a role in determing standing crop.

The other effect of time is through the aging of individuals comprising the population. As individuals grow through the stanzas of their life history their metabolic abilities change, as discussed above. The result is that the ratio of assimilation/consumption decreases as animals mature, and

* The units generally used are calories. The content of 1 g of phytoplankton carbon is about 15.8 kcal, 1 g of dry weight phytoplankton is about 5.3 kcal. Consumption of 1 ml of O_2 by an animal provides about 3.4–5 cal; taking the latter value, and an *R.Q.* of 1, 1 mg of dry food is about 5.5 cal. Energy content is also expressed in joules (J), and 1 cal = 4.19 J.

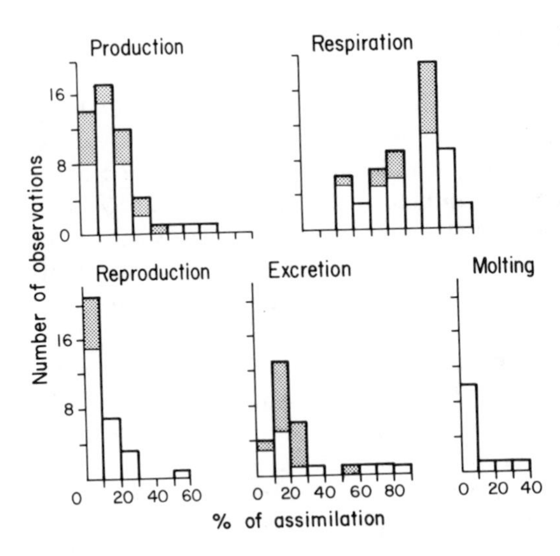

Figure 7-22. Partition of assimilated food into the various parts of the energy budget of populations of marine invertebrates (open part of bars) and fish (stippled part of bars). Many of the measurements of respiration do not include values for active expenditures. The excretion values for fish include both feces and urine. From data compiled by Conover (1978), Sushchenya (1970), Parsons et al. (1977), and Baird and Milne (1981).

relatively more of the assimilation goes to respiration at the expense of production (Fig. 7-24). In addition to respiration, excretion also increases at the expense of production. Larger animals also defecate larger proportions of consumption. For example, a 35-mg mussel (*Mytilus californiensis*) egests an amount of feces equivalent to 15.2% of assimilated energy,

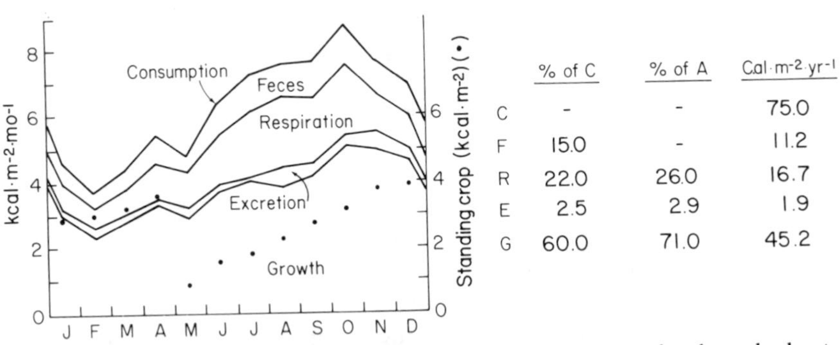

	% of C	% of A	Cal·m-2·yr-1
C	-	-	75.0
F	15.0	-	11.2
R	22.0	26.0	16.7
E	2.5	2.9	1.9
G	60.0	71.0	45.2

Figure 7-23. Partitioning of energy consumed through the year by the polychaete *Nereis virens*. Reproduction is included in growth. To the right are the percentage of total consumption (C) or assimilation (A) that is devoted to F, R, E, or G. The values for respiration seem low, probably because the calculations were done assuming a diet of high quality prey rather than detritus; further, metabolic expenditures for activity may be low if this polychaete were sedentary in behavior. This polychaete is also capable of anaerobic metabolism and of uptake of dissolved organic compounds, so both losses and consumption may thus be different than calculated. The average standing crop through the year is 28 kcal m^{-2}. From data of Kay and Brafield (1973).

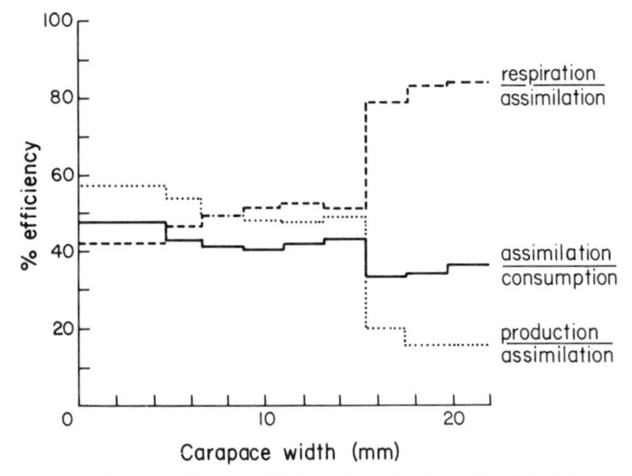

Figure 7-24. Changes in metabolic efficiencies during the life history of the fiddler crab *Uca pugnax* in Great Sippewissett Marsh, Massachusetts. Data from Krebs (1976).

while a 10-mg mussel only releases 4.5% (Elvin and Gossor, 1979). The net result of all this is that specific production is lower for larger individuals, and if older cohorts predominate in a population the level of production by the population may change.

Production of biomass (P) in a given trophic link relative to food consumed by that link (C) cannot be very high, since there is so much dissipation of energy by metabolic demands (Fig. 7-1). The ratio of P to C is called "ecological efficiency" and is important because it yields some notion of what proportion of consumed energy is available to the next link in the food web, whether it is another predator or a fleet of fishing vessels. This is a fundamental property of food chains, and is of interest to fisheries and in understanding how marine ecosystems work. Steele (1974a), following Ryther (1969) concluded that ecological efficiencies in the water column should be about 10–20%. Actual measurements of ecological efficiencies in many types of animals range broadly, but most measurements are lower (Fig. 7-25). There is a scatter of high values, some too high to be

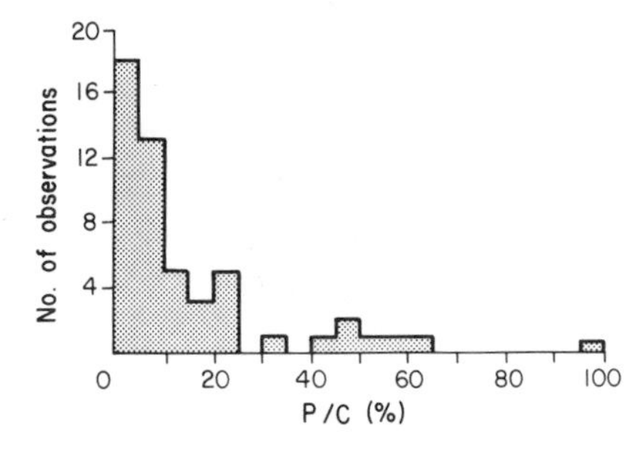

Figure 7-25. Ecological efficiencies (consumption by consumer/production by consumer) for invertebrates. Data were obtained from Conover (1978), Shafir and Field (1980), Tenore et al. (1973), Krebs (1976), Reiswig (1974), Teal (1962), Petipa et al. (1970), and Baird and Milne (1981). Data include a wide variety of taxa, trophic habitat, and size.

reasonable, in view of what we know of respiration and egestion. In general, the ecological efficiency of a trophic link is most frequently less than 10%, some values reach 25%, and there is a scattering of not totally convincing larger values.

7.7 Energy Budgets for Ecosystems

We have so far been concerned with the fate of organic matter and energetics of specific populations, but it is possible to conceive of the flow of energy through all the producers and then through all the consumers in an ecosystem. The amount of primary and secondary production and the patterns of energy flow are ecosystem properties rather than populational properties.

Energy flow budgets for whole ecosystems were first put together for freshwater systems by C. Juday and R. Lindemann. The first rough energy budgets for marine systems were put together by Clarke (1946) for Georges Bank and Harvey (1950) for the English Channel, followed by Odum and Odum (1955) for a coral reef and Golley et al. (1962) for a mangrove swamp. Below we concentrate on better developed energy budgets for a salt marsh and estuarine ecosystems, and also examine a compilation for planktonic systems.

7.71 Salt Marshes

Teal (1962) took advantage of the many studies of salt marsh organisms carried out over the years at Sapelo Island, Georgia, to construct what is still one of the best-documented energy budgets of a marine ecosystem. More recent research has revised many parts of such budgets (Nixon and Oviatt, 1973; Pomeroy and Wiegert, 1981) and has shown the importance of anaerobic processes (Howarth and Teal, 1980) (cf. Chapter 10), but Teal's budget of aerobic energy use is still informative.

In Georgia, about 600,000 kcal m^{-2} year arrive at the marsh surface as light energy, of which primary producers in the marsh used only 6.1%—this is actually somewhat better than the 0.1–3% typical of other producers. Teal used estimates of respiration and growth of *Spartina alterniflora* above-ground and algae to calculate gross primary production, 36,400 kcal m^{-2} $year^{-1}$, of which 77% was respired, leaving 8,200 kcal m^{-2} $year^{-1}$ as the net production. This is the input into the consumer food web. Herbivores consumed only 4–6% of the net production of marsh grass, and probably more of the algal production. In this ecosystem most of the fixed carbon and energy therefore flow through the food web as detritus.

By applying various methods and assumptions, Teal estimated consumption by detritus feeders—including aerobic bacteria—and predators. Thus, of the 8,200 kcal m^{-2} $year^{-1}$ of net above-ground production, 47% were respired by microbes, 7% by detritus feeders or herbivores and 0.6% by predators. The importance of microbial respiration in a salt marsh is clear, and it also might be important in many other marine environments. The herbivore–detritivore partition of energy probably depends on the quality of the biomass available as food for consumers. Salt marsh vegetation produces litter of poor nutritional quality (cf. Chapters 6 and 10). In planktonic food webs, the phytoplankton biomass is more assimilable; more of the primary production flows first through herbivores, and organic matter enters detrital food webs mainly via egestion of fecal pellets and cast molts.

The various consumers in the salt marsh studied by Teal were responsible for the dissipation of 55% of the energy of net production. This left 45% unaccounted for, probably exported by tidal flushing to tidal creeks and perhaps to coastal waters. There has been a continuing controversy as to the fate or reality of this export (Nixon, 1980; Valiela, 1983), but the export does highlight one important matter: adjoining ecosystems may influence each other (cf. Chapter 15), at least by exchanges of organic materials and energy. Allochthonously produced subsidies may be important in many producer communities, from salt marshes to subtidal environments (Chapter 9) and to the deep sea (Chapter 10). These relationships among adjoining ecological systems are discussed more fully in Chapter 15.

7.72 Estuarine Ecosystems

Another comprehensive study of ecosystem energy flow is that done on the Ythan estuary in Scotland (Baird and Milne, 1981). Only about 2.7% of photosynthetically active radiant energy available to plants is converted into primary production by phytoplankton, benthic microalgae, and macrophytes. Primary production reaches about 630 g C m^{-2} $year^{-1}$, of which macrophytes contributed about 72%.

The secondary production by invertebrates totals about 69 g C m^{-2} $year^{-1}$, of which the zooplankton, meiofauna, and macrofauna contribute about 1, 28, and 71%, respectively. Secondary production by consumers is about one order of magnitude smaller than primary production. Grazers consume a small fraction (7%) of plant production; other major groups of consumers feed on detritus and on other particles. The production of the meiofauna is about 52% of the production of the deposit-feeding macrofauna, even though the biomass of smaller organisms (2.1 g C m^{-2}) is only 13.6% of the biomass of the macrofauna (15.1 g C m^{-2}). This reiterates our earlier statements on relatively higher production associated with smaller body size.*

The meiofauna consume about 69 g C m^{-2} $year^{-1}$, of which 38% is egested as feces and 46% is respired. The major suspension feeder is *Mytilus edulis*. This mussel consumes about 66 g C m^{-2} $year^{-1}$ from the water column, of which 50% is respired, and 25% is egested. From the standpoint of the mussel-predator flow of energy this is about a 12% ecological efficiency, well in the range of our earlier review.

The macrobenthic deposit-feeders consume about 400 g C m^{-2} $year^{-1}$, respire 29% of this, and defecate 61%. Presumably the large amounts of feces are due to the low quality of food available as detritus. The macrobenthos net production is about 38 g C m^{-2} $year^{-1}$, of which birds, fish, and crabs eat 24.7, 22, and 4%, respectively.

The efficiency of energy transfer within the estuarine ecosystem can be measured by the consumption efficiency, the consumption by trophic link n relative to the net production of trophic link $n - 1$ (recall that ecological efficiency was $[P_n/C_n] \times 100$). In the Ythan estuary, fish, birds, and crabs consume suspension feeders and macrofauna with consumption efficiencies of 71 and 48%, respectively. The prey organisms therefore suffer rather significant annual losses to predators, amounting to about 50–75% of the yearly crop of biomass. Predators thus are very likely to be a major structuring feature of estuarine environments.

7.73 Open-Water Ecosystems

Complete energy budgets for planktonic ecosystems are not available, but recent advances in methodology have prompted a reassessment of earlier ideas of the fate of energy in pelagic ecosystems (Pomeroy, 1974; Williams, 1981). Earlier views, represented by Riley (1946) and Steele (1974a), were that the fate of phytoplankton was to be eaten by grazers, and it was believed that if the algal-grazers relation was stipulated, the system could be modeled.

Grazers, however, rarely seem to consume all the biomass produced by phytoplankton. For example, Petipa (1979) estimates that herbivores (including microplankton) in the top 12 m of the Black Sea use about 59% of the energy available in phytoplankton. The herbivores located between 12 and 100 m use about 24% of the phytoplankton energy. In the tropical

* There are some other estimates of the relative activity of macro- and meiofauna. In a tidal mudflat in the Wadden Sea, macrofauna biomass is 20 times that of the sum of meiofauna, microfauna, and bacteria (Kuipers et al., 1981). Yet, since the metabolic rate of a nematode, for example, is about 21 times higher than that of an average macrobenthic specimen, 74% of the organic matter is consumed by the small organisms. In other soft-bottom marine habitats respiration of the macrofauna ranges from 2 to 34% of the total respiration of the community (Pamatmat, 1968; 1977; Banse et al., 1971; Smith et al., 1972; Smith, 1973; Davies, 1975).

Pacific the consumption is about 32% (Petipa, 1979). Other examples of the inability of grazers to consume phytoplankton are compiled in Table 8-1. Thus, the earlier view needs revision.

Even if all the phytoplankton were consumed, there is still need for a revision of earlier ideas, because substantial amounts of carbon fixed are exuded from phytoplankton cells and are available to microbes. Williams (1981) has produced a best-estimate budget of energy in a representative reconstruction of a temperate pelagic system during the summer. Williams assumed that all the phytoplankton cells produced are consumed by grazers; 30% of net production is released (exuded) by live phytoplankton before they are grazed. Further, in view of recent work, Williams also assumed that 15% of the algal biomass taken by grazers is lost before ingestion due to fragmentation while feeding. Other assumptions were that assimilation, excretion, and egestion are about equal, and microbial growth efficiency is about 70%. These are more or less reasonable guesses corroborated by the material reviewed earlier in this chapter and in Chapter 10, where many of these concepts are discussed in more detail.

The import of Williams' (1981) calculations is that perhaps 56% of the primary production is channeled to the microbes, even if we just consider the release of dissolved organic matter due to algal exudation and grazer excretion. If we then add the organic matter consumed by predators, we find that about 60% or so of primary production passes through the heterotrophs. Vinogradov et al. (1977) estimated that 70% or so of the energy flow passed through the bacteria in a tropical pelagic system. Azam et al. (1983) estimate that 10–50% of carbon fixed by photosynthesis passes through bacteria. These calculated flows of energy to microbes depend primarily on estimates of rate of algal exudation. This process is still not well understood, but is discussed further in Chapter 10.

The next important point raised by Williams (1981) is that we do not really know the subsequent fate of the large amount of energy fluxing through the microbes. Some bacteria are consumed by micro- and macroplanktonic grazers (King et al., 1980; Fenchel and Jorgensen, 1977; Fenchel, 1982). This grazing is probably intense and may control the abundance of bacteria (Azam et al., 1983). Most bacteria are probably consumed by microzooplankton, including ciliates and flagellates (Fenchel and Jørgensen, 1977; Landry and Hassett, 1982; Azam et al., 1983).

All these consumers process organic matter, release CO_2, and mineralize waste products (ammonia, phosphate, etc.). In the oceanic water column excretion by microplankton, rather than decay by bacteria, may be responsible for the bulk of the mineralization of organic matter. Bacteria may accumulate rather than release nutrients in seawater. For example, bacteria grown on kelp detritus incorporate 28% of the carbon in the detritus, while nitrogen is taken up at 94% conversion efficiency (Koop et al. 1982); bacteria thus are not likely to be involved in release of inorganic nitrogen, while the microzooplankton are.

The protozoa are probably eaten by larger zooplankton, but since one or more trophic steps have intervened, the yield to the larger zooplankton should be less than if the primary producers had been eaten directly. In the pelagic ecosystem, then, much like the coastal system described earlier, a substantial part of photosynthetically fixed energy passes through the microbes and associated microzooplankton.

The marked loss of fixed energy in the various steps of food chains evident in the coastal energy budgets is also true for pelagic ecosystems. This phenomenon puts constraints on the amounts of protein harvestable from the sea. Ryther (1969) calculated that about 24×10^7 tons of fish were produced in the oceanic, coastal, and upwelling areas of the sea. His calculation* was based on rates of primary production, estimates of the number of trophic links and assumed ecological efficiencies in each region (10–20%). Oceanic regions, as befits biological deserts, were estimated to produce only about 16×10^5 tons, in spite of making up 90% of the surface area. The upwelling areas, even though only 0.1% of the area, were calculated to yield 12×10^7 tons. The coastal zones produced another 12×10^7 tons of fish, and make up 9.9% of the area of the sea. Clearly there are many uncertainties in each of the elements of the calculation. For example, level and patchiness of primary production, role of gelatinous plankton, ecological efficiency, lack of proper coverage of the Antarctic sea, among other aspects, are all still inadequately known. Ryther's estimate is close to the 200 million tons of fish estimated to be produced annually in the sea by Schaefer (1965).

This 200–240 million tons of wet weight of fish cannot be completely harvested by man, however. A certain proportion of the annual production has to be left to provide for maintaining of the population, plus other top predators in the sea. Recall, for example, that predators consumed 50–75% of the production of prey in the Ythan estuary.

It is hard to estimate the sustainable level of harvest of fish from the sea, but perhaps 100 million tons per year is about the maximum we can expect. In 1967, about 60 million tons of fish were harvested worldwide, and this value may have increased since. Although there is perhaps some room for further increase in fish harvest, the fish resources in the sea are not going to solve the world's need for protein. To make matters worse, much of the current fish harvest is made into fish meal, which is then fed to poultry and other animals. Although poultry are fairly efficient at food conversion, averaging about 20%, the feeding of the fish catch to poultry just adds another trophic link to the food web, which means a substantial loss of the energy caught as fish.

---

* Ryther calculated that 20×10^9 tons of carbon were produced per year in the sea by primary producers. This is lower than more recent estimates of 31×10^9 tons of carbon year^{-1} by Platt and Subba Rao (1975). Ryther also did not consider the microbial pathways of energy flow that have received recent attention. It is not clear if these omissions cancel each other; Ryther's estimates should therefore be taken as just that: estimates.

Part III

Interaction of Producers and Consumers: Competition and Consumption

Sea urchins (*Diadema*) in sea grass bed. Photo courtesy of R. Phillips.

Competitive interactions can result in the presence, absence, or changes in abundance of a species, as evident from the discussion in Chapter 4, but in nature a population also faces consumption by predators. Considerable space has been devoted in Chapters 5 and 6 to the process of consumption by predators. The next two chapters present evidence of the relative importance of competition and predation in marine ecosystems and assess whether these processes determine the abundance of species.

The impact of consumers on producers is not different from that of consumers on prey animals as will become obvious. Nonetheless, Chapter 8 is devoted to competition among producers and consumption by herbivores, and Chapter 9 to competition and predation among consumers. This separation is prompted mainly by the examples available, rather than to significantly different processes.

Chapter 8

The Consequences of Competition Among Producers and Consumption by Herbivores

In Chapter 4 we considered how competition could in theory and in nature affect abundance of organisms. In this chapter we will briefly introduce models and concepts that predict the effects that consumers might have on their food species especially producer species. Then we discuss some specific case histories that permit an assessment of the impact of grazing on marine producers.

8.1 Models and Theory of Consumer-Food Interaction

The consequences of eating or being eaten should be reflected in the abundance of food and consumers. The simplest model of these relationships was derived by Lotka and Volterra, and considers that the rate of growth of a prey population (dN_1/dt) depends on the intrinsic growth rate ($r_1 N_1$) minus the rate of fatal encounters with a predator

$$\frac{dN_1}{dt} = r_1 N_1 - p_2 N_1 N_2, \qquad (8\text{-}1)$$

where r_1 and N_1 are the rate of increase and abundance of the prey organism 1, and p_2 is the predation rate on species 1 by the predator species 2. For the predator population, the rate of population growth (dN_2/dt) equals the rate of successful predation encounters ($p_2 N_1 N_2$) minus the death rate ($d_2 N_2$),

$$\frac{dN_2}{dt} = p_2 N_1 N_2 - d_2 N_2, \tag{8-2}$$

where d_2 is the death rate of predators without prey.

Solutions of Eqs. (8-1) and (8-2) can be shown as graphs in which N_1 and N_2 fluctuate over time, with predators increasing their number when prey are abundant, then dying when prey are scarce. Simulations using Eqs. (8-1) and (8-2) show the phase of the fluctuations of the predator population lagging behind the fluctuations of the prey population, and results in oscillations of unchanging amplitude for both populations. The model assumes no time lags, no differences among individuals, and that the amount of interaction is proportional to the product of prey and predator numbers. The Lotka–Volterra equations obviously apply to a very simplified situation.

We can look at field data to see an example of such a consumer–food cycle. Fenchel (1982) measured the abundance of bacteria and flagellates in Limfjord off Denmark. Flagellates are presumed to graze on the bacteria. The latter fluctuate in abundance, and the abundance of the flagellates roughly follows the pattern expected in a predator–prey cycle (Fig. 8-1, top left), with peaks in abundance of flagellates lagging behind peaks in bacterial abundance.

To inquire whether the pattern of abundance of bacteria and flagellates could be the result of a predator–prey interaction, Fenchel (1982) constructed a model. The Lotka–Volterra equations were too simple to be

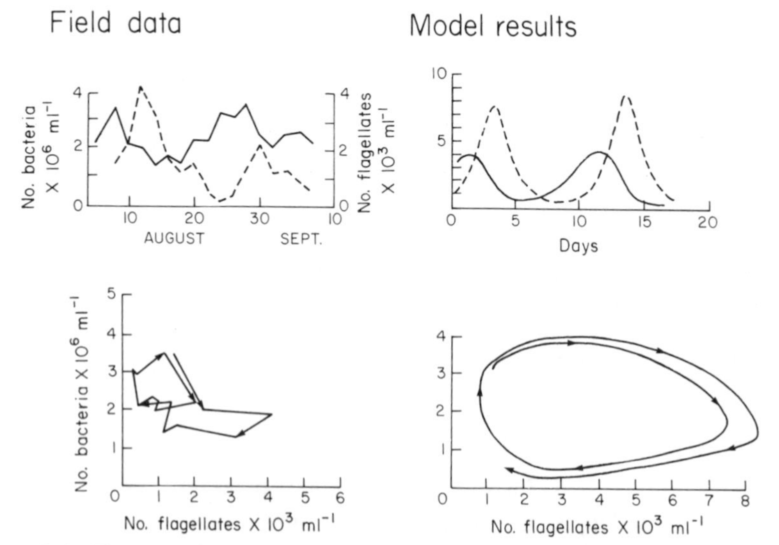

Figure 8-1. Field data and model results on the abundance of bacteria and flagellates in surface waters in Limfjorden. The units on the vertical axis are the same for both field and model results. Adapted from Fenchel (1982).

used directly; other factors discussed in Chapters 5 and 7 plus some realistic initial conditions had to be included in the model. Experimental evidence suggested that the rate of maximum consumption of bacteria by flagellates was 60 bacteria hr^{-1}, and the growth of flagellates was accomplished with a growth efficiency of 3×10^{-3} flagellates per bacterium consumed. The functional response of the flagellates to bacterial abundance was type II, and could be described by a Michaelis–Menten equation with a half-saturation constant of 5×10^6 bacteria ml^{-1}.

Fenchel also had to make educated guesses that the bacteria grew logistically with $r = 0.03$ hr^{-1} and $K = 10^7$ bacteria ml^{-1}, and that flagellates were subject to a density-independent death rate of 0.045 flagellates hr^{-1}. The model results produced fluctuations in abundance of bacteria and flagellates (Fig. 8-1, top right). As in the field data (Fig. 8-1, top left), the grazer abundance in the model simulation lagged behind the abundance of bacteria. With enough further adjustment of the values of various components of the model, a much better fit of the simulated values to the field data could be obtained. The major point here, however, is that fairly simple assumptions about this grazer-bacterial system could provide a not unreasonable depiction of the interaction.

The prey–predator cycle predicted by the Lotka–Volterra model is also provided by Fenchel's more complex model (Fig. 8-1, bottom right). The graph of simulated flagellate vs bacterial abundance is not, however, a very close representation of the actual data (Fig. 8-1, bottom left). The discrepancies mean that either the assumptions of the model need revision, that values of variables need adjustment, or that new factors need to be added. Checks on the assumptions and values of variables can be carried out by further simulation in which the assumptions or variable values are changed; similarly, other factors can be introduced into the model and the results verified versus actual data. This points out one of the primary benefits of such use of models, their role as tools to identify areas that need further study. The construction of models forces a clear account of assumptions made and makes clear what is missing in current knowledge.

In general, analysis of the implications of simulation studies using models is done by mathematical manipulations; another approach is to use graphical analysis as done by Holling (1965), Rosenzweig and MacArthur (1963) and others. Noy-Meir (1975) surveyed graphical analyses used in relation to grazer–producer systems, and made predictions as to the consequences of competition and grazing on producer populations. The basic approach used is to compare two curves, one that depicts rate of growth of the food organisms, and a second that portrays consumption by the grazer.

A graph of rate of population growth versus abundance of a population often results in a humped curve—the Allee effect of Section 4.3, determined by intraspecific competition. This curve is a measure of the ability of the prey or food population to grow and hence provide food for the consumer populations.

The shape of the consumption curve can be represented, for relatively short time intervals, by the functional response discussed in Chapter 5. Of course, the actual rate of consumption would be higher at higher densities of the consumer; this is depicted by the higher or lower position of the consumption curves in Figure 8-2.

With a type II functional response (Fig. 8-2, top), consumption by high densities of consumers always exceeds growth of food and drives the food species to extinction. At lower densities of consumers there are equilibrium points (E_i, E_l) where consumption equals growth. Consider the consumption by consumers at an intermediate density. To the right of the left-most E_i, growth exceeds consumption, and the prey increases in abundance. If the population for some reason happens to grow beyond the density of the right-most E_i, consumption exceeds growth and numbers of prey will decline. Thus within the values of E_i, there is control of the prey population by the consumer. At very low densities of consumers, consumption only exceeds growth at densities that produce such intense intraspecific competition that growth is extremely low (Fig. 8-2, top). The equilibrium (E_l) is therefore reached at very high densities of prey.

Where the consumer requires some minimum abundance (R) of food organism to start feeding (Fig. 8-2, middle) consumption by high densities of consumers is exceeded by growth of the food species, and the latter increases in abundance until E_h. Conversely, at prey densities higher than E_h, prey numbers are reduced to E_h. The minimum feeding abundance thus produces a refuge below which it can grow at least up to a density of E_h. At lower consumer densities the equilibrium points (E_i and E_l) occur at much higher prey densities.

If the consumer employs a type III functional response (Fig. 8-2, bottom), growth of food species may exceed consumption and increase in density up to some equilibrium point (E_h, E_i) so that the prey increases in density. At high density of consumers, consumption exceeds growth beyond this first equilibrium point, and the abundance of food species diminishes to E_h. Thus the equilibrium in this case occurs at low prey density. At intermediate densities of consumers there may be a series of equilbrium points (E_i) all along the various prey densities. At the lower densities of consumers, the equilibrium due to consumption and growth of prey is only found at high prey densities.

The curves of Figure 8-2 are hypothetical and extremely simplified. For instance, there is no accounting of interspecific competition for both the consumer and prey species. The graphical models also are not cast in long enough time scales to include the impact of lowered food abundance on consumption rate by the grazer. If consumer populations were exposed to low prey abundance long enough, there would doubtless be reductions in consumer abundance and hence in consumption.

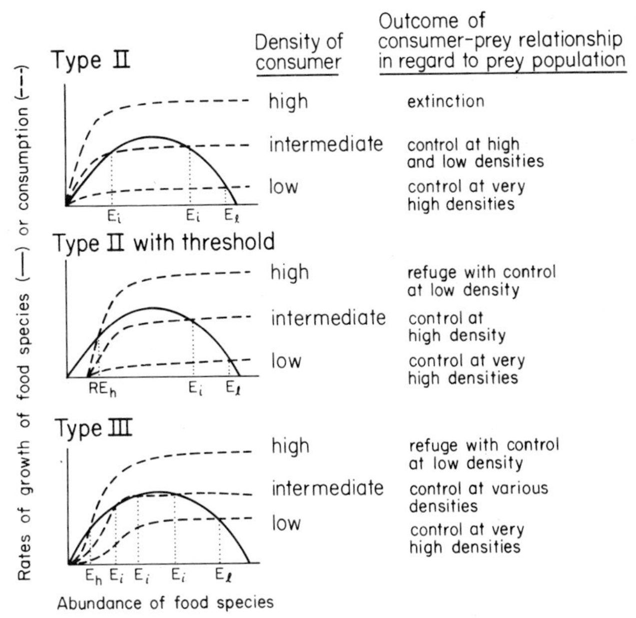

Figure 8-2. Graphical analysis of possible outcomes of consumer-food relationships where abundance of the food and consumer species vary. "E" refers to various equilibrium points discussed in text. Adapted and modified from Noy-Meir (1975).

The message from the graphical models of Figure 8-2, nonetheless, is that with a few assumptions about the consumer–food relationship, a scheme can be put together that predicts circumstances in which consumption of a food organism will lead to extinction or to control of the prey species. The density of both consumer and prey are critical factors, as is intuitively obvious.

The models discussed above focus attention on fairly delicately balanced equilibrium points. Realistically, it seems unlikely that in natural systems such equilibria will be long-lived. Caswell (1978) argues that the impact of predators (or any disturbances that remove prey) occurs in patches, and that in such units of environment extinction can take place, while the prey persist in other units. We thus have transient, nonequilibrium conditions, since migration can reestablish prey in empty patches. Such local extinctions due to consumption are common. Equilibrium ideas such as discussed above apply to individual patches of environment. Consumer–prey models in which transient local extinction is included can account for long-term coexistence of competitors in different patches under the influence of motile predators (Caswell, 1978). In fact, much of consumer–food dynamics may take place as non-equilibrium, transient extinction of local patches.

The above arguments, together with the material in Chapters 4 and 5, suggest that competition and consumption can be important in affecting the abundance of species, and hence the structure of communities. Before examining the actual evidence on the role of competition and predation in marine environments, there is one other theoretical approach, derived for terrestrial situations, that merits some discussion. Terrestrial ecology has had the benefit of a seminal paper (Hairston et al., 1960) that inquired into the mechanisms that determine the abundance of populations of species in different trophic levels. Hairston et al. (1960) made a series of observations and drew conclusions as to how abundances of producers, herbivores and carnivores are determined. Even though some of their points are now outdated, it is still of interest to go through their arguments:

1. Since the accumulation of organic matter in environments is negligible in comparison to the amount produced by photosynthesis, decomposers must be limited by their food supply. If they were not, organic matter would accumulate.
2. Depletion of green plants or their destruction by natural catastrophes is exceptional. Producers are therefore not limited by herbivores or catastrophes. They must be limited by their own exhaustion of nutrients.
3. Where carnivores have been excluded by extrinsic factors, herbivores do deplete vegetation. Herbivores are therefore limited by predators rather than by their depletion of their plant foods.
4. If carnivores limit herbivore populations they are limiting their own resources, and therefore carnivores must be food limited.

These enormously broad generalizations attracted their full share of attention and criticism (Murdoch, 1966; Ehrlich and Birch, 1967; Slobodkin et al., 1967; Wiegert and Owen, 1971). They also served to stimulate a large amount of research. Much of the heated argument that ensued dealt with nuances of interpretation and semantics, but several aspects of Hairston et al. (1960) do need updating.

The classification of consumers into decomposers, herbivores, and carnivores is easy to make on paper but difficult when dealing with real animals. The typology stems from the idea of trophic levels, an idea that has heuristic value but is not easily applicable to natural assemblages. This is because most detritus feeders are facultative predators, most predators can feed occasionally on dead organic matter, and many herbivores can use animal prey at certain times.

Biotic and abiotic mortality factors seldom act separately, and resources and predation may also work in tandem to limit population growth. Hairston et al. (1960) were aware of these difficulties but isolated the effects as a polemical device to focus the arguments.

The fact that on land the "world is green," as put by L. Slobodkin, does not mean that herbivores have an unlimited food supply. Recent work, in part stimulated by Hairston et al. (1960), has shown that plants produce a multitude of compounds that can act as antiherbivore agents (cf. Section 6.321). Further, there is evidence that plant tissues, although plentiful, can be nutritionally inadequate (McNeill, 1973; Caswell et al., 1973; Mattson, 1980) for herbivores. Thus either because of poor quality or because of the presence of chemical defenses, green plants may not be adequate or usable food for terrestrial herbivores. This new information requires modification of the statement by Hairston et al. (1960): terrestrial herbivores can be limited by the quality of their food resources.

This modification of the original formulation helps solve an awkward difficulty in Hairston et al. (1960) having to do with the fact that they had to conclude that limitation due to restricted food supply—in other words, competition for food—would not be important to herbivores, since plant food was not limiting. This seems unreasonable in view of the many examples of sympatric species of terrestrial herbivores that clearly partition food resources (Bell, 1970, for example). The revised version allows for competition among all groups of consumer, which corresponds to what seems to take place in nature.

There have also been criticisms of some of the data on which Hairston et al. (1960) based their statement that herbivores can be controlled by carnivores (Caughley, 1966), but other examples still support that conclusion.

Is any of this relevant to aquatic ecosystems or to the sea? In what follows of this and in the following chapter we evaluate available information on the role of competition and predation in the control of producers and consumers in marine environments. Consideration of the idea that the accumulation of organic matter is negligible and that decomposers are thus food limited is deferred, since it is more convenient to discuss that topic after we have gone into the subject of cycling of organic matter in the sea (Chapter 10).

Most of the discussion in this and the next chapter will revolve around biological relationships, specifically interactions of herbivores, carnivores and their food. Physicochemical features are neglected, but there is no doubt that they are important. For example, desiccation clearly determines the upper limits at which marine organisms are found in the rocky shore. The impact of waves in exposed shores may remove mobile predators, leaving only sessile species on rock sufaces; if intense enough, wave action may in fact allow the presence of very few large organisms. Low light intensities prevent algae from using available nutrients below the mixed layer.

These examples suffice to make the point that abiotic variables may determine the limits of the distribution of a species over time or space. This is not to say that abiotic features are not important in the more

Table 8-1. Percentage of Primary Production Consumed by Herbivores in Marine and Terrestrial Environments[a]

	Percentage of prod. eaten by herbivores	Number of trophic steps	Source
Terrestrial environments			
Deciduous forests			
Tennessee, USA	2.6	3	Reichle et al. (1973)
Ontario, Canada	1.5–2.5[b]	3	Bray (1964)
Old fields			
S. Carolina, USA (1–7 yrs)	12	3	Odum et al. (1962)
Michigan, USA (20 yrs)	1.1	3	Wiegert and Evans (1967)
Tennessee, USA	12.5	3	Pfeiffer and Wiegert (1981)
African grasslands			
Serengeti, Tanzania	14.5–38.2	3	Sinclair (1975)
American grasslands[c]			
USA	2–8	3	Pfeiffer and Wiegert (1981)
USA	0.1–19	3	Scott et al. (1979)
Moors and wet meadows			
U.K., Poland	9–30[d]	3	Pfeiffer and Wiegert (1981)
Grazing lands			
Colorado, USA; U.K.	5.6–26.7	3	Pfeiffer and Wiegert (1981)
Coastal environments			
Vascular plants			
Eelgrass, North Sea	4	3	Nienhuis and van Ierland (1978)
Salt marsh, Georgia, USA	4.6	3–4	Teal (1962)
Salt marsh, North Carolina, USA	58		Smith and Odum (1981)
Mangrove swamp, Florida, USA	9–27[e]		Onuf et al. (1977)

Phytoplankton			
Long Island Sound, USA	73[f]	4	Riley (1956)
Narragansett Bay, USA	0–30[g]	4	Martin (1970)
Cochin Backwater, India	10–40		Qasim (1970)
Beaufort Sound, USA	1.9–8.9		Williams et al. (1968)
Offshore California	7–52 (ave. 23)		Beers and Stewart (1971)
Peruvian upwelling	92, 54–61	3	Walsh (1975), Whitledge (1978)
Open Seas (all phytoplankton)			
Georges Bank	50–54	4	Riley (1963), Cohen et al. (1981)
North Sea	75–80	4–6	Crisp (1975)
Sargasso Sea	100	5	Menzel and Ryther (1971)
Eastern Tropical Pacific	39–140 (ave. 70)[h]	5	Beers and Stewart (1971)

[a] Annual consumption except where indicated otherwise. These values are rough but best possible estimates based on many assumptions and extrapolations.
[b] Leaves only; 0.5–1.4% of total production is consumed by herbivores (Bray, 1961).
[c] This considers grass–cattle–man as the food chain.
[d] Includes above- and below-round production and consumption.
[e] Leaves and buds only.
[f] This is an estimate of consumption of organic matter in the water column. Larger zooplankton consume about 20%, microplankton and bacteria an additional 43%. In the bottom, benthic animals use an estimated 31% of net primary production.
[g] Of standing stock of algae.
[h] Includes only microzooplankton that passed through a 202 μ mesh. The biomass of these small species was about 24% of that of the larger zooplankton. Total consumption could easily be larger than reported if any of the larger species are herbivorous.

favorable season or part of the environment. Abiotic properties are important and interact everywhere with the density-dependent biological mechanisms that limit density of populations.

Returning to the issue of grazing, however, what evidence is there that grazers do control producer populations? The extent of consumption of primary producers by marine grazers is extremely variable (Table 8-1). This variability reflects largely the architecture and, as was seen in Section 6.32, chemical composition of the producers. In a few environments, especially in oceanic situations, grazers of algae consume large proportions of annual primary production. This contrasts with terrestrial environments where producers have complex, multicellular support structures. Herbivores consume only small proportions of the net annual production of forests and mangrove swamps; environments dominated by grasses (including salt marshes) suffer somewhat larger grazing pressure. The data of Table 8-1, although suggestive, are not sufficient to establish the importance of grazing since, even where grazers consume most of the production, other limiting factors may have determined the level of primary production or the composition of the plant community. It is necessary, then, to examine some case studies of the producer–herbivore interaction in several marine environments to gain some perspective on the role of grazers.

8.2 Herbivores of Macroalgae and Marine Vascular Plants

8.21 Two Case Studies

8.211 Herbivores on Rocky Intertidal Shores

The vegetation of the upper half of the rocky intertidal in New England is a mosaic of various species. In tide pools the extremes vary from almost pure stands of the opportunistic green alga *Enteromorpha intestinalis* or of the perennial red alga *Chondrus crispus* (Irish moss), to situations where many different types of algae coexist.

The most obvious herbivore in the upper intertidal zone is the snail *Littorina littorea*, and its density varies from pool to pool. *L. littorea* has clear-cut preferences among the algal foods available in the New England shore: *Enteromorpha* is a highly desirable food, while *Chondrus* is not (Fig. 8-3, bottom). *Enteromorpha* is abundant where there are few snails and *Chondrus* is the dominant species in pools with lots of snails. To see if these differences in plant composition were caused by the different herbivore abundances, Lubchenco (1978) carried out experimental alterations of snail density (Fig. 8-3). All snails were removed from a pool where *Chondrus* was the dominant plant, while snails were added to a pool

Figure 8-3. Experimental manipulation of a grazer snail (*Littorina littorea*) in tide pools in the higher reaches of the New England rocky intertidal zone. Adapted from Lubchenco (1978). © University of Chicago, reprinted by permission.

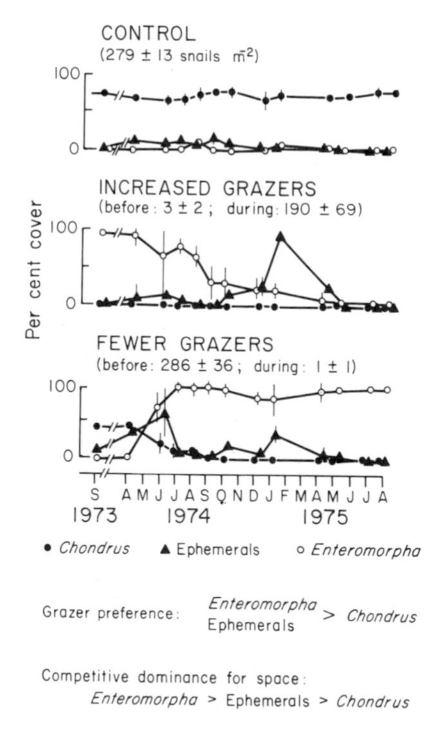

dominated by *Enteromorpha*. In an untreated control pool, the abundance of *Chondrus* remained high through the 1 1/2 years of the study (Fig. 8-3, top).

In the pool where snails were abundant, their grazing gradually reduced the abundance of *Enteromorpha*. Some "ephemerals" (fast-growing, short-lived algae) became abundant in winter, the time when snails were less active.

In the pool with no snails, *Enteromorpha* and several ephemeral species settled quickly, but after a short period of time *Enteromorpha* became the most abundant species in the pool. *Enteromorpha* actually settled on *Chondrus*, grew and shaded the fronds of the Irish moss, and after one summer of the experimental manipulation only the holdfasts of *Chondrus* remained.

These experiments show that grazers control the composition of the producer community in tidal pools. The first-order effect of grazers is the virtual elimination of preferred species such as *Enteromorpha* and other ephemerals. The second-order effect of grazers is that by removing *Enteromorpha*, the best competitor for space, other species may become the dominant components of vegetative cover.

The effect of herbivores is therefore considerably more complicated than merely a removal of a more or less important proportion of the plant biomass.* Selective feeding by the herbivores leads to differential grazing pressure on the various species of producers present in the environment, a feature of fundamental importance in determining what species make up the community. The amount of biomass of the surviving species, the ones not preferred by herbivores or competitively dominant, is then determined by other factors, principally light and nutrients.

8.212 Herbivores in Kelp Forests

The subtidal kelp forests off California consist of the canopy-forming giant kelp (*Macrocystis pyrifera*), and two understory kelp species (*Laminaria dentigera* and *Pterygophora californica*). A luxuriant growth of foliose red algae covers the rock surfaces below the kelp canopy. Along the seaward margin of the forest is a band of the brown alga *Cystoseira osmundacea* and the canopy-forming bull kelp *Nereocystis leutkeana*.

Sea urchins are widespread grazers of marine macroalgae (Lawrence 1975), including kelp. *S. franciscanus* prefers *Macrocystis* over all other species (Leighton, 1966), but other species of kelp are eaten when *Macrocystis* is not available. If no kelp at all are available (years 1974–1975, Fig. 8-4) urchins feed on detritus, microscopic plants, newly settled juvenile plants, or whatever else may become available.

A natural experiment that showed the importance of grazer control of producers took place in 1976 on the seaward edge of a giant kelp forest when an unknown disease decimated the population of the sea urchin *Strongylocentrotus franciscanus* with remarkable consequences for the vegetation (Fig. 8-4). Soon after the near-disappearance of urchins off California, the density of *Macrocystis* increased markedly (Fig. 8-4, second row), and by 1977 only about 1% of the light at the surface reached the bottom. This light intensity is about the lower limit at which *Macrocystis* can achieve positive net photosynthesis (Neushul, 1971), so that intraspecific competition through self-shading was probably responsible for the decline in numbers of *Macrocystis* during 1976–1977. The number of fronds on the surviving kelp increased, so that the total biomass of *Macrocystis* remained significantly higher after the removal of sea urchins. Shading by the large established kelp prevented further recruitment of young kelp.

The two understory kelp species (*Laminaria* and *Pterygophora*) increased rapidly after the urchins were gone (Fig. 8-4, third and fourth rows). In subsequent years, both these species decreased in abundance, with few live individuals left by 1977. The decrease in *Laminaria* and

---

* Herbivores may have many additional effects on vegetation, including influencing plant morphology. For example, some algae of the upper intertidal zone in rocky shores exist as upright morphs during the part of the year where grazing pressure is low, while crustose or boring morphs are dominant when grazers are more active (Lubchenco and Cubit, 1980).

Figure 8-4. Changes in vegetation of kelp forest after collapse of sea urchin population off California. The vertical dashed lines show the time of occurrence of the sea urchin mortality. Adapted from Pearse and Hines (1979).

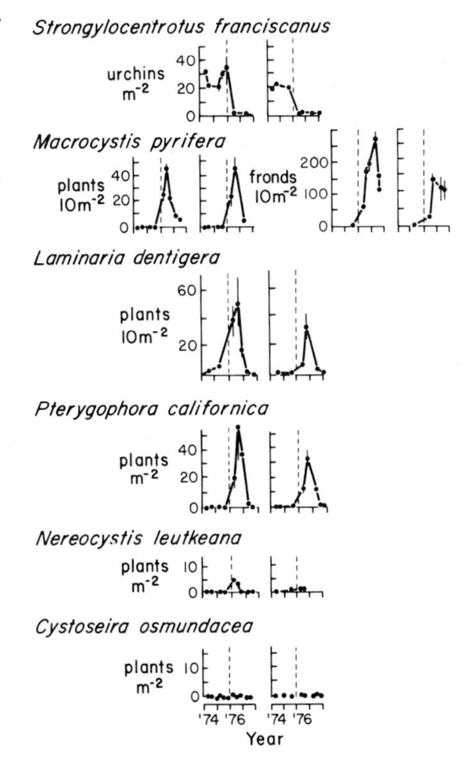

Pterygophora was most probably due to shading by the *Macrocystis* canopy; to test this hypothesis, a plot of 200 m^2 was cleared of large *Macrocystis* and the vegetation was examined 3 months later. The increased light due to the clearing allowed increases in density of *Laminaria* and *Pterygophora* as well as of juvenile *Macrocystis*. The biomass of foliose red algae also increased. As the kelp forest recovers following the disturbance, the biomass per unit area of *Macrocystis* will presumably be lowered to the average of about 70 fronds · 10 m^{-2} of undisturbed kelp forests. This lower density may allow enough increased light penetration so that the understory kelp species can reestablish themselves.

Sea urchins are not usually found grazing on the upright kelp, since wave action dislodges them from the fronds. Instead the urchins feed on kelp fronds kept down near the sea floor by the combined weight of many clinging urchins. The ability of urchins to affect kelp forests is related to urchin abundance. In kelp forests off Nova Scotia, sea urchins* have to exceed a density of 2 kg m^{-2} before these grazers can consume enough kelp to reduce the area covered by kelp forests (Breen and Mann, 1976). If the urchins are abundant enough, they can graze out a kelp bed (Bernstein

* In Nova Scotia, the main kelp is *Laminaria longicrurus* and the sea urchin is *Strongylocentrotus droebachiensis*.

et al., 1981). Then it is rather difficult for the kelp community to reestablish itself, since the urchins remain, feeding on detritus and young algae. The remaining urchins slowly lose weight and newly recruited urchins may not obtain maximum size; dense populations of stunted urchins may occupy the area of a former kelp bed for many years.

The specific species and details of the grazer–producer interaction may vary from site to site. Off the coast of California, *Nereocystis* and *Cystoseira* were not influenced by the absence of urchins or competition among algae (Fig. 8-4, bottom two rows), although elsewhere they may be subject to both (Paine and Vadas, 1969; Foreman, 1977). Nonetheless, although the specific *result* of grazing in rocky shore and kelp forests may differ, the role of grazing and competitive *processes* evident in the kelp case history are very similar to that of the rocky intertidal shore. Grazing pressure may be intense enough that if applied to competitively dominant species of producers may result in competitive release of other producer species. This conclusion may be general enough to apply to other marine environments.

8.22 The Structure of Macrophyte Communities: Palatability to Herbivores and Competitive Ability

As seen in the above case histories, the preference by the grazer for one species of producer over another affects the composition of producer communities and raises the important question as to how certain plants deter feeding by herbivores.

Deterrents are involved in establishing feeding preferences. Terrestrial plants protect themselves by means of spines, hairs, and chemical defenses (Section 6.32). Marine macroalgae have not been studied as thoroughly, but as discussed in Section 6.32, many species do have a variety of chemical defenses against herbivores.

The quality of certain marine vascular plants as food for herbivores is also important. Enrichment studies in salt marshes and mangrove swamps (Section 6.32) show that there is close coupling between the nutritive quality of plants and palatability to herbivores. The biomass of herbivores and the effects of herbivores on the plants also are influenced by nutritive quality. Increased nitrogen content of salt marsh plants* (Fig. 8-5, top) leads to increased biomass of herbivores (Fig. 8-5, middle and bottom).

* *S. alterniflora* is a species that shows C_4 metabolism. The name C_4 derives from the four-carbon compound that is the first product of carbon fixation rather than the three-carbon compound typical of the more usual Calvin cycle metabolism. Such C_4 species have a number of remarkable biochemical, physiological, and ecological properties (Black, 1971, 1973). One property of C_4 plants relevant here is that they are relatively free of herbivores (Caswell et al., 1973), and one reason for this is their relatively low nitrogen content (Fig. 8-5, top). The experimental fertilization increases the percentage nitrogen of *S. alterniflora* to that of the C_3 plants, and the ensuing response of the herbivores shows one reason why C_3 plants are more attractive to herbivores.

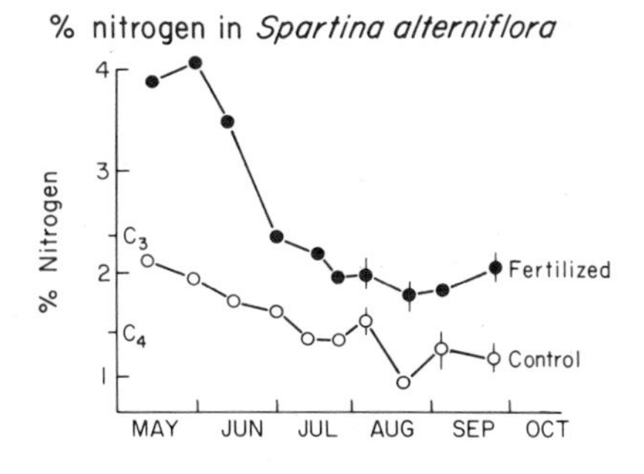

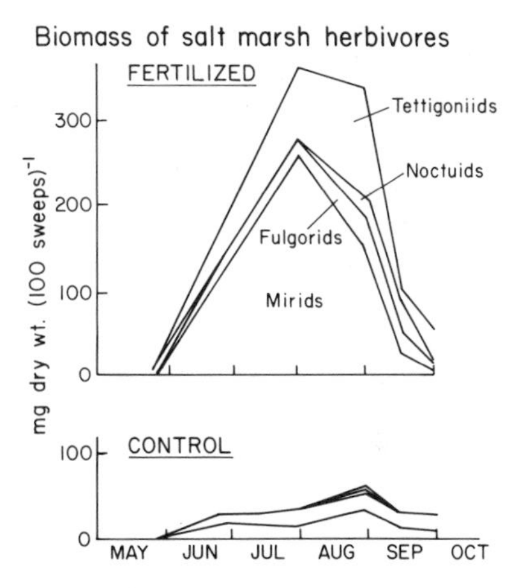

Figure 8-5. Effect of increased nitrogen in plants on the herbivore fauna in a Massachusetts salt marsh. Top: Percentage nitrogen through the year in control and experimentally nitrogen-enriched plots of cordgrass (*Spartina alterniflora*). C_3 and C_4 refer to average concentration of nitrogen in C_3 and C_4 plants at harvest, comparable to September values for cordgrass. Middle and bottom: Biomass of herbivores in fertilized and control plots. Unpublished data. I. Valiela and S. Vince.

Herbivores do not respond merely to the nitrogen content of the food plant. In these enrichment experiments, the fertilization also decreased the concentration of certain secondary plant substances (phenolic acids), making the plants more palatable to herbivores. The relatively low consumption of salt marsh plants (Table 8-1) by grazers is the result of the low nutritive quality of these plant tissues as well as by the relatively high concentrations of secondary plant substances. In salt marshes and in mangroves, which also provide relatively poor food, herbivores do not usually deplete vegetation; they are limited not by the quantity, but probably by the quality of their food plants.

The same holds for many marine grazers that feed on algae growing in kelp forests and intertidal pools. These grazers can eliminate the edible species of macroalgae. The remaining vegetation is composed of a few species—such as Irish moss—that are not consumed by grazers, perhaps because of chemical deterrents. Herbivores in such environments can and often do completely consume the highly palatable or edible food supply; the remainder is undesirable food, so that these herbivores face a limiting supply of food. In both the kelp forests and intertidal rocky shore case studies we lack information as to the particular factors that determine the amounts of biomass of the plants that successfully escape grazers. These details will most likely involve nutrients or light, but further work on this is needed.

In the kelp forest case history, we have a species, *Macrocystis*, that for some reason has not evolved effective phytochemical defenses and is therefore susceptible to overgrazing by herbivores, much like *Enteromorpha* on the rocky shore. It turns out that *Macrocystis* is a species of fundamental importance to the entire kelp forest ecosystem, so that the impact of grazers is critical in determining what the community looks like. Just what is so particular about the phytochemistry of *Macrocystis* or the ability of urchins to ignore what feeding deterrents may be present in this plant is not known (see Section 9.2 for some speculations).

In both our macroalgal case histories, the producer–herbivore interaction is primarily influenced by the relative palatability of the array of producer species, which is in turn determined by the chemistry of the tissues. The impact of herbivores is greatest on the most palatable species. Once palatable species are removed, the success of the less palatable producers is determined by competitive interactions (for light in the case of the kelp forests). The interplay of these features, plus restrictions due to wave action, and other factors of local importance, determines the make-up of the producer community.

8.3 Grazers of Microalgae

8.31 Grazers of Phytoplankton

8.311 Palatability of Phytoplankton

A large proportion of phytoplankton standing crops seems available and palatable to herbivores, even though there is some evidence from freshwater studies that certain phytoplankton such as blue-green bacteria are unpalatable (Arnold, 1971) or undigestible (Porter, 1973; 1976) for zooplankton. Certain dinoflagellates such as *Amphidinium carterae* produce choline-like substances that may be grazing deterrents (Wangersky and Guillard, 1960) and are rejected by their tintinnid grazers (Stoecker et al., 1981). There are many examples of preference by grazers for one phytoplankton species over another, but there is little information on the cues used by grazers to establish preferences. Most algal cells seem to be relatively palatable to some herbivore.

The standing crop of algae is always much lower than that of terrestrial plants, although production rates may be comparable (Chapter 1). In the English Channel and in the South Atlantic, standing crops may be on the order of 2–3% of the annual production. In sharp contrast to the green terrestrial world, the ocean is blue, largely devoid of green cells. In part, the low standing crop of phytoplankton typical of seawater is due to consumption by grazers (Table 8-1).*

8.312 Impact of Grazers of Different Size

Consumers and their particulate prey occur in the water column in a very wide variety of sizes. To deal practically with this very large size range, a series of categories has been proposed, as discussed in Section 1.112: ultrananoplankton, mainly bacteria, smaller than 2 μm; nanoplankton, chiefly small plankton between 2 and 20 μm; microplankton, consisting of large phytoplankton and small zooplankton about 20–200 μm; macroplankton, made up of large zooplankton ranging in size between 200 and 2000 μm; and megaplankton, the remaining plankton larger than 2000 μm.† There are consumer species in every one of these categories, and each consumer species eats smaller species. Research has focused on grazing by macroplankton since measurements are easier, but grazing by nanoplankton and microplankton may be important (Pomeroy, 1974; Williams, 1981).

There are few estimates of the grazing impact of nanoplankton. Fenchel (1982a) measured the abundance of six species of flagellates in waters of Limfjord off Denmark. The flagellates ranged in size from about 0.5 to 10 μm. Using estimates of the volume of water cleaned per organism, Fenchel calculated that, on average, 20% of the water of Limfjord was filtered by the microflagellates per day. The flagellates thus could consume 20% of the bacteria daily. There are few measurements of the growth rate of bacteria in marine plankton. In the Baltic bacterial generation times may vary between 10 to 100 hr (Hagström et al., 1979). Fenchel (1982a) calculates that if the flagellates consumed 20% of the bacteria daily, and had a gross growth efficiency of 35%, the bacteria could support a growth rate of the flagellate population of about one doubling per day. This is a high rate and, if true, implies that grazing has to be an important factor affecting bacterial abundance.

A few available measurements of feeding activity of microplankton are given in Table 8-2 as percentage of production of phytoplankton. The consumption of producers can be substantial, so that microzooplankton surely convey significant amounts of organic matter through planktonic

* In terrestrial plants extensive support and vascular systems are needed to keep photosynthetic tissues exposed to adequate light and transport nutrients; in the water column water movement supplies these functions. The extensive, nongrowing tissues of terrestrial plants slow turnover rates (Table 2-3). This contrast is also important in chemical defense, since many of the deterrents are usually chemically associated with structural tissues.

† The larger size categories (2 cm to 20 m) often consist of active swimmers (nekton) in addition to the more passively drifting plankton. Such distinctions are not always clear, and different classifications have been forwarded (Sieburth et al., 1978).

Table 8-2. Estimates of Phytoplankton Consumption by Microzooplankton, Compiled by Landry and Hassett (1982)[a]

Site	Percentage of primary production	Percentage of standing crop day^{-1}	Source
Southern California Bight	23(7–52)[a]		Beers and Stewart (1971)
Long Island Sound	Up to 43[b]		Riley (1956)
Long Island Sound		Up to 41	Capriulo and Carpenter (1980)
Coastal waters off Washington	17–52[a]	6–24	Landry and Hassett (1982)

[a] Range of values given in parentheses.
[b] Of daily production.
[c] Of annual production.

food webs. Microzooplankton may be less able to limit populations of their food organisms than nanoplankters since, at least in the available measurements, their grazing rates are low relative to rates of production by phytoplankton.

The impact of macroplanktonic herbivores can be impressive on certain occasions in coastal or shallow waters. Daytime grazing in the surface coastal waters off California may remove 40–50% of the algal populations, so that the phytoplankton have to divide overnight to maintain their abundance (Enright, 1969).* Riley (1946, 1963), in a pioneer study of the seasonal fluctuation of phytoplankton and herbivores on Georges Bank in the North Atlantic, estimated the partitioning of the carbon fixed by the plants into respiration by the algae and grazing losses throughout the year. His calculations showed that grazing by copepods consumed about 50% of the net production of phytoplankton throughout the year. Actually Riley only included data for large *Calanus* in his model; more recent data show that biomass of smaller macrozooplankton may exceed that of *Calanus*, so that the grazing pressure by macroplankton on phytoplankton is probably much more important (Davis, 1982).

Such estimates of high consumption of phytoplankton by macrozooplankton in some environments can be contrasted by many instances where consumption is lower. For instance, the macrozooplankton of many shallow inshore waters and estuaries consume only small proportions of the annual primary production (Table 8-1) (Deason, 1980). Even

* Recall Table 1-2, where the average phytoplankton in shelf waters turned over about 300 times a year. Other marine waters may not support such fast reproduction.

where macroplanktonic grazers are very dense, the phytoplankton may not be depleted: concentrations of 3,000 mussel larvae liter^{-1} in waters off Denmark consumed less than half the daily production of planktonic flagellates, their dominant food item (Jørgensen, 1981).

The intensity of grazing pressure thus varies markedly from one planktonic environment to another, but it is not clear why this is so. The species of phytoplankton of rich coastal waters and estuaries, where grazing losses are low, are not dominated by species known to be endowed with herbivore deterrents. Blue-green bacteria, for example, are not very common except for very small forms probably eaten mainly by very small zooplankton such as some ciliates and protozoans (Waterbury et al., 1980). The palatability of most algae, especially estuarine and inshore forms, need further study, but other explanations for the low consumption by herbivores may exist. In some cases most of the plant production may occur during colder parts of the year, when the zooplankton may be less able to respond either functionally, numerically, or developmentally (cf. Chapter 7). In nutrient-rich environments algal growth may proceed so fast that their growth rate overwhelms the grazing capacity of extant zooplankton (cf. Section 13.2).

It should be evident from this examination of the grazing by plankton of various sizes that the actual knowledge is in a far more primitive state than required to assess even the most simplistic models discussed earlier in this chapter. The models, however, should provide a guide to the kind of information that will be needed to assess the role of grazing.

Much foregoing material, especially Chapter 5, has been an introduction to mechanisms involved in grazing. To assess the importance of grazing, we could take a reductionist approach, in which we would compile a complete description of each mechanism, with faith that once this itemization is available, an explanation of the entire process would be available. In fact this has not been feasible, for at least two reasons. First, there are, as is obvious by now, many mechanisms and possible interactions. No one could practically afford the time and funds to complete the required itemization. Actually, the complete list may not be needed, because in specific cases some processes may be far more important than others. Second, there are some processes that are characteristic of a certain level of biological integration. A rate of population growth, for instance, is only appropriate for the populational, not the individual, level of biological organization. There are some relationships involving, say, producers, herbivores, and predation that, as we shall see below, only emerge when interactions of all three components of a marine community are considered. These difficulties have made a component approach to consumer-food interaction unfeasible for whole systems. A variety of other approaches have been applied to this problem, with results described below.

8.313 Approaches to Study of the Relation of Grazers and Phytoplankton

8.3131 Experimental and Seasonal Studies

Even if we could say that most of the production by phytoplankton was consumed by grazers, we still could not claim that zooplankton controlled phytoplankton abundance. To assess this, we need experiments where grazers are excluded and the effect on producers measured, and assessments of the rates and timing of grazing pressure in relation to those of primary production.

There have been some simple experiments where removal of grazers from a sample of 20 liters of seawater resulted in 10-fold increases in chlorophyll concentration after 4 days (Thomas, 1979), so that zooplankton seemed effective in limiting algal crops. This is only a preliminary result done on the deck of a ship. More experimentation of this sort, done *in situ* and with appropriate controls, is needed to examine the role of grazers in oceanic waters. Exclosure experiments in the open sea are logistically extremely difficult, and the measurement of *in situ* grazing and production rates through the year is costly due to the needs for ship time. Seasonal studies (surveyed in Chapter 14) where grazer activity and phytoplankton production are measured together are also expensive. The available data suggest that zooplankton grazing may be important but are by no means conclusive.

8.3132 Modelling Studies

Simulation studies have been the most practical approach to gain insight into the relation of phytoplankton and grazers. Models have been constructed in which grazing is a major determinant of algal biomass over the year (Riley, 1947; Steele, 1974a,b; O'Brien, 1977). There are still disagreements as to the key elements of these models. The threshold in phytoplankton abundance (discussed in Section 5.21) below which grazers do not feed is an important feature of a model by Steele (1974a,b), since it provides the simulated phytoplankton with a way to avoid extirpation by grazers. Others find, however, that populations of phytoplankton may be maintained in their models by the inclusion of alternative mechanisms. For instance, Landry (1976) obtained long-term stability in his version of Steele's model by including predators that eat grazers.* Wroblewski and O'Brien (1976) found that the inclusion of spatial heterogeneity (about which more is found in Chapter 13), and some physical mixing to disperse

* This is an instance of a property that emerges only when consideration is given to the community level, including algae, grazers, and predators. There is a growing awareness that there are significant interactions, not only between producers and herbivores but also among producers, herbivores, and carnivores, and that these three-way interactions can be crucial. This is the message, for example, of the review by Price et al. (1980) of interactions among terrestrial plants, insects, and their predators.

phytoplankton cells made it unnecessary to appeal to a grazing threshold for stability.

The way in which the models are initially formulated has a significant effect on the results, so that a great deal of subjectivity is added *a priori* in any model (Steele, 1976). Since there is still so much uncertainty as to the mechanisms that need to be included in models of the planktonic food web, we will not examine these models in any more detail. We can summarize by saying that these models do a fair to adequate job when verified against real numbers, as in, for example, the classic early model results of Riley (1963).

In nature and in the models the influence of grazers is most clear-cut only during times of year or in regions where nutrient supply is very low—and therefore growth of algae is low—and warmer temperatures allow high activity by the grazers. Light and nutrients take over the limiting of algal growth in other circumstances. Thus, although grazers may consume a large part of annual production by the phytoplankton, the results of modeling studies and actual observations suggest that most of the time factors other than grazing by zooplankton control biomass or growth of phytoplankton. We have to note here these studies only consider macrozooplankton, so that inclusion of smaller consumers in future models may lead to other conclusions.

8.3133 Correlational Studies

Another approach applied to study the relation of grazers to phytoplankton is to conduct surveys in areas where the abundance of grazers and of phytoplankton varies. Such empirical field surveys lead to the conclusion that in fact grazers are more abundant where phytoplankton production is larger (also see Tables 2-4 and 2-5). Included in this general conclusion are observations that the proportions of herbivorous species of zooplankton increase where upwellings provide nutrients for phytoplankton (Table 9-2). Standing crops of herbivores are positively correlated with chlorophyll levels (Beers and Stewart, 1971) and primary production (Taniguchi, 1973; Blackburn, 1973). Although these are only correlations, they are a substantial body of data that suggests that the abundance of grazers may depend on rather than control the abundance or activity of the producers.

If the latter interpretation of the correlation studies is correct, densities and growth of phytoplankton are determined primarily by the rates at which the physics of the water column supplies nutrients rather than through the impact of grazing. In Section 14.2, where we discuss seasonal cycles in marine environments, this conclusion is expanded and the argument is made that at different times of year light, nutrients, and grazing in turn influence primary production, with grazers able to exert a controlling role only while algal growth rates are reduced for other reasons. There is, however, a real need for experimental tests of the controlling factors at different times of year in various marine planktonic systems. Until results from such manipulative research are forthcoming, it will be difficult to put together a convincing theory of control of plankton communities.

Most correlational studies have dealt with macroplankton, precisely the size class of grazers that have shown the least likelihood of being able to limit producers. Studies with nano- and microzooplankton are needed.

8.32 Grazers of Attached Microalgae

Single-celled algae that grow on surfaces are found in the sediments of the shallow coastal zone, as epiphytes on floating surfaces and on surfaces of rocky seashores. The abundance and distribution of attached microalgae are very markedly affected by grazers.

An example of the impact of grazers can be found in the intertidal zone of Southern Oregon, where dark brown carpets of attached diatoms grow on the sandstone surfaces during the colder months. When the weather warms up, the diatoms are removed by grazing herbivorous snails (*Littorina scutulata*). Castenholtz (1961) installed wire-mesh covers over artificial tide pools and varied the density of the littorines within these cages. As the density of grazers increased, the carpet of diatoms was reduced in percentage cover, until it disappeared above a threshold density of grazers. A density of about 3 snails dm^{-2} kept an area free of diatoms, while algal patches developed below this density of grazers. In the summer natural densities of grazers were equal to or several times the threshold values so that during the warm season grazers eliminated algal patches.

The major effect of grazers on attached microalgae may be due to several factors. The elimination of attached microalgal producers by grazers suggest a lack of grazing deterrents in the attached algae. In addition, while phytoplankton cells are widely dispersed, attached microalgae offer many cells in close proximity to grazers. The effort (search time, energy expenditures) involved in feeding is therefore much smaller in the case of attached microalgae, and this may be enough to allow consumption rates by grazers to exceed growth rates by populations of attached microalgae, and hence limit abundance of these producers.

8.4 Basic Impacts of Herbivores on Marine Producers

The examples discussed above highlight several important aspects of the herbivore-producer interaction. First, feeding preference by the herbivore, mediated by the chemical composition, is a key element. Both deterrents in the producer, and the quality of the producer biomass as food, often determine how much and which of the producer species will remain in the environment.

Second, the differential effect of herbivores on different species of producers alters competitive relationships. The activity of grazers determines which array of producers—and other associated species—are present, depending on whether the preferred food is one species or another, as evident in the cases of the nonpreferred *Chondrus crispus*, and in the preferred *Macrocystis pyrifera*.

Third, the effect of herbivores is often dependent on quantitative factors. To control producers, there must be enough herbivores, and their feeding rate must be faster than the growth of producers; otherwise, the producers will escape control by herbivores and be limited by some other factors. In the case of kelp forests, a minimum density of urchins was needed to cause kelp forests to recede; in the plankton, the number and feeding rates of zooplankton may be insufficient to cope with reproductive rates of phytoplankton, at least during blooms. Where algal cells are close together, as in benthic algae, feeding is expedited, and control by herbivores may take place. Every consumer can potentially exhaust its food supply; whether this happens depends on the rates of consumption and renewal of the food supply.

The above three points update the ideas of Hairston et al. (1960) on the control of producers by grazers, but still are far from a complete statement of interaction between herbivores and their food. There is still much to be learned in this field, both as to actual consequences of herbivory in different environments and in regard to herbivore–producer coevolution. For example, Owen and Wiegert (1976, 1981), based on some observations on terrestrial cases, hypothesize that there is a complex mutualism between grazers and plants advantageous to both; this has been disputed (Silvertown, 1982, Herrera, 1982), but there are many other examples of coevolution between herbivores and their food plants (Rosenthal and Janzen, 1979). One such example is the apparent induction of higher concentrations of secondary compounds in leaves of oak trees that were previously defoliated by larvae of gypsy moths (Schultz and Baldwin, 1982). In fact, Mattson and Addy (1975) argue that the long coevolutionary history of consumers and plants has resulted in feeding rates inversely proportional to the productivity of forest vegetation. This relationship may allow herbivores to actually regulate production in forests, in spite of the low consumption typical of forest herbivores. Most of the active research in this field is being done in terrestrial situations, but many aspects are extendable to marine environments. In both terrestrial and marine situations much remains to be learned about producer–herbivore interactions.

Chapter 9

The Consequences of Competition Among Prey and Consumption by Predators

There are many studies that demonstrate that the abundance and species composition of species in an environment are determined by a combination of competitive interactions among the species and the effect of predators on those species. As in the case of herbivores, the impact of predation may not be straightforward, as we can see in the following case histories.

9.1 The Impact of Predators in Kelp Forests

We saw in the last chapter that grazing by sea urchins can destroy kelp forests if urchin densities are high enough. Removal of these herbivores by experimental manipulation (Paine and Vadas, 1969), by oil spills (Nelson-Smith, 1968), or by disease (cf. Section 8.212) results in rapid recolonization by marine vegetation. The presence or absence of predators also has important consequences for urchin populations and for kelp.

The sea otter (*Enhydra lutris*) once occupied a range from northern Japan through the Aleutian Islands and south to Baja California. At present, the sea otter occurs mainly in certain islands off Alaska, as a remnant population in Central California, and in small populations that have been reintroduced to the coasts of Oregon, Washington, and British Columbia. The near extinction of the sea otter was due to the hunting pressure for the highly prized pelts and occurred after the coming of Europeans to the northwest of North America.

At Amchitka Island (Aleutian Archipelago, Alaska), sea otters are present at high densities (20–30 individuals km^{-2}) (Estes and Palmisano,

1974; Estes et al., 1978). On the average, an otter weighs 23 kg and consumes 20–30% of its body weight in food daily. Otters feed mainly on benthic invertebrates and some fish, and as a result of their high consumption rates, have a major role in structuring the near shore community. The population in Amchitka has evidently recovered from the near extirpation of the early 20th century, but recolonization has not yet taken place on Shemya Island, 400 km to the West. At Shemya, where sea otters are absent, there is no subtidal algal cover, and the density of sea urchins (*Strongylocentrotus polyacanthus*) is much higher than that in Amchitka (Fig. 9-1, top and bottom left). Other herbivorous invertebrates are also scarcer and smaller (Simenstad et al., 1978). As usual with many of the predators we have examined, otters prefer to eat the largest urchins (over 32 mm diameter). Only small urchins remain where otters are present (Fig. 9-1, bottom right).*

In Amchitka algae almost completely cover the substratum (Fig. 9-1, top left), while in Shemya space is occupied by urchins, chitons, mussels, and barnacles. Climate, seat state, tidal ranges, and substrates were similar in both sites so that the marked differences in the two communities are likely due to the presence or absence of seat otters.

Confirmation of the importance of otters is also evidenced at other sites where otters have become reestablished (Table 9-1). Otters were reestablished in Surge Bay, Alaska following transplantation into adjacent areas (Duggins, 1980). After some time, the otter population became very numerous and presumably reduced urchin populations (cf. Deer Harbor data, Table 9-1). This led to large enough increases in fast growing annual kelp species that local fishermen reported the area so congested with the annual kelp *Nereocystis* that some boats were unable to make their way through the kelp beds. In Torch Bay, where no otters were present, *Nereocystis* and several other annual kelp became the dominant primary producers one year after Duggins (1980) experimentally removed urchins from an area devoid of kelp. The annuals were subsequently outcompeted and replaced in the experimental quadrats by the perennial *Laminaria groenlandica*. These experimental results agree nicely with the data of Table 9-1 and demonstrate that the presence of otters fundamentally alters the nearshore community.

The impact of the otters on the kelp forest community has several other probable consequences. Estes and Palmisano (1974) and Simenstad et al. (1978) believe that the low abundances or lack of certain fish (rock greenling (*Hexagrammos lagocephalus*), harbor seals (*Phoca vitulina*), and

* The fact that large urchins are present in high densities in Shemya seems unusual. Other studies referred to earlier show that stunted populations were more often the result of crowded, overgrazed conditions where the key predator was not present. Perhaps in Shemya there is enough horizontal transport of food from the rich rocky intertidal zone into the area where urchins are found (Simenstad et al., 1978) or cannibalism of young urchins so that stunting is prevented.

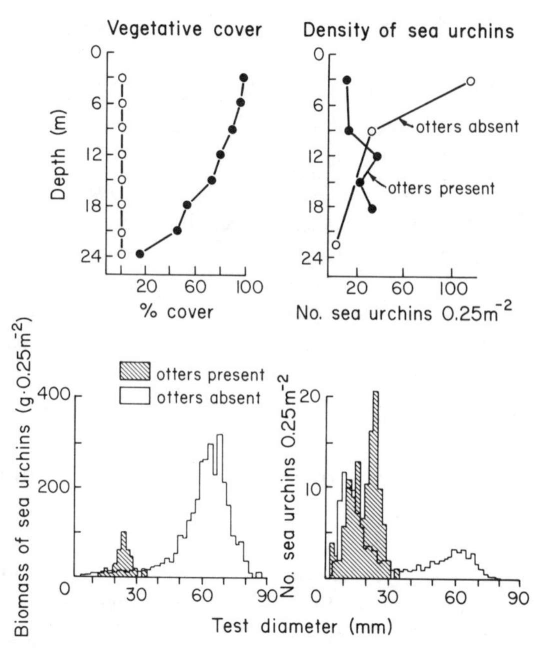

Figure 9-1. Interactions among sea otters, sea urchins, and vegetative cover in kelp beds off the Alaskan coast. Sea otters are present in Amchitka Island and absent in Shemya Island. Symbols on top left are the same as top right. Size of sea urchins is shown as the diameter of the test not including spines. Adapted from Estes and Palmisano (1974). © AAAS, reprinted by permission.

bald eagles (*Haliaetus leucocephalus*)) at Shemya are related to the significant lack of algal cover, although the specific linkages to sea otters are not clearly known. Islands with sea otters have near-shore fishes that depend on and use sublittoral macroalgae for cover and spawning. These fish feed on epibenthic organisms such as mysids and amphipods that consume detrital macroalgal particles (Simenstad et al., 1978). There are fewer near-shore fish at islands lacking otters, and the species that are present are mostly species that feed on the open water. These manifold consequences of the activity of sea otters demonstrate that they are what Paine (1969) calls keystone predators, that is, species whose activities structure the surrounding community. Otters also feed on the older, less productive age classes of fish such as the rock greenling (Simenstad, unpublished data). Thus, otters not only structure the near-shore community but also may increase the secondary production by the effect of their selective feeding.

Removal of the keystone predator in kelp forest communities may have occurred more than once, as shown by animal remains in kitchen middens of aboriginal Aleuts and pre-Aleuts at Amchitka (Simenstad et al., 1978).

Table 9–1. Mean Number m^{-2} (± Standard Deviation) of Key Species in 1-m^2 Quadrats in Southeastern Alaskan Kelp Forests[a]

	No otters (Torch Bay)	Otters present less than 2 yrs (Deer Harbor)	Otters present more than 10 yrs (Surge Bay)
Urchins			
S. franciscanus	6 ± 7	0.03 ± 0.03	0
S. purpuratus	4 ± 13	0.08 ± 0.06	0
Kelp			
Annual species[b]	3 ± 7	10 ± 5	2 ± 5
Laminaria groenlandica	0.8 ± 5	0.3 ± 0.6	46 ± 26
Percentage samples with no kelp	69	0	0

[a] From "Kelp beds and sea otters: An experimental approach" by D. O. Duggins (1980). © Ecological Society of America, reprinted by permission.
[b] Annual kelp include *Nereocystis leutkeana*, *Alaria fistulosa*, *Cymathere triplicata*, and *Costaria costata*. These algae are opportunistic species that temporarily use space free from urchin grazing.

During time intervals when sea otters, harbor seals and fish remains were abundant in the middens (about 580–80 B.C. and a period around 1,080 A.D.), sea urchins and limpets were not, and vice versa. The Aleuts may therefore have supplanted the otter as the keystone species. The Aleuts had the technological know-how with which to hunt or even exterminate sea otters* but for whatever reason Aleut hunting of otters varied in intensity over long periods of time. When hunting pressure increased, the otters became scarce and the kelp forest community reverted to its sea urchin-dominated state.

Elsewhere in the world where there are kelp forests, other predators such as crabs, lobsters, or fish may perform the keystone role. As mentioned in Chapter 8, kelp forests off Nova Scotia have declined in area, with reduction of up to 70% of the habitat in St. Margaret's Bay (Breen and Mann, 1976; Wharton and Mann, 1981). At the same time, the lobster catch per unit fishing effort declined by 50% between 1959 and 1973, probably due to over-fishing of lobsters. Subsequently, sea urchins have increased and have destroyed kelp beds at a high rate. The picture of lobsters as the keystone predator controlling sea urchin numbers is not entirely clear because there are substantial numbers of crabs and fish that

* During the enslavement of Aleut hunters by Russian fur traders, the Aleut nearly eliminated the sea otter population (Kenyon, 1969). Thus, when they had to, these hunters could bring about marked a decrease in abundance of sea otters.

prey on urchins and the fish have declined in number (Wharton and Mann, 1981; Bernstein et al., 1981). The crabs in turn are highly preferred by lobsters over urchins (Evans and Mann, 1977). There may also be important effects of diseases or parasites (as we have seen in Section 8.212) in changing urchin abundance.

9.2 Interaction of Predation and Competitive Exclusion in the Rocky Intertidal Zone

The relative ease with which manipulative experiments can be carried out in the rocky intertidal has made this environment one of the best understood in regard to competitive and predatory interactions among the constituent populations. We have already dealt with results on competition and herbivory. Work done primarily on the Pacific Coast of North America has documented that predation (Paine, 1966; Connell, 1972) and physical disturbances (such as waves and floating logs) (Dayton, 1971) are important as causes of mortality and maintain competing species of prey at relatively low densities. In such situations, competitive interactions are less intense and competitive exclusion is usually avoided. Since predation and other disturbances periodically provide open space (the resource most limiting in the rocky intertidal environment), recruitment of new individuals can take place.

A good example of these interactions is provided by experimental studies by Lubchenco and Menge (1978) in the rocky intertidal zone of New England, where the predatory intertidal whelk, *Thais lapillus*,* and the sea star, *Asterias forbesi*, feed on the barnacle *Balanus balanoides* and the blue mussel *Mytilus edulis*, as well as affect their competitor for space, the macroalgae *Chondrus crispus*. The experiments included clearing areas on the rock surface and following the course of events in the areas (Fig. 9-2, top four graphs). In cleared areas with no further manipulations (called "control" in Fig. 9-2) *Mytilus* settled but did not survive predation for very long, and *Chondrus* became the dominant species in the site. Where stainless-steel cages excluding *Thais* and *Asterias* were attached to cleared rock surfaces, *Mytilus* survived and was able to outcompete *Balanus* in a short time. In other cleared areas with cages, the *Mytilus* were removed from the cages; this allowed *Balanus* to settle and eventually to grow into the most abundant species. In cages where *Mytilus* were removed and *Balanus* were reduced in density, *Chondrus*, after some initial coexistence with *Balanus*, became dominant. Thus, at least

* The genus of these snails has recently been changed to *Nucella*. We retain the earlier usage for convenience.

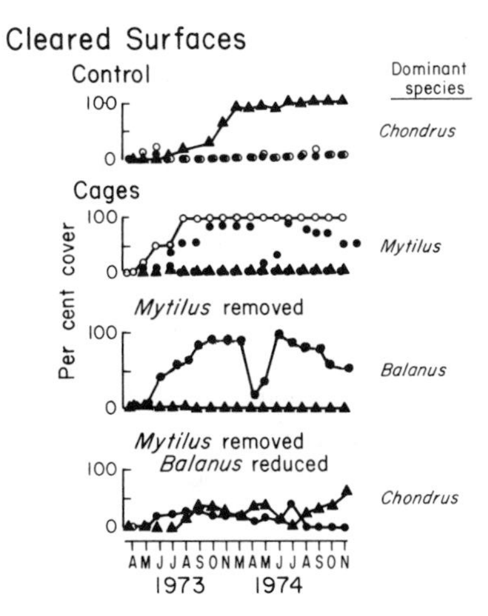

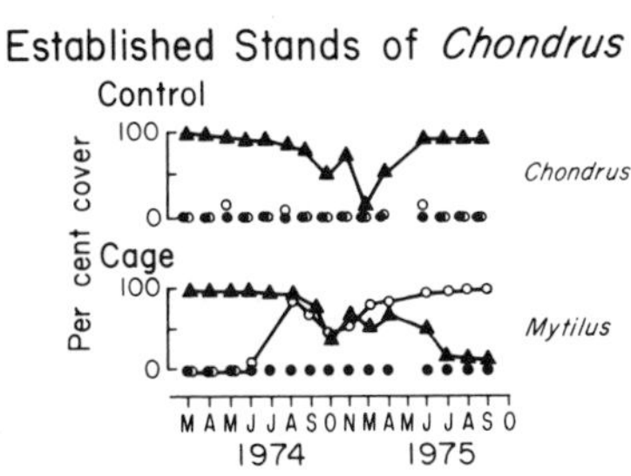

Figure 9-2. Experiments on the role of predation and competition in the lower rocky intertidal in New England. The abundance of each of the three major species, *Mytilus edulis, Balanus balanoides,* and *Chondrus crispus,* is expressed as percentage cover. One set of experiments (top four graphs) was carried out starting with experimentally cleared surfaces, while the second set (bottom two graphs) was done on already well-established stands of *C. crispus.* Cage controls (not shown) consisted of cages with no sides and show results similar to controls. Adapted from Lubchenco and Menge (1978). © Ecological Society of America, reprinted by permission.

as colonizers of bare substrates, the competitive hierarchy is (1) *Mytilus*, (2) *Balanus*, (3) *Chondrus*.

There are places where predators are naturally absent, such as in sites exposed to severe wave action, and in such habitats *Mytilus* is the most abundant species. In protected sites, predators are not swept away or damaged by waves, and their presence prevents their preferred prey (*Mytilus* and *Balanus*) from monopolizing space. *Chondrus* can then colonize and establish itself.

Another set of experiments was started in already-established stands of *Chondrus crispus* (Fig. 9-2, bottom two graphs). The exclusion of predators allowed recruitment of *Mytilus* into the stand, and growth of the mussels eventually resulted in exclusion of *Chondrus* from the experimental site. Then, regardless of whether *Chondrus* was established or not, mussels, the dominant competitors, replaced the alga, but only if predators were absent. Notice that the mechanisms involved, preferences by the consumer, and competitive dominance are very much the same as discussed for herbivores in Chapter 8.

The outcome of predatory and competitive interactions can be markedly modified by physical-chemical features. The clearing experiments of Figure 9-1 were done at 0.3 m above mean low water. Similar experiments done higher in the intertidal showed that *Chondrus* does not cover all the space, probably because it cannot survive long exposures to drying.

The results of these experiments suggest how the rocky intertidal community of New England is structured. In habitats protected from severe wave action, predators can severely limit the abundance of the dominant competitor (mussels). Feeding by predators in protected areas produces open space that is colonized by algae. Competition among the various species of algae ensues, and, as we saw in Chapter 8, *Chondrus* eventually excludes other algal species. In exposed habitats the intense wave action may remove predators, since they are not permanently anchored in the substrate. In such habitats the competitively superior mussels monopolize available space.

The results of work on intertidal rocky shores by Connell, Paine, and Lubchenco and Menge provide an appealing scheme of how community structure is determined, one that has been applied to a wide variety of other environments. The scheme may not, however, be entirely generalizable to other marine environments, or even perhaps to all hard substrates. The outcome of competitive relationships among assemblages of invertebrates on the underside of foliose corals seem determined by competition rather than by predation, and competitive exclusion is prevented by reversals in competitive superiority from one site to another (Jackson, 1977a). Predation also does not seem to determine the structure of fouling communities that settle on hard surfaces suspended in the water column (Sutherland and Karlson, 1977). In spite of such exceptions, the rocky intertidal experiments have produced the basic model from which hypotheses have been derived to be tested in other environments.

9.3 Predation and Competition in Soft Sediment Benthic Communities

Predators, particularly fish, are major influences on the competition and abundance of organisms of soft sediments. There is correlational evidence for this: the abundance of benthic invertebrates in the Baltic Sea, for example, increased after overfishing depleted flatfish during the 1920s (Persson, 1981). Better evidence can be obtained by caging experiments where an area of sea floor is protected from predation. Early experiments showed that such caging led to 60-fold increases in the densities of prey on the sea floor (Blegvad, 1925). Experiments using cages or exclosures are now commonplace, and many of them have been done on soft sediments.*

* Recently it has become fashionable to question the validity of caging experiments in soft sediments, since caging is claimed to lead to artifacts, including changes in the sediments and fauna within the cages (McCall, 1977; Virnstein, 1977; Eckman, 1979). There is no

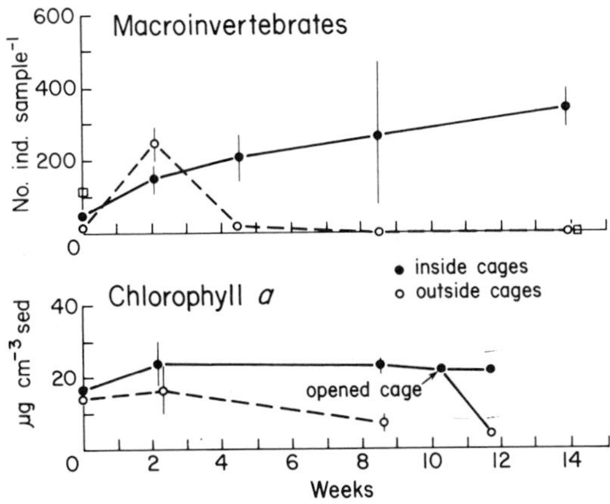

Figure 9-3. Experiments in which the density of macroinvertebrates (top) and the concentration of chlorophyll *a* (bottom) were measured in soft salt marsh creek sediments. The experimental treatments included two replicates of control plots and two plots provided with a mesh cage to prevent the entrance of fish and crabs. The two sets of data were not from the same experiments, but they were both carried out from early to late summer. In bottom graph, one of the two replicate cages was opened during week 10. The access of fish and crabs reduced chlorophyll to levels similar to those outside cages. The open squares represent cages with an open side to control for cage effects. Adapted from Wiltse et al. (in press) and unpublished data of Kenneth Foreman.

In soft organic sediments of intertidal creeks of New England salt marshes the major macrofaunal taxa include amphipods, an oligochaete (*Paranaius litoralis*), hydrobiid snails, an anemone (*Nematostella vectensis*) and a diverse group of annelids, including *Streblospio benedicti, Capitella* spp., *Manayunkia* spp., and *Hobsonia florida*. During warm months of the year, fiddler (*Uca pugnax*) and green (*Carcinus maenas*) crabs and fish (principally the killifish *Fundulus heteroclitus* and *F. majalis*) forage through the surface sediments and detritus, ingesting a variety of foods, including the fauna. Experiments in which cages were used to exclude predators (Fig. 9-3, top) show that in the absence of predators, the biomass of benthic macroinvertebrates remains high through the growing season or increases a bit. Where predators have access—in natural sediments and in control cages—biomass of macroinvertebrates decreased markedly after an initial increase during spring. Predation by fish and crabs thus seems responsible for removal of a very large proportion of

---

question that there are cage effects, but many of these changes are no doubt the result of changes in the fauna brought about by the lack of predators since organisms can markedly change the density and physiochemical properties of sediments (Aller, 1980a). The caging approach is too powerful a tool to dismiss too readily, especially when not many alternatives are available to directly study the effect of consumers. Cages should be used with careful attention to minimizing artifacts, and cage effects should be assessed using partial cages.

prey biomass (Werme, 1981; Wiltse et al., in press). Recruitment and growth of macroinvertebrates take place in early summer but the predators quickly remove the added biomass, so that by August or so the standing crop has been reduced by at least one order of magnitude. It is clear that the predators are indeed key members of this community and that their activity structures the benthic community.

Similar caging studies have also been carried out in a variety of lagoons and estuaries with unvegetated soft sediments. In all these studies, protection from predation increased the density of the macroinvertebrate prey two- to three-fold (Peterson, 1979). This suggests that, as in the other marine environments we have examined, predation can control composition and density of the fauna of soft sediments.* We should note that none of these experiments distinguish between the effect of ingestion of prey by the predator from mortality due to the disturbance created by the activity of the "predators." Disturbance of sediments by large animals can disrupt the life activities of benthic animals and may, if severe enough, kill or smother smaller species.

Caging studies on soft sediments, in contrast to the studies on rocky substrates, show little evidence of competitive exclusion once predators are removed (Peterson, 1979). In fact, more species of invertebrates were often found within the cages. The conclusion from experiments done in rocky intertidal environments that predation, by preventing competitive exclusion fostered the presence of multiple species, is therefore not completely extrapolable to soft sediments.

The effectiveness of predators in determining abundance of prey species varies from one habitat to another. A major feature that affects the success of predators is the physical complexity of the habitat where predation takes place (cf. Section 6.6). This complexity is provided by topography, rock fragments, cobblestones, reefs, or vegetation. Experiments using cages to exclude predators have also been carried out in estuarine or lagoon sediments on which vegetation was present. The results of these experiments (Peterson, 1979) are that the exclusion of predators has very slight effects on the macroinvertebrate community of such habitats, since the vegetation already excludes large predators or permits prey to hide. The fauna in vegetated habitats is usually more numerous compared to similar but unvegetated sediments and far more species are present. In these relatively predator-free habitats there is also no tendency for competitive exclusion.

The experimental work on soft sediments thus fails to show competitive exclusion, but the mechanisms that prevent exclusion are not clear (Peterson, 1979). Perhaps one explanation is that soft sediments do not

* The supposition is that small predators about the same size as the macroinvertebrate prey are not effective as predators, as discussed in Chapter 5. This assumption needs some more careful examination, especially in soft sediments and in plankton communities.

provide the hard substrate where competitors can be crushed, pried off, or overgrown (cf. Chapter 4). Thus, the more obvious mechanisms of interference competition, leading to direct elimination of one species by another, may be lacking in soft sediments. Animals that live in soft sediments may compete mainly by exploitative competition. Another explanation may be that competitive advantages of one species over another are slight (E. Gallagher, personal communication) so that it may take a relatively long time for one species to exclude another. If it takes longer to achieve exclusion than the average time in which the environment changes or disturbances occur, competitive exclusion may seldom take place. In soft sediments there are frequent small scale local disturbances that alter competitive advantages. Such disturbances may thus prevent competitive exclusion and maintain multi-species assemblages. There are many possible mechanisms that may disturb patches of soft sediments, including physical and biological mechanisms (Thistle, 1981). Van Blaricom (1982), for example, showed that bat rays (*Myliobatis californica*) repeatedly disturb the sea floor by digging pits in the sand. This markedly affected the fauna of the disturbed pits and resulted in a sequence of various species colonizing the disturbed patches. On average, about 23% of the sea floor was in some phase of recovery following disturbances by bat rays; during periods when rays were very active this figure reached 100%.

Another alternative explanation may be that perhaps the three-dimensional nature of sediments as habitats for organisms allows for the continued presence of several species. Steep gradients in redox, for example, may provide refugia for species tolerant of reducing conditions. Such stratified habitats may be partitioned by a greater number of species than could be accommodated in a two-dimensional rock surface.

Another reason for the lack of exclusion may lie in the nature of "predation" in soft sediments. In many instances, as in the salt marsh example of Figure 9-3, the "predators" are really omnivores. Salt marsh fish and fiddler crabs, for example, ingest detritus, algae, and meiofauna as well as macrofauna. Exclusion of the fish and crabs results in more abundant algae (Fig. 9-3, bottom). If these algae are food for the prey macrofauna within the cage, competition among prey may be lessened so that exclusion is less likely to occur. Some other limiting factor besides food may then prevent any one species of macrofauna monopolizing resources.

Another possible mechanism that prevents saturation, and hence exclusion, may be a lack of sufficient larvae available to settle and saturate available space. Whatever the factors involved, the coexistence of many competing species is a major unresolved issue in the understanding of communities in soft bottoms.

The lack of competitive exclusion seen in short-term experiments does not mean that competition has not occurred on different spatial or time scales. There is suggestive, indirect evidence of character displacement, resource, and habitat separation among benthic organisms (Rees, 1975; Fenchel and Kofoed, 1976; Levinton, 1979; Frier, 1979, 1979a).

Although there is room for considerably more study, it seems from the evidence on soft sediments, foliose coral, and fouling communities that the entire picture of the role of predation and competition may not be directly transferable from the rocky shore to other environments.

9.4 Speculations as to Control of Prey Populations by Predators in the Marine Water Column

9.41 The Kinds and Sizes of Predators in Pelagic Food Webs

Planktonic predators often feed on each other, and have usually very variable diets. There are also marked changes in plankton communities even over relatively short periods of time. These features make it difficult to define planktonic food webs. Just as there are herbivorous species in most size groups of zooplankton (Section 8.3), there are a plethora of predaceous species of every size. In the microzooplankton, for example, there are ciliated protozoa, including tintinnids and oligotrichs, that consume naked ciliates, as well as graze on dinoflagellates (Stoecker and Evans, in preparation). Macro- and megaplankton, and nekton, including copepods, euphausiids, bony fishes, sharks, mammals, birds, squid, and some of the larger gelatinous plankton are usually nearer the top of pelagic marine food webs.

The small plankton consume even smaller particles and are themselves the prey for still larger predators. Oligotrophic waters may have more size classes of zooplankton than eutrophic waters. Predation is size-dependent (Chapters 5,6), so the more size classes of plankton that there are, the more kinds of predators that can be present. It is therefore possible to have a larger number of links in oligotrophic oceanic than in coastal or more eutrophic food webs (Table 8-1). The greater proportions of species of carnivorous zooplankton in oceanic compared to coastal waters (Table 9-2) may reflect the larger number of trophic links.*

Pelagic food webs tend to be "unstructured" (Isaacs, 1973) because there are many predaceous species that are only slightly larger or smaller than other potential prey species. Most zooplankton have the capacity to feed on most everything within a certain size range. Since few species of marine zooplankton are therefore free of some predation, it may be diffi-

* Additional evidence is that the proportion of carnivorous zooplankton is much larger in depauperate tropical waters than in relatively rich subpolar waters (Vinogradov, 1970).

Table 9-2. Gradients in Herbivory and Carnivory in Zooplankton[a]

	Areas of stable water layers	Areas somewhat influenced by upwellings	Areas of intensive upwelling
	Tropical Pacific surface water	California coastal water	Baja California upwelling
Herbivores	46–47	50–52	63–78
Mixed feeders	4–12	16–24	5–25
Carnivores	41–48	26–34	12–17
	South Equatorial Current, Indian Ocean	Divergence between South Equatorial counter-current	South Equatorial Current, Indian Ocean
Herbivores	17	18	50
Mixed feeders	26	41	24
Carnivores	57	30	26

[a] Values are percentage of the numbers (Indian Ocean data) and biomass (Pacific data) of net zooplankton in environments with varying influence of upwelling. Adapted from Longhurst (1967) and Timonin (1969).

cult for any one species to build up its own abundance in response to the abundance of its prey. Thus, the complex feeding relationships among consumers themselves may prevent planktonic predators from completely determining the number and kinds of prey, because not many predators are themselves free of predation pressure.

In contrast, freshwater food webs do not often consist of as diverse a fauna as marine planktonic food webs, and tend to be more unidirectional (more "structured"). A freshwater predator may be able to respond more clearly to changes in prey density and, rather than be limited by its predators, it is more likely limited by its own food supply.

There are many major unanswered questions about the role of the various groups of planktonic predators. In particular there is a notable lack of information on the abundance and impact of nano- and microzooplankton, squid, and gelatinous zooplankton, and on their importance as consumers. The problem is mainly the lack of adequate methods to estimate densities of, for example, gelatinous zooplankton. Observations during *in situ* diving by R. Harbison and L. Madin and associates (Harbison et al., 1978) indicate that gelatinous zooplankton may be as important consumers as copepods. In some areas, jellyfish are major predators of zooplankton. J. Purcell (unpublished data) found that a siphonophore (*Rhizophysa eysenhardti*) consumed over 28% of the available planktonic prey per day in a shallow cove in the Gulf of California. Other siphonophores in more open waters were less effective predators, consuming only a few percent of available prey daily. Experiments with large plastic enclosures set in a coastal inlet of British Columbia (Grice et al.

1980) demonstrate that ctenophores (*Pleurobrachia* and *Bolinopsis*) can effectively consume smaller zooplankton, and their predation pressure leads to very low densities of prey populations. All this is only fragmentary information; in spite of the long history of research in the sea, there is very little known about the role of most invertebrate planktonic consumers.

Bony fishes are the most studied of all marine predators, primarily due to the impetus of commercial fisheries. Marine birds and mammals are also important and obvious predators that have attracted attention and some data on these groups of consumers are available. In the sections that follow we examine what is known of the role of fish, mammals, and birds as potential predators in pelagic food webs. We treat the fish separately from the mammals and birds because, as will be seen, most fish use a reproductive stategy very different than mammals and birds. This fundamental difference (typically large numbers of tiny young in fish versus few, large young in mammals and birds) leads to very different results in their role as predators.

9.42 The Importance of Predation by Fish

9.421 Some Relevant Evidence from Freshwater Environments

There are several examples of the impressive influence of predation on plankton communities in freshwater environments. Smith (1968) records the remarkable changes that have taken place in the fish fauna of Lake Michigan as a result of changes in intensity and selectivity of predation, including introduction of sea lamprey and changes in fishing effort and gear. Brooks and Dodson (1965) document the consequences of introduction of alewives (*Alosa pseudoharengus*) in Connecticut, U.S. lakes: the larger zooplankton species were eliminated after the appearance of the alewife. Zaret and Paine (1973) describe the results of the invasion of *Cichla ocellaris*, a predatory fish, in Gatún Lake, Panama, where its feeding changed the food webs of the lakes, with a marked reduction of prey organisms and the appearance or increase of nonprey species. In the case of *Cichla*, the depletion of prey also meant that densities of other predators that fed on the same prey as *Cichla* were also much reduced. Since, as we know, predation is generally size specific, the prey requirements of the new fish species led to a very different community of herbivorous zooplankton and presumably to a very different distribution of sizes and species of phytoplankton.

Similar conclusions emerge from experimental manipulation of fish densities in replicated freshwater ponds by Hall et al. (1970), who also manipulated densities of invertebrate predators. Invertebrate predation did not seem to have nearly as important an effect as vertebrate predators. The fish were larger in size than the invertebrate predators, more mobile, and evidenced better-developed functional and developmental responses. Nutrient concentrations were also experimentally manipulated but did not seem to have as much of an effect on the composition of the zooplankton as did fish but did lead to higher growth of the fish predators.

The theory that emerges from these studies (Hall et al., 1976) is that as fish density increases, size-selective predation leads to decreases in the size of the dominant zooplankton. The resulting community of abundant, but small zooplankton species may, however, be just as productive as the community dominated by the larger species. In the absence of fish, the smaller species cannot outcompete the larger species because the larger species have higher grazing and food processing efficiencies (cf. Chapter 5).

Freshwater communities thus seem to be thoroughly dominated by the top predators. The supply of nutrients and phytoplankton may have less of an effect on the composition and production of freshwater zooplankton than do the fish. Most freshwater phytoplankton are available to grazers, although some species may not be readily palatable or assimilable. Many freshwater benthic macrophytes, for example, do have what may be chemical defenses, such as alkaloids in water lilies. The many different compounds found in a variety of species are summarized in Hutchinson (1975).

Grazers are much more vulnerable to fish predation in the water column than in habitats with benthic vegetation. Since the protection provided by the structure of macrophytes is lacking, the impact of predation on planktonic species is not damped as it may be in vegetated benthic habitats.

We have no data with which to evaluate the impact of predation of pelagic marine fish comparable to those available for freshwater. The kind of manipulative experiments that allowed the role of fish in lakes to be studied are very hard to do in the open sea. Cases of invasions or extinctions of well-defined parcels of ocean are also not easily studied. Thus a scheme about how predation by fish may effect pelagic marine communities can only be speculatively pieced together using various other lines of evidence.

9.412 Fish as Carnivores in Marine Pelagic Communities

Fish can be effective consumers, but it is difficult to say whether they control the number and size of prey in the sea. Fishery biologists traditionally believe that stocks of adult fish rarely overexploit their food resources (Beverton, 1962; Gulland, 1971; Cushing, 1975). For example, the average-year class of haddock (*Melanogrammus aeglefinus*) is thought to consume only a fraction of its potential food supply, and only when the occasional very strong year class appears is the potential food resource used more extensively. The 1962 and 1967 age classes of haddock in the

North Sea were about 20 times as large as those of other years. Such large increases in abundance must to some extent have taxed food supply, as indicated by the small decrease in the average size of fish entering the fishery (Jones and Hislop 1978). The 1958 autumn age class of herring in the Gulf of St. Lawrence was 44 times as large as the smallest age class recorded (1969) and the 1959 spring age class was over 60 times as large as the 1969 age class. These contrasting abundances provide some idea of the tremendously different recruitment success from year to year, reflecting how variable water column conditions and plankton communities can be over time. There is little evidence of how the peak abundances due to strong recruitment relate to the carrying capacity of marine water columns. It is clear, though, that primary production does not change by 60X from year to year.

To really assess how close populations of marine fish are to the carrying capacity of their food supply, we need an idea of the food resources and demands by fish. There are very few instances where both abundance of food and requirements for growth of fish have been measured. Burd (1965) compiled data on the size of adult (3-, 4-, and 5-year-old) herring in the southern North Sea and the average density of *Calanus*, a favored food item for herring. Herring mature at about 3 years of age. The size of adult herring is *not* related to food availability at the time of collection of the sample, but, if size of the 3-year-old herring is plotted versus the average density of *Calanus* during the 3 years of life of the fish (Fig. 9-4), a response in growth during early life to prey abundance is evident. These data are certainly not conclusive, since many other factors could have affected fish length, and because older fish grow more slowly than younger fish. The example does corroborate the evidence seen earlier (Table 4-3) that growth in young fish is more closely coupled to abundance of food than is growth of adult fish.

The importance of a growth response to prey density such as seen in Figure 9-4 is that larger young fish have better survivorship, mature faster, and become more fecund adults (Sections 4.4, 7.4). Thus, the

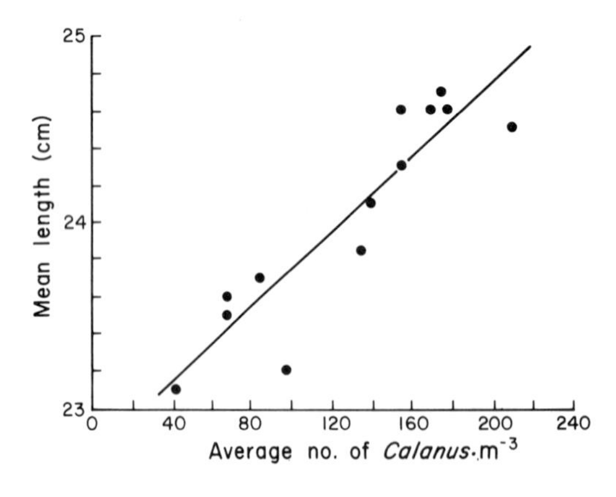

Figure 9-4. Length of 3-year-old herring in the southern North Sea in relation to the abundance of *Calanus*, a favored food item. Data from Burd (1965).

developmental response discussed in Section 5.4 is a major potential mechanism whereby population numbers may be adjusted to food supply. This growth response has been consistently observed in freshwater, where fish predators effectively control prey populations (Le Cren, 1965). In marine environments the abundance of adult fish seldom seems to affect the development response, implying a looser coupling of numbers of pelagic marine fish to their food resources.

Pelagic bony fish have generally adopted a reproductive strategy very similar to the complete metamorphosis of terrestrial insects: their larvae are very numerous and use very different resources than the adults. Other large marine predators, of course, use other strategies. Mammals, sharks, skates, and birds produce few, large offspring that by size and feeding habit enter the ecosystem at a similar position to the parent. Gelatinous zooplankton use both the few-large and many-small strategies, depending on the species.

The reproductive strategy used by pelagic bony fishes has a major drawback: producing lots of larvae necessarily means that they will be small, in the order of 1 mm or so. The problem with being small in the plankton is that the steps in the food chain are primarily based on size of predator and prey [cf. Section 6.31 and Ursin (1973)], and a growing fish larva essentially has to grow its way up the food web, and is exposed to a gauntlet of predators at each stage.

A further problem is that the larvae are true plankters and drift with the water. During the period of larval drift, the small larvae are very susceptible to haphazard hydrographic events which may take the larvae to unsuitable areas. These multiple detrimental factors not only lead to a marked mortality of young fish but also impose what appears to be a random variability to rates of larval survival. The variability is large and haphazard enough to overwhelm the influence of density of parent and cohort stocks at some point along the survival curve of the cohort.

Fish larvae live in a habitat populated by numerous other larvae and zooplankton of similar size, all feeding on similar food items of appropriate sizes. The influence of cohort density is significant in early life, as evidenced by the density-dependent mortality and growth of young fish (Section 4.4). Density-dependent competition for food is therefore a dominating feature of larval life (Section 4.4). The larvae that are most successful at feeding grow faster, stay healthier, are more active, and more sucessfully escape predators.

Densities of larvae fall very markedly as the larvae age and competition may become less intense. The density dependence lessens rapidly with age in most fish species due to the random nature of the predation and hydrographical events that affect mortality and dispersal. This uncoupling of survival of young individuals from adult density makes it difficult to find density-dependent recruitment to adult populations in most pelagic species (cf. Section 4.44). The lack of such a relationship suggests that adult fish stocks rarely over-exploit their food resources. There is also no clear relationship between growth rates and densities of adult fish (Section 4.4).

Another telling fact that suggests that adult fish populations are generally low relative to resources is that many fish stocks are maintained mainly through sporadic but exceptionally strong year classes (Cushing, 1975). As pointed out earlier, in general, perhaps only 10 out of every million fish larvae survive to adult size. There are so few larvae that make it through the tight bottleneck of larval and juvenile mortality that adult stocks are seldom abundant relative to resources. On occasion the severe mortalities that affect young fish may be slightly and fortuitously relaxed, and even a small increase in survival rate may result in a very strong cohort. Strong-year classes can be achieved even with minimal egg production; for example, the large size class of herring of 1958-1959 in the Gulf of the St. Lawrence (Fig. 4.9, left) arose out of a very small egg production.

Recruitment of strong-year classes is, by all appearances, a common but haphazard event that depends on unknown but probably climatic and hydrologic factors. Often strong-year classes of particular species may occur simultaneously over much of a large area such as the North Atlantic, suggesting that a large-scale climatic event is responsible.

Stunting of individual fish within a strong cohort is rare and, in fact, strong cohorts usually are characterized by faster growth of young individuals (Gulland, 1971). This is unusual since mortality and growth of larvae and young fish are generally density-, and by extension, food-dependent. It is as though larval mortality due to predators or other factors had been reduced at the same critical time that larval resources increase. The common occurrence of the strong-cohort recruitment pattern and high growth rate of the cohort throughout the subsequent years suggests that there may be a surplus of food not used by the extant, average-year fish stock.

The strong-cohort reproductive pattern may be a result of what might be called an *r*-selected, exploitative, or "many-small" reproductive strategy (cf. Section 3.4) and of life in water columns subject to high variability. Conditions in the water column are extremely heterogeneous both in space and in time. Fish release as many larvae as possible into this hazardous and variable environment, and every so often circumstances allow greater survival of a cohort. Since the extant population is usually below carrying capacity, the strong cohort can proceed to grow well.

9.413 Contrasts in Role of Fish in Marine and Freshwater Environments

The observations discussed in the previous paragraphs suggest that substantial increases in abundance of marine fish do not lead to substantial changes in mortality, growth, recruitment, or fecundity of adult stages. Some indication of incipient resource limitation such as reduced growth are evident only in environments where very high densities are found. The body of evidence as a whole suggests that these predators may be limited by their food resources as young fish but that adults are not limited by and therefore do not limit their food resources. This contrasts with the marked effectiveness of freshwater fish in controlling prey populations.

Oceanic food webs often have five steps or so (Table 3-1) while in freshwater three steps may be more typical. This is an as yet unexplained but significant qualitative difference between the two environments. We may speculate that freshwater fish larvae are therefore exposed to fewer types of predators, and are perhaps less subject to predatory mortality. In freshwater there may therefore be a better chance for the effect of parent density to be reflected on recruitment, and therefore a greater likelihood of a density-dependent response by the total fish population to changing abundance of the prey. In part this may be why freshwater fish are more effective at controlling the size of prey populations.

The contrasts in performance between marine and freshwater fish might more likely be due to the large differences in density of food and fish between these two habitats. It is not easy to make comparisons of the representative densities of marine and freshwater fish, and, in fact, there are few methods available to do so directly. Horn (1972) estimated in a very rough way the number of species and average number of individuals per species in fresh and seawater. The final point of his calculations was that each individual marine fish had about 10 to 10,000 times more volume of water to deal with than each freshwater fish. Clearly this calculation needs to be taken lightly due to the assumptions needed to make the calculations, but it does suggest that marine waters are dilute not only in the nutrients, algae, and grazers, but also in regard to fish.

We then have a situation in the sea where both fish and their food are in low abundance compared to freshwater environments. Yet fish, at least adult stocks, seem, in general, not to be food limited. Why don't marine fish tax their food supply? It may be that the crucial bottleneck is the larval stage. Mortality of larval fish in the sea may be so severe due to catastrophes, competition for food, and predation that, as said above, there is seldom a chance for the numbers of adults to accumulate.

Why do marine fish not make more eggs? Actually marine fish do produce more eggs than freshwater fish. The modal marine bony fish produces about 10^6 eggs per spawning (with a range of 100 to 3×10^8), while freshwater bony fish produce a mode of about 10^4 eggs per spawning (with a range of 10 to 10^7) (Altman and Dittmer, 1972, pp. 149–150). Since marine and freshwater fish are not too different in size (marine fish *on average* are probably only a bit larger than freshwater species in spite of swordfish, tuna, marlin, and other giants), the production of more eggs means that the eggs are smaller. Perhaps there is a lower limit to the size of a viable egg; smaller eggs may just have too many trophic steps to grow through, or perhaps there are anatomical and physiological constraints.

Alternatively, why do pelagic marine fish not evolve better larval survival ability? Some have: sharks, for example. Bony fish may have taken irreversible evolutionary steps in embryology so that larger larvae are not feasible. In addition, perhaps planktonic conditions are so unpredictable that the *r* strategy of opportunism has proven favorable in the long term.

Freshwater fish may be freer to respond to food abundance because larval mortalities in freshwater fish are lower, perhaps because there are fewer trophic steps and less variability. If this is true it might explain why there is a greater density dependence of growth and of age of first reproduction in freshwater than in marine fish (Cushing, 1981).

This section was headed as "speculations," and most of the above discussion was just that. Much critical research remains to be done on this subject. We need to add that fishes of certain marine communities (coral reefs, salt marshes, estuaries) may achieve very high densities, and may be less exposed to variable hydrography. Many fishes of coastal environments tend to take better care of their eggs, either by actual brooding or care in selection of a potential site for oviposition. Not all species engage in these activities, but in species that do so, larval mortality is less random or intense. The marine fish of some coastal environments perhaps are thus more likely to limit prey abundance (cf. Section 9.3). In the open ocean, the evidence available suggests that populations of adult fish are not likely to control abundance of their prey.

9.42 Competition and Predation by Mammals and Birds in Pelagic Communities

Mammals and birds, in contrast to most fish, use a *K*, saturation selected, or few-large reproductive strategy (cf. Section 3.4), and this may have fundamentally different consequences for the role of those animals in the exploitation of their food resources. Unfortunately, there is little concrete information on the way in which pelagic communities may be structured by mammal and bird activity. The only data available by which we can infer some of the effects of mammals and birds as predators on pelagic communities are provided by the consequences of the intensification of whale harvesting in the southern oceans since the late 1920s (Fig. 9-5) due to the development of the exploding harpoon, faster whaling vessels, and factory ships.

The whalers behaved as a typical predator, first harvesting the largest whales accessible to their gear, focusing their attention in the early years on the blue whale. When stocks of blues began to dwindle, the whalers turned ("switched"), again in classical predator fashion, to two other large whales, the fin and the humpback. In the 1960s, as the larger whales became very scarce, the whale catch fell steeply and the remaining whaling fleets turned to the considerably smaller sei and even to minke whales.

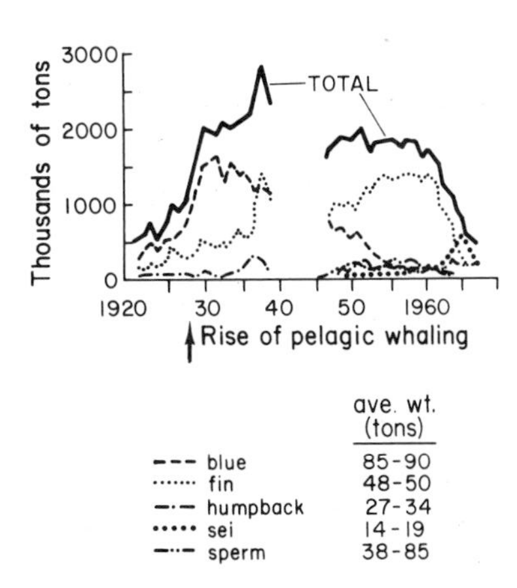

Figure 9-5. Tonnage of Antarctic whale catches over time. The whale species include the blue (*Balaenoptera musculus*), fin (*B. physalus*), sei (*B. borealis*), minke (*B. acuturostrata*), humpback (*Megaptera novaeangliae*), and sperm whale (*Physater catadon*). The latter feeds on squid and fish, a diet rather unlike the copepod, krill, and small fish eaten by the baleen whales represented by the other two genera. Adapted from MacIntosh (1970); weights from Laws (1977a,b) and Lockyer (1976).

The stock of plankton-feeding whales was sharply reduced after only 30 years of industrialized whaling.

By the late 1960s stocks of baleen whales had been lowered in the Southern Ocean to about a third of their former numbers. Blue and humpback whales were the most affected, having been reduced to 3 and 5% of the estimated initial stock (Laws, 1977a,b). The sperm whale was thought to be half as abundant as formerly.

Such large changes in abundance of major consumers altered the apportioning of biomass in the Antarctic pelagic food web (Table 9-3). The total tonnage of whales is now perhaps one fifth of what it was. Before the pelagic fishery reduced whale stocks, the annual migration of whales in and out of the Southern Ocean to more northern waters drained over six times the amount of biomass away from the Antarctic compared to

Table 9-3. Rough Estimates of Whale and Seal Stocks, Migratory Losses, and Food Consumption in Prepelagic Whaling and in Recent Times[a]

	Whales			Seals
	Initial	Recent	Percentage left unconsumed	Recent
Stock	46	8	83	3.5
Loss from Antarctic Ocean	19	3	84	Not migratory
Food consumption				
Krill	190	43	77	64
Fish	4	1	75	6
Squid	12	5	58	7

[a] Values are in millions of tons. Data on food consumption by penguins, other birds, and fish are not available. Adapted from Laws (1977a,b).

present losses. This loss is due to the fact that whales lose weight while away from the rich feeding areas of the Southern Ocean (Hinga, 1979).

If production and biomass of krill, fish, and squid are about the same over the years, the reduction in whales leaves considerable amounts of food unconsumed (Table 9-3). There is, unfortunately, no direct information about whether the densities of krill, squid, and copepods have actually increased in response to increased food. We need, therefore, to look for indirect evidence of such a response by predators still present in Antarctic waters.

Available data on current whale stocks suggest that availability of food per whale has increased. In contrast to the fish populations discussed earlier, whale populations seem very closely coupled to their food resources, as shown by changes in variables known to be density and food dependent (cf. Chapter 4). The reduction of whale densities has, in fact, resulted in increased growth rates of the remaining individuals (Laws, 1977a,b). The percentage of the blue, fin, and sei females that are pregnant has also increased since the 1900s; there is evidence of earlier maturity in fin and sei whales. The halt in whaling during the Second World War allowed increased survivorship and lowered food availability for a short time, reducing the reproductive performance of fin and blue whales. The hiatus in whaling during the war may also have resulted in a brief period of reduced growth of average individuals, since smaller fin whales were taken while whales were more abundant (1945–1946) than during a period of lower whale standing stocks (1962–1963) (Fig. 9-6).

The increase in growth and maturation rates when more food is available per capita* have also been recorded on other environments for harp, fur, and elephant seals (Sergeant, 1973), so it seems that in general, marine mammals are food limited.

All the above changes in the whale populations suggest that whales were limited by food resources prior to industrial whaling. Where there were fewer whales, more food was available, and this led to faster growth, earlier maturity, and successful pregnancies. These increased pregnancy rates turned into more recruitment of young since survival rates of young are so much greater than those of most fish (Chapter 3). Such reproductive changes in the sei whale population are especially convincing because the response took place before the whalers began to exploit sei populations extensively (Gambell, 1968). The general conclusion from all these observations is that whales remaining after exploitation responded as if there were an increase in food abundance.

---

* Such improvements in the "standard of living" after density has been lowered seem to be a rather general phenomenon. In the Baltic Sea, growth of flatfish increased after those populations were thinned out markedly in the 1920s (Persson, 1981). The phenomenon is not limited to marine organisms: food supply, wages, and quality of life in general improved for European people as a whole during 1350–1550, a period following severe outbreaks of the Black Death (Braudel, 1981, p. 193).

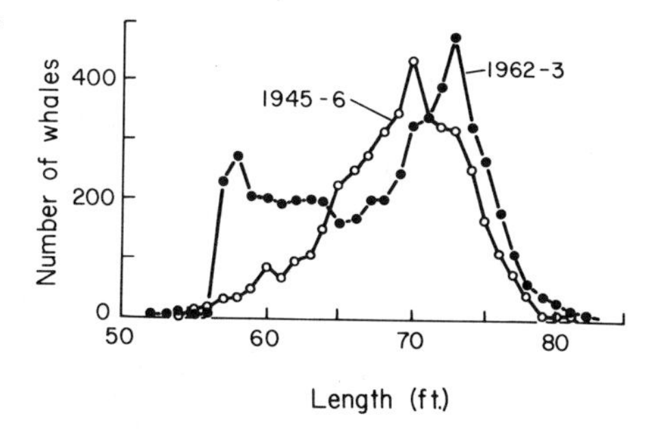

Figure 9-6. Length composition of female fin whales caught in the Antarctic in the 1945–1946 and 1962–1963 seasons. The small mode in the 1962–1963 data is an artifact of the well-known "elastic" tape measure used by whalers on smaller whales after the introduction of the 57-ft lower size limit. Adapted from Gulland (1971). © Academic Press, reprinted by permission.

Other vertebrate predators in the southern oceans also seem to have responded to the apparent increase in availability of food (Table 9-4). Data on the several penguin and six seal species present in Antarctica are very difficult to get because penguins and seals are sparsely distributed, often are difficult to see and count among the ice floes, and spend much time diving. Surveys of nesting colonies show that four of the five major penguin species in Antarctica have increased in abundance. The fifth (the emperor) is a species that feeds inshore and probably was seldom in competition with whales. There is information on two primarily krill-

Table 9-4. Recent Changes in Penguins and Seal Populations in Antarctica[a]

	Principal foods	Changes in population
Penguins		
Emperor (*Aptenodytes forsteri*)	Fish	No significant increase
King (*A. patagonica*)	Fish	Marked increase
Adelie (*Pygoscelis adeliae*)	60% krill, 40% fish and other	Local increases in whaling areas
Chinstrap (*P. antarctica*)	Krill	Marked increases, extended range
Gentoo (*P. papua*)	Benthic fish, some krill	Some increases
Seals		
Crabeater (*Lobodon arcinophagus*)	94% krill, 3% fish, 2% squid	Earlier maturity, presumed increase in numbers
Fur (*Arctocephalus gazella*)	34% krill, 33% fish, 33% squid	Population explosion especially in overlap with range of baleen whales; appearance of new colonies

[a] Data from Conroy (1975), Stonehouse (1975), Laws (1977a,b), Payne (1977), Øritsland (1977), and Hinga (1979).

feeding seal species (Table 9-4), and for both there is some evidence of increased numbers, especially for the fur seal. The seals presently in the southern oceans have increased to the point where their estimated consumption of krill, fish, and squid exceed that of whales (Table 9-4).

Although the above observations are not very convincing individually, in the aggregate they do suggest that other open-ocean predators currently in the southern oceans have responded to the greater food availability due to removal of whales by increasing their own abundances. The species involved (mammals, birds) do not use the "many-small" reproductive strategy of fish, and perhaps thus avoid the loose coupling to resources typical of populations of adult bony fish. The Antarctic surveys imply that prior to industrialized whaling the abundance of whales had increased to the level of being limited by and limiting their food supply. After the advent of industrialized whaling, other predators (seals, penguins) used the no-longer-consumed food resources, and their populations, in turn, have increased.

Scarcity of whales, public opinion, and the exigencies of economics, especially fuel costs, have caused the whaling effort to dwindle, and soon there may be little if any large-scale hunting of whales. Will they return to their earlier abundance? Perhaps some, such as blue whales, have too few individuals left to do so,* but other species certainly can recuperate earlier abundances. Human ability to affect natural environments, however, never ends: Russian efforts to harvest krill in the Antarctic are already under way (Omori 1978). With so many protein-short human populations, it seems inevitable that krill will be exploited.

Whether whales can recover their earlier abundance in the face of a new, efficient competitor for krill, we can only speculate at present, but it is certain that a new balance will take place. There is a very pertinent comparison available in the interactions among birds, men, and anchovies in the Peruvian upwelling region (Fig. 9-7). Sporadic natural episodes of El Niño† usually lead to massive reductions in anchoveta. The lack of fish causes populations of the fish-eating guano birds to crash, as occurred in 1957 and 1965. While the birds recovered readily from the earlier crash, competition from men produced a different result after 1965.

By the 1960s the human harvest of anchoveta had become quite large, about 7.5×10^6 metric tons, compared to $0.2–0.3 \times 10^6$ tons commercially harvested during the late 1950s. The birds in the late 1950s ate about 2.5×10^6 metric tons of fish. In the late 1960s the sustainable yield of anchoveta

* Actually blue whales are not rare in certain parts of the ocean (W. Watkins, personal communication), and since they have notable ability to move for very long distances, there is a good chance of recolonization of locally depleted areas if whaling pressure stops.

† El Niño (The Child) episodes start toward the end of the calendar year (hence the connection to the Nativity), when warmer water intrudes into coastal areas of Peru, preventing upwelling of cold, rich water. The result is markedly lowered primary production and massively reduced abundance of anchoveta. The hydrodynamics of El Niño are connected to events going on at surprisingly distant areas of the tropical Pacific (Wyrtki, 1975).

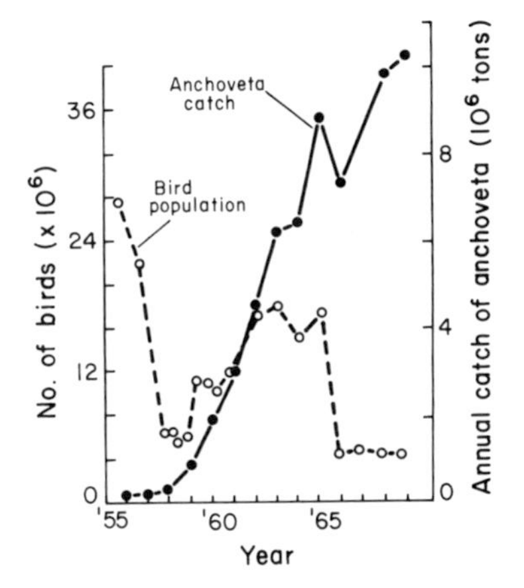

Figure 9-7. Commercial catch of anchoveta (*Engraulis ringens*) off Peru and census of guano birds during the same span of time. Adapted from Schaefer (1970).

was calculated to be about 10×10^6 metric tons; the birds took 0.7×10^6 metric tons. The fishermen had by then developed the capacity to take at least the remaining 9.3×10^6 metric tons. The result is that guano birds did not therefore have the surfeit of fish on which to base a return to earlier abundances. This may be the situation that may await the whales if the krill fishery becomes well established.

The Peruvian upwelling is a second example of a pelagic ecosystem where the species occupying the upper trophic positions seem to be closely coupled to food availability. As in the case of the whales in the southern oceans, the predators that control prey abundance in the Peruvian upwelling are mammals or birds. These predators, as mentioned already, use a K-selected reproductive strategy which may be best suited to control populations of prey or to increase numbers to closely match available resources.

All species of marine mammals have single offspring, even though they belong to four different orders whose nearest terrestrial relatives have variable and larger clutch sizes (Estes, 1979). In marine species the strongly developed parental care reduces mortality of the young, especially from predation. This reproductive strategy may have set up a situation where the numbers of marine mammals would be limited by resources, rather than by predators. If marine mammals are resource-limited, they in turn are likely to have a strong influence on the structure of the communities of organisms on which they feed.

Whales, seals, penguins, and most marine birds thus seem to exist near the carrying capacity of their environment, and show *K* or saturation selected traits such as iteroparity, few offspring, parental care, delayed age of first reproduction, and long lives. Such life history characteristics make a population easily susceptible to over-exploitation since there is little slack in the demography of such species to absorb predatory losses [cf. Chapter 4 and Estes (1979)]. Whale stocks were lowered by whaling, as were the exploited stocks of elasmobranchs, another group of *K* strategists, described in Section 3.4. Exploitation of sea otters led to extinction over most of their range (Estes et al., 1982); exploitation of the Great Auk, a marine bird ecologically very similar to penguins, led to its complete extinction.

All of our examples in this section on large planktonic predators come from rich waters, none from less productive tropical areas. Perhaps in such nutrient-poor water we can expect much closer adjustments of consumers to their food supply, perhaps even in fish, but there is little evidence of this.

We should be cautious to note that virtually all the evidence we have examined comes from exploited populations. Fishing has reduced populations of fish and other top predators over much of the sea. Fishing mortality, for example, can be 58% of all mortality in cod (Daan, 1975). The many-tiered food webs of the marine water column have at present relatively reduced populations of many top predators. For example, during and after the late fifties the tuna stocks were drastically reduced by fishing in the North Sea (Tiews, 1978). The annual consumption of fish by bluefin tuna in certain areas of the North Sea during the maximum *recorded* tuna abundance (1952) has been roughly estimated at 210,000–256,000 metric tons of mackerel and herring (Tiews, 1978). If bluefin tuna consume 75% herring, this means that about 150,000–180,000 tons of herring were eaten by tuna during their annual stay in the North Sea. In 1952 the stock of herring in the area in question was about 140,000 tons (Burd, 1978), so that consumption of herring by tuna could have removed the entire stock of herring. The reduction of tuna abundance, due to fishing, may have therefore been one reason for the very marked increase in herring (Daan, 1978) and other pelagic fish in the North Sea during the 1950s.

Such comparisons are very rough but do bring up an important notion. What we see today in the food web of the North Sea—and elsewhere in the sea—is severely affected by the ultimate top predator, the fishing fleet. Virtually all large top predators in the area have been much reduced by fishing and man may now have replaced these as the keystone predator of marine food webs. We are in the same position that a sentient otter may be in interpreting what might be going on in the coastal communities of Amchitka Island. It may be hard to separate our effects from those of other control mechanisms.

The fishing activity of man has been clearly instrumental in affecting the community structure of every marine environment. The increased stocks after the respire of World War II, the lack of herring in recent years in the North Sea, the near disappearance of tuna and whales, are among the many examples of the real impact of fishermen as the top predator. It is not yet clear how this affects our conclusion as to the role of pelagic fish in regard to their food supply.

The studies reviewed in this section provide some tantalizing glimpses at some processes that organize open water marine communities. This entire field needs innovative approaches, especially including experimentation and study at the multispecies and ecosystem levels. In coming decades, exploitation of resources in the sea will increase, and the evaluation of the consequences and the management of marine resources will require considerably improved knowledge as to the dynamics and organization of marine communities. The speculations in this section merely probe the surface of the matter.

Part IV

Organic Matter, Decomposition, and Nutrient Cycles

Fecal pellet of a copepod, photographed at three different magnifications to show shape and contents. Photos courtesy of Susumo Honjo.

The last few chapters have dealt with transformations of organic matter and energy in individuals, populations, and in food webs, and have expanded into physiological and behavioral aspects. It has been made clear that the chemical composition of organic matter is critical in many ways. Nutrients are involved in virtually all other aspects of marine ecology. To understand how marine ecosystems work, it is of vital importance to consider the role of essential nutrients. The specific mechanisms that link nutrients to primary production, consumers, and microbes have been documented in earlier chapters. In the next two chapters we delve into properties and transformation of just a few of the most important chemical elements involved in organic matter in the sea: carbon (Chapter 10), phosphorus, nitrogen, and sulfur (Chapter 11).

To be entirely consistent with the orientation to processes in this book, the material ought to have been organized in terms of nutrient uptake, regeneration, sedimentation, and so on. Instead, each element is treated separately, simply because the information available for each process and for each element is not equivalent and also because each element has idiosyncracies of its own.

The separate treatment of elements should not obscure the intimate coupling among elemental cycles. This coupling is especially tight in the formation and decay of organic matter, since element contents and their transformations, as seen in earlier chapters, tend to follow certain ratios. Nitrogen, carbon, phosphorus, and sulfur are also linked by various assimilative microbial processes. The point is that although for explanatory purposes the elemental cycles are here separated, this is by no means true in nature.

Chapter 10

Transformations of Organic Matter: The Carbon Cycle

10.1 Inorganic Carbon

The sea contains a substantial amount of carbon compared to that present in the atmosphere* or in terrestrial organisms (Table 10-1); only rock deposits contain more carbon, but these resources are not involved in the carbon cycle at ecologically meaningful time scales.

Most of the carbon in the sea is present as dissolved inorganic carbon (Table 10-1), which originates from complex equilibrium reactions of dissolved carbon dioxide and water

$$H_2O + CO_2(aq) \rightleftharpoons H_2CO_3 \rightleftharpoons H^+ + HCO_3^- \rightleftharpoons 2H^+ + CO_3^{2-}$$

The properties and consequences of the CO_2-carbonate equilibrium depicted by Eq. (10-1) are of fundamental importance to the biology and geochemistry at the sea, but an adequate treatment of carbonate equilibrium is beyond the scope of this book. Reviews of the topic are avail-

---

* The relation of atmospheric CO_2 to aqueous CO_2 in the sea has received a lot of attention recently due to the continued increase in CO_2 in the atmosphere attributable to the burning of fossil fuels (Keeling et al., 1976), and the consequent potential for disturbance of the world's climate. Recall that CO_2 is a major absorber of energy (Fig. 2-1), especially infrared radiation. Increases in CO_2 might increase the temperature of the earth's atmosphere by trapping heat in the so-called "greenhouse effect." The ocean holds about 50 to 60 times as much CO_2 as the atmosphere and is able to take up more CO_2. It is not certain just how much of the CO_2 added to the atmosphere will be taken up into the larger pool or what the effects of the increased temperature will be. In fact, some believe that the higher temperature will increase evaporation and will result in increased interception of energy such that temperatures will drop worldwide. The only conclusion agreed upon is that the effects of our altering the CO_2 content of the atmosphere will have major consequences, and that the ocean will play a large role.

Table 10-1. Approximate Size of Major Natural Carbon Reservoirs

Reservoir	gC × 10^{20}
Atmosphere (about 1973)	0.000675
Ocean	
Inorganic carbon	0.38
Organic carbon	0.01
Detrital carbonates	0.0129
Terrestrial	
Organisms	0.0164
Organic C in sedimentary rocks	68.2
Carbonate rocks	183

[a] Compiled by Skirrow (1975) from several sources.

able in Riley and Chester (1971) and Broecker (1974). More advanced treatments are provided by Stumm and Morgan (1981) and Skirrow (1975).

One major aspect of the CO_2-carbonate system is the buffering capacity it provides. Carbon dioxide or rather, carbonic acid, its aqueous form, is present mainly as the bicarbonate (HCO_3^-) in seawater, at concentrations of about 25 mg liter^{-1}. CO_2 is removed by photosynthesis and added by respiration. The amounts of CO_2 in seawater generally exceed photosynthetic demand, so the changes in pH that occur by Eq. (10-1) due to photosynthetic removal of CO_2 are relatively small. Aqueous solutions of weak acids and bases such as Eq. (10-1) resist pH changes, so that the pH of seawater is buffered between 7.5 to 8.5 by the CO_2-carbonate system, at least for time scales of tens to hundreds of years. The CO_2-carbonate system is not the sole buffering element in seawater; borate for instance, also contributes to buffering. Temperature and pressure also affect carbonate equilibria.

In anoxic water columns or in interstitial water within sediments, anaerobic metabolic reactions by microorganisms generate carbon dioxide in far larger concentrations than in aerobic seawater. These waters also have very concentration dissolved salts. In anoxic situations, pH does not therefore depend solely on the CO_2-carbonate system.

Our too-brief mention of the CO_2-carbonate system here is just one way to show that organisms are intimately linked to the carbon cycle. Because of the chemical consequences of such biological control of the CO_2-carbonate equilibrium, affecting pH, buffering capacity, and the equilibrium reactions of many other elements, much of the chemistry of the oceans is mediated biologically. There are other important biogeochemical processes, such as the deposition of carbonates in reefs and sea floor, that are also biologically influenced.

The uptake of inorganic carbon from seawater by organisms produces organic carbon, and this leads to complex transformations discussed below. In addition, there are transports of organic carbon by physical processes, transports that can have major consequences. The carbon cycles of aerobic and anaerobic environments are described separately below because many of the biological processes differ significantly in these two circumstances.

10.2 The Carbon Cycle in Aerobic Environments

10.21 Concentration of Organic Carbon in Seawater

Organic carbon in seawater occurs in dissolved forms (DOC) and in living or dead particulate forms (POC). DOC is taken to be the organic carbon that passes through certain filters (usually of a pore size of 0.2–0.45 μm) while POC is retained. Somewhat different methods of analysis are used for these two fractions. The separation into POC and DOC is not clearly justified, since there is a continuous size distribution of organic matter in seawater from 10^{-3}–10^{3} μm (Sharp, 1973). Some of the DOC is also colloidal rather than truly dissolved, but here we use the term DOC to comprise both.

The concentrations of dissolved organic matter in seawater are much larger than those of particulate organic carbon (Table 10-2); there is usually about a 100 : 10 : 2 ratio among DOC, dead POC, and living organic carbon (Parsons, 1963). The DOC is made up of many classes of compounds; metabolically some of the most valuable are low-molecular-weight carbohydrates, sugars, and amino and fatty acids. These compounds generally are present at concentrations of about 1–100 μg liter^{-1}.

Table 10-2. Range of Concentrations of Dissolved (DOC) and Particulate Organic Carbon (POC) and of Dissolved Organic Matter (DOM) in Marine and Lake waters[a]

	DOC (mg C l^{-1})	POC[a] (mg C l^{-1})	DOM (mg l^{-1})
Open ocean, surface	0.4–2	0.1–0.5	0.5–3.4
Open ocean, deep water		0.01	
Inshore water			1–13
Lake	Trace–10	Trace–10	1–50

[a] POC includes living and detrital particulates. Adapted from Parsons et al. (1977), Fenchel and Blackburn (1979), and Wetzel (1975).

[b] The proportions of POC made up by living zoo- and phytoplankton can be found in Table 1-2. Bacteria average about 6.6×10^5 cells ml^{-1} in coastal water and make up 4–25% of the total plankton carbon (Ferguson and Rublee, 1976). Bacteria are less abundant in nearshore waters (2.1 cells ml^{-1}) and offshore (0.3 cells ml^{-1}) (Ferguson and Palumbo, 1979), so that the proportion of POC due to bacteria in open ocean is very small.

In most environments much of the particulate organic carbon (POC) is dead material. This is especially true in the plankton (cf. Table 1-3), although it is difficult to identify clearly the detrital or live nature of particles, in part because detrital particles have attached microorganisms not easily separated from non-living particles.

10.22 External Sources of Organic Carbon

Carbon is introduced into the sea by various mechanisms (Fig. 10-1 and Table 10-3). In terms of the entire World Ocean, photosynthesis is the most important mechanism by which organic carbon is formed by about two orders of magnitude (Duce and Duursma, 1977; Deuser, 1979).

Fluvial and groundwater transport (Fig. 10-1) are also important but smaller carbon sources (Table 10-2). The organic matter carried by rivers, often of vascular plant origin, is relatively decay resistant. Such terrestrially originated and fairly refractory material, POC, then finds its way to the sea floor and may be a quantitatively important source of sedimentary carbon in coastal areas. Studies on several chemical tracers—hydrocarbons (Gardner and Menzel, 1974; Farrington and Tripp, 1977; Lee et al., 1979), ratios of stable isotopes of carbon (Parker et al., 1972), and degradation products of lignin (Hedges and Mann, 1979)—show that there is considerable horizontal transport of terrestrial, estuarine, and coastal organic matter to deeper sediments. There is, however, evidence that some of the very resistant compounds—humic and fulvic acids—found in sea floor sediments are formed *in situ* (Nissenbaum and Kaplan, 1972).

Horizontal transport from land can be very large in certain locations. For example, in certain areas off the state of Washington the input of terrestrial matter may constitute 95% of the total transport of organic carbon to the sea floor (Hedges and Mann, 1979). Tidal flushing of estuarine rivers and creeks transports organic matter between coastal marshes and adjoining coastal waters, as discussed in Section 7.71. There has been

Table 10-3. Estimates of the Amounts of Organic Carbon Transported to the Oceans by Various Mechanisms[a]

	g C $yr^{-1} \times 10^{14}$
Primary production by phytoplankton	200–360
Rivers and streams	3–3.2
Ground water flow	0.8
Plant volatiles and airborne particles	1.5–4
Petroleum hydrocarbons	0.046

[a] Adapted from Handa (1977), Duce and Duursma (1977), and Farrington (1980).

some dispute as to the direction of the export (Nixon, 1980; Valiela, 1983), but recent evidence suggests that marshes export carbon. Hopkinson and Wetzel (1982), for instance, found that a source of organic matter, most likely from adjacent salt marsh estuaries, was needed to balance carbon budgets of nearshore environments off Georgia. The heterotrophic activity of coastal organisms is considerably higher within 10 km from shore (Kinsey, 1981), and measured exports of organic matter and nutrients from estuarine salt marshes in Georgia seem large enough to support activity by the auto- and heterotrophs present (Turner et al., 1979).

Other mechanisms by which organic materials are transported to the sea include wind and water-borne coastal and terrestrial carbon and man-made hydrocarbons (Table 10-2) (Duce and Duursma, 1977; Handa, 1977; Wakeham and Farrington, 1980). Carbon may also be transported laterally from one marine environment to another. For example, lateral injections of organic carbon into a water column increase the oxygen consumption by heterotrophs in the affected layers of the water column, and lead to otherwise unexplainable zones of oxygen depletion (Craig, 1971; Skopintsev, 1972).

Another example of lateral transport is the movement of detritus from coastal plants to deeper waters. The occurrence of detritus of seagrasses and other coastal plants is locally common in the deep sea (Wolff, 1976; Wiebe et al., 1976; Staresinic et al., 1978). This source of carbon to the deep sea is very small, considering the area of ocean relative to that of coastal zones. The amounts of carbon needed by deep-water organisms are far too large to be supplied by terrestrial and coastal export (Hinga et al., 1979).

Within any one water mass it is the death of the producers that yields most of the nonliving POC (Fig. 10-1). Death and defecation by animals, and fragmentation of food during feeding, all add or modify organic particles, but of course these particles are derived from carbon fixed by plants. The aggregation and adsorption of dissolved organic carbon onto large complexes (cf. Section 10.1221) are further internal sources of POC (Fig. 10-1).

10.23 Internal Sources of Dissolved Organic Carbon

Dissolved organic carbon is excreted by consumers, released from broken cells during feeding, released by leaching of soluble organic carbon and by hydrolysis due to microbial extracellular enzymes, and exuded by primary producers (Fig. 10-1).

Excretion by consumers (Fig. 10-1) can account for perhaps 3–10% of the carbon fixed photosynthetically in coastal waters off California (Eppley et al., 1981). Excretion by zooplankton is perhaps of greater importance in regeneration of nutrients than of carbon (cf. Chapter 11).

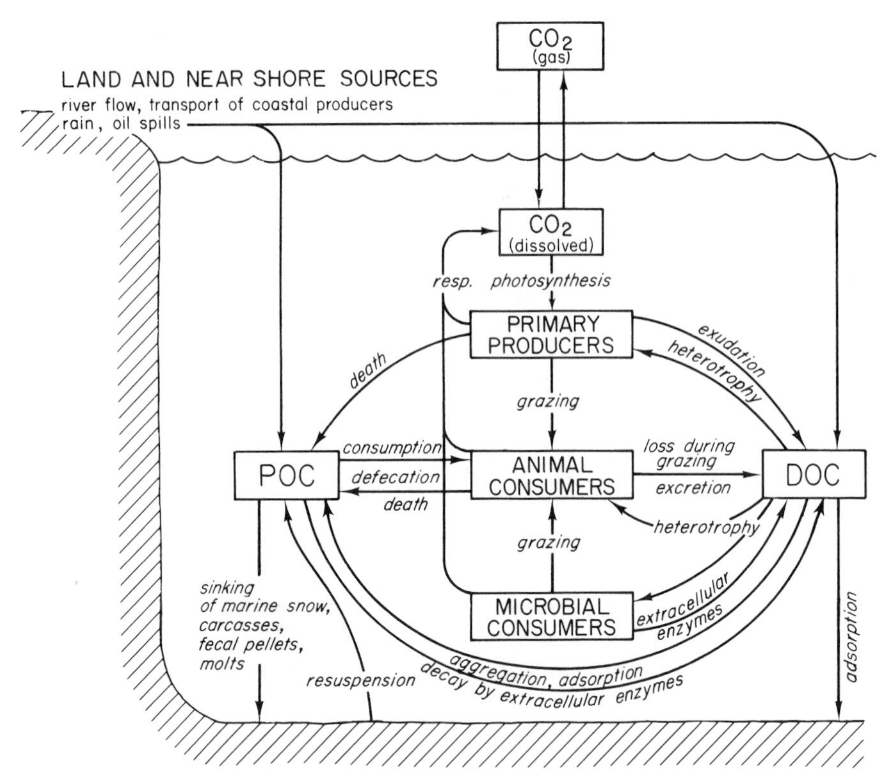

Figure 10-1. The transfer of carbon in aerobic marine environments. The boxes are pools while the arrows represent processes. The inorganic parts of the cycle have been simplified. ''Marine snow'' refers to organic aggregates and debris. There is probably some resuspension of DOC from sediments onto overlying waters not shown in diagram.

Loss of DOC from cells damaged during grazing may amount to 15–20% of the consumed carbon (Fig. 10-1). Lasker (1966) used ^{14}C-labeled carbon to show that about 16–21% of the carbon was released into the water during feeding by an euphausiid. Copping and Lorenzen (1980) used similar techniques to calculate that grazing by a copepod released 17-19% of the ingested labeled carbon.

The release of DOC from dead particles is, of course, substantial. Up to 40–60% of the initial weight can be leached as DOC from recently dead organic matter in aerobic situations (Figs. 10-1 and 10-9).

The largest single source of DOC in aerobic situations may be release (exudation) by producers (Fig. 10-1). Primary producers release DOC as a result of excretion of metabolic waste, secretion of compounds inhibitory to competitors or predators, or as unavoidable losses through membranes in an aqueous medium. Both single-celled and multicellular producers release some of the assimilated carbon, but there is some controversy as to how large this release is. Estimates of exudation by phytoplankton vary widely, from 7 to 62% of photosynthetically fixed carbon (Hellebust, 1967; Anderson and Zentschel, 1970; Choi, 1972; Smith et al., 1977; Lancelot, 1979; Larsson and Hagström, 1979). The *in situ* rates of exudation by natural phytoplankton assemblages may lie toward the low end of the range. Williams (1975) estimated that perhaps 10% of the net phytoplankton production is exuded in the open ocean, and Smith et al. (1977) calculated that about 9% of fixed carbon was excreted in the upwelling off northwest Africa. The estimated rates of exudation by macroalgae are as variable as those of microphytes. They range from 1–4% (Majak et al., 1966; Brylinsky, 1977) to up to 35–40% of the carbon fixed (Sieburth, 1969; Sieburth and Jensen, 1969; Hatcher et al., 1977). In coastal waters release of mucilage can amount to 25% and 50–60% of carbon fixed photosynthetically by macroalgae and corals, respectively (Linley et al., 1981; Crossland et al., 1980).

The extracellular products released as exudates by producers are mainly carbohydrates, but include various nitrogenous compounds, organic acids, lipids, phenolics, and many other kinds of organic compounds (Hellebust, 1974). The amount of DOC released may depend on growth conditions, light intensity, and species present.

The growth condition of producers affect the release of DOC, although it is not yet possible to generalize. In cultures, cells in log growth phase release less organic carbon than those in a stationary growth phase. *Thalassiosira fluviatilis*, for example, releases only 5% of photoassimilated carbon when growing exponentially, while over 20% is released from cells in a stationary phase. Senescence or poor growth conditions, where cells are metabolically active but unable to divide, often lead to release of organic carbon. Thus, early in a bloom or in succession (cf. Chapter 15) in nutrient-rich water, the release of DOC per cell is lower than later when nutrients are depleted (Ittekkot et al., 1981). On the other hand, Lancelot (1979) finds that phytoplankton producing at high rates release *more* DOC. This relation also applies when DOC release is expressed on a per chlorophyll unit basis, so that in high-production environments, more cells may be "leaky" than those where production is low.

Light seems to have an impact only at high intensities. The release of DOC is proportional to the amount of photoassimilated carbon in cells over a wide range of light intensities (Fogg et al., 1965). At very high light intensities, where inhibition of photosynthesis occurs, there are unusually high releases of DOC.

The variability in rates of exudation from producers may be in part due to species-specific characteristics. For example, eelgrass exudes 1.5% of its net production (Penhale and Smith, 1977). In salt marshes, where cordgrass dominates, 9% of the net above-ground production is released into the water as dissolved organic matter (Turner, 1978).

Measurements of exudation such as reported above are rough, since the compounds released by exudation tend to be easily metabolized by microbes. Many measurements of exudation are hampered by the presence of bacteria and their rapid uptake of the released DOC.

10.24 Losses of Organic Carbon

10.241 Sedimentation

Algal cells are subject to sinking. Animals release fecal pellets, cast molts, shells, skeletal parts, and carcasses, all of which also fall down the water column (Fig. 10-1). This process of vertical movement of particles is crucial to one of the most debated problems in marine ecology—the source of organic matter for consumers living far below the photic zone.

Measurements of particulate carbon in seawater done in the 1960s showed that in many parts of the ocean concentrations decreased sharply below the surface layers (Fig. 10-2), and that the low concentrations continued unchanged to very great depths (Menzel and Ryther, 1970). Such vertical profiles suggested that labile carbon compounds were degraded by heterotrophs near the surface, and that only residual nondegradable carbon compounds remained at depth. In fact, Menzel and Goering (1966) incubated particulate matter from various depths and found that bacterial growth took place on POC from surface water but not on POC from 200 to 2000 m. These results suggested that there was little if any vertical flux of organic carbon and that the carbon present at depth consisted mainly of biologically unavailable compounds. How, then, do the deep water and benthic faunas obtain food?

The water samples studied by Menzel and colleagues were obtained by collecting water with water samplers that closed at a desired depth and collected some tens of liters of water. This method of collection samples the small abundant particles that are more or less suspended in the water. It does not adequately sample the large, rare particles that sink through the water column, yet these large particles are largely responsible, as will

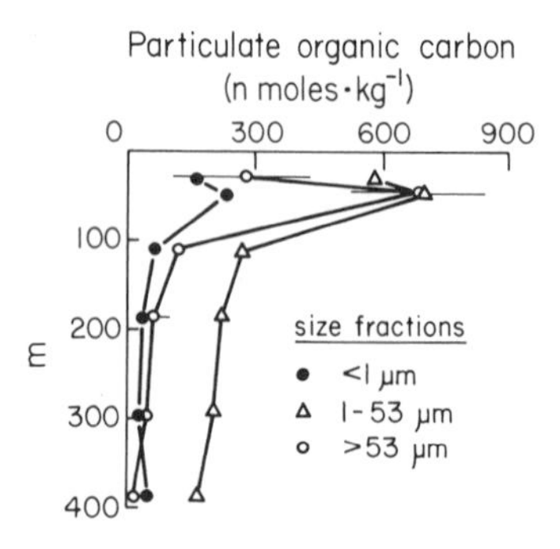

Figure 10-2. Organic carbon profile for three particle size fractions. Samples obtained with a large volume *in situ* filtration system in the tropical Atlantic. Adapted from Bishop et al. (1977). © Pergamon Press Ltd., reprinted with permission.

be seen, for the carbon fluxes that supply the needs of the deep water consumers (Fournier, 1971, and many others).

Various approaches to measure vertical flux of large particles have been used. The various approaches have included comparison of production in the photic zone with consumption in the water column and benthos; others have calculated flux from vertical profiles of organic matter in the water column. Another approach has been to make direct measurements using sediment traps.* There is substantial variation in estimates of POC leaving surface waters using these various methods even in similar areas, for example, in measurements within the Equatorial Atlantic and the Pacific gyre (Table 10-3). Some of this variability is regional due to horizontal differences in productivity, or vertical due to lateral advection of water masses containing different amounts of particulate matter. Unfortunately, due to the cost of the deployment of traps, replication is rare, so it is hard to assess the variability within a site.

The range of values for export of carbon from oceanic photic zones to deeper water obtained using sediment traps is 1–100 mg C m^{-2} day^{-1}. The flux in nutrient-rich coastal waters is greater, ranging from 30 to 600 mg C m^{-2} day^{-1}. The intensity of production on the surface therefore affects the extent of carbon flux, as can be seen in Figure 10-3, where values of primary production are plotted against the resulting carbon fluxes from the surface layer.

Since intensity of production varies seasonally (cf. Chapter 14), so does the flux of carbon to the deep sea. Near Bermuda (Fig. 10-4) the flux varies from 20 to 60 mg of total particles m^{-2} day^{-1} (Deuser et al., 1981). Such pulses provide one of the few variations that may be experienced by the fauna of the otherwise very constant deep sea environment. The more

* Sediment traps are of remarkably variable design (Blomqvist and Hakanson, 1981). In general they are containers of various sizes and shapes open at the top and are moored at the desired depth to collect particles for certain periods of time. Decay of the collected organic matter is possible, especially at warm temperatures in shallow water, so poisons are often added to the traps to prevent decay. Animals that eat detritus may also be attracted to traps and may consume detritus; in poisoned traps, the attracted consumers may be killed, so that measurements of accumulation may be increased. The design and deployment procedures for sediment traps present several problems. The opening-to-height ratio changes the effectiveness of traps (Hargrave and Burns, 1979; Gardner, 1980, 1980a). The raising of traps to the ship may result in loss and alteration of the captured particles. Some sediment traps of different shapes and sizes may be especially susceptible to error in strong currents, since the containers themselves alter hydraulic flow lines such that under and over trapping can result and the faster the current the larger the effects. Particles of different size react differently in any given velocity of current, so it is not a simple matter to apply a correction factor to the trap based on flow velocity. One way to reduce the effects of currents is to use drifting traps that move along with a water mass, collecting particles as they go (Staresinic et al., 1978). A preliminary report of recent intercalibrations (Spencer, 1981) show that different traps gave very different measures of flux for one site in the Panama Basin of the Pacific, ranging from a few to about 200 mg m^{-2} day^{-1} of material. The size and shape of traps seemed less important than other unidentified sources of variation.

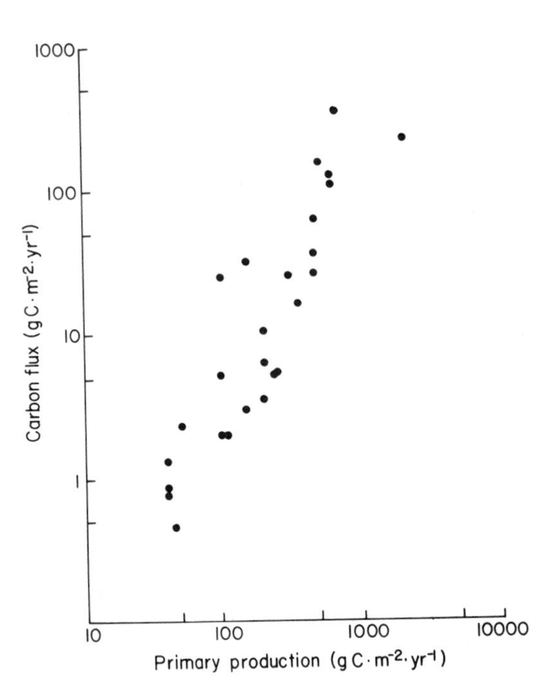

Figure 10-3. Rates of carbon from the surface layers to deeper waters in relation to the rate of annual production in the photic zone. Data from many authors compiled by Hargrave (in press).

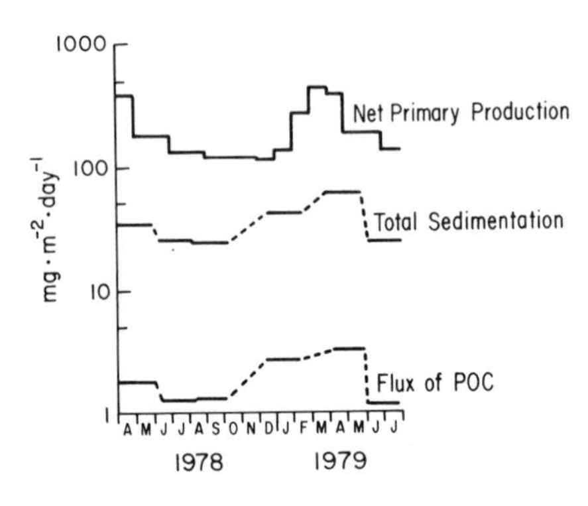

Figure 10-4. Seasonal pattern of net primary production [from Menzel and Ryther (1961) for the years 1957–1960], total sedimentation and POC flux into traps located at a depth of 3,200 m and moored at 1,000 m above the sea floor in the Sargasso Sea. Adapted from Deuser and Ross (1980). © 1980 Macmillan Journals Ltd., reprinted by permission.

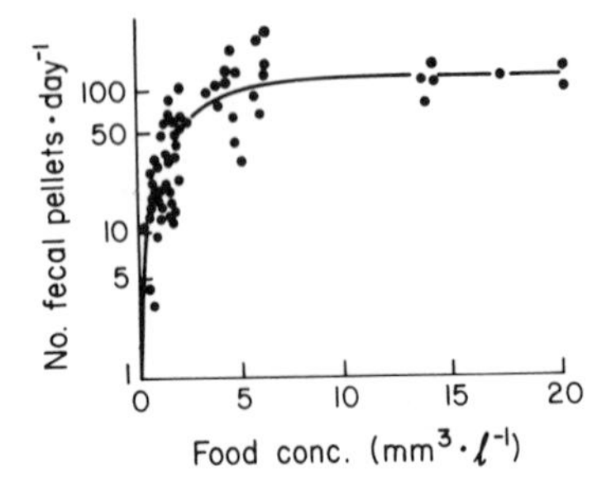

Figure 10-5. Egestion of fecal pellets by adult female *Eucalanus pileatus* feeding on various concentrations of phytoplankton. Adapted from Paffenhöfer and Knowles (1979).

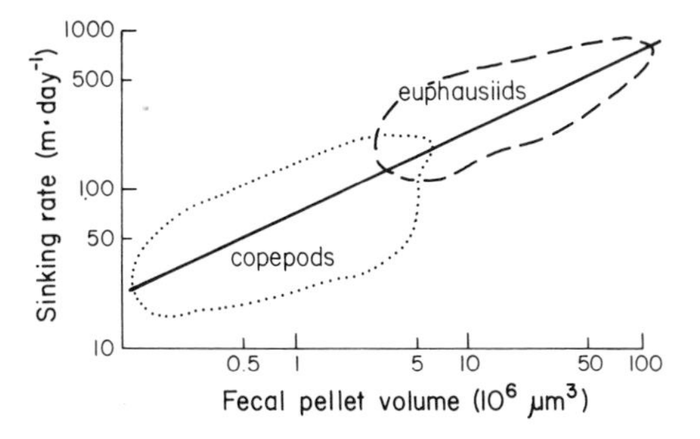

Figure 10-6. Relation of sinking rate to volume for fecal pellets of copepods and euphausiids in the Ligurian Sea. Adapted from Small et al. (1979).

production, the more food available to zooplankton, and therefore the higher the rate of egestion of fecal pellets, up to a threshold beyond which feeding rate (Fig. 10-5) and release of fecal pellets do not increase further. Most of the time algal production rates are low and cell concentrations are such that increased production increases the release of fecal pellets.

Large particles such as fecal pellets may be only a small proportion of the suspended organic carbon in surface water [4% in the tropical Atlantic (Bishop et al., 1977)], but the *flux* of large particles may be large. The relatively greater specific gravity of large particles increases their fall velocity (Fig. 10-6); for example, a fecal pellet containing several coccolithophorid cells may sink 160 m day^{-1}, while a single coccolith sinks at 0.14 m day^{-1} (Honjo, 1976). The export of POC out of the photic zone due to the sinking of fecal pellets can be large; Honjo (1980) estimated that 92% of the coccolithophorid production in a station in the Equatorial Pacific was transported to the sea floor.

Larger particles (mm to cm in diameter) fall even faster through the water column.* Such large organic aggregates may contain thousands of fecal pellets and fall fast (43–95 m day^{-1}) through the water column. The fall of such particles may remove 3–5% of the POC and 4–27% of the particulate organic nitrogen daily from surface water (Shanks and Trent, 1980). Thus sinking of fecal matter and relatively large aggregates can lead to very significant losses of organic matter from the water column (Fig. 10-1).

There are other indirect lines of evidence that suggest that fast sinking by large particles is a major pathway by which organic matter reaches the sea floor. This evidence includes the synchrony of seasonal pulses of primary production at the surface of the sea and the flux of organic matter at great depth (Fig. 10-4). The very fast dispersal of recently manufactured industrial chemicals (polychlorinated biphenyls) to great depths

* Some of this so called "marine snow" does not sediment out of the water column and is perhaps the site of substantial amounts of primary production (G. Knauer, unpublished data). In addition to algal cells such aggregates are also enriched microenvironments for bacterial and protozoan populations, which are five orders of magnitude more abundant in marine snow than in the surrounding water (Caron et al., 1982).

(Harvey et al., 1973) also argues for the transport by adsorption onto fast sinking large particles.

10.242 Consumption

10.2421 Ingestion of Particulate Carbon

Most of the organic matter within the photic zone and that leaving the photic zone is consumed (Fig. 10-1). Some calculations show that at least 87% of the organic carbon, 91% of the nitrogen, and 94% of the phosphorus is recycled above 400 m in the Equatorial Atlantic (Bishop et al., 1977). Degens and Mopper (1976) estimate that 93–97% of the flux of organic matter is recycled within the oceanic water column.

Grazers in different layers use particles that fall from the layer above. In deep water the "green" fecal pellets containing phytoplankton are far less abundant than "red" fecal pellets containing mainly mineral sediments (Honjo, 1980). This suggests that the zooplankton that produce pellets at different depths have very different foods available, and that only shallow-living zooplankton have access to phytoplankton. The peritrophic membranes surrounding fecal pellets slowly disintegrate as the pellets sink, probably due to microbial attack. The degradation is faster at warmer temperatures (Honjo and Roman, 1978; Iturriaga, 1979). The pellets are also consumed by the water column fauna.

Consumption of particles falling through the water column by microbes and fauna has not been measured directly but must be large, and is responsible for the high degree of recycling of organic matter and nutrients within the water column. These consumption processes produce an exponential decrease in the flux of carbon as a function of depth below the photic zone (Fig. 10-7). The decrease in POC down the water column is reflected in the markedly lower rates of oxygen consumption at 5,000 m (about 0.01–0.1 μM liter^{-1} year^{-1}) compared to surface water (about 1–100 μM liter^{-1} year^{-1}) (Deuser, 1979). In shallow water about 5–50% of the primary production reaches the deepest traps while only 0.8–9% does so in oceanic waters (Table 10-2). Although this may partially depend on

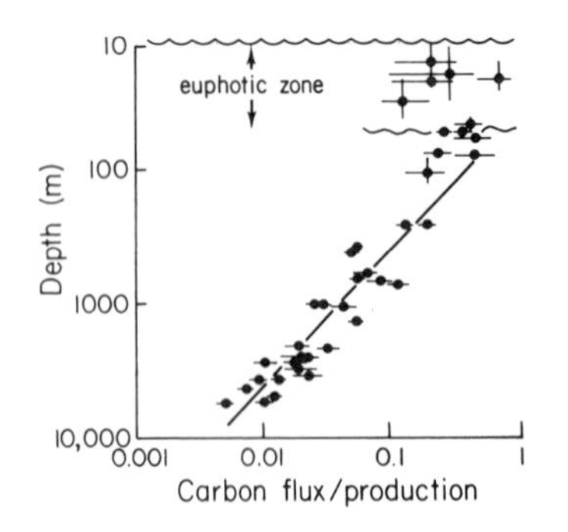

Figure 10-7. Carbon flux standardized to production in the photic zone in relation to depth, for a series of observations on a wide variety of areas. The equation that fits the points is $C_{\text{flux}(z)} = C_{\text{prod}}\,(0.0238z + 0.212)$, for depth z, and $z \geq 50$, $r^2 = 0.79$. Adapted from Suess (1980). © 1980 Macmillan Journal Ltd., reprinted by permission.

differences in production in the photic zone, the reduction is mainly a function of depth (cf. Fig. 10-7, where flux is normalized to production). Depth can be equated to time exposed to consumption by animals and degradation by microbes, so the deeper the water column, the greater the disappearance of particulate organic carbon.

One result of differential sinking rates by particles of different size is that if zooplankton communities are dominated by smaller species their smaller pellets sink relatively more slowly and the sedimentation to the sea floor is smaller. For instance, 67% of primary production can be consumed within the water column over shallow shelf sites in the southeastern United States, but if small species of copepods are dominant, their small* pellets sink slowly, and only 0.2% of the primary production reaches the sea floor (Hoffmann et al., 1981). This low rate of delivery of organic matter to the sea floor probably accounts for the sparse benthic fauna of the southeastern shelf of the United States (Hoffmann et al., 1981). This is an example of the structure of one community—the plankton—affecting the function of a very different but dependent community—the benthos.

There has been much discussion on the origin of the organic matter that supports benthic production. Some of the carbon fixed in the photic zone reaches the sea floor, regardless of depth, as anticipated long ago (Agassiz, 1888). Estimates of the flux of carbon to the sea floor are variable, as can be expected by the nature of the methods used to obtain the measurements, but the estimated carbon fluxes are enough to account for a tenth to seven times the respiration of benthic infauna (Table 10-4). These are only very rough comparisons that do not always include the mobile epifauna, carbon losses to sediment, and other components of consumption besides respiration. Respiration, however, can be correlated to and is a large fraction of total metabolic demands (cf. Chapter 7). Using the respiration values as a lower estimate of carbon demand, we can see that the estimates of flux and consumption are about the same order of magnitude. Sinking of organic particles thus can probably provide enough carbon to support benthic communities.

The variability of the flux estimates, due primarily to the limitations of the ways in which we measure flux, masks what in fact may be a fairly tight coupling between production in the photic zone and activity and abundance of consumers in the benthos. We have some evidence of this in the results of the studies of the relation between zooplankton and benthos on the shelf off the southeastern United States referred to earlier. More generally, the abundance of benthos can be related to the crop of phyto-

* Fecal pellets of small copepods such as *Paracalanus* and *Oithona* nauplii may be 2×10^3 μm^3, while a large zooplankton such as the euphausiid *Meganyctiphanes norwegica* makes pellets of 10^5 μm^3 (Paffenhöffer and Knowles, 1979).

Table 10-4. Estimates of Sedimentation of Particulate Organic Carbon (POC) in Various Marine Environments and Their Ability To Support Respiration[a]

	POC leaving surface waters (mg C m^{-2} day^{-1})	Percentage of primary production reaching deepest point of measurement (depth)	Percentage of benthic respiration accounted for by POC arriving at bottom[b]	Sources
Low to moderate productivity				
Tongue of the Ocean	36.5		14	Wiebe et al. (1976b)
Equatorial Atlantic	6.8	0.8 (5,000 m)	133–667	Honjo (1980)
Equatorial Atlantic	41.2[c]		470	Bishop et al. (1977)
Equatorial Atlantic			75–160	Gardner (1977)
Equatorial Atlantic			13–334	Hinga et al. (1979)
Sargasso Sea		1(3,200 m)		Deuser and Ross (1980)
Pacific Gyre	3.6	1.5 (5,600 m)		Honjo (1980)
Pacific Gyre	91.2	9 (1,050 m)		Knauer et al. (1979)
Gulf of Alaska	34.3	2.1 (3,500 m)		Lorenzen and Schuman (unpub. data)
W. North Atlantic	5.5–16.5	4–6 (2,170–3,132 m)		Rowe and Gardner (1979)
Oligotrophic gyres (average)	1.6	6.2		Hargrave (in press)
Coastal California	30[d]	3–4 (1,500 m)		Knauer and Martin (191)
Panama Basin	103[c]	12 (1,300 m)		Bishop et al. (1980)
Panama Basin	—	2 (3,791 m)		Spencer (1981)
Panama Basin	—	2 (1,700 m)		Spencer (1981)

Moderate to high productivity				
Peruvian upwelling	533[e]	10 (50 m)		Staresinic et al. (1978)
California Current				
during upwelling	576	53 (700 m)		Knauer et al. (1979)
no upwelling	122	34 (700 m)		Knauer et al. (1979)
Dabob Bay, Washington	192	24 (110 m)		Lorenzen et al. (1981)
New York Bight	299[f]	59		Garside and Malone (1978)
Inshore and neritic waters	37–168[g]	30–46		Hargrave (in press)
Inshore and neritic waters	8–4,611	8–60 (2–65 m)		Zeitschel (1980)

[a] The depths for the first column of numbers vary from 50 to 400 m; values were selected as close to the bottom of the photic zone as the data allowed. Burial of organic matter in sediments of the deep sea is generally much smaller than consumption by organisms (cf. Gardner, 1977, for example), although not so for shallow seas (Deuser, 1979).

[b] This entry is generally calculated by comparison to measurements of O_2 consumption of the infauna (Smith and Teal, 1973; Smith et al., 1979).

[c] Obtained using a large sample filtration system where water is pumped out from the appropriate depth.

[d] About 22% of the carbon fixed by primary producers sank out of the photic zone.

[e] On average, about 19% of the daily production of POC sank out of the photic zone.

[f] Calculated on basis of O_2 consumption and vertical distribution of POC.

[g] Calculated from estimates of nitrate assimilation and use of Redfield ratio to convert to production, from Eppley and Peterson (1979).

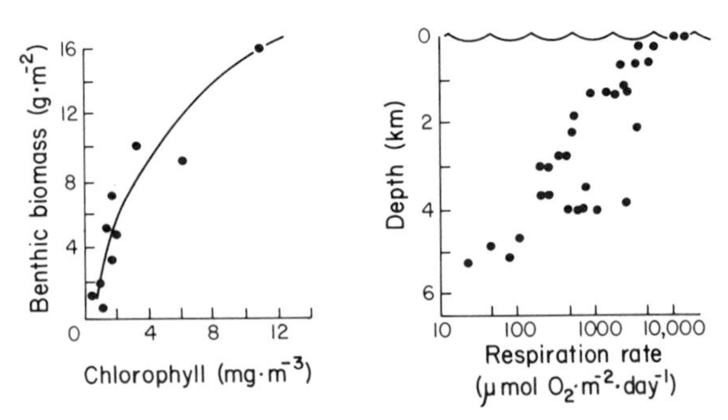

Figure 10-8. Left: Biomass of benthic macrofauna in relation to the chlorophyll concentration in the photic zone during March–May, in various areas. After Mann (1978) from data of B. T. Hargrave and D. C. Peel. Right: Respiration of infaunal bottom organisms at various depths. Compiled by Hinga et al. (1979) from various authors.

plankton during the spring bloom (Fig. 10-8, left). The respiration of the benthos, furthermore, is greatest at shallow depths (Fig. 10-8, right) because the flux of carbon to the sediment is higher at shallow depths (Fig. 10-7). Since respiration is higher (presumably because biomass of organisms is higher), more of the carbon flux is consumed, so even where the particulate carbon flux is high, there is little of the carbon that arrives at the sea floor that is not consumed by some heterotroph.*

One mechanism that assures this very high degree of consumption even in shallow waters is the intense reworking of sediments done by benthic organisms. In Buzzards Bay, a shallow, relatively rich water body off Cape Cod, Massachusetts, Young (1971) found that the bivalve *Yoldia limatula* was capable of reworking the annual deposition of sediments 2.5 times per year. This is more impressive if we realize that *Yoldia* make up only 0.25% of the animals in the sediments. The reworking maintains the surface layer in a flocculent state so that particles can be easily resuspended. Resuspension results in renewed colonization of particles by bacteria and the particles thus enriched by growth of microbes are reingested by animals.† This recycling of organic particles via resuspension amounts to 98–99.5% of the annual sedimentation.

* It is possible for the sedimentation rate to be so high (Table 10-4) that some organic matter is buried before being completely decomposed (Müller and Suess, 1979). The material that is buried, even in such cases, is probably mainly organic compounds highly resistant to microbial attack.

† This potential improvement of the quality of particulate organic matter due to microbial regrowth or colonization is important to many detritivores. For example, hydrobiid snails do not ingest fecal pellets until enough time has elapsed to allow microbial colonization (Levinton and Lopez, 1977).

10.2422 Direct Uptake of Dissolved Carbon

Some of the dissolved organic matter exuded by producers or leached from detritus is readily usable by microbes (Fig. 10-1). Ogura (1975) found that about 10–30% of the DOC from phytoplankton is labile and is degraded rapidly, with a turnover rate of 0.045–0.25 per day. Brylinsky (1977) found that 20–30% of released carbon was assimilated by heterotrophic microbes within 2 hr. Others have found even faster assimilation of exudates (Larsson and Hagström, 1979; Lancelot, 1979). Twenty-nine percent of the DOC released from kelp is converted into bacterial biomass while only 11% of the particulate debris is used by bacteria (Stuart et al., 1981).

Some of the exuded DOC is fairly resistant to microbial attack. For example, the DOC released by brown algae such as *Fucus, Ascophyllum*, and *Laminaria* contains phenolics and is very similar to the refractory yellow humic substances (gelbstoff) found in fairly high concentrations in coastal waters (Craigie and McLachlan, 1964; Sieburth and Jensen, 1969; Kalle, 1966). Much of the DOC in seawater is made up of high-molecular-weight polymers including proteins, lignins, and humic acids (Williams, 1975), the bulk of which are not readily degraded by microbes.

The annual input of labile DOC released by plants and other sources is probably small relative to the total DOC pool (Larsson and Hagström, 1979; Wiebe and Smith, 1977). Since most of this total DOC is made up of the refractory fraction, the turnover of this carbon may be very slow, up to 3400 years in some waters (Williams et al., 1979).

There is as yet no consensus on the importance of the producer:DOC: microbe pathway relative to other fates of fixed carbon. We have already discussed the rough estimates by Williams (1981) in regard to the magnitude of flow of energy through the decomposer pathways in open ocean communities (Chapter 7). In coastal water Iturriaga and Hoppe (1977) calculated that microbial activity was high enough to metabolize the carbon exuded daily by phytoplankton. Turner (1978) measured carbon lost by salt marsh grass and found the amounts were high enough to sustain the respiration of plankton. In coastal waters off Stockholm, Larsson and Hagström (1979) estimated that bacteria assimilate 49 g C m^{-2} of the total phytoplankton production of 110 g C m^{-2}, so that about 45% of primary production is channeled through the bacteria.* Andrews and Williams (1971) obtained similar results.

---

* Bacterial production was about 29 g C m^{-2} per year, so the growth efficiency was about 60%, a high value compared to most heterotrophs (cf. Fig. 7-9). For bacteria using organic matter derived from kelp, Robinson et al. (1982) calculated growth (or conversion) efficiencies of 28 to 91%, while Hoppe (1978) believes that the range is most often between 60 and 70%

Smith et al. (1977) calculated that in the upwelling area off northwest Africa bacteria could remove carbon exuded by algae fast enough to maintain a low concentration of DOC. The production by heterotrophs may be as much as two orders of magnitude lower than primary production, so that there is not much bacterial biomass available to other consumers in the food web of the sea.

Rhee (1972) believes that in nature bacteria are limited by the supply of organic carbon (recall the arguments of Chapter 8), while phytoplankton are nutrient-limited, as discussed in Chapter 2. Bacteria thus appear to have the potential to determine the chemical composition of seawater, at least in regard to DOC. Although the calculations of Larsson and Hagström (1979) and Smith et al. (1977) are only preliminary, they establish the potential importance of exudation and uptake of DOC as a significant pathway of carbon and energy flow. Future models of marine food webs will need to consider DOC dynamics.

10.243 Aggregation and Adsorption of Dissolved Organic Carbon

The action of surface charges and microbial activity prompts the formation of flakes of marine snow, a pervasive feature of the marine water column. Such flakes probably adsorb further dissolved compounds (Fig. 10-1). An earlier suggestion that particles could be formed from dissolved compounds by the action of bubbles or other natural surfaces (Baylor and Sutcliffe, 1963; Riley, 1963b) has not been consistently demonstrated to be important in further work (Barber, 1966). Aggregation processes do occur in the sea (Johnson and Cooke, 1980), but there is little evidence of the extent to which DOC is transformed to POC via this pathway.

10.224 Decomposition

10.2441 Phases of Decomposition

The decay of organic matter (POC and DOC) (Fig. 10-1) is a complex of several processes. The processes that affect the rates of loss of nonliving organic matter include leaching of soluble compounds, microbial degradation, and consumption by other heterotrophs.

Strictly speaking, leaching is not a mechanism of decomposition but a way in which soluble material is transported. We have already referred to consumption by consumers. We include these processes in this section because in most cases they are very hard to separate from microbial degradation and thus have been studied together. Leaching, degradation, and consumption act at all times, but these processes may not be equally active during different phases of decay.

The phases of decay for organic matter derived from phytoplankton (Otsuki and Hanya, 1972), macroalgae (Tenore, 1977) and vascular plants

(Wood et al., 1967; Godshalk and Wetzel, 1978) have been described. First, there is a short-lived phase during which leaching rapidly removes materials that are either soluble or autolyzed after death of cells (Fig. 10-9, top graph). The dissolved material lost from dead organic matter (detritus) is readily available to microbial heterotrophs for uptake and mineralization into CO_2 and inorganic salts (Otsuki and Hanya, 1972; Godshalk and Wetzel, 1978; Newell et al., 1981). The rates of loss during this early leaching phase are high and are most likely not mediated by microbes since losses occur at similar rates in sterile and nonsterile conditions (Cole, 1982). The leaching phase may last from minutes to a few weeks, depending on the detritus involved.

A second stage, the decomposition phase, with lower rates of loss, is longer lasting and occurs after the leaching phase. During this second phase (Fig. 10-9, first graph), microbial activity is the prime process that degrades organic matter (Figs. 10-1 and 10-9). Microbial degradation takes place mainly through hydrolysis by enzymes released from microbial cells, with subsequent uptake of the solubilized compounds by the microbes. Part of the partially broken-down organic matter is used by microbes for growth and part is respired as CO_2. Some of the detritus converted to DOC by the extracellular enzymes is lost by leaching away from the immediate vicinity of the microbes (Fig. 10-1).

Some compounds—sugars, some proteins—contained in detritus are susceptible to microbial attack, and are depleted (Stout et al., 1976; Velimirov et al., 1981; Cauwet, 1978) during the decomposer phase. Other compounds—cellulose, waxes, and especially certain phenolic compounds such as lignins—are less easily degraded, and thus during the decomposer phase the chemical composition of POC changes as a result of differential degradation (Fig. 10-9, bottom graph). The percent lignin, for instance, increases during decay. Such resistant compounds are present in vascular plants, but may be absent or less abundant in algae; hence, algal detritus disappears faster than vascular plant detritus (Tenore et al., 1982). The differences in rates among detritus of different taxonomic origin are not due to differences in the processes or phases of decay but rather in the chemical composition of the organic matter. The chemistry of decay of phytoplankton-derived detritus needs more detailed chemical study (Cauwet, 1978; Lee and Cronin, 1982).

Eventually, the remaining detritus contains primarily compounds that are degraded slowly or very slightly. This third or refractory phase may occur within several weeks in the case of phytoplankton detritus (Otsuki and Hanya, 1972) or may last months to years in detritus from vascular plants such as *Spartina* (Fig. 10-9, top graph) or *Zostera* (Godshalk and Wetzel, 1978a). The end result of the decay of both algae and vascular plants is organic matter with high contents of fulvic and humic acids. These are very refractory phenolic polymers and complexes (Nissenbaum and Kaplan, 1972), which eventually form what may be called marine

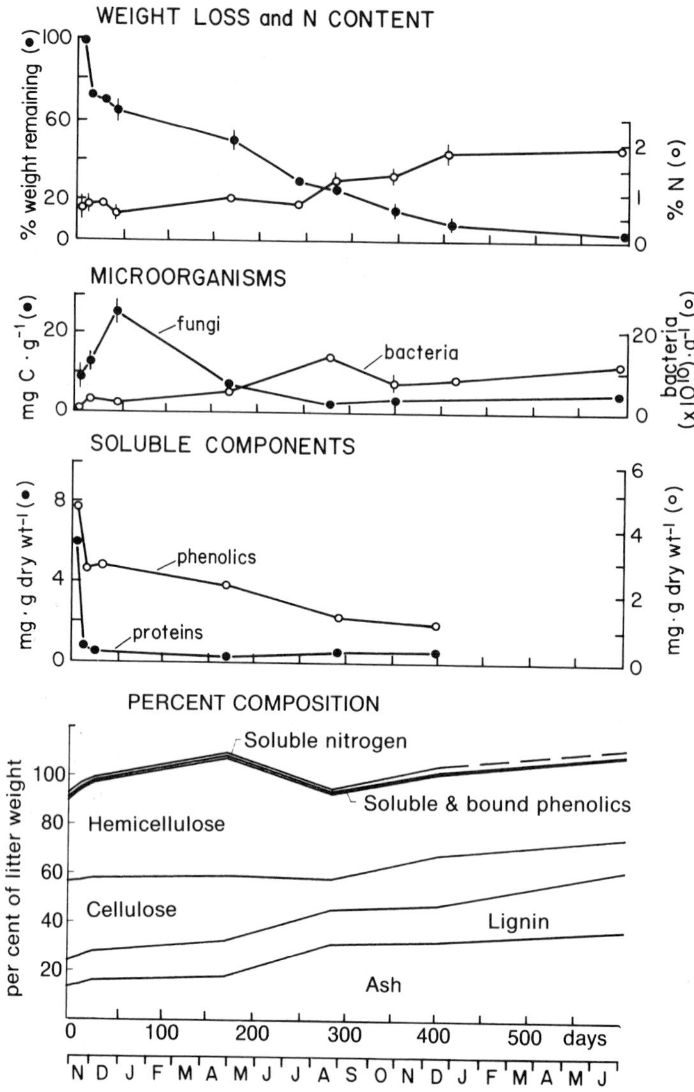

Figure 10-9. Decomposition history of biomass of *Spartina alterniflora* in a Massachusetts salt marsh. Data obtained by means of litter bags set out in November and collected at intervals. The two soluble components are small relative to other chemical components, and so the pattern of the third graph does not show in the bottom graph. Cellulose is a straight-chain polymer of glucose, and hemicellulose is a base-soluble cell wall polysaccharide closely associated with cellulose. Lignin is an amorphous phenolic heteropolymer. Data of I. Valiela, John Teal, John Hobbie, Tony Swain, John Wilson, and Robert Buchsbaum.

humus or "gelbstoff." The differences in initial composition of algal and vascular plant detritus lead to subtle differences in the chemistry of the derived fulvic and humic acid (Gagosian and Stuermer, 1977; Stuermer and Payne, 1976). It might be thus possible to use these differences as indicators of the source of the gelbstoff. This last refractory phase may occur principally, but not exclusively, in sediments.

10.2442 Chemical Changes During Decay and Internal Controls of Decay Rates

It is evident from the above discussion of the phases of decay that the relative abundance of certain types of organic compounds in organic matter can affect rates of decay. The more lignin initially present, for example, the slower the decay. The mineral elements in detritus also affect decay rates. During the course of decay of organic matter, bacteria and fungi take up mineral elements, principally nitrogen, but also phosphorus. The source of nutrients can be internal (contained in the tissues) or external (supplied by water). Increased internal nitrogen in detritus (Fig. 10-10,

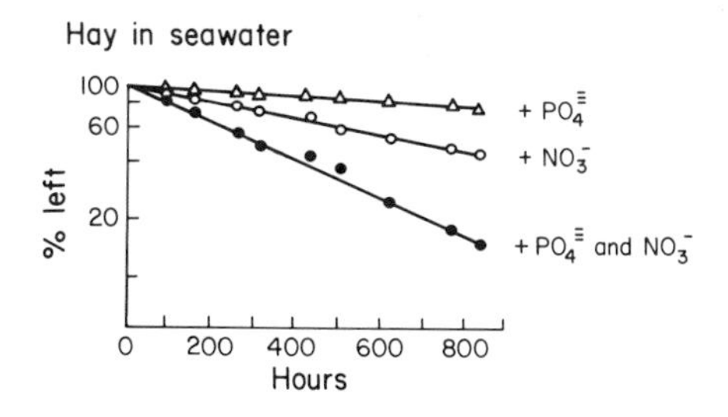

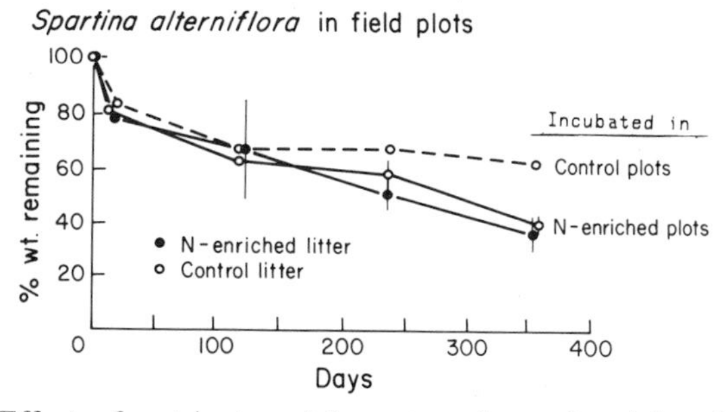

Figure 10-10. Effect of nutrient enrichment on loss of weight of dead organic matter. Top: ^{14}C-Labeled hay incubated in seawater in the lab with additions of nutrients (adapted from Fenchel 1973); bottom: *Spartina alterniflora* litter incubated in mesh bags in nitrogen-enriched and control salt marsh plots. Litter from plants grown in enriched plots has higher nutrient content (cf. Fig. 6.11); enriched litter incubated in control plots (not shown) decayed at the same rates as when incubated in enriched plots. Unpublished data of I. Valiela and John Teal.

bottom) or nutrients supplied externally both may increase microbial activity and decay of salt marsh detritus (Fig. 10-10, top and bottom).

As we have seen above, particulate organic matter falling through the oceanic water column is to a large extent degraded by the time it reaches the deep sea floor, with significant changes in organic and mineral composition, including nitrogen content. For example, in some parts of the open ocean the POC in surface water may have C/N values of 14, while below 1,000 m the C/N may reach 29 (Knauer et al., 1979). Much of the organic matter in the deep sea may be refractory. Labile dissolved organic matter (such as provided by exudation) may constitute perhaps 15–25% of total organic matter. This "young" organic matter degrades rapidly, while the older fraction (perhaps 4,000 to 6,000 years old) are complex substances refractory to degradation (Degens and Mopper, 1976).

The rates of decay of the POC that does reach great depths may be slow for other reasons rather than just the chemical composition of detritus. This was brought into focus serendipitously by the accidental sinking of the research submarine *Alvin* in 1968. *Alvin* was recovered 10 months later with the crew's lunches largely intact (Jannasch et al., 1971). This evidence of obviously slow decay in the deep sea prompted a substantial effort (Jannasch and Wirsen, in press; Morita, 1980) to identify the factors that inhibited decay rates in the deep sea. Enrichment studies (Jannasch and Wirsen, 1973) show no large increment in the rate of *in situ* decay of various organic substrates, even at high concentrations. Other factors have to be involved. The uniformly low temperatures (2–4°C) in the deep sea slow growth of even cold-adapted bacteria, and the high hydrostatic pressures further reduce bacterial growth. Temperature and pressure cannot be the complete answer, since there is considerable bacterial activity found in the guts of deep sea cold-blooded animals, where nutrients are significantly concentrated but where temperature and pressure are the same as in the water. In addition, it is not clear why pressure would so hamper procaryote activity while apparently not really decreasing the activity of eucaryotic organisms. Perhaps free-living bacteria in the deep sea are adapted in a fundamental way to living at very low substrate concentrations. Such bacteria have been unfortunately named "oligotrophs" by microbiologists—needlessly confusing a term otherwise used by ecologists. Such bacteria, perhaps better called "oligocarbophilic," may be characterized by rather slow metabolic and growth rates (Poindexter, 1981) and may be very poorly adapted to using substrates in high concentrations, hence the negative results of enrichment experiments. Perhaps the deep sea microflora is so dominated by oligocarbophilic species that the low rates of decay of organic matter associated with these microbes prevail in the deep sea. The finding of such slow decay of organic matter even when substrate may be added has applied importance, since organic wastes that find their way to the deep ocean are therefore likely to be degraded very slowly.

While the percentage carbon content in most kinds of dead organic matter generally decreases with age during decay as a result of respiratory losses of CO_2, in detritus of vascular plants the percentage nitrogen may increase. During the course of decay of vascular plant detritus bacteria and fungi immobilize nitrogen by incorporation into new cells, and this is in part responsible for the increased nitrogen content of detritus as it ages (Fig. 10-9 top). Microbial nitrogen is not enough to account for all the nitrogen in aged detritus (Cammen et al., 1978; Odum et al. 1979). Fungal biomass accounts for 12–22% of the nitrogen in salt marsh detritus, while bacteria amount to a tenth of the fungal contribution (Marinucci et al., 1983). Nitrogen may also be immobilized by the accumulation of microbial extracellular protein and nitrogen-containing exudates (Glenn, 1976; Hobbie and Lee, 1980). The third way by which nitrogen may be immobilized is through binding of proteinaceous substances by phenolic compounds in the detritus (Leatham et al., 1980; Tenore and Rice, 1980; Rice, 1982).

The actual amounts of immobilized nitrogen accounted for by these mechanisms has not been measured. As a result of immobilization of nitrogen, the carbon to nitrogen ratio of vascular plant detritus is progressively reduced. At C to N ratios about 30 : 1 the rates of immobilization approximate the rates of mineralization; below 15 : 1 or 10 : 1, nitrogen in litter can be mineralized since such ratios provide N in excess of that needed by the microbes for cell building (Alexander, 1977).

In phytoplankton detritus the percent nitrogen does not increase during decay (Otsuki and Hanya, 1972a). In dead phytoplankton fungi are scarce while in vascular plant detritus fungi make up one of the principal pools of nitrogen that increases over time. Further, the phenolic compounds that bind proteins are less abundant in phytoplankton than in vascular plants. The result is that immobilization of nitrogen may not be very important in phytoplankton detritus.

10.2443 The Roles of Different Groups of Decomposers

Fungi and bacteria are the principal competitors for organic substrates. The fungi contain relatively less nitrogen relative to carbon than bacteria (Table 10-5) and do better on detritus that has a higher C to N ratio. Fungi are more efficient than bacteria at transforming carbohydrates into living tissue and may assimilate 30–40% of the carbon of organic matter, while bacteria only assimilate 5–10% (Alexander, 1977). Bacteria, in addition, may cause faster loss of carbon, since they require only 1–2 units of nitrogen for the decomposition of 100 units of carbon, while fungi require 3–4 units of nitrogen (Alexander, 1977). Fungi therefore tend to immobilize nitrogen more strongly than bacteria, so that mineralization may therefore be slower where fungi predominate. Organic matter such as derived from salt marsh grass, mangroves, or eelgrass has high C/N val-

Table 10-5. Approximate Carbon to Nitrogen Ratios in Some Terrestial and Marine Producers[a]

	C/N
Terrestrial	
Leaves	100
Wood	1000
Marine vascular plants	
Zostera marina	17–70
Spartina alterniflora	24–45
Spartina patens	37–41
Marine macroalgae	
Browns (*Fucus, Laminaria*)	30 (16–68)
Greens	10–60
Reds	20
Microalgae and microbes	
Diatoms	6.5
Greens	6
Blue-greens	6.3
Peridineans	11
Bacteria	5.7
Fungi	10

[a] Data compiled in Fenchel and Jorgensen (1977), Alexander (1977), Fenchel and Blackburn (1979), and unpublished data of I. Valiela and J. M. Teal.

ues (Table 10-5) and fungi are prominent. With detritus derived from phytoplankton, C/N values are lower (Table 10-5) and bacteria may be more important.

Further evidence of the role of fungi in immobilizing nitrogen is given by the low nitrogen content in anaerobic sediments such as found in salt marshes. Fungi are aerobic and are rare in marsh sediments. Organic matter in sediments has low nitrogen content, suggesting that mineralization of nitrogen has taken place.

Fungi are more important than bacteria in the degradation of cell wall material, especially in regard to complex molecules such as lignin, but the breakdown of such compounds generally requires aerobic conditions. Since fungi are rare or inactive under anaerobic conditions, lignins are not degraded in such situations.

Detritivores may directly assimilate some of the available organic matter, since cellulases and other needed enzymes have been detected in their guts (Elyakova, 1972; Hylleberg-Kristensen, 1972; Kofoed, 1975); most of these cases may involve symbiotic floras (Fong and Mann, 1980). Direct assimilation of ingested detritus by animals usually accounts for only a small proportion of decomposition (Yingst, 1976; Lopez et al., 1977; Levinton, 1979) (see also Section 7.221). There are, however, important indirect effects of detritivore feeding, including the stimulation of microbial activity and changes in the microbial flora.

Feeding by large consumers stimulates microbial activity through two mechanisms: the fragmentation (comminution) of detrital food particles and the harvesting of microbial cells attached to particles. Comminution alters the surface to volume ratio of particles; where amphipods are allowed to feed on detritus there is a significant reduction of size of detrital particles (Fig. 10-11, left), with a 2.3-fold increase in surface area. The increased surface area leads to more microbes and larger standing crops of organisms that consume microbes, such as ciliates and flagellates (Fig. 10-11, middle). The larger surface area also enhances activity of the biota, as measured by respiration rates (Fig. 10-11, right). The presence of amphipods increased the oxygen consumption of the entire system (detritus, amphipods, and microorganisms) two-fold in less than 4 days compared to the respiration of detritus without the amphipods. The increase was about 110% of the respiration due to the amphipods, so the increase must have been due to microbial respiration.

The harvest of microbial cells off particles by animals may maintain microbial populations in a more actively growing state. This may occur through either removal of relatively more abundant, older and metabolically slower cells, or by relative removal of species that are slower growing and metabolically less active.

The major consumers of microbes are small organisms—ciliates, microflagellates, nematodes, and foraminifera. The presence of these small organisms stimulates degradation of organic matter. Feeding by flagellates and ciliates can increase the percentage loss of weight of detritus of eelgrass by 46% over the loss that take place when microbes alone are present, even though bacterial populations are reduced by one or two orders of magnitude (Harrison and Mann, 1975). A mixed protozoan

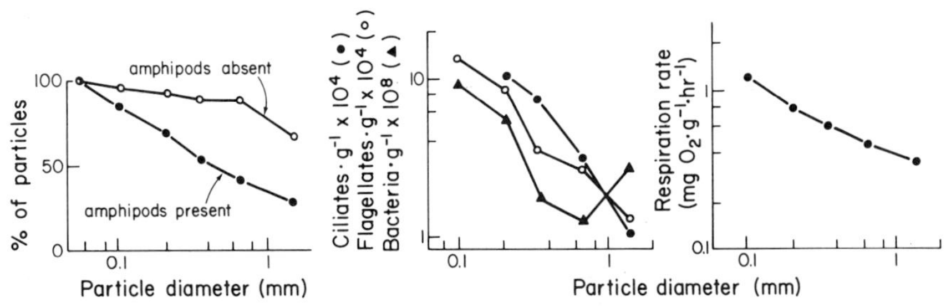

Figure 10-11. Left: Comminution of particulate organic carbon derived from the seagrass *Thalassia testudinum* by feeding of a detritivore, the amphipod *Parhyalella whelpleyi*. Middle and right: Effect of size of particulate organic carbon on abundance of organisms (middle) and on respiration by the biota (right). Adapted from Fenchel (1970).

fauna decreased the bacterial fauna of eelgrass detritus by 70% while increasing the rate at which plant matter decayed (Fenchel and Harrison, 1976). In these cases the bacterial population is most likely turning over at a rapid rate, even though the standing population may be low.

There are important interactive effects of activity by large and small detritivores on rates of decay. Meiofauna may enhance use of detritus by large detritivores. Where ciliates are present, the polychaete *Nereis succinea* incorporates significantly more carbon from detritus of a red algae than from detritus with no ciliates (Briggs et al., 1979). When meiofauna are present 30–400% more ^{14}C from detritus of *Spartina* and *Zostera* is incorporated by polychaetes (Tenore et al., 1977). Such increases in uptake of detrital carbon are due both to enhanced microbial activity and to the consumption of small animals by the large detritivores.

Feeding by large detritivores may, in turn, greatly affect abundance and activity of small consumers of microbes. In field situations, the abundance of meiofauna can be mediated by the cropping pressure due to macrofaunal predators (Bell, 1980). The predation pressure by large species on small ones may be so large that the microbes increase in abundance (Morrison and White, 1980, Lopez et al., 1977) when macrofauna crop meiofauna. The increased abundance of microbes is probably due to the reduction of their principal predators, the micro- and meiofauna, to comminution, and to the mechanical stirring of detritus by the large animals. This stirring probably makes nutrients released by the mineralization process more available to the microbes. There are therefore important second- and third-order effects that link large detritivores, small microbe-feeding species, microbes, and rates of decay of organic matter.

There is also the possibility that feeding by detritivores affects the relative abundance of fungi and bacteria. Evidence from terrestrial soils shows that animal feeding on detritus reduces fungal biomass (Hanlon and Anderson, 1980; Hanlon, 1981). Apparently fungi may not grow fast enough to compensate for losses due to cropping by detritivores. Bacteria may grow faster, and the cropping thus results in a shift from a fungal to a bacterial microbial community. Morrison and White (1980) found such a shift in detritus cropped by coastal amphipods in the laboratory. So far such a shift has not been demonstrated in the field but detritus-feeder activity may be responsible for the shift from fungi to bacteria over time evident in Figure 10-11 (second graph). The importance of this effect of detritivore feeding lies in the changes in chemistry and decay rates that may accompany a shift from fungi to bacteria. There may be less immobilization of nitrogen and, if lignin-containing detritus is involved, slower decay rates.

Throughout this section we have been discussing examples from coastal and marsh systems. Comparable data are not available for open-water environments, but we can hypothesize that the principles apply. What we have said need only be modified to account for the much more dispersed nature of detritus in the plankton, and for the differences in chemical composition between phytoplankton and coastal and marsh producers. The implications of these modifications need some concerted study.

10.245 Burial in Sediments

Some of the carbon fixed by producers is buried in sediments and lost from the biological carbon cycle. The rates of consumption and decomposition of organic matter by heterotrophs are such that in general only a small proportion of the organic matter produced by autotrophs accumulates in the sediments over most of the sea (Table 10-6). In the case of Buzzards Bay sediments mentioned above, permanent accretion to sediments amounts to about 0.5–2% of the annual sedimentation. In exceptional cases of very high production and shallow depths, such as in the Baltic Sea and the Peruvian upwelling (Table 10-6) considerable proportions of annual primary production may accumulate in pelagic sediments. Even in salt marshes, only where accretion rates of sediment are very high do large proportions of the organic matter produced accumulate to form peat.

In general, over most marine environments (Table 10-6), little carbon accumulates. Organic matter does accumulate in anoxic sediments where decomposition is not carried to completion by consumers, but only specialized consumers exist in these environments. Further, the organic matter that does accumulate in aerobic or anaerobic environments probably consists mostly of refractory compounds unsuitable for assimilation by

Table 10-6. Accumulation of Organic Matter in Marine Sediments[a]

	Percentage of primary production accreting in sediments	Sedimentation rate (mm yr^{-1})
Oceanic Sediments		
Abyssal plain, average of several sites	0.03–0.04	0.001
Central Pacific	<0.01	0.002–0.006
Off N.W. Africa, Oregon, and Argentina	0.1–2	0.02–0.7
Peru upwelling, Baltic	11–18	1.4
Salt Marsh Sediments		
Cape Cod, USA	5.3	1
Long Island, USA	37	2–6.3

[a] Pelagic data compiled by Müller and Suess (1979) and Wishner (1980). Salt marsh data from various references in Valiela (in press).

consumers (Degens and Mopper, 1976). The proposition of Hairston et al. (1960) (Chapter 8) that organic matter generally does not accumulate in terrestrial environments thus seems to be true also for most marine systems. Since there is little excess organic matter available and what is available may be refractory, we conclude, as Hairston et al. (1960) did for terrestrial detritivores, that marine consumers that feed on dead organic particles are limited by the quantity and quality of their food supply.

10.3 The Carbon Cycle in Anoxic Environments

10.31 Occurrence of Anoxic Conditions

Reducing conditions occur in marine environments where the rate of consumption of oxygen exceeds the supply. The dissolved oxygen used in the oxidation of live and dead particulate matter is derived from photosynthesis and the atmosphere. The supply of oxygen from the water column is thus dependent on events within the surface layer of the ocean but there may be lateral movements of oxygen or oxygen-consuming organic matter. Such local phenomena ultimately may create a larger or smaller concentration of oxygen or carbon, but generally most of the oxidation that takes place in the sea goes on near the surface. The combination of supply and consumption rate at various depths results in characteristic oxygen profiles with a minimum concentration at some intermediate depth. Below the oxygen minimum consumption rates are lower, and oxygen concentrations rise again up to perhaps 6 ml O_2 per liter (Deuser, 1975).

Where there is some impediment to water flow, such as a physical restriction or a stratification due to changes in density of different layers of water, water below the mixed layer may be devoid of oxygen. Such situations occur in the Black and Baltic Seas, the Cariaco Trench, and in some coastal lagoons and fjords.

There may also be low oxygen waters below photic zones with high rates of primary production. This occurs in certain large oceanic areas north and south of the Equator in the eastern Pacific (Deuser, 1975), regions where divergences create upwellings of nutrients.

The other major kind of situation where oxygen consumption exceeds supply is within sediments below productive water columns. This applies to most coastal sediments. In most coastal zones much organic matter reaches the sea floor, and since flux of interstitial water is slow, the renewal of O_2 is very restricted. Deep-sea sediments receive much smaller amounts of organic matter, so consumption of oxygen may not exceed supply rates, and the interstitial water may remain aerobic.

10.32 Sources of Organic Carbon in Anoxic Environments

The fixed carbon of anoxic environments is primarily of allochthonous origin. The major source is of course the rain of organic matter from above, but there are two other mechanisms that can produce small amounts of autochthonous organic carbon. As mentioned in Section 1.2, carbon can be fixed by anoxigenic photosynthesis carried out by bacteria (cf. Section 1.2). Another alternative source is the chemosynthetic reaction in which hydrogen gas is oxidized by bacteria that use CO_2 as the electron acceptor. The result of this reaction is the release of methane (Table 10-7).

10.33 Losses of Organic Carbon in Anoxic Environments

10.331 Microbial Decomposition Processes

Carbon compounds can be degraded by dissimilative reactions in which the energy contained in reduced compounds is used by bacteria. In these reactions the reactants are not assimilated. Some chemosynthesis may be carried out based on inorganic reduced compounds (chemolithotrophy), including reactions such as the oxidation of H_2 with SO_4^{2-} or NO_3^-, or the oxidation of HS^- with NO_3^- (Table 10-7).

More relevant to our topic here are reactions in which reduced organic compounds are the source of energy (chemoheterotrophy). Examples of these reactions are fermentation, nitrate reduction, denitrification, and sulfate reduction (Table 10-7). In the latter three reactions some oxidant (NO_3^-, SO_4^{2-}) must be transported into a reduced environment from an oxidized environment. This usually means that the microbial reactions take place near the interface between oxidized and reduced environments.

Fermentation and anaerobic reduction reactions are the means used by bacteria to reoxidize the compounds—such as NADPH—that transfer reducing power in living cells when oxygen is not available. Both these types of reactions can be carried out with a variety of specific substrates (Fenchel and Blackburn, 1979; and Hamilton, 1979). Below we examine some general types of reactions involved in the degradation of organic matter.

10.3311 Fermentation

Fermentation is one of three processes by which organisms obtain energy. Photosynthesis and respiration, the other two, are discussed in Section 1.2. Fermentation evolved first, before the appearance of oxygen

Table 10-7. Some Representative Reactions Illustrating Pathways of Microbial Metabolism and Their Energy Yields[a]

		Energy yield (kcal)
Aerobic respiration	$C_6H_{12}O_6$ (glucose) + $6O_2 = 4CO_2 + 4H_2O$	686
Fermentation	$C_6H_{12}O_6 = 2CH_3CHOCOOH$ (lactic acid)	58
	$C_6H_{12}O_6 = 2CH_2CH_2OH$ (ethanol) + $2CO_2$	57
Nitrate reduction and denitrification	$C_6H_{12}O_6 + 24/6\ NO_3^- + 24/5\ H^+ = 6CO_2 + 12/5\ N_2 + 42/5\ H_2O$	649
Sulfate reduction	$CH_3CHOHCOO^-$ (lactate) + $1/2\ SO_4^= + 3/2\ H^+ = CH_3COO^-$ (acetate) $+ CO_2 + H_2O + 1/2\ HS^-$	8.9
	$CH_3COO^- + SO_4^= = 2CO_2 + 2H_2O + HS^-$	9.7
Methanogenesis	$H_2 + 1/4\ CO_2 = 1/4\ CH_4 + 1/2\ H_2O$	8.3
	$CH_3COO^- + 4H_2 = 2CH_4 + 2H_2O$	39
	$CH_3COO^- = CH_4 + CO_2$[b]	6.6
Methane oxidation	$CH_4 + SO_4^= + 2H^+ = CO_2 + 2H_2O + HS^-$	3.1
	$CH_4 + 2O_2 = CO_2 + 2H_2O$	193.5
Sulfide oxidation	$HS^- + 2O_2 = SO_4^= + H^+$	190.4
	$HS^- + 8/5\ NO_3^- + 3/5\ H^+ = SO_4^= + 4/5\ N_2 + 4/5\ H_2O$	177.9

[a] Energy yields vary depending on the conditions, so different measurements may be found in different references. The values reported here are representative. Adapted from Stumm and Morgan (1981), Fenchel and Blackburn (1979), and Martens and Berner (1979).

[b] This reaction is sometimes considered fermentation.

in the atmosphere. It has been retained in microorganisms living in anoxic environments and also occurs within many other higher organisms.

The specifics of fermentation in marine systems are poorly known. In this process, energy is transferred by the oxidation of part of an organic compound and the reduction of another part; there is no net oxidation of the substrate. The fermentative reactions are considerably less efficient at harnessing energy than aerobic respiration (Table 10-7).

The most common fermentation reactions involve sugars or other high-molecular-weight carbohydrates, with a variety of end products, including acetate, propionate, CO_2, and alcohols, among others (Table 10-7). Fermentation is important in anoxic environments, because it is the major mechanism by which organic matter synthesized by primary producers is broken down into organic compounds of lower molecular weight readily usable by other groups of microbes (Fig. 10-12).

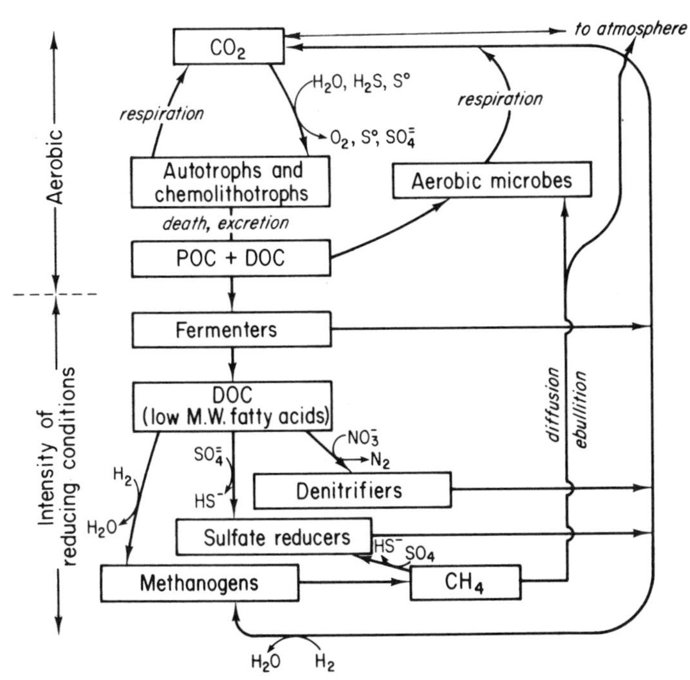

Figure 10-12. Carbon transformations in the transition from aerobic to anaerobic situations. The top part of the diagram is a simplified version of Fig. 10.1. The gradient from aerobic to anaerobic can be thought of as representing a sediment profile, with increased reduction and different microbial processes deeper in the sediment. Boxes represent pools or operators that carry out processes; arrows are processes that can be biochemical transformations or physical transport. Elements other than carbon are shown, where relevant, to indicate the couplings to other nutrient cycles. Some arrows indicate oxidizing and some reducing pathways.

10.3312 Nitrate Reduction and Denitrification

When oxygen is depleted, bacteria use compounds other than oxygen as terminal electron acceptors. The energy yielded from the oxidation of organic matter by the reduction of nitrate (NO_3^-) or nitrite (NO_2^-) is comparable to the yield by aerobic respiration (Tables 10-7, 10-8) and NO_3^- and NO_2^- are thus readily reduced in anoxic environments. The further reduction through nitric (NO) and nitrous (N_2O) oxides to nitrogen gas (N_2) completes the process of denitrification (cf. Fig. 11-6). Denitrifiers generally obtain energy from organic compounds of low molecular weight (Fig. 10-12), but some also use H_2 and reduced sulfur.

The reactions that oxidize organic matter at the expense of nitrate can result in either complete conversion of the organic nitrogen to N_2 or to release of both N_2 and NH_3 (Table 10-8). In coastal sediments of Japan, for example, denitrification to N_2 accounted for 27–57% of the nitrate consumption, so both reactions may be important.

The relative magnitude of the oxidation of carbon compounds by nitrate and oxygen in marine environments is not well known (Fenchel and Blackburn, 1979). In part the difficulty lies in that it is hard to measure small changes in production of N_2 into water that already contains a very large pool of dissolved N_2. Further, it is not clear how much nitrogen is lost as N_2O, or the extent to which nitrite is denitrified. Depending on these factors, 2–5 carbon atoms could be oxidized for every 4 nitrogen atoms reduced. In a few anoxic marine environments where nitrate is abundant, nitrate reduction to nitrite can consume up to 50% of the photosynthetically fixed carbon (Richards and Broenkow, 1971). Such cases are the exception rather than the rule, and in most instances the oxidation of organic carbon by nitrate is small. In a coastal sediment off Denmark, Jørgensen (1980) calculated that 2% of the carbon was oxidized by denitrification. Denitrification in a New England salt marsh consumed less than 1% of readily available carbon (Table 10-9). We can conclude, then, that heterotrophic consumption of DOC via nitrate reduction and denitrification is usually unimportant as a mechanism of oxidation of organic matter (Table 10-9), even though the energy yields are quite favorable (Table 10-7). It simply turns out that in anaerobic environments there is little available nitrate and nitrate reduction and denitrification are therefore limited by the amount of oxidized nitrogen.

10.3313 Manganese and Iron Reduction

Reactions involving the oxidation of organic matter by Mn^{4+} and Fe^{3+} (Table 10-8) doubtless go on, although it is not certain whether microorganisms directly reduce these metals. The reduction to Mn^{2+} and Fe^{2+} may be indirectly mediated by microbes, since the anaerobic conditions created by bacterial activity may reduce and solubilize Mn^{4+} and Fe^{3+} (Fenchel and Blackburn, 1979).

Table 10-8. Theoretical Stoichiometry of Major Reactions Involved in the Oxidation of Organic Matter (OM, defined as $C_{106}H_{236}O_{110}N_{16}P$) Through Reduction of Oxygen, Nitrate, Manganese, Iron, and Sulfate[a]

Reaction	Energy yield (kcal · mole OM^{-1})
$138\ O_2 + 1$ unit OM $\longrightarrow 106\ CO_2 + H_3PO_4 + 122\ H_2O + 16\ HNO_3$	13,334
$94.4\ HNO_4^- + 1$ unit OM $\longrightarrow 106\ CO_2 + H_3PO_4 + 177.2\ H_2O + 55.2N_2$	12,665
$84.8\ HNO_4^- + 1$ unit OM $\longrightarrow 106\ CO_2 + H_3PO_4 + 148.4\ H_2O + 42.4N_2 + 16\ NH_3$	11,495
$236\ MnO_2$[b] $+ 1$ unit OM $+ 472\ H^+ \longrightarrow 106\ CO_2 + H_3PO_4 + 366\ H_2O + 8N_2 + 236\ Mn^{2+}$	12,206–12,916[c]
$212\ Fe_2O_3$[b] $+ 1$ unit OM $+ 848\ H^+ \longrightarrow 106\ CO_2 + H_3PO_4 + 520\ H_2 + 16NH_3 + 424\ Fe^{2+}$	5,559–5,894[c]
$53\ SO_4^{2-} + 1$ unit OM $\longrightarrow 106\ CO_2 + H_3PO_4 + 106\ H_2O + 16NH_3 + 53\ S^{2-}$	1,588

[a] Adapted from Froelich et al. (1979). The exact stoichiometry can be expressed in different ways (cf. Stumm and Morgan 1981; Emerson et al., 1980, for example), and the energy yield depends on the way the equation is written.
[b] The Mn^{4+} or Fe^{3+} compounds may differ; for example, 424 FeOOH may react with the same compounds, and 742 H_2O would be produced.
[c] The actual energy yield varies depending on the specific mineral phase of the reactants.

Table 10-9. Amounts of CO_2 Released by the Various Pathways of Mineralization of Organic Matter in Sediments of a New England Salt Marsh [a]

Pathway	Rate of CO_2 release (mole C m^{-2} yr^{-1})	% of total mineralization
Aerobic respiration[b]	30.1	44.9
Nitrate reduction[c]	0.4	0.6
Sulfate reduction[d]	36	53.7
Methanogenesis[e]	0.5	0.7
Total	67	

[a] Data from Howes et al. in press and in press a.
[b] Even though the sediments are anoxic, the plants present have internal air ducts; there is also some movement of aerated water into and out of the sediment. These mechanisms support some aerobic respiration. Value obtained by subtracting other terms from total.
[c] Based on Kaplan et al. (1979).
[d] The sulfate reduction is based on fermentation products, and the efficiency of use of these products by sulfate reduction is very high. Fermentation does not therefore appear as a separate entry in this table.
[e] Based on $CH_3COO^- = CH_4 + CO_2$ (Table 10-7).

The reduction of manganese can co-occur with nitrate reduction, since the energy yield of these reactions is similar (Table 10-8), and both these processes may take place in sediments that are not completely anoxic (Froelich et al., 1979; Emerson et al., 1980, among others). After the labile nitrate and manganese are consumed, iron is reduced. The energy yield of reduction of ferric iron is lower than that of nitrate and Mn^{4+} (Table 10-8).

10.3314 Sulfate Reduction

Bacteria capable of using SO_4^{2-} as terminal election acceptors are strictly anaerobic (Fig. 10-12). Sulfate reduction takes place mainly in anoxic sediments, in anoxic microzones within aerobic sediments, or in anoxic waters. Even in aerated water columns there may be anoxic microzones within particles where diffusion of O_2 may be slow enough to produce anoxic conditions within the particles. Jørgensen (1977a) calculated, for example, that a detrital particle of 2 mm diameter was large enough to be anoxic at the center, even if it was surrounded by oxygenated water.

The energy yield of sulfate reduction is much less than that of nitrate reduction (Tables 10-7, 10-8), so that sulfate is consumed by microorganisms in anoxic sediments only after nitrate and perhaps after Mn^{4+} and Fe^{3+} are depleted. Since there is little nitrate in coastal marine sediments, sulfate reduction becomes the dominant process by which organic carbon is oxidized. In some coastal sediments about half of the oxidation of carbon is due to sulfate reduction (Jørgensen, 1977). In salt marsh sediments about half the carbon dioxide that results from degradation of organic matter may be consumed by sulfate reduction and fermentation (Table 10-9). Since sulfate is plentiful in seawater but is rare in freshwater, sulfate reduction is far more prominent in the sea than in freshwater ecosystems.

Organic substrates of relatively low molecular weight usually serve as electron donors for sulfate reduction, and the sulfate reducers depend on fermenters to provide them. Lactate, acetate, hydrogen, and methane are oxidized, with H_2S, CO_2, and H_2O the end products (Table 10-7). The activities of sulfate reducers, fermenters, and methanogens are thus closely coupled. The H_2S and organic sulfides, primarily dimethyl sulfide, are the source of the "rotten-egg" odor of anoxic marine sediments.

10.3315 Methanogenesis

Methanogens are strict anaerobes whose metabolic activity generates methane (Fig. 10-12). We have seen above that methanogenic bacteria can be a source of organic carbon (as methane, Fig. 10-12). To make matters confusing, some bacteria produce methane by degrading organic compounds, so some methanogens also act as decomposers of organic matter. The principal reactions involved in this latter role are the reduction of CO_2 (Deuser et al., 1973; Claypool and Kaplan, 1974) or the fermentation of acetate (Atkinson and Richards, 1967; Cappenberg, 1974, 1975) (Table 10-7). Methanogens use metabolic end products such as CO_2 or acetate released by the activity of other groups of bacteria (Table 10-7). This consumption of metabolic end products of other decomposers is important because accumulation of end products could otherwise inhibit activity of the other bacteria.

The energy yield of methanogenic reactions (Table 10-7) is low, and other potential electron acceptors (O_2, NO_3^- and $SO_4^=$) are used first (Oremland, 1975; Balderston and Paine, 1976). Methane production generally occurs in environments with no oxygen or nitrate (Fig. 10-12), such as anaerobic muds rich in organic matter, and in the guts of animals.

Methane concentration in reduced sediments are usually low where sulfate concentrations are high (Fig. 10-13 left). It has been suggested that this is because there is competition for substrates by sulfate reducers (Rudd and Taylor, 1980), or by inhibition of methanogens by sulfide produced by the sulfate reducers. Recent evidence suggests that the lack of methane is due to oxidation of methane by sulfate reducers (Table 10-7, Fig. 10-12). This oxidation may be responsible for the concave shape of the upper part of methane profiles (Fig. 10-13) (Martens and Berner, 1974). The methane that escapes the reduced part of the sediment may then be subject to aerobic oxidation. Methane is energy rich and provides

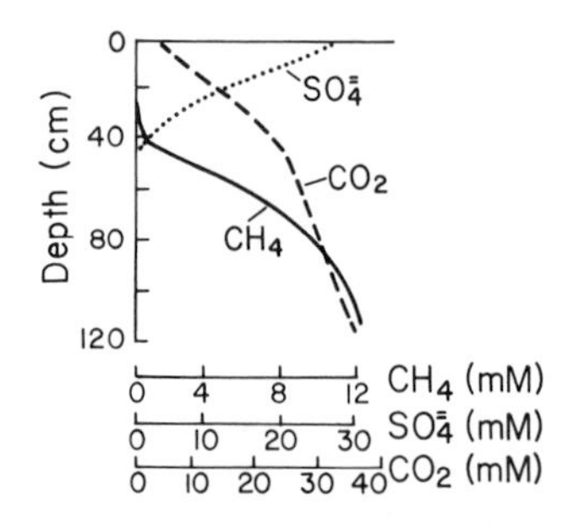

Figure 10-13. Schematic diagram of vertical profiles of sulfate, methane, and carbon dioxide in a reduced coastal marine sediment. Adapted from Rudd and Taylor (1980), data from Reeburgh (1976).

a favorable energy yield for bacteria capable of its oxidation (Fig. 10-12, Table 10-7). Thus, very little of the methane produced within sediments escapes to overlying waters (Table 10-9).

In environments with high rates of production of organic matter and anoxic sediments, such as those of upwellings and marshes, methane from sediments can be found in overlying waters (Scranton and Farrington, 1977). The mixed layer of the open ocean, however, is so distant from the sea floor that transport of methane from sediments does not appear feasible, yet methane concentrations 48–67% in excess of equilibrium concentrations with the atmosphere have been measured (Scranton and Brewer, 1977). There is also a subsurface maximum where the methane concentration may exceed equilibrium concentration by more than twofold. This methane must be generated within the water column, either as a by-product of algal metabolism or in the guts of animals. Although the latter source at first sight may seem trivial, it is not: about 20% of the biologically produced methane that enters the atmosphere is produced in animal digestive tracts (Enhalt, 1976).

The rates at which methane is produced in freshwater systems are much larger than those in marine environments (Table 10-10). Freshwater systems have considerably less sulfate than marine systems (about 0.1 mM compared to more than 20 mM, respectively), so competition by sulfate reducers, inhibition of methanogens by sulfides, and anaerobic reduction of methane by sulfate reducers are less important in freshwater than in marine environments (Reeburgh and Heggie, 1977). The lack of sulfate thus allows methanogens to mineralize a larger proportion of the primary production. The large amounts of methane produced in freshwater muds are not anaerobically oxidized and may escape into the water, where aerobic oxidation takes place, or into the atmosphere.

Some very rich marine and brackish sediments resemble freshwater environments, with occasional depletion of sulfate near the surface of the sediment due to sulfate reduction (Martens, 1976). One of the reasons this can take place is that during the times of year where the sea floor is anoxic the benthic fauna are killed (Sansone and Martens, 1978). Tube building, feeding, and pumping of seawater by the benthic animals are prime mechanisms that transport seawater—and the dissolved sulfate and oxygen—

Table 10-10. Approximate Rates of Methane Production in Various Aquatic Environments[a]

	Methane production ($\mu M\ m^{-2}\ hr^{-1}$)
Marine environments	
Seagrass beds	0.2–2
Coral reefs	0.02–0.5
Cariaco Trench	18
Santa Barbara Basin	12
Salt Marsh	3–380
Freshwater environments	
Paddy soils	570
Lakes	33–4,580

[a] Condensed from various authors in Rudd and Taylor (1980), and King and Wiebe (1978).

into surface sediments. With no fauna there is little available sulfate, reduced activity by sulfate reducers, and the bubbling of methane out of anaerobic sediments then becomes a major source of methane to overlying water.

In seagrass beds, mangroves, and salt marshes the vegetation provides a large, intricate network of oxygen-containing roots (Oremland and Taylor, 1977; Teal and Kanwisher, 1966a). The oxidation associated with roots may provide for a significant amount of aerobic respiration of methane in these very productive environments; this may be part of the reason why marsh profiles of sulfate and methane differ from both marine and freshwater sediments in that while they are very productive and rich in sulfate, and there is enough oxidation by sulfate reducers that little methane leaves the sediment. The pathwork of oxidized and reduced microzones could also probably allow the simultaneous occurrence of aerobic respiration and sulfide oxidation.

10.332 Incorporation Into Anoxic Sediments

Only a small fraction of the organic matter produced by autotrophs accumulates in sediments (Table 10-6), due to consumption by animals and both aerobic and anaerobic decay. Where production is very high, such as in upwelling regions, salt marshes, and mangroves (Table 10-6), environments that characteristically have anaerobic sediments, organic matter or peat do accumulate. A certain proportion of organic material is resistant to decay under anaerobic conditions. Fungi are rare, as discussed earlier, so the abundant lignin fraction in detritus from vascular plants is not degraded and peat accumulates. This may explain the accumulation of

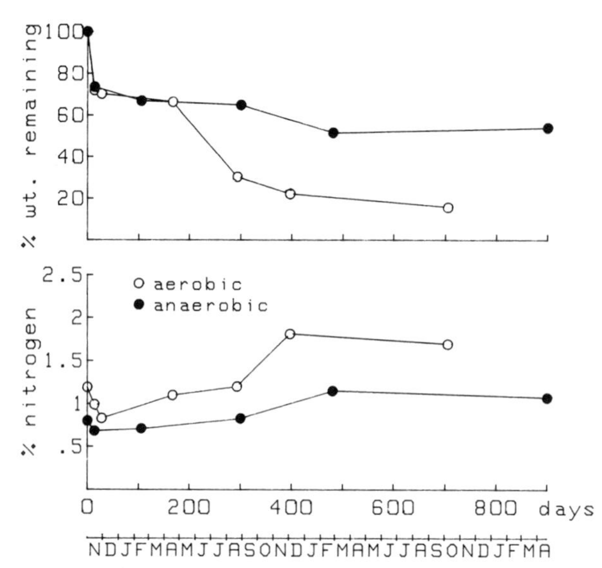

Figure 10-14. Course of decomposition of biomass and N % content of *Spartina alterniflora* exposed aerobically (litter bags on the marsh surface) and anaerobically (buried 5–10 cm below the marsh surface). Data of I. Valiela and John Teal.

organic matter in marshes, where decay of detritus under anaerobic conditions is slower than under aerobic conditions (Fig. 10-14).* This explanation may not apply where the organic matter originates as phytoplankton, since lignins and fungi are rarer in phytoplankton.

Based on analyses of chemical composition of water masses, Richards (1970) concluded that there was no evidence for slower decay of organic matter under reducing compared to oxidizing conditions. Actual experiments in which dead phytoplankton were incubated under anaerobic and aerobic conditions provide evidence that in fact anaerobic decay is somewhat slower than aerobic decay of phytoplankton detritus (Otsuki and Hanya, 1972, 1972a). Perhaps the accumulations of toxic end products such as sulfides, ammonium, alcohols, or acids are responsible. The comparison of anaerobic and aerobic decay in phytoplankton needs further study.

Mineralization and release (''regeneration'') of nutrients contained in organic matter (Table 10-8) take place more readily in anoxic environments than under aerobic decay. Fungi are rarely present to aid the immobilization of nitrogen. Further, nutrients in detritus are released in large quantities by sulfate reducers since they must oxidize more organic matter to obtain a given amount of energy (Table 10-7). This results in a

* This discussion applies to decay prompted by microbes during the decomposer phase; note in Figure 10-14 that the leaching phase is more or less the same under both conditions.

higher carbon-to-nitrogen ratio of detritus under anaerobic conditions than under aerobic conditions. The percentage of nitrogen of anaerobic detritus is about half that of aerobic detritus (Fig. 10-14, bottom). Less nitrogen has been immobilized during decay, and more nitrogen is mineralized and available for transport out of the organic matter. This is probably one mechanism that produces the very high concentrations of inorganic nitrogen (principally NH_4^+) in waters of salt marshes, mangroves, and estuaries.

Chapter 11

Nutrient Cycles: Phosphorus, Nitrogen, and Sulfur

Nutrients have been discussed throughout this book in terms of limiting factors for producers, consumers, and decomposers, as electron donors for microbial decomposers, and in a number of other roles. It is clear, then, that nutrients are inextricably linked to almost all other ecological processes. In this chapter we look at the transformations, exchanges, and general dynamics of three nutrients: phosphorus, nitrogen, and sulfur. The dynamics of these three elements illustrate how nutrients interact with sediments and water, and the great significance of oxidation state and of the interactions with organisms.

11.1 Phosphorus

11.11 Chemical Properties of Phosphorus

In aerobic environments inorganic phosphorus occurs virtually exclusively as (ortho-)phosphates* in which the P atom has an oxidation state of +5. The orthophosphate ions include phosphoric acid (H_3PO_4) and its dissociation products ($H_2PO_4^-$, HPO_4^{2-}, PO_4^{3-}), and in ion pairs and complexes of these products with other constituents of seawater (Gulbrandsen and Roberson, 1973). HPO_4^{2-} is the major ion in seawater by about one order of magnitude over PO_4^{3-}, which in turn is one order of magnitude more abundant than $H_2PO_4^-$, while H_3PO_4 is negligible. Although poly-

* Orthophosphate refers to any salt of H_3PO_4, phosphoric acid, which can also be described as $3H_2O \cdot P_2O_5$. Water molecules can be removed from orthophosphate; the various resulting condensed compounds are named polyphosphates (Van Wazer, 1958).

phosphates are abundant within cells, only traces are found naturally in seawater (Hooper, 1973); high concentrations of polyphosphates are often used as an indication of pollution with waste waters. Bacteria can reduce phosphate to phosphine (PH_3) and phosphide (P^{3-}) but such reduced compounds are of rare occurrence (Hutchinson, 1975); phosphate is by far the dominant form of phosphorus. The range of concentrations of phosphates in seawater is given in Fig. 2-18.

A number of organic compounds that contain phosphorus can be found dissolved in seawater. These are primarily phosphate esters that originate from living cells and can be hydrolyzed to phosphates once released into the environment.

There are at least two aspects of the chemistry of phosphate that are important in maintaining low concentrations of the dissolved forms and greatly complicate the dynamics of phosphorus: the facility for adsorption, and the propensity to form insoluble compounds with certain metals.

Phosphate adsorbs readily under aerobic conditions onto amorphous oxyhydroxides, calcium carbonate, and clay mineral particles (Krom and Berner, 1980; Stumm and Morgan, 1981). Adsorption onto clay particles takes place by bonding of phosphate to positively charged edges of clays and by substitution of phosphates for silicates in the clay structures (Stumm and Morgan, 1981). Because of the ease of adsorption under aerobic conditions, phosphate seldom travels far in sediments, except where transported by movement of particles. Phosphate ions also precipitate with cations such as Ca^{2+}, Al^{3+}, and Fe^{2+}. On a geological time scale phosphate in seawater is in equilibrium with phosphate minerals such as francolite ($Ca_{10}[PO_4CO_3]_6Fe_2$) or hydroxyapatite ($Ca_5OH[PO_4]_3$), but for ecological purposes these phosphate compounds can be considered almost insoluble.

The chemistry of phosphorus changes markedly in anaerobic environments. In anaerobic soils certain microorganisms may reduce phosphorus to oxidation states of +3 and +1 (HPO_4^{2-} and HPO_2^{2-}) (Silverman and Ehrlich, 1964), but these oxidation states have not been reported in anaerobic marine sediments.

As far as phosphorus is concerned, a much more important aspect of anaerobic sediments is that bacteria and the H_2S present reduce ferric iron (Fe^{3+}) to ferrous iron (Fe^{2+}). Ferrous iron is much less effective at adsorbing phosphate than ferric iron (Krom and Berner, 1980), and the reduction of iron thus results in greater availability of dissolved phosphate in anaerobic environments. The reduction of ferric to ferrous iron causes the amorphous ferric oxyhydroxides to dissolve, making them unavailable to adsorb phosphates. This mechanism is the principal cause of the solubilization of phosphate, aided perhaps by inhibition of adsorption on clay surfaces due to the buildup of a coating of the abundant organic matter typical of anoxic sediments (Krom and Berner, 1980). The result of all this is that in reduced sediments the ratio of adsorbed phosphate (by weight) to the equilibrium concentration in the interstitial water is 1–5, while in oxic sediments the ratio is 25–5,000 (Krom and Berner, 1980).

Dissolved phosphate may leave anoxic sediments, but some of the phosphate may reprecipitate as $FePO_4$ at the oxic–anoxic interphase, and most is probably adsorbed onto amorphous ferric oxyhydroxides (Krom and Berner, 1980a). The phosphate adsorbed to ferric hydroxides is readily exchangeable and available as phosphate (Shukla et al., 1971; Golterman, 1973). In any case, the release of phosphate from anaerobic sediments is faster than the reoxidation and immobilization, and phosphate is regenerated from anoxic sediments into overlying water.

11.12 Transformations and Exchanges in the Phosphorus Cycle

Phosphorus in seawater is found in living organisms or as dissolved inorganic phosphorus (DIP), dissolved organic phosphorus (DOP), and particulate phosphorus (part. P).* In most aquatic environments the amount of part. P is much greater than that of DOP, which in turn is larger than DIP. In lakes the phosphorus is partitioned into about 60–70% part. P, 20–30% DOP, and 5–12% DIP. In seawater, where cell concentrations are lower than in lakes, the dissolved organic fraction (which comes from cell exudates) may be smaller than the inorganic fraction. In the English Channel most of the phosphorus present in winter is DIP, but in summer, when cells are most abundant and senescing, there is more DOP than DIP (Harvey, 1955). In eutrophic marine environments, such as the Baltic Sea, the amount of DIP may be 91% of the dissolved phosphorus (Sen Gupta, 1973).

The sources, losses, and pools of particulate, dissolved inorganic, and dissolved organic phosphorus are schematically represented in Figure 11-1, which should be referred to for the remainder of this section.

Uptake by primary producers and bacteria is responsible for the low phosphate concentration typical of surface waters (Fig. 2-17; the kinetics of uptake are discussed in Section 2.21). Some coastal macrophytes in salt marshes, mangrove swamps, and eelgrass beds take up phosphorus from the sediments, as do some freshwater macrophytes (Carignan and Kalff, 1980). This pathway is not shown in Figure 11-1 because in the sea the amounts of P involved are trivial relative to other exchanges. Some phosphate is excreted by bacteria, and DIP is also provided by microbial hydrolysis of the esters of DOP. This lysis is a very rapid process that

* The standard reference for methods of analysis for nutrients in seawater is Strickland and Parsons (1968), which includes procedures for the various forms of phosphorus as well as nitrogen and carbon. A critique of methods of measuring phosphorus is given by Chamberlain and Shapiro (1973), and for measurement under anaerobic conditions see Bray et al. (1973) and Loder et al. (1978).

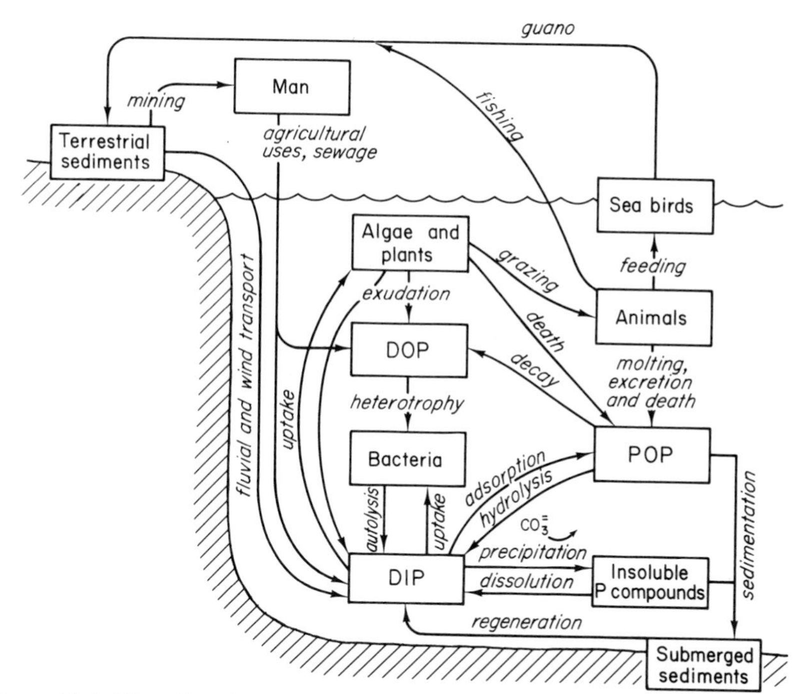

Figure 11-1. The phosphorus cycle. Boxes indicate pools of P, arrows show processes, including transport or transformations.

limits the persistence of DOP to a few hours (Van Wazer, 1973). Abiotic hydrolysis also occurs (Johannes, 1965), but is 10^3–10^4 times slower than where bacteria are active. The release of phosphate directly from macroalgae is very low (Kirkman et al., 1979; Carignan and Kalff, 1980) and the amounts produced by hydrolysis of DOP are very small. By far, phosphate is principally regenerated by the decay of particulate (POP) and by animals. The zooplankton in the Central North Pacific Ocean, for example, may release 55–183% of the daily requirements of phosphorus by the phytoplankton (Perry and Eppley, 1981).

The processes that provide and remove DIP are rather fast; the residence time of dissolved phosphate in oligotrophic waters, for example, is just a few minutes (Hayes and Phillips, 1958; Pomeroy, 1960), mainly due to rapid uptake by algae and bacteria. The rapidity of uptake is a major reason why P in seawater is found principally incorporated in living particles. This can be demonstrated b considering the exchanges between the DIP, DOP, and particulate P as equilibrium reactions (Fig. 11-2). The rates of transformation are such that the concentration of particulate P exceeds those of DIP and DOP.

Death, shedding, or molting of organisms plus adsorption of phosphate onto particles produce particulate organic phosphorus (POP) (Fig. 11-1). Some of the POP is released as DIP as particles decay in the water

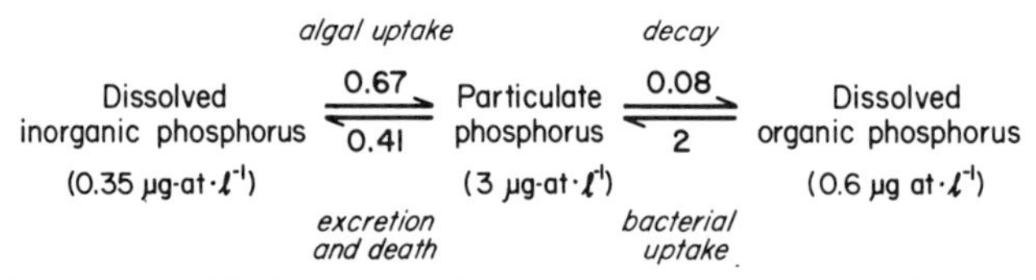

Figure 11-2. Relationship between DIP, Particulate P, and DOP in coastal waters off Nova Scotia. The concentrations of the three species of P are shown in parentheses. The arrows indicate the exchange rates in μgram atoms P liter^{-1} day^{-1}. Adapted from Watt and Hayes (1963).

column, but some settle onto sediments. In the sediments further degradation of settled organic P to DIP can take place and some of this DIP is precipitated or adsorbed.

There is an extremely slow part of the phosphorus cycle in which submerged sediments are uplifted over geological time to provide phosphate rocks on land. The weathering of this rock is the geological process that supplies P back to the oceans by river run off and wind transport. Phosphate rocks are also mined, and much of the phosphate used as fertilizer finds its way down estuaries to the sea, transported on particles. After 1 year following use as fertilizer, 12% of the total P mined in the United States finds its way to aquatic environments, and 40% within 4 years (Griffith, 1973). Guano birds who feed on fish and defecate on land during the nesting season provide the only natural mechanism whereby some small amount of the P in the sea returns to land on a short-term basis (Fig. 11-1). Because of the chemistry of phosphorus, it is therefore clear that, at least for human use, phosphorus must be treated as a nonrenewable resource in limited supply.

11.13 Phosphorus Dynamics in the Marine Water Column and Sediments

In surface waters phosphate is usually very low (usually less than 1 μg atom P liter^{-1}, cf. Fig. 2-17), primarily due to uptake of phosphate by algae and bacteria. Algae are most likely the primary agent of phosphate removal (Fuhs et al., 1972; Cole, 1982) since their K_s concentrations are low (0.6–1.7 μg atom liter^{-1}) and comparable to ambient concentrations (cf. Fig. 2-18). Values of bacterial K_s range from 6.7 to 11.3 μg atom liter^{-1}, very high values compared to seawater concentrations. The ability to store nutrients also makes algae a more important reservoir than bacteria. The importance of the uptake by algae is reflected by the sharp reduction of phosphate in surface layers of temperate coastal waters during the spring bloom. Algal growth produces an increase in particulate phosphorus and, as the algae senesce, release of dissolved organic phosphorus (Harvey, 1955; Strickland and Austin, 1960).

In the open ocean phosphate concentrations increase below near-surface waters to a peak near the permanent thermocline (Fig. 2-24). Peak concentrations are usually associated with the layer of minimum O_2 and maximum CO_2, where microbial and zooplankton activity regenerates both dissolved phosphorus and carbon. We have already seen that carbon fixed in the oceanic photic zone is regenerated within the water column (Chapter 10) and so is phosphorus. Measurements of the rate of release of phosphate (Pomeroy et al., 1963; Johannes, 1965) suggest that zooplankton may regenerate substantial amounts of the nutrients fixed by algae in oceanic waters. Butler et al. (1970) calculated that in *Calanus* about 17% of the P ingested was retained for growth, 23% egested as fecal pellets, and 60% excreted as dissolved phosphorus. The excreted P is accessible to organisms in the water column. Since most of the organic matter, including fecal pellets, that leaves the photic zone is degraded within the water column (Chapter 10), recycling of the contained P is very active. In coastal waters of California, for example, about 68–87% of the P used in surface production is regenerated above 75 m; above 700 m about 91–95% of the P is regenerated (Knauer et al., 1979). In oceanic waters of the Pacific, 87% of the P was regenerated above 75 m and 99% at 1000 m (Knauer et al., 1979).

The rate at which zooplankton regenerate phosphorus is not constant. When algae are abundant, a larger amount of P is converted to eggs (cf. Chapter 3) and is stored as phospholipids. When little food is available, the stored lipids are used as an energy source and the phosphorus is excreted (Martin, 1968), so that relatively more phosphorus is regenerated into the water.

If the water column is deep enough there is a more or less uniform concentration of phosphorus below the thermocline, probably maintained by vertical mixing, and at some depth nearer the bottom there is usually an increase in phosphate. Such gradients suggest that deeper waters are a common source of phosphate. In the Gulf of Maine, for example (Fig. 11-3, left), the concentrations are higher near the bottom, suggesting that vertical diffusion from the sediments supplies the water column. There are examples of horizontal advection of nutrient-rich water, but by and large the pattern of Figure 11-3 is general.

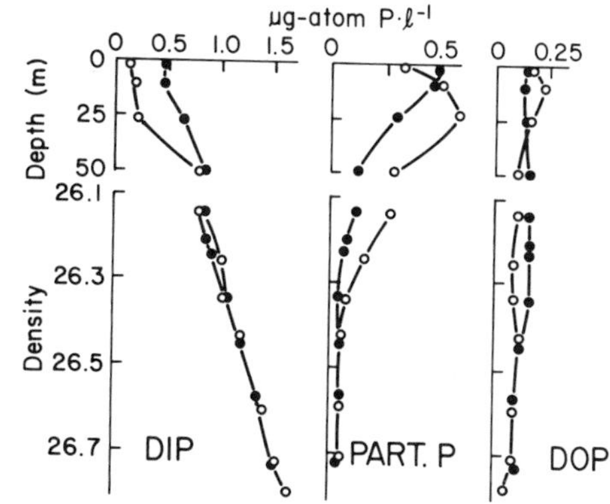

Figure 11-3. Vertical profiles of dissolved inorganic, particulate, and dissolved organic phosphorus in the water column of the Gulf of Maine. Below 50 m depth is expressed as density of the water. Adapted from Ketchum and Corwin (1965).

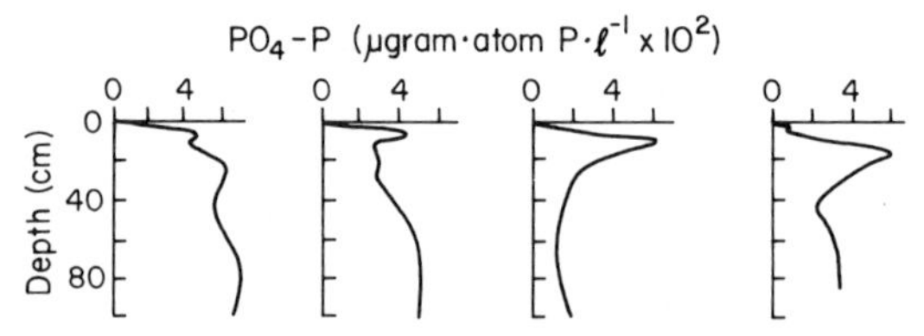

Figure 11-4. Vertical profiles of phosphate in the interstitial water of sediments at different stations within Chesapeake Bay. Adapted from Bray et al. (1973). © AAAS, reprinted by permission.

Phosphate concentrations in sediments are quite variable. In anaerobic sediments there is a peak at some relatively shallow depth, presumably where reduced conditions combine to make phosphate salts most soluble (Fig. 11-4). Above the maximum there is a pronounced decrease toward the surface, suggesting that there is a significant diffusion of phosphate out of reduced sediments to the overlying water. This can be the source of DIP that diffuses upwards in Figure 11-3, as shown by Klump and Martens (1981) for organic-rich coastal sediments. The gradients in the top few centimeters of the sediment of Figure 11-4, if they are due to diffusion, could provide about 5% of the total P content of the water column above per week, a significant flux of phosphate.

More evidence that regeneration of benthic phosphate affects water column concentrations is given by the relation of the seasonal pattern of phosphate in water and the flux of phosphorus from sediments (Fig. 11-5). Regeneration of phosphate is temperature dependent, increases in the spring, and is followed by increases in phosphate in the water, at least in Narragansett Bay. The rates of release of phosphate by a wide variety of marine sediments range from -15 to 50 μmoles m^{-2} hr^{-1} (Nixon, 1981; Fisher et al., 1982). Regeneration from sediments in Narragansett Bay can provide enough phosphate to support 50% of the primary production in the water column (Nixon et al., 1980). An additional 20% or so may be provided by DOP also released by the sediments. In various other coastal

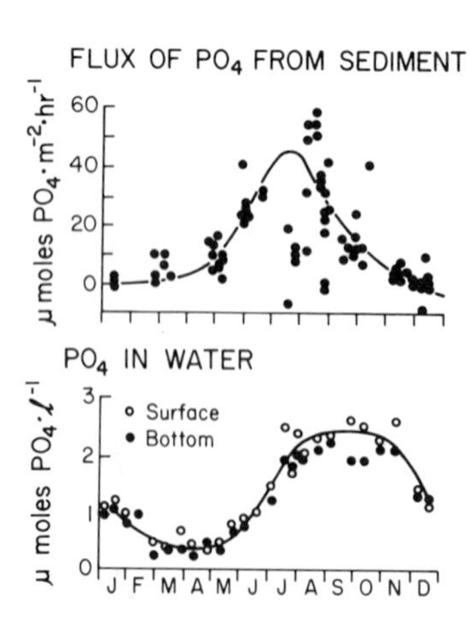

Figure 11-5. Flux of phosphate from sediments over the year and phosphate in the water of Narragansett Bay, 1972–1973. Lines drawn by eye. Adapted from Nixon et al. (1980). © University of South Carolina Press, reprinted by permission.

sediments the rates of regeneration of phosphate provides an average of 28% of phytoplankton requirements (Fisher et al., 1982). Regeneration of PO_4 by sediments is thus important in coastal water.

The regeneration process is so active that only a very small proportion of the P in coastal ecosystems is lost permanently to the sediments by burial. These losses to sediments may be balanced by inputs from terrestrial sources, including wind-borne dust particles and fluvially carried particles (Fig. 11-1).

Our discussion of phosphate in water and sediments demonstrates that its concentration is to a very large extent determined by biological activity. As a result, for example, oxygen uptake by organisms in water is well correlated to phosphorus (and nitrogen) concentrations (Stumm and Morgan, 1981), since nutrients are released during aerobic respiration of organic matter [Eq. (1-2) and Table 11-1]. All the dissimilative reactions (Table 10-7) result in regeneration of phosphorus.

Biological activity by microbes and animals also controls the distribution of phosphorus and other elements in organic matter by altering the redox of sediments, and hence the chemical properties of phosphorus.

The stoichiometric equations of Table 10-7 emphasize the point that cycles of individual nutrient are never isolated; the biological transformations link phosphorus, nitrogen, sulfur, and all elements present in organic matter. The reduction reactions involved in oxidation of organic matter are also important in that they not only alter redox conditions but

Table 11-1. Percentage of Total Nitrogen Inputs Contributed by Nitrogen Fixation in Marine and Other Environments

	Percentage of inputs	Source
Estuaries		
Narrangansett Bay, R.I.	0.4	Nixon (1981)
Rhode River, Chesapeake Bay	5	Marsho et al. (1975)
Salt Marshes		
Great Sippewissett, MA	9	Valiela and Teal (1979)
Sapelo Island, GA	23	Haines et al. (1977), Whitney et al. (1981)
Open Ocean		
Sargasso Sea	3.9×10^{-6}–7.3×10^{-7}	McCarthy and Carpenter (1984)
North Pacific gyre	1.6×10^{-6}–3×10^{-9}	McCarthy and Carpenter (1984)
Terrestrial Environments		
Hardwood forests[a]	31	Bormann et al. (1977)
Tundra	75	Barsdate and Alexander (1975)

[a] Estimated by difference to balance losses.

also change alkalinity* (Emerson et al., 1980), and this buffers the pH of the medium. Sulfate reduction (Table 10-7) increases alkalinity, as does reduction of nitrate, Mn^{4+}, and Fe^{3+}. Since alkalinity has important consequences for chemical reactions, it is clear that the chemical status of water and sediments is thoroughly affected by biological activity. The CO_2-carbonate system is of course a major influence on water chemistry, but it is also affected by biological activity.

11.2 Nitrogen

11.21 Transformations of Nitrogen

Unlike the phosphorus cycle, the nitrogen cycle is dominated by a gaseous phase, and microbial transformations involving changes in oxidation state play important roles.

Nitrate (NO_3^-) is the most oxidized form of nitrogen; it is taken up in aerobic environments by algae, bacteria, and plants, and once within the cells, reduced by assimilation processes involving several enzymes [including nitrate reductase (Packard et al., 1971)] to the amine form, which is used in metabolic processes. The uptake kinetics can be described by Michaelis–Menten kinetics (Section 2.21).

Nitrate is also used as a terminal electron acceptor and is reduced by dissimilatory processes in a series of steps. The first step is the reduction of nitrate to nitrite (NO_2^-), then NO and N_2O, and finally to N_2 gas (Fig. 11-6). This pathway is called denitrification and requires a supply of organic compounds (i.e., reduced carbon) which is concomitantly oxidized (Table 10-7). Denitrification also requires anaerobic conditions (Reddy and Patrick, 1976), is most active at pH of 5.8–9.2 (Delwiche and Bryan, 1976) and is very temperature dependent (Focht and Verstraete, 1977; Kaplan et al., 1977). Denitrifiers may reduce nitrate to ammonium via hydroxylamine intermediates (Fig. 11-6) but this is probably not a large source of ammonium. This source of NH_4^+ is probably not as important as the degradation of organic nitrogen.

Ammonium can be removed from water by uptake by plants, algae, and bacteria (cf. Section 2.2). Ammonium may also be oxidized by nitrifying bacteria (Fig. 11-6). Nitrification is a bacterially mediated process in which the first step is the oxidation of ammonium to nitrite by bacteria, usually *Nitrosomonas*. Nitrification continues with the further oxidation of nitrite to nitrate by *Nitrobacter* and other genera. Both reactions re-

---

* Alkalinity is a measure of the bases that are titratable with strong acid. It can be thought of as the acid-neutralizing capacity. It is measured in milliequivalents, and in seawater is due primarily to the presence of bicarbonate (HCO_3^-). Various forms of borate and any other base in seawater add to the alkalinity. Edmond (1970) Morel and Morgan (1972) discuss advanced methods to determine alkalinity in seawater.

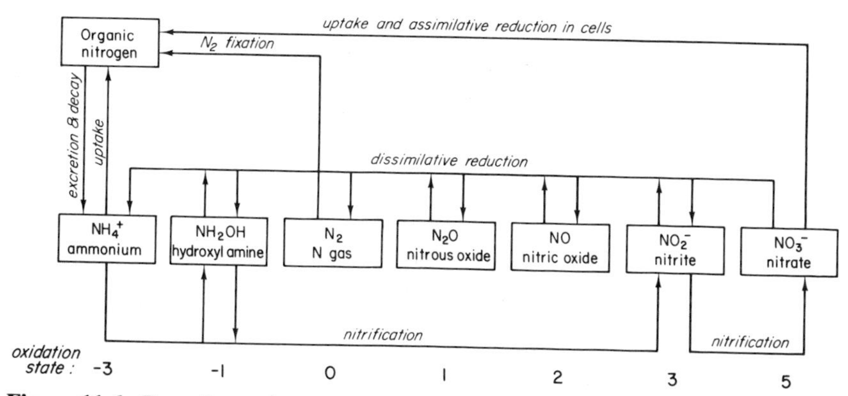

Figure 11-6. Transformations of nitrogen. The boxes show the various nitrogen species and arrows indicate the processes of transformation. Adapted Fenchel and Blackburn (1979). © Academic Press Inc. (London) Ltd., reprinted by permission.

quire oxygen and are energy yielding. These two genera of bacteria are autotrophic, in that they use the energy from reduced nitrogen compounds to fix carbon. Nitrification rates are often limited by supply of oxygen. Increase in aeration of sediments can increase nitrification, but nitrification can occur even at low levels of O_2 (Billen, 1975), due to the high affinity of nitrifiers for O_2. The importance of this is that some nitrate is produced even in sediments or water low in oxygen. The production of nitrate then allows denitrification to take place.

The cycle of nitrogen also includes a pathway by which nitrogen in gaseous form can be fixed into organic compounds by certain bacteria, including blue-greens (Fig. 11-6). The microbes that fix nitrogen can be free-living (Carpenter, 1973; Marsho et al., 1975; Gotto and Taylor, 1976; Potts and Whitton, 1977), or symbiotic with primary producers (Head and Carpenter, 1975; Zuberer and Silver, 1978; Teal et al., 1979; Capone et al., 1977) or with animals [shipworms, Carpenter and Culliney (1975), and sea urchins, Fong and Mann (1980)].

Nitrogen fixation is much reduced if free ammonium is available (Goering et al., 1966; Van Raalte et al., 1974; Carpenter et al., 1978). Uptake of ammonium should be energetically favored over fixation since the assimilative reduction needed within cells when N_2 is fixed is an energy-demanding process. Anaerobic conditions are required by fixers because the enzymes involved in N_2 fixation are sensitive to oxygen. Most blue-greens that fix nitrogen have structures called heterocysts in which anaerobic conditions are maintained and in which fixation may take place even in aerobic water. Some cyanobacteria like the pelagic *Oscillatoria* lack such structures; they may create anoxic zones within bundles of associated cells (Carpenter and Price, 1976).

Nitrogen fixation requires energy, which in shallow sediments may be provided by the abundant organic matter. In the water column the DOC may be very low, so other energy sources are used. Fogg and Than-Tun (1960) postulated that perhaps light energy could drive fixation; this agrees with results that show that fixation in the water column by the blue-green *Oscillatoria* stops in the dark (Dugdale et al., 1961).

Large amounts of organic nitrogen are released after death of animals (Fig. 11-6). Most of these organic nitrogen compounds are probably decomposed by heterotrophs, ultimately to ammonium.

Excretion by animals also releases dissolved nitrogen. Zooplankton excrete free amino acids and ammonia, with urea making up some of the remainder (Corner and Newell, 1967). Fish excrete relatively more urea and other organic compounds. We have already discussed the potential importance of animal regeneration of nutrients in regard to primary production (Chapter 2). This is a topic needing further study. Even less is known about the release and transformations of organic than inorganic nitrogen compounds in seawater.

11.22 Distribution of Nitrogen Species

11.221 Water Column

Nitrate is generally the primary form of inorganic nitrogen in seawater. It is usually most abundant in winter (Fig. 11-7) when it is not taken up by producers. Nitrate may disappear from the photic zone during algal blooms. Since active uptake by phytoplankton is restricted to the photic zone, nitrate concentrations are greater at depth (Fig. 2-17). Where there is a breakdown of stratification of the water column, vertical advection of deeper water replenishes nitrate (and other nutrients) in surface waters (cf. Section 14.3).

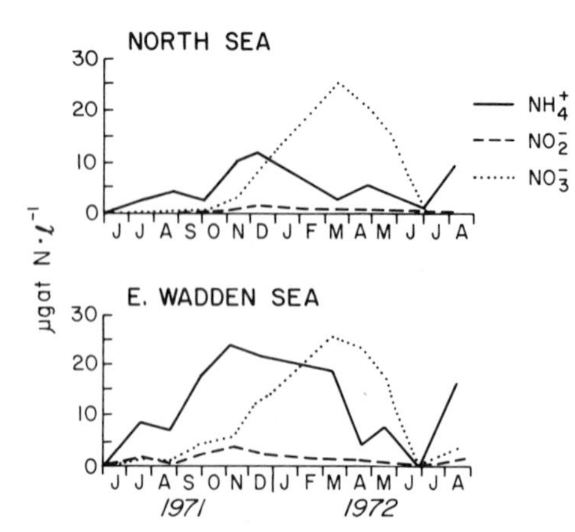

Figure 11-7. Seasonal changes in the concentrations of ammonium, nitrite, and nitrate in the North Sea and eastern Wadden Sea off Holland. Adapted from Helder (1974).

The amount of nitrite present in the water column is low (Fig. 11-7). Ammonium may be abundant in very productive shallow environments (Fig. 11-7) and since it is generated in large measure from decay of organic nitrogen, it peaks about the time of senescence in the producer growth cycle. Concentrations of ammonium range from traces up to 1.9 μg atoms N liter^{-1} in oceanic water and much higher in coastal areas. In salt marsh water for instance, 1–8 μg atoms N liter^{-1} are not uncommon (Valiela et al., 1978).

The amount of dissolved organic nitrogen (DON) is usually much larger than the amount of inorganic nitrogen. The concentrations of DON range from 5 to up to more than 30 μg atoms N liter^{-1} (McCarthy, 1980). Amino acids and urea comprise only a small component of this DON. Amino acids range from 0.025 to 1.4 μmole liter^{-1}, while urea concentrations reach 8.9 μg atoms N liter^{-1} but are usually much less (McCarthy, 1980). The remainder, or rather most of the DON, is not nutritionally suitable for oceanic phytoplankton (Thomas et al., 1971) or for estuarine microbes (Asiz and Nedwell, 1979). Mass balance and other calculations show only slight degradation of DON (Valiela and Teal, 1979; Nixon, 1981). Thus much of the DON may be resistant to decomposers. Most of the DON in seawater is still chemically uncharacterized and its chemical and biological properties are poorly known.

11.222 Sediments

The concentrations of nitrogen compounds in interstitial water of sediments are much higher than in water (Fig. 11-8). This is due to degradation of the large concentrations of organic matter, low rate of percolation, and to the presence of active exchange surfaces in sediments.

The redox state of a sediment determines the relative abundance of inorganic nitrogen compounds [NH_4^+, NO_2^-, and NO_3^-], and their vertical distribution in sediment. In aerobic sediments (Fig. 11-8, left) organic matter accumulates near surface due to the rain of particles from the water column; this leads to active decay of organic matter near the surface, so that most ammonium is released near the sediment surface. The benthic fauna are more abundant near the surface and their excretion also contributes to high ammonium near the sediment surface. Nitrification converts the ammonium into nitrate, and tube building and feeding activity of the benthos carry nitrate-bearing water down into the sediment column; both these processes create the nitrate gradient.

In anaerobic sediments (Fig. 11-8, right) ammonium is far more abundant than in aerobic situations due to the stoichiometry of anaerobic decay (cf. Table 10.7). The vertical profiles of ammonium in intertidal water show that ammonium diffuses upward to the overlying water. Oxidized inorganic nitrogen is seldom abundant (Fig. 11-8, bottom right) in anaerobic profiles, since denitrification, and less importantly, ammonifi-

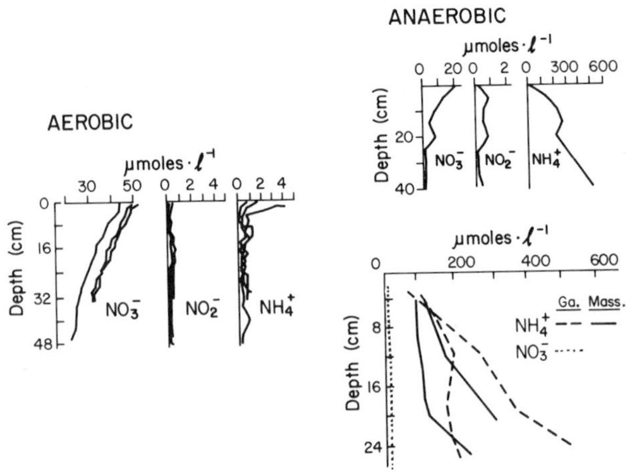

Figure 11-8. Vertical profiles of dissolved inorganic nitrogen in sediments. Left: Aerobic sediments in the East Pacific Rise [adapted from Emerson et al. (1980)]. Right top: Anaerobic sediments of Puget Sound [adapted from Grundmanis and Murray (1977)]. Right bottom: Anaerobic salt marsh sediment in two sites each in Massachusetts and Georgia. Concentrations of nitrite were very low. Adapted from Giblin (1982) and Haines et al. (1977).

cation, remove nitrate. In some anaerobic sediments the activity of animals may transport nitrite and nitrate-bearing waters into the sediments (Fig. 11-8, top right). Models of vertical distribution of nitrogen species constructed on the basis of diffusion and the stoichiometry of reduction reactions (Table 11-1) may not completely describe the actual distribution. Stirring of sediment by animals (Fig. 11-8, top right) or ebullition of bubbles of gases up animal tubes may enhance vertical fluxes (Emerson et al., 1980; Klump and Martens, 1981). Dissolved and exchangeable ammonium are released and flux out of sediments to overlying waters at rates between 13 to 710 μmole m^{-2} hr^{-1} (Nixon, 1981). Thus, not only do biological reactions provide nutrients in proportions determined by the biochemical stoichiometry, but also biological activity by animals affects the distribution and transport of the products of these reactions.

Of the ammonium produced by decay, some remains dissolved in interstitial water, and one to two times as much can be adsorbed to sediments (Rosenfeld, 1979). Ammonium is adsorbed by two means: exchangeable ammonium is attached to surface of clays and organic matter by ion exchange; fixed ammonium is adsorbed within layers of the clay structure where it is not readily replaced by other cations.

11.23 Cycle of Nitrogen in the Sea

Nitrogen is transformed and transported in a complex pattern in marine environments (Fig. 11-9). We will not consider all components of the cycle but rather focus on the major sources of new nitrogen, mechanisms

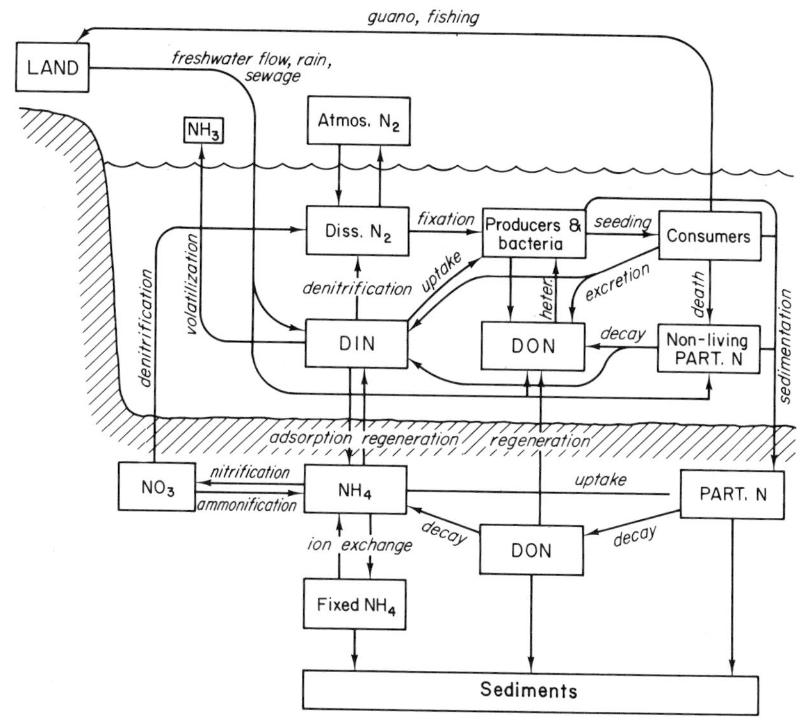

Figure 11-9. Simplified scheme of major transformations and transport of nitrogen in marine environments.

of uptake and regeneration of nitrogen, and on budgets of inputs and outputs. Referral to Fig. 11-9 will help through this discussion.

11.231 Sources of New Nitrogen

11.2311 Nitrogen Fixation

Gaseous nitrogen dissolved in seawater is generally in equilibrium with the atmosphere (except in special circumstances where water exchange is restricted (fjords, Black Sea, interstitial water in sediments). The fixation of gaseous nitrogen (N_2) is a source of new nitrogen for marine ecosystems. The acetylene reduction method* has been used to provide measurements of N_2 fixation in many environments but has drawbacks; N_2 fixation is best measured by a ^{15}N tracer technique (Painter, 1970; Hauck and Bremner, 1976).

Nitrogen fixation can provide a significant input of nitrogen into certain marine environments. In the Sargasso Sea (Carpenter and McCarthy,

* In this method the reduction of acetylene to ethylene is measured, and the rates are transformed to N_2 fixation using a 3 : 1 ratio. This assumes that the enzymes used to attack the triple bonds in acetylene (HCCH) and N_2 are the same.

1975) the amount of N_2 fixed by the cyanobacterium *Oscillatoria* is trivial compared to other sources, but in the Caribbean (Carpenter and Price, 1977) N_2 fixation can provide about 20% of the nitrogen required to support phytoplankton production.* Fixed nitrogen can be released reasonably quickly by exudation or by death of cells, so that the new nitrogen is quickly incorporated into the pool of recycled nitrogen.

Another way to assess the relative importance of fixation is by comparison with total nitrogen inputs (Table 11-1). In coastal and estuarine environments there are other significant sources of nitrogen (Tables 11-2, 11-3), so fixation plays a relatively minor role. In oceanic ecosystems with stratified upper layers the water column is seldom mixed, so that even small amounts of "new" nitrogen may be significant. In all environments, however, fixation is a minor source of external nitrogen (Table 11-1).†

11.2312 Vertical Mixing

The second source of new nitrogen for the photic zone is vertical advection or mixing of nitrate-rich deeper waters. We have already mentioned how upwelling of nutrient-rich water or the breakdown of thermo- or haloclines can provide a source of new nutrients to the photic zone. This is by far the major source of nutrients for most pelagic systems (McCarthy and Carpenter, 1984), and is discussed further as a major influence on seasonal patterns of production in Chapter 14. The hydrographic provi-

* The cyanobacteria are the principal organisms that fix nitrogen in aquatic environments. Blue-greens are most abundant in lakes with N:P = 29 (Smith 1983). The scarcity of blue-greens in seawater relative to fresh water is curious, since the low supply of nitrogen in seawater would make fixation advantageous. Recent work has shown an abundance of small cyanobacteria in seawater reaching densities of 10^3–10^5 cells ml^{-1} (Waterbury et al., 1980; Johnson and Sieburth, 1979). These small blue-greens, however, do not fix nitrogen (J. Waterbury, personal communication), so the higher fixation of nitrogen in freshwater compared to seawater still remains to be explained. Several possibilities have been suggested. Blue-green bacteria are most effective at obtaining CO_2 at low concentrations (high pH) (King, 1970), do best at taking up PO_4^- at high PO_4^- concentrations (Shapiro, 1973), do best in warm waters (Clendenning et al., 1956), and at high densities may secrete enough inhibitory substances to deter growth of other phytoplankton (Keating, 1978). These properties do not seem well suited for life in the sea, where pH is buffered at about 8, nutrient concentrations are low, waters are as a rule colder, and cell densities are lower than in freshwaters. One other possibility is that nitrogen fixation by marine blue-greens is limited by the availability of molybdenum, an element that is part of the enzymes involved in nitrogen fixation. R. Howarth and J. Cole (personal communication) suggest that the abundant sulfate in seawater interferes with the uptake of molybdenum by fixers, and therefore reduces their abundance and the rates of nitrogen fixation in seawater. All these explanations need substantiation.

† The term "new" nitrogen was first used by Dugdale and Goering (1967) to differentiate allochthonously provided nitrogen from "old" or recycled nitrogen. The idea that these are separable is perhaps applicable only to deep water, since some nitrate is regenerated from sediments in shallow water (Fisher et al., 1982).

Table 11-2. Budget of Inputs and Regeneration of Nitrogen and Phosphorus for Narragansett Bay, Rhode Island[a]

	Annual inputs (10^6 g-atoms yr^{-1})	
	Nitrogen	Phosphorus
Inputs		
Fixation	0.2	—
Rain	2.8	0.19
Runoff	16.2	0.8
Rivers	235	17.3
Sewage	278	21.7
Total	532	39.9
Regeneration		
Menhaden	0.8	0.1
Ctenophores	8.1	0.8
Net zooplankton	98.5	—
Benthos	264	41.1
Total	371	42

[a] Adapted from Nixon (1981).

Table 11-3. Nitrogen Budget for Great Sippewissett Marsh Itemizing the Contribution of Each Mechanism of Input and Output[a]

	Inputs	Outputs	Net exchange
Precipitation	380		380
Groundwater flow	6,120		6,120
N_2 fixation	3,280		3,280
Tidal exchange	26,200	31,600	−5,350
Denitrification		6,940	−6,940
Sedimentation		1,295	−1,295
Other	9	26	−17
	35,990	39,860	−3,870
Percentage of exchanges via biotic mechanisms	9	18	
Percentage of exchanges via physical mechanisms	91	82	

[a] Units are in kg N yr^{-1} for the entire marsh. Below the table are the calculated percent exchanges due to biological and physiological mechanisms, which emphasize the importance of non-biological processes. Negative signs indicate net export to coastal waters. Adapted from Valiela and Teal (1979) and Valiela (1984).

sion of nutrients has already been discussed in Chapter 2 as the feature that generally determines the overall level of primary production.

11.2313 Horizontal Transport of Nitrogen

In estuarine, coastal, and oceanic areas, there is considerable horizontal transport of nutrients, both in solution (nitrate and ammonium) and adsorbed to particles (ammonium). Shelf waters off Georgia, for example, are enriched by intrusion of nutrient-rich waters from the Gulf Stream (Hanson et al., 1981). Another example is provided in Section 14.312, where the importance of river outflow for phytoplankton production in coastal waters is discussed, but there are many other examples. The surface waters within the plume discharging from the Columbia River in the Pacific northwest of the United States contains up to 1.2 μg atoms liter^{-1} NH_4^+ even many miles off shore. Beyond the plume, ambient oceanic surface waters average about 0.1 μg atoms liter^{-1} (Jawed, 1973). Much of the nitrogen transported by rivers originates as fertilizer nitrogen. In lakes and rivers of the midwestern cornbelt of the United States 55–60% of the nitrate in surface waters is due to agricultural use (Kohl et al., 1971); a substantial portion of this can flow to the sea.

While river transport is an obvious source of nutrients, there is increasing evidence of considerable intertidal and subtidal flux of groundwater carrying nutrients from land to coastal waters (Johannes, 1980; Bokuniewicz, 1980; Valiela, 1984, and Table 11-3). Although this topic remains to be studied, it is likely that the transport of nitrogen by groundwater is significant for coastal ecosystems.

The disposal of sewage can also convey large amounts of nitrogen. The New York City metropolitan area contributes so much dissolved nitrogen to the western end of Long Island Sound that ammonia concentrations are as high as 45–100 μg atoms liter^{-1}. Away from the city, waters of the Sound carry only 0–5 μg atoms liter^{-1}. Nitrate is similarly high near the city, 8–20 μg atoms liter^{-1}, while the rest of the Sound has 0.5–8 μg atoms liter^{-1}. Near the city the nutrients are high enough not to limit growth of algae, as we saw in the enrichment experiments discussed in Section 2.2332.

Transport from land can carry massive amounts of nitrogen to coastal environments. Runoff, rivers, and sewage provide 99.5% of the nitrogen received by Narragansett Bay, and 99.7% of the phosphorus (Table 11-2).

The nitrogen budgets of salt marshes are dominated by physical transport processes. In Great Sippewissett Marsh, flow of groundwater and tidal exchange provide 90% of the nitrogen inputs (Table 11-3). Environments where inputs of nitrogen are physically determined have no way to respond to sudden increases in nitrogen loadings. Biologically active processes such as fixation decrease if the supply of DIN increases. Such mechanism may furnish some control of the inputs of nitrogen. They could be important where fixation is a large proportion of nitrogen inputs.

11.2314 Precipitation

The amount of nitrogen conveyed to marine environments by rain and snow is usually small relative to other inputs (Table 11-3). Near land and especially near areas of industrial and agricultural activity, this source of nitrogen increases (Eriksson, 1952), but since other inputs are also large, precipitation is less important. For example, precipitation provides 12% of the nitrogen entering the mixed layers of the Baltic Sea, (Sen Gupta, 1973), a body of water whose shores are heavily industrialized. Only 0.5% of the nitrogen entering Narragansett Bay (Nixon, 1981), a body of water surrounded by a less urbanized region, is due to precipitation (Table 11-2).

Precipitation may be more important in open, stratified oceanic water, such as the subtropical gyres where there are few other external sources of nitrogen. There is a significant correlation, for example, between the amount of rainfall and the concentration of ammonium in surface water in the Sargasso Sea (Menzel and Spaeth, 1962). Perhaps only in such oligotrophic areas are nutrients borne by rain important in sustaining primary production (Carpenter and Price, 1977).

Industrial contamination of the atmosphere has increased the nitrate content of rainwater, and marine waters downwind of industrial regions must receive increased doses, although this has not yet been clearly documented.*

11.232 Uptake of Nitrogen

The dynamics of uptake of dissolved nitrogen has been discussed in Chapter 2, and are reviewed by Collos and Slawyk (1980) and McCarthy (1980). Most marine algae preferentially take up NH_4^+ over NO_3^-, and take up nitrate when ammonium is depleted.† Some producers, especially estuarine and coastal species, may use organic sources of nitrogen, for instance, methylamine by kelp and phytoplankton (Wheeler, 1979, Wheeler and Hellebust, 1981), and methyl ammonium in phytoplankton (Wheeler, 1980). A variety of amino acids and urea are also used by phytoplankton, but this uptake may only amount to a fraction of the nitrogen used by primary producers (McCarthy, 1980).

---

* The increased acidity of rainfall associated with industrial regions is due to the sulfuric and nitric acids released into the air. Freshwater is usually not well buffered and its pH can be changed by external addition of CO_2 or other acids; thus acid rain is a serious problem, and many lakes and ponds in North America and Europe have been acidified enough to lose most of their fauna. Acid rain is a far less severe problem in the more strongly buffered seawater.

† The mechanism behind preference for ammonia is not well known. Either actual uptake or reduction by nitrate reductance may be inhibited by high ammonium concentrations. The ultimate reason for the preference for ammonium may be that energy is required to reduce nitrate to the amino form. Uptake of ammonia thus avoids the need to spend energy, since the reduced nitrogen taken up as NH_4^+ can be directly incorporated into proteins.

Particulate nitrogen is consumed by animals, sinks to the sediments, decays, and is regenerated as dissolved inorganic nitrogen, much like particulate carbon.

11.233 Regeneration of Nitrogen

The inputs of new nitrogen in many marine environments are not sufficient to support the requirements of primary production. As in the case of phosphorus, nitrogen incorporated in particles is recycled by the release from bacteria, zooplankton, and fish, and by the regeneration of dissolved inorganic nitrogen from the benthos (Fig. 11-9). Measurements of these sources of regenerated nitrogen (Table 11-4) show that each is potentially capable of providing substantial amounts of nitrogen relative to the amounts assimilated by producers.

11.2331 Regeneration in the Water Column

Most animals and microbes in the water column contribute significantly to the regeneration of nitrogen (Table 11-4). Smaller organisms release relatively more nutrients than larger species (Johannes, 1964; Glibert, 1982). In coastal waters off California 90% of the mineralization of NH_4^+ is accomplished by organisms smaller than 35 μm; 40% was done by cells less than 1 μm in diameter (Harrison, 1978). These two size classes were 40 and 10% of the total particulates. The abundance of small and large plankton vary so much, however, that bacteria do not always release more total ammonium than the large zooplankton (Table 11-4). Gelatinous zooplankton may be important in regeneration of nitrogen: in the Sargasso Sea just one average-sized siphonophore per cubic meter of water could excrete enough ammonium to satisfy 40–60% of the requirements of phytoplankton (Biggs, 1977). If the fish are abundant enough they may also be important releasers of ammonium, as in the Peruvian upwelling ecosystems (Table 11-4).

Nitrogen is regenerated by heterotrophs mainly as ammonium, the form of nitrogen preferentially taken up by phytoplankton. Even when nitrate concentrations are very high in Antarctic waters (17–31 μg atoms liter^{-1}) the phytoplankton still use very substantial amounts of ammonium, although the concentrations of ammonium may be only 0.1–2.5 μg atoms liter^{-1} (Glibert et al., 1982). Thus, it is obvious that the regeneration of ammonium is important for algal growth.

Regeneration of nitrogen (and phosphorus) by zooplankton increases as more food is available to the zooplankton (Eppley et al., 1973). *Calanus* excretes 2.6% of the nitrogen in its body per day in winter, when phytoplankton is scarce, but in spring the excretion of nitrogen by *Calanus* reaches 4.6% of body N per day (Butler et al., 1970).

The importance of regenerated versus allochthonously provided "new" nitrogen has not yet been directly assessed. Eppley and Peterson

Table 11-4. Sources of Regenerated Nitrogen in Various Marine Environments[a]

<table>
<tr><th></th><th>Bacteria</th><th>Zooplankton</th><th>Fish</th><th>Benthos</th></tr>
<tr><td>New York Bight, inshore[b]</td><td>13</td><td>6</td><td></td><td>21</td></tr>
<tr><td>New York Bight, shelf break[b]</td><td></td><td>61</td><td></td><td></td></tr>
<tr><td>Southern Calif. Bight[c]</td><td>70</td><td>100</td><td></td><td></td></tr>
<tr><td>Saanich Inlet, B.C.[c]</td><td colspan="2">--------------100--------------</td><td></td><td></td></tr>
<tr><td>Pacific Ocean, off Oregon[d]</td><td></td><td>36</td><td></td><td></td></tr>
<tr><td>Pac. Oc., plume of Columbia R.[d]</td><td></td><td>90</td><td></td><td></td></tr>
<tr><td>N.W. Africa upwelling[e]</td><td></td><td></td><td></td><td></td></tr>
<tr><td>shelf less than 200 m</td><td></td><td>24</td><td>9.5</td><td>27</td></tr>
<tr><td>shelf over 200 m</td><td></td><td>18</td><td>6</td><td></td></tr>
<tr><td>Peru upwelling[e]</td><td></td><td>15</td><td>22</td><td></td></tr>
<tr><td>Long Island Sound[f]</td><td colspan="3">--------------------50--------------------</td><td>53</td></tr>
<tr><td>Narragansett Bay[g]</td><td></td><td colspan="2">------------10------------</td><td>26</td></tr>
<tr><td>Bering Sea, coastal[h]</td><td></td><td>2–16</td><td></td><td></td></tr>
<tr><td>Bering Sea, open sea[h]</td><td></td><td>13</td><td></td><td></td></tr>
</table>

[a] The values are expressed as percentage of the nitrogen required by measured rates of primary production. N.Y. Bight values based only on use of NH_4^+ by producers; others based on DIN use.
[b] Conway and Whitledge (1979).
[c] Harrison (1978).
[d] Jawed (1973).
[e] Whitledge (1978), Rowe et al. (1977): In this case upwelled water provided an additional 117% of the nitrogen required by the phytoplankton production.
[f] Bowman (1977).
[g] Nixon (1981).
[h] Dagg et al. (1982).

(1979) obtained rough estimates of these two kinds of nitrogen by assuming that uptake of ammonium by phytoplankton corresponds to use of regenerated nitrogen, while uptake of nitrate is based on new nitrogen made available by mixing. Using this distinction they calculated that in neritic (open coastal) and inshore water 54 and 70% of the production is based on regenerated nitrogen, while in the open ocean 82 to 87% of production is supported by regeneration. In subtropical gyres 94% of the nitrogen used by producers is regenerated by other organisms; the phytoplankton of these depauperate waters are thus almost completely dependent on recycled nitrogen.

11.2332 Regeneration in the Benthos

Organic nitrogen is mineralized (converted to an inorganic form, usually NH_4^+) in sediments. The inorganic nitrogen in interstitial waters can then be released to the overlying water; this is a major pathway by which regenerated nitrogen is made available to the overlying plankton (Table 11-4). The rate of release of ammonium depends on biological activity (Fig. 11-10, left). The greater the rate of aerobic decay (or oxygen con-

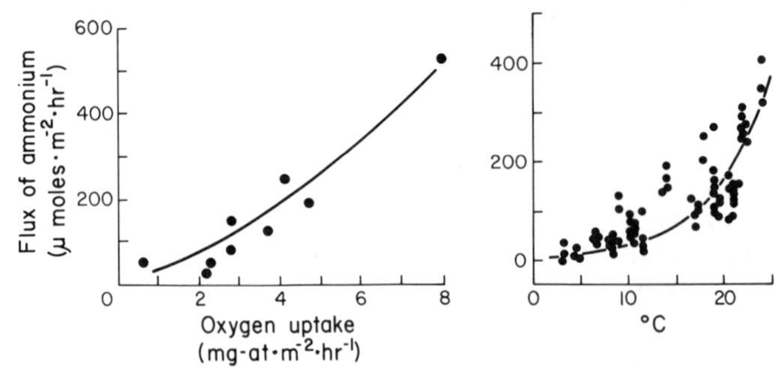

Figure 11-10. Effect of biological activity (left) and bottom water temperature (right) on the release of ammonium from sediments in Narragansett Bay, Rhode Island. Adapted from Nixon (1981) and Nixon et al. (1976).

sumption) the more organic nitrogen that is mineralized to ammonium. Since aerobic respiration rates in microbes and animals are strongly temperature dependent, the regeneration of NH_4^+ also depends on temperature (Fig. 11-10, right). Anaerobic decay also regenerates ammonium in large amounts and is temperature dependent.

We have earlier referred to the Redfield atomic ratios of 16 N–1 P usually found in particulate matter. This ratio is fairly constant over much of the ocean, but is generally lower in coastal areas. A comparison of the nitrogen and phosphorus generated from coastal sediments (Fig. 11-11) shows that there appears to be a consistent depletion of nitrogen compared to phosphorus. This loss may be the result of denitrification. The nitrogen limitation commonly found in coastal waters may be due to remineralization of 25–50% of the organic matter in the sediments, where nitrogen may be lost by denitrification. Bioturbation and burrowing by benthic animals mix the organic matter into the surface layers, where mineralization takes place. In the sediments some of the ammonium released from the organic matter is nitrified to nitrate, and denitrification then converts a substantial proportion of the nitrogen to N_2.

The large gradients (100- to 1,000-fold differences) in concentration of nitrogenous compounds between interstitial and overlying water lead to

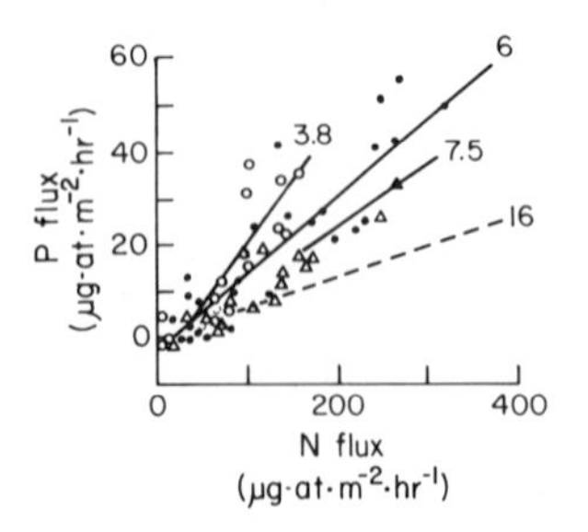

Figure 11-11. Relationship of flux of phosphate and dissolved inorganic nitrogen (mainly ammonium) in three stations in Narragansett Bay. The three stations showed fluxes with N/P of 3.8, 6, and 7.5, with correlation coefficients of $r^2 = 0.69$, 0.76, and 0.83, respectively. The Redfield ratio (N/P = 16) is included for comparison. Adapted from Nixon et al. (1980).

Table 11-5. Average Rates of Regeneration (Mean ± Standard Error) of Forms of Inorganic Nitrogen (Nitrate, Nitrite, Ammonium, and Total Dissolved Inorganic Nitrogen) from 10 Coastal Sediments to the Overlying Water[a]

Nitrogen compounds	Exchange rates ($mmole\ m^{-2}\ day^{-1}$)
NO_3^-	0.73 ± 0.42
NO_2^-	0.09 ± 0.06
NH_4^+	2.2 ± 0.6
Total DIN	3 ± 0.9

[a] From Fisher et al. (1982). © Academic Press Inc. (London) Ltd., reprinted by permission.

movement of nitrogen out of sediment (Table 11-5). Measured rates of regeneration of inorganic compounds in coastal sediments such as those of Table 11-5 are enough to supply an average of 35% (± SE of 8.7%) of the nitrogen demands of phytoplankton.

11.24 Nitrogen Budgets for Specific Ecosystems

To understand the dynamics of nutrients in specific ecosystems, we need budgets of the inputs and losses, but very few budgets are available. Comparison of the magnitudes of the various sources of nitrogen and of recycling are therefore possible only in a few environments.

In the shallow coastal system of Narragansett Bay (Table 11-2) rivers and sewage are the major sources in the nitrogen and phosphorus budgets. Animals regenerate 70% of the N and 128% of the P that enter the bay. The activity of small animals and microbes, while not measured, probably adds considerably to the recycling of N and P. Note that the inputs have a N:P = 13.3, somewhat lower than the ideal Redfield ratio, perhaps in part creating a nitrogen-limited situation. The loss of nitrogen by denitrification in sediments of Narragansett Bay was equivalent to about half the input of nitrogen from sewage and rivers (Seitzinger et al., 1980).

The Narragansett Bay budget (Table 11-2) shows that in shallow coastal systems, external sources of nutrients are mainly due to horizontal transport, nitrogen recycles slower than phosphorus and is depleted (due to nitrate reduction and denitrification) relative to phosphorus. The

Narragansett Bay budget also demonstrates that anthropogenic sources of nutrients can be very large compared to other sources (Table 11-2). Both agricultural fertilizer and waste water contribute to the nitrogen loadings of coastal waters. Since nutrients, as we have seen, are so intimately linked to virtually every process important in marine ecology, very significant changes can be potentially brought upon marine environments by eutrophication due to human activity. The nitrogen in culturally eutrophied waters is of special significance since we have seen that coastal primary production is nitrogen limited.

Another nitrogen budget where losses as well as inputs are reported is available for a salt marsh (Table 11-3). Most of the inputs and outputs of nitrogen in a salt marsh are conveyed by physical mechanisms (flow of groundwater and tidal exchanges), while biotic mechanisms (N_2 fixation and denitrification) provide smaller inputs and outputs. As we have seen, perhaps in very stable, nutrient-poor oceanic waters fixation may be more important, but most nitrogen transport in marine environments is probably dominated by physical mechanisms.* Considering the variation of each of the entries of Table 11-3, the relatively small deficit of nitrogen is not significant; this ecosystem is roughly in balance with regard to nitrogen.

Another way to make use of nutrient budgets is to tally the inputs, fates, and outputs of each of the major species of nitrogen so as to depict the net transformations that go on within the ecosystem (Fig. 11-12). The exact amounts in each entry are not the important issue, since these are only rough estimates. Instead, notice that the internal transformations within the marsh ecosystem can be very large, often exceeding annual inputs and outputs. This is invariably the case in many ecosystems. For instance, consider the particulate nitrogen at the bottom of Figure 11-12. Particulate nitrogen enters the marsh by transport and by fixation. About half of the particulate nitrogen is consumed by animals, of which about two thirds is released as particulate nitrogen in feces. Producers add another substantial amount of particulate nitrogen. The particulate nitrogen released by the marsh to coastal waters exceeds the input; within the marsh major qualitative transformations have taken place, and the nitrogen has been extensively reused while in the ecosystem.

Similar complex patterns of use and reuse can be followed for any of the forms of nitrogen in Figure 11-12. The point here is that once in an ecosystem, nutrients recirculate in complex ways; knowledge of the inputs and outputs provides only partial understanding of how the salt marsh works or how active processes are.

It is still instructive, however, to compare the chemical state of the inputs vs outputs of nitrogen. The salt marsh essentially converts NO_3 to

---

* Notice that this discussion deals with inputs and outputs to the ecosystem. Internal transformations due to biological activity—for example, benthic regeneration in Narragansett Bay (Table 11-2)—can be larger than inputs or outputs.

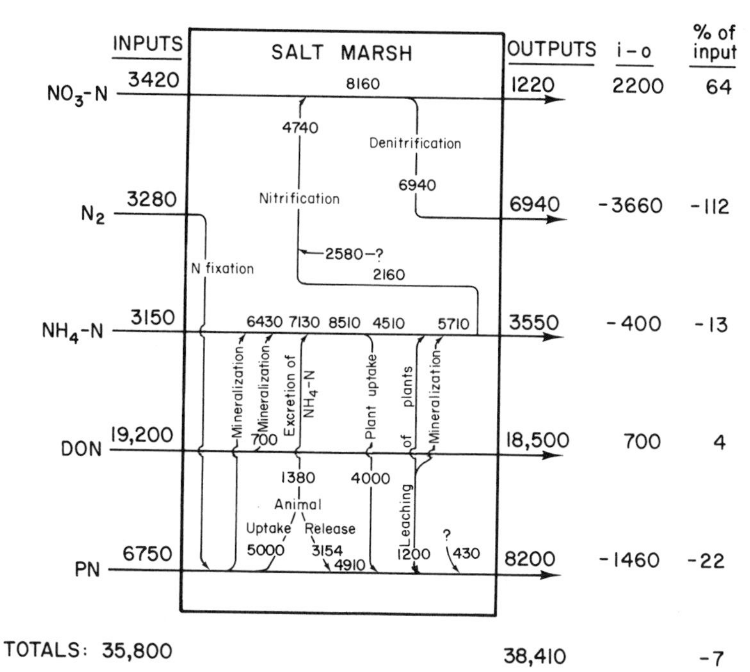

Figure 11-12. Nitrogen budget for Great Sippewissett Marsh for a year. Inputs and outputs are indicated along the margins of the box; the two right-most columns show the difference between inputs and outputs and the percentage of the input for each nitrogen species. Inside the box representing the marsh are indicated some of the major transformations. From Valiela (1983), data from Valiela and Teal (1979), Jordan and Valiela (1982), and unpublished data.

N_2 gas, ammonium, and particulate nitrogen (Fig. 11-12). The large amounts of DON seem to be flushed back and forth with the tides, and may remain relatively unchanged. Perhaps many of the compounds making up DON are not easily degradable (cf. Section 11.221). The N_2 exported from the marshes is not useful to coastal organisms but ammonium and particulate nitrogen are energy rich and accessible. Export of these reduced compounds (already mentioned in Sections 7.71 and 10.12) could provide energy, nitrogen, and carbon to the adjoining coastal waters. Thus, the end products of processes in one environment can be transported and can affect assemblages of organisms in adjoining ecosystems.

11.3 Sulfur

The dynamics of sulfur include aqueous, sedimentary, and atmospheric phases, as well as a wide range of oxidation states. As a result the sulfur cycle is very complex. We have already covered aspects of sulfur dynamics involved in the decomposition of organic matter. Microbiological reactions of sulfur also involve the nitrogen cycle, as well as many other

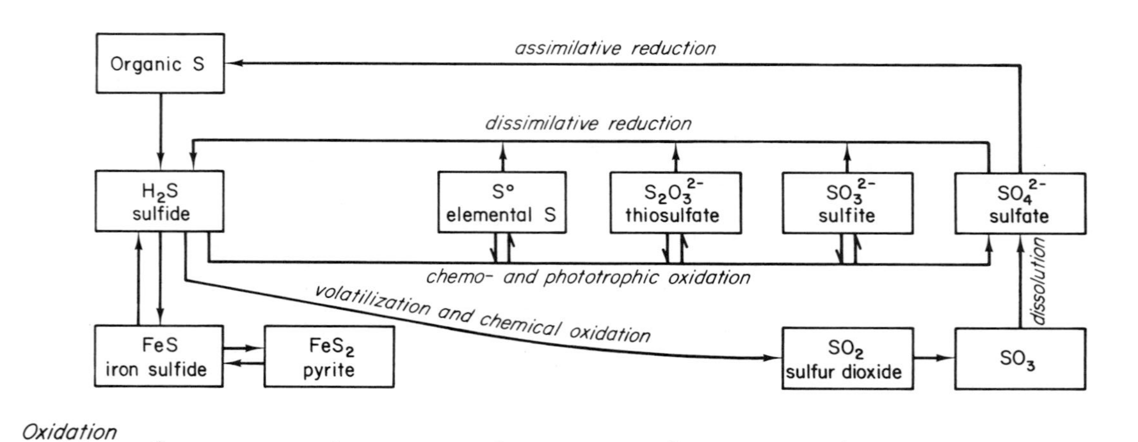

Figure 11-13. Scheme of microbial and chemical transformations of sulfur. The oxidation state of S in pyrite is −1 on average; there are probably mixtures of sulfur atoms in 0 and −2 states. Adapted and expanded from Fenchel and Blackburn (1979). © Academic Press Inc. (London) Ltd., reprinted by permission.

elements. The cycle of sulfur thus illustrates the couplings among nutrient cycles, production, and decay processes.

11.31 Chemical Transformations

In aerobic environments, sulfate (SO_4^{2-}) is by far the most abundant form of sulfur. Although uptake and assimilative reduction of sulfate by bacteria and algae occur (Figs. 11-13, 11-14), these processes do not materially change concentration of sulfate because this ion is so abundant in seawater. There is sulfur within cells, including methionine (–S–), cystine (–S–S–), or cysteine (–SH), all reduced organic sulfur compounds. Sulfolipids and sulfate esters are also important. Marine producers contain 0.3–3.3% sulfur, but the decay of this organic sulfur is usually not an important source of sulfide in marine environments, accounting for only a few percent of the sulfide production (Jørgensen, 1977; Deuser, 1979).

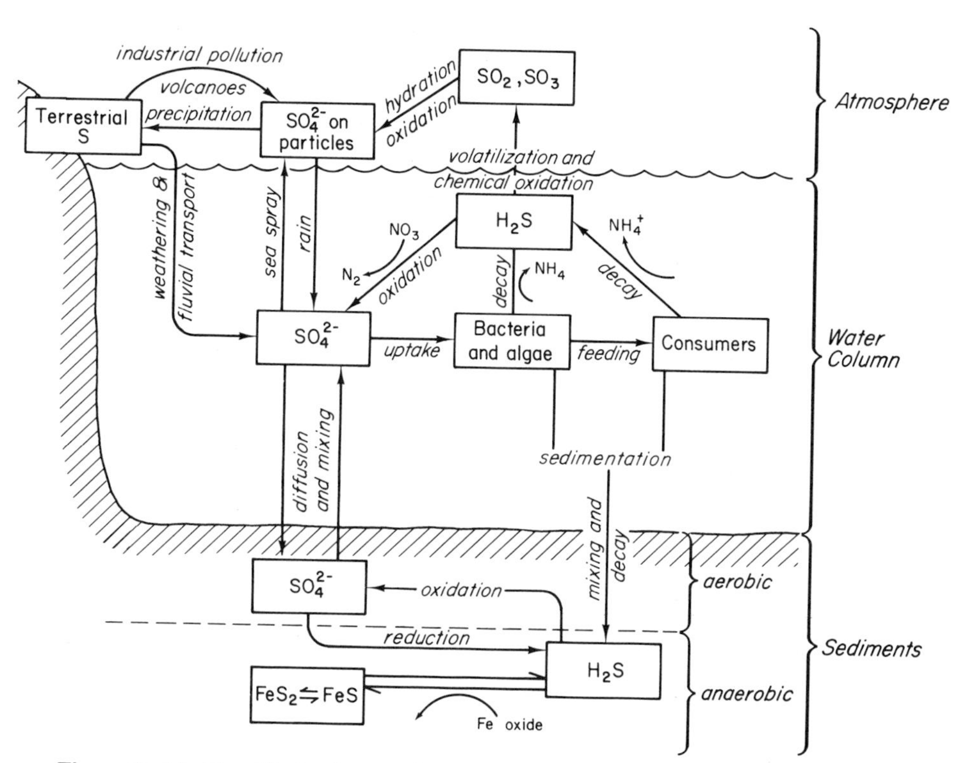

Figure 11-14. Simplified scheme of sulfur dynamics. Couplings to nitrogen cycle are indicated on the arrows that represent processes of transfer or transformations. This is representative of a coastal environment; in the deep sea sediments are usually not reduced. The diagram does not include the volatilization of dimethyl sulfide, probably released by producers, and whose flux is probably larger than that of H_2S.

Dissimilative reduction of sulfate is the principal source of sulfide in anaerobic environments (Tables 10-7, 10-8). Sulfate reduction rates range from 0.01 to 10 μmole SO_4^{2-} ml^{-1} day^{-1} (Jørgenson, 1982). The rates depend on the mechanisms that convey sulfate into the anaerobic environment, such as diffusion, mixing by animals, and on the availability of organic matter. Sulfate reducers also depend on fermentation of complex, polymeric organic matter (Table 10.7) by fermenting microorganisms. Such fermentation produces low-molecular-weight substances which in turn are respired by sulfate reducing bacteria (Jørgensen, 1982; Howarth and Teal, 1979). *Desulfovibrio* and *Desulfomatuculum* are the principal taxa involved in sulfate reduction.

Sulfide [actually dimethyl sulfide is the major volatile sulfur compound (Andreae and Raemdouck, 1983)] can diffuse out of a reducing environment* (Fig. 11-14) and be oxidized biologically and chemically into more oxidized sulfur compounds (Fig. 11-13), including elemental sulfur, thiosulfate, sulfite, and sulfate. Oxidation of sulfide occurs where O_2 or nitrate is available, and at the same time sulfide diffuses out of an anaerobic zone; sulfide oxidizers therefore occur in a narrow layer or plate, either in sediments (Fig. 11-15) or in the water column or around anaerobic microzones. The use of nitrate links the nitrogen, sulfur, and carbon cycles, as does the production of ammonium during sulfate reduction (Table 11-1) and aerobic decay (Fig. 11-15).

Some of the sulfide in anaerobic sediments reacts with dissolved iron or oxidized iron minerals to form FeS and FeS_2 (Figs. 11-14, 11-15) (Rickard, 1975; Fenchel and Blackburn, 1979; Luther et al., 1982). Pyrite (FeS_2) can be oxidized by *Thiobacillus* if there is oxygen or nitrate available, and the reaction results in production of acid (this is the process that produces acid mine wastes).

A number of forms of sulfur of intermediate oxidation state (elemental sulfur and sulfite, Fig. 11-14), and polythionates ($S_{3+n}O_6$, n = 0–3) are produced by different bacteria or are oxidized chemically. These are usually found in low concentrations or have very fast turnover in seawater.

11.32 Dynamics of Sulfur in Marine Environments

Although the qualitative outlines of the sulfur cycle have long been known, quantitative knowledge of sulfur transfers in natural environments is rare.

Studies in Solar Lake, a hypersaline coastal pond on the edge of the Gulf of Elat, Red Sea, have demonstrated an anaerobic, hypersaline

* Anaerobic conditions occur in sediments or in the water column; Figure 11-14 is more representative of a coastal environment, where anaerobic sediments are common. In the Black Sea (cf. Fig. 1-10), fjords, and in certain coastal ponds, water below the mixed layers is anaerobic.

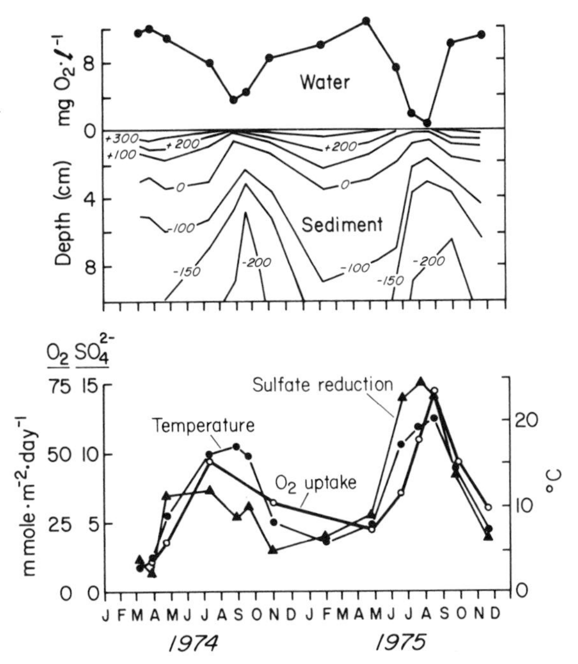

Figure 11-15. Top: Vertical and seasonal profiles of oxygen and redox in the water and sediments, respectively, in Limfjorden, Denmark. Positive redox potentials show oxidized conditions, while negative values show reduced conditions. Bottom: Surface temperature and rates of sulfate reduction and dark O_2 uptake in top 10 cm of sediments. Adapted from Jørgensen (1977).

lower layer of water in which there is anoxygenic photosynthesis primarily carried out by cyanobacteria (Cohen et al., 1977, 1977a, 1977b). In Solar Lake, bacterial activity controls the vertical gradients of sulfur and other elements. Sulfide and oxygen, for example, have profiles that look much like those of the Black Sea (Fig. 1-10), although of course the depth of Solar Lake is just a few meters. Oxygen is depleted down the water column, and sulfide increases below the depth where oxygen disappears.

Another comprehensive study of the dynamics of sulfur was done by Jørgensen (1977) on Limfjorden, Denmark. Limfjorden is a shallow brackish water body with an aerobic water column that exhibits a seasonal fluctuation in dissolved oxygen (Fig. 11-15, top). Even though there is seasonal variation in the reduction of the surface sediments, below about 5 cm the sediments are permanently reduced. Near the sediment, surface temperature varies seasonally and seems to drive both the dark uptake of O_2 and the rate of sulfate reduction in the top 10 cm of the sediment (Fig. 11-15, bottom). Sulfate reduction takes place even in winter when the sediments are most oxidized, because of the presence of reduced microzones 50–200 μm in diameter (Jørgensen, 1977a). In such sediments sulfate reducers (*Desulfovibrio*) and sulfide oxidizers (*Beg-*

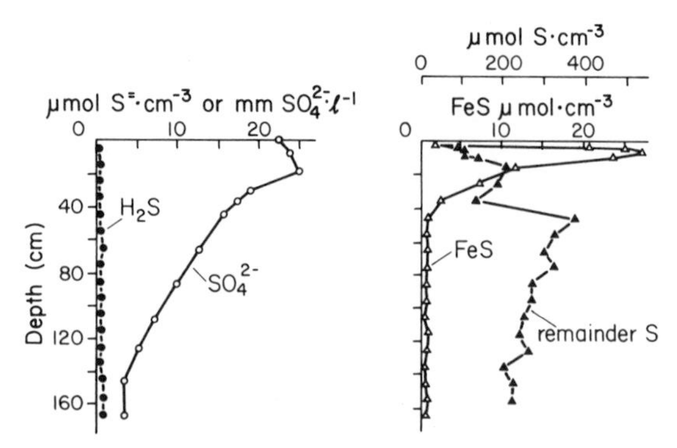

Figure 11-16. Vertical profiles of sulfur compounds in sediments of Limfjorden, Denmark. Remainder S is shown as total S minus H_2S, FeS, and SO_4^{2-}, and probably consists of pyrite (FeS_2). Adapted from Jørgensen (1977).

giatoa) occur close together, and so the sulfide diffusing out of anaerobic microzones is oxidized; in addition, some of the sulfide may be converted to iron sulfides. The result of both these processes is that little H_2S accumulates in the sediment (Fig. 11-16).

The vertical profiles of the various forms of sulfur in sediments are due to various processes. In the upper 10 cm, the sulfate concentrations are like those of the overlying waters (Fig. 11-16, left). This is probably the effect of irrigation caused by the activities of tube-dwelling animals, as in the case of vertical profiles of nitrogen compounds (Section 11.2332). Below the zone inhabited by animals, bacterial reduction decreases sulfate (Fig. 11-16, right). Ferric iron in the sediment is released to ferrous iron and precipitates with sulfide to form iron monosulfides and pyrite (Luther et al., 1982) Pyrite is not a permanent sink for S. Although there are disagreements as to rate of formation and oxidation of pyrite (Howarth, 1979; King, 1983; Howes et al. in press), there are seasonal changes in the net oxidation of pyrite.

A considerable portion (170 g organic C m^{-2} $year^{-1}$) of the primary production of the phytoplankton and eelgrass flats reaches the sediments of Limfjorden and its decay supports the sulfur cycle. Only 12% of this carbon accumulates, and the whole carbon pool in the sediment can be turned over in 3–5 years, by both O_2 respiration and sulfate reduction. In Limfjorden about 53% of the organic carbon is consumed by sulfate reducers. In other coastal environments there are similar high proportions (50% or so) of consumption of organic matter by sulfate reduction (Jørgensen, 1982). In Great Sippewissett Marsh, annual rates of sulfate reduction are high enough to consume over half the annual accumulation of organic carbon by living salt marsh plants (Table 10-9). In such anaerobic environments, therefore, dissimilative sulfate reduction is a very major pathway by which organic carbon is degraded.

In deeper coastal waters the depth of the aerobic layer of sediment increases. The amount of organic matter reaching the anaerobic layers decreases and the refractory nature of the organic matter increases, so that the organic substrate available for sulfate reducers is less abundant. In deeper waters, then, rates of sulfate reduction are lower and the proportion of organic matter mineralized by sulfate reducers may only amount to 35% of that consumed by aerobic respiration (Jørgensen, 1982).

Part V

Structure of Marine Communities

Previous chapters have provided information on the physiological and populational levels. We have touched on the matter of how marine producers and consumers interact and how these populations may be controlled. We also have seen how materials cycle in marine ecosystems and how both nutrients affect populations, and organisms affect nutrient dynamics.

The next and last part of this book uses the information of previous chapters in an account of the structure of communities. By structure we mean what and how many organisms are present, and how they are distributed in space and time. We devote the last chapters to the three components of community structure: species composition (Chapter 12), spatial pattern (Chapter 13), temporal changes (Chapters 14 and 15).

It will be obvious in these chapters, especially in Chapters 14 and 15, that it is fatuous to refer to structure separately from function. In fact, the emphasis will be on how the functioning of these communities changes structure over time.

Chapter 12

Taxonomic Structure: Species Diversity

12.1 Introduction

Terrestrial ecologists long ago remarked on the richness of the floras and faunas of tropical environments relative to colder climates. The diversification within many specific taxonomic groups is clearly greater in the tropics than in temperate latitudes, both in terrestrial and in marine environments (Fig. 12-1). Low numbers of species are also typical of severe and disturbed habitats. Such observations have spawned an abundant and contentious body of publications that have dealt with three major problems: first, how to quantify the clearly observable differences in diversity, second, how are such differences in taxonomic richness of communities generated and maintained, and third, what do such differences mean ecologically.

12.2 Measurement of Diversity

12.21 Diversity Within a Habitat

The assessment of number of species in a habitat is not a straightforward matter, for it is hard to know when to stop sampling additional individuals. Rare species will be missed if large numbers of individuals are not included. To solve this uncertainty, the number of species (S) can be graphed against the total area sampled (A) (Fig. 12-2, top) or the number of individuals included in the sample (Fig. 12-2, bottom). Such a curve is described by

$$S = CA^z, \tag{12-1}$$

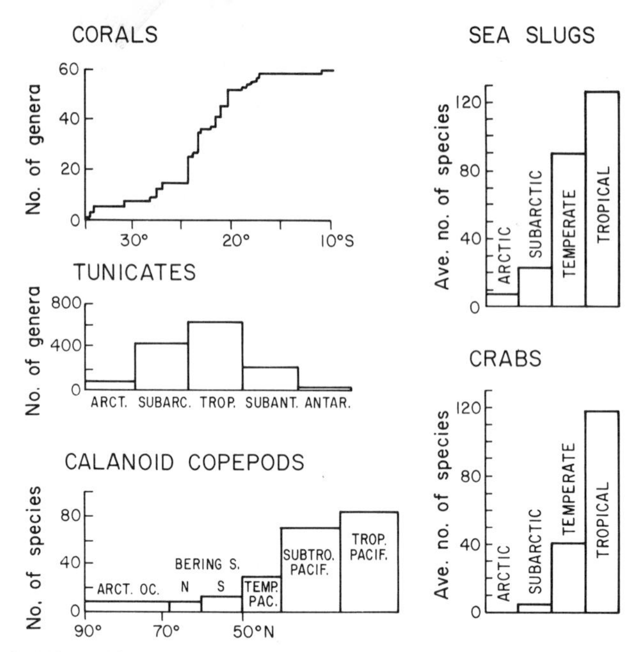

Figure 12-1. Diversification in selected marine taxa along geographical gradients. Corals from the Great Barrier Reef, Australia; the copepods are from the Pacific and its northward extensions, while the remaining data are from all oceans. Adapted from Thorson (1957) and Fischer (1970).

where C and z are constants that can be estimated if a series of values of S and A are available and a regression using logarithmic transformations of the numbers of species and individuals is used:

$$\log S = \log C + z \log A. \tag{12-2}$$

The number of species in any area can then be obtained. Such species–area curves generally depict deceleratingly increasing numbers of species as area or number of specimens increase. The asymptote eventually reached gives some idea of the total number of species in the environment.

Such an estimation of the total number of species is still not a completely satisfactory way to measure diversity, since the distribution of individuals among species also needs to be considered. For example, a sample containing two species, each represented by 50 individuals seems intuitively more diverse than one consisting of 99 individuals of one species and 1 of another species. Diversity, then, has two components, species richness and equitability of abundance of the species. To deal with these two components a number of approaches using either probability or information theory have been made. We deal with these only briefly; anyone interested in measuring diversity should consult the reviews by

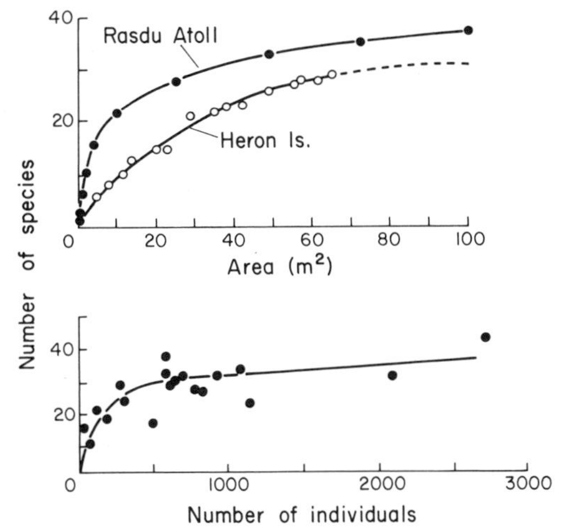

Figure 12-2. Top: Species-area curve for corals in coral reefs on Rasdu Atoll, Maldives, and on Heron Island, Great Barrier Reef. Adapted from Scheer (1978). UNESCO, reprinted by permission. Bottom: Relation of number of species and number of individuals in a sample, based on twenty samples of benthic invertebrates collected from Buzzards Bay, Massachusetts, during a 2-year period. Adapted from Hessler and Sanders (1967). © Pergamon Press Ltd., reprinted by permission.

Poole (1974), Peet (1974), and Pielou (1975), the major sources of the following discussion.

Simpson's index. Simpson (1949) suggested the use of an index of diversity that was based on the probability that two individuals drawn from a population of N individuals belong to the same species. This index, modified more recently (Pielou 1969), is

$$D = 1 - \sum_{n_i} n_i\,(n_i - 1)/N(N - 1), \tag{12-3}$$

where n_i is the number of individuals of the ith species and N is the total number of individuals. D can also be used as the probability of interspecific encounters in random meetings of the individuals of the population.

Logarithmic series. Fisher et al. (1943) proposed that the frequency of species with a given number of individuals was described by a logarithmic series, and derived an equation for S, the number of species in a sample:

$$S = \alpha \log(1 + [N/\alpha]), \tag{12-4}$$

where α is a fitted constant that can be used as an index of diversity.

Log-normal distributions. If a variable is affected by many independent factors, the distribution of the variable will be normal. Since the occurrence of individuals is affected by such an array of factors, in a community there ought to be mainly intermediately abundant species, with few abundant or rare species as in normal distributions. This abundance pattern can be fitted by a log-normal distribution where data are plotted as the number of species in steps (called octaves) of one to two individuals, two to four individuals, etc. (Fig. 12-3). The steps are equivalent to taking logarithms of the abundances to the base 2. Such log-normal distributions fit many populations (Preston, 1948, 1962). The number of species (s) in the Rth octave is

$$s = s_0 e^{-(aR)^2}, \tag{12-5}$$

where a is value estimated from the data itself and s_0 is the number of species in the modal octave. The total number of species (S) in the environment from which the range was taken can be obtained from

$$S = s_0\sqrt{2\pi\sigma^2} = 2.5\ \sigma\ s_0, \tag{12-6}$$

where sigma is the standard deviation of the log-normal distribution expressed in octaves. The accuracy of this calculation of S is not as good as desirable. Further, S is an index of species richness and does not reflect equitability. To include both richness and equitability, Edden (1971) suggests the use of the ratio s_0/σ. Because of Eq. (12-6), the ratio equals $S/\sqrt{2\pi\sigma^2}$. The ratio, then, has both the elements of species richness (S) and equitability, since the variance reflects the frequencies with which

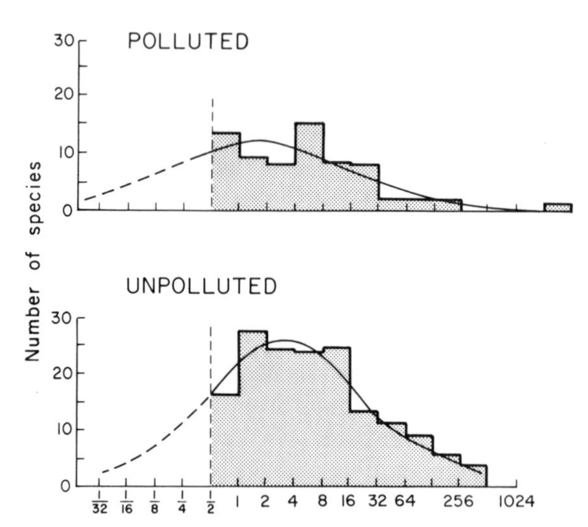

Figure 12-3. Frequency distribution of species of diatoms settling on glass slides in a polluted and an unpolluted stream. The histogram shows the number of species in each of the octaves (doubling intervals) of abundance. The line shows the idealized log-normal distribution as if it were a continuous distribution. Since frequencies of less than one individual cannot occur, the dotted part of the distribution cannot be seen left of the dashed vertical "veil line" located left of the first octave. Adapted from Patrick et al. (1954).

individuals are distributed. Values of Edden's index ratio calculated for phytoplankton correlate fairly well with those calculated using the Shannon–Weaver index (see below).

The use of the log-normal requires that the mode be clearly identified, so that the distribution should be moved to the right as far as possible (Fig. 12-3). For the mode to move one octave to the right, a doubling of the number of individuals is needed, so the use of the log-normal requires quite large censuses.

Information theory indexes. The use of diversity indexes derived from information theory were first suggested by Margalef (1957). If we are examining a collection of specimens from an environment, and we find that all the individuals examined so far belong to species A, we feel reasonably sure that the next individual will also be an A; we therefore gain little information by looking at that specimen. If, in contrast, the sequence so far has been A, N, B, D, X, Z, we cannot easily predict what species will be represented by the next individual, and its identification does produce new information about the collection. An increase in the number of species and a more equitable distribution therefore increase uncertainty, and the average uncertainty can be calculated in various ways.

The information theory index most often used is the Shannon-Weaver expression*

$$H' = -\sum_{i=1}^{S} P_i \log p_i, \tag{12-7}$$

where S is the number of species and $p_i = n_i/N$, for the ith species. Actually Eq. (12-7) is a biased estimator, and at least one correction term $[(S - 1)/2N]$ should be subtracted from the right-hand side of Eq. (12-7) (Hutcheson, 1970). The variance of H' can be obtained from the series

$$\mathrm{Var}(H') = \frac{\sum_{i=1}^{S} p_i \ln^2 p_i - \left(\sum_{i=1}^{S} p_i \ln p_i\right)^2}{\mathrm{N}} + \frac{S - 1}{2N^2} + \cdots \tag{12-8}$$

For large samples the first term of Eq. (12-8) is usually sufficient (Poole, 1974). The calculation of Eq. (12-8) makes it possible to subject the values of H' to statistical tests (Hutcheson 1970).

In some instances involving sessile, colonial, or vegetatively growing organisms (bryozoans, tunicates, salt marsh grasses are examples) it is not possible to define individuals. For such cases Pielou (1966) suggests that the area of study should be divided into many quadrats, and the diversity of a randomly chosen quadrat calculated using Brillouin's index

* The choice of the base of logarithm is arbitrary; if base 2 is used, the units will be in bits; if base e is used, the units are bels.

$$H = 1/N \log \frac{N!}{N_1!H_2!N_3!...N_S!}, \tag{12-9}$$

where N is the total number of individuals, and N_1, N_2 ... N_S are the abundances of each species.

Brillouin's index applies where all individuals in a collection or a community can be identified and counted. Since H measures the entire community it has no standard error, and values of H calculated for different communities that are not the same are therefore significantly different.

The area covered by a species or perhaps the weight of a plant within a quadrat are taken as N_i. The procedure is then repeated but using values of N_i from the first quadrat pooled with those of a second randomly chosen quadrat. Then a third quadrat is added to the first two, and H estimated for the pooled data, and so on. The cumulative values of H are graphed versus the number of pooled quadrats, and there will be a point where the calculated pooled diversity reaches an asymptote. An estimate of the pooled diversity (H'_p) of the whole area is

$$H'_p = \frac{1}{z - t + 1} \sum_{k=t}^{z} \frac{M_k H_k - M_{k-1} H_{k-1}}{M_k - M_{k-1}}, \tag{12-10}$$

where z is the total number of quadrats sampled, t is the number of quadrats at which the pooled diversity reaches an asymptote, M_k is the summed area or weight of all species in k quadrats, and M_{k-1} is the summed area or weight of all species in $k - 1$ quadrats.

Many other indexes of diversity exist; none, including the ones listed above, have a well-documented underlying biological mechanism; all the justifications given above are analogies or logical speculation. All have some drawbacks (Hurlbert, 1971; Taylor et al., 1976). Further, they each respond somewhat differently to aspects of diversity, for example, to changes in the importance of the rarest or most abundant species (Peet, 1974; May, 1975; Taylor et al., 1976). The result is that there may be a great deal of discordance in the trends shown by the different indices even when applied to the same data (Hurlbert, 1971). It should always be kept in mind that the diversity we more or less quantify using indices is our construct. There is no evidence that diversity has any meaning to the organisms involved.

Many other problems or difficulties need to be considered, not the least that indices used should be independent of sample size. The avalanche of new indices and papers on diversity has currently subsided, and the attitude prevails that perhaps the number of species or a simple index of species richness (R), such as Margalef's (1951)

$$R = (S - 1)/\log N \tag{12-11}$$

will suffice for most purposes.

One of the reasons diversity and diversity indices became so popular in the 1960s and early 1970s was the growing need to provide some assay of the effect of pollution of natural communities. It was obvious that the number of species decreased in polluted circumstances (Fig. 12-3 and Table 12-1) and that the more severe the level of pollution the greater the reduction (Fig. 12-4). An index such as H' contains information on the abundance and species composition of an array of species (Fig. 12-4). It also depicts the dominance by an opportunistic, tolerant species (the polychaete *Capitella capitata* in Fig. 12-4 bottom) in more polluted waters, and the more diversified assemblage away from the source of pollution. Applied scientists thus seized on the apparently comprehensive, easy-to-calculate indices as a way to subsume a lot of complex data.

As another example, a comparison of the number of species and variance of the distribution of Figure 12-3 shows that there were fewer species and that these fewer species were relatively more abundant in a polluted than in an unpolluted environment. Pollution simplifies the assemblage of diatoms and prompts the numerical dominance of a handful of species. The Shannon–Weaver index, because of ease of calculation, has been a favorite (Fig. 12-4). In addition to an ambiguous ecological interpretation, this use of diversity indices delivers less than it promises. Assemblages of species are variable, and so are the indexes calculated from samples of such assemblages. This variability, plus the statistical idiosyncracies of the index used, may prevent the indices from being sensitive enough to detect incipient pollution, a role that would be of great benefit. By the time a diversity index reflects changes in the assemblage of species, there are other clear and more dramatic symptoms of pollution. Judicious use of the indexes can, however, serve as a shorthand way to show change in species composition, and the results should be used keeping in mind their limitations.

Table 12-1. Effects of Experimental Eutrophication on the Diversity of Higher Plants and Benthic Diatoms of Salt Marsh Plots[a]

	Higher plants	Diatoms		
	No. of species	No. of species	H′ (bits ind^{-1})	Percentage *Navicula salinarum*
Control plots	9–11	129–196	3.8–4.1	5–9
Fertilized plots	5	92–117	3.1–3.4	20–23

[a] Fertilized plots received additions of nitrogen or mixed fertilizers throughout the growing season. H' is the diversity index of Shannon–Weaver. The right-most column is the percentage of the total number of cells made up of *Navicula salinarum*. Adapted from Van Raalte et al. (1976a) and unpublished data.

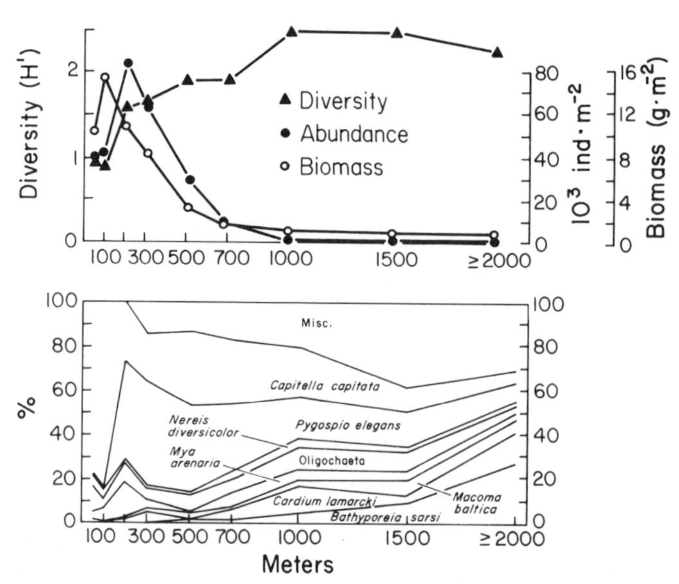

Figure 12-4. Benthic fauna at a depth of 3 m in the Baltic Sea. Top: Abundance, biomass, and diversity in samples collected at various distances away from an outfall releasing untreated sewage. Bottom: Percentage composition of benthic species along the same transect. Adapted from Anger (1975).

12.22 Diversity Between Habitats

If larger and larger samples of organisms are taken there will be increases in the diversity of the total census that are due to the inclusion of different habitats with different faunas (MacArthur, 1965). This can be seen in Figure 12-5, where the number of species from environments that contain a variable number of habitats is shown. This between-habitat component of diversity has also been termed beta-diversity, in contrast to alpha- or within-habitat diversity (Whittaker, 1972). The distinction of within- and between-habitat diversity is to an extent ambiguous since it depends to a large degree on the size of the organism and the size of habitat patches.

A variety of indices can be used to measure the compositional similarity of assemblages of species from different sites (between-habitat diversity), including some from information theory (Field, 1969; Williams et al., 1969). A commonly used index of overlap among samples containing two arrays of species is that of Morisita (1959)

$$C_D = \frac{2 \sum_{i=1}^{s} x_i y_i}{(D_x + D_y)XY}, \tag{12-12}$$

where species i is represented x_i times in the total X from one sample, while y_i times in the total Y from another sample. D_x and D_y are Simpson's

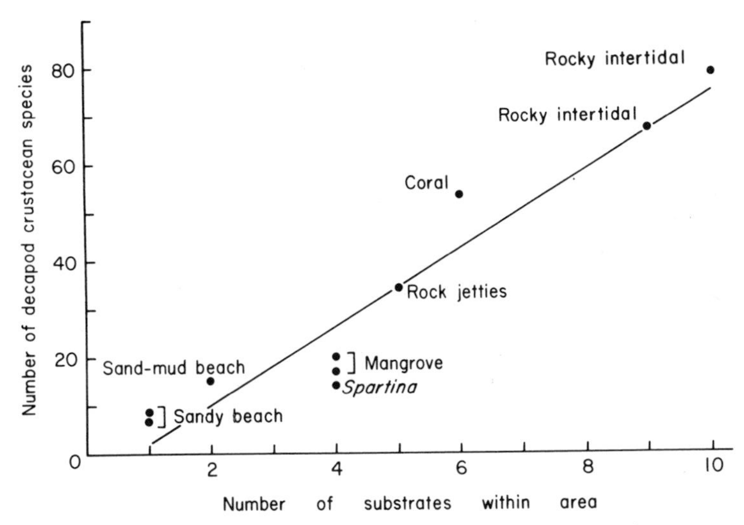

Figure 12-5. Effect of between-habitat diversity on decapod species present. The number of different substrates in each of the habitats was more or less arbitrarily designated from experience. The species counts are probably underestimates in all cases but serve for comparative purposes. From Abele (1982).

index [Eq. (12-3)] for the two samples. When the samples are completely different $C_D = 0$. The value of C_D increases to about one when species are in the same proportions. Some pitfalls of similarity indices are discussed by Bloom (1981). Simplified versions of Morisita's index and others adapted from information theory are provided by Horn (1966). Plant ecologists have developed many other simple empirical indices of similarity that can be used for such comparisons (Goldsmith and Harrison, 1976). Much more complex analyses are available for the identification of relations among assemblages of organisms from many different sites (Whittaker, 1973).

12.3 Factors Affecting Diversity

The tropical–temperate gradient in species diversity mentioned above has been taken as an example of the observation that diversity varies from one environment to another and has thus attracted an enormous amount of attention. There are many hypotheses put forward to try to explain the gradient and differences in diversity. In very general terms, the various hypotheses can be summarized as follows (Sanders, 1968):

The time hypothesis. Communities diversify with time, and older communities are therefore more diverse.

The spatial heterogeneity hypothesis. The more heterogeneous and complex an environment the more diverse the flora and fauna.

The competition hypothesis. In benign environments selection is controlled by biological interactions rather than physical variables. Interspecific competition is therefore important, species have evolved narrower niches and hence there is higher diversity because more species can partition the available resources.

The environmental stability hypothesis. The more stable the environmental variables, such as temperature, salinity, oxygen, etc., the more species are present, since it is less likely that species will become extinct due to vagaries of environmental variables.

The productivity hypothesis. All other things being equal, the greater the productivity, the greater the diversity, since more resources allow more species.

The predation hypothesis. Predators crop prey populations and thus lower competition that may lead to exclusion of many species-

Detailed evaluations of each of these are available in Pianka (1966), Ricklefs (1979), Pielou (1975), and Huston (1979). Most of the papers that proposed these hypotheses dwell on the evolutionary origin of species diversity and often disagree on factors that might affect species diversity. This is a field where logical argument exceeds factual content and, as such, is far from settled in spite of great effort.

We do not have space to scrutinize the evidence for each hypothesis, but material relevant to the spatial heterogeneity hypothesis was discussed in the previous section in terms of between-habitat diversity. Some elements of the productivity and predation hypotheses are discussed later in this section.

The various hypotheses listed above are a mix of evolutionary considerations and factors that affect diversity on a local scale. We will discuss these two aspects separately.

12.31 Evolutionary Considerations

Sanders (1968) combined features of the several hypotheses listed above into an integrated statement that he called the time-stability hypothesis. This hypothesis incorporates evolutionary considerations, and was based on the most thorough study of marine benthic organisms available. Sanders (1968) concluded that variable, stressful shallow habitats had lower diversities than the more constant, benign abyssal environments. To account for these differences the time-stability hypothesis holds that diversity in communities increases during evolutionary time. Competitive pressures result in competitive displacements; the assemblage of species becomes relatively more specialized over time. The increase in diversity, however, is restricted by the level of physiological stress and environmental variability that the assemblage of species has to endure. Thus, diversity is the net balance of a process in which the more specialized an assemblage of species, the more subject it is to vagaries of the environment.

The ambiguities and difficulties of schemes to explain diversity are represented by those of the time-stability hypothesis (Peters, 1976). Unfortunately it is impractical to follow the diversity of communities over evolutionary time, making it difficult to test directly whether diversity does indeed increase over time, the first part of the time-stability hypothesis. The second element of the hypothesis, specialization, is very difficult to define in most situations, since species may have narrow preferences for one dimension of their needs, while showing broad tolerances for another. Branch (1976) put together a subjective index, based on the occurrence of a species in one or more habitats, and on morphological traits, and applied it to an assemblage of limpet species found on rocky shores of South Africa. In this one example there is a clear relation of specialization of the assemblage and the number of species (Fig. 12-6). This agrees with the prediction of the time-stability hypothesis, but it does not necessarily mean that the relationship arose as per the hypothesis.

Because of the difficulty of measurement over time and difficulty in the definition of specialization, the tests of Sanders' hypothesis have to focus on stress, the remaining element. Stress, however, is usually defined by its effect so there is danger of circular reasoning, and it is not easy to state a falsifiable hypothesis about the effect of stress. Thus, the time-stability scheme does not have the clear predictive quality that defines scientific hypotheses. Perhaps tests can be carried out if stress can be thought to be related to the extent to which variables fluctuate in different environments. Comparison of communities of similar age but differing amplitude of variation could then test of the effect of "stress" on diversity. Coral reefs of approximately the same age (Abele, 1976) contain more species of decapod crustaceans (55 vs 37) in reefs with more variable conditions

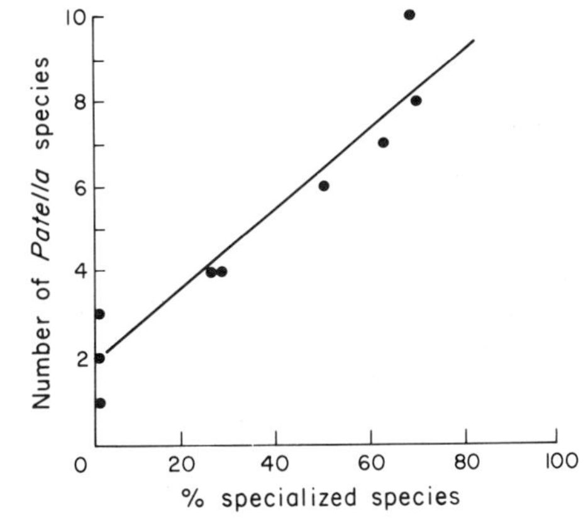

Figure 12-6. Relation of number of *Patella* species in a South African shore and the percentage of specialized species. Adapted from Branch (1976).

(temperature ranges of 17–29°C vs 28–29°C, salinity ranges of 22–36‰ vs 31–32‰). This, although a very inadequate test, does not seem to support the time-stability hypothesis in regard to higher diversity in constant, benign environments.

The high diversity of benthic fauna found by Sanders (1969) in the deep sea floor and the very uniform conditions at such depths inspired the time-stability hypothesis. The high diversity no doubt was facilitated by the long evolution in relative constant conditions. Part of the explanation, however, may be attributable to other factors. The increased diversity of the deep-sea benthos compared to shallower areas may to a large extent depend on the inclusion of a series of heterogeneous habitats in the very lengthy tows used to sample the larger areas in the deep sea. The data may thus include substantial between-habitat diversity. There may also be sampling artifacts due to the use of different collection devices (Osman and Whitlach, 1978; Abele and Walters, 1979). Regression analysis of the species–area curve for some of the data collected by Sanders (1968) and colleagues showed that perhaps 80% of the variation in the species numbers down to about 2,000 m could be accounted for by the effect of area alone. More extensive analysis of benthic data including data down to about 5,000 m shows that the diversity of several taxonomic groups decreases below about 3,000 m (Rex, 1981) (Fig. 12-7). Since the largest area of sea floor is at or below this depth, the effect of area alone cannot explain the gradients in diversity. Nevertheless, the devices used to sample the deeper parts of the sea floor, such as epibenthic sleds and anchor dredges, are towed over the sea floor for several kilometers, while shallow stations were sampled in Sanders' data by coring devices only a few centimeters in diameter. The sampling is thus at different scales and must affect the number of species obtained. Newer approaches to deal with such spatial problems are discussed by Jumars (1976, 1978).

The difficulties in interpretation and methods associated with the time-stability hypothesis are representative of other evolutionary arguments to

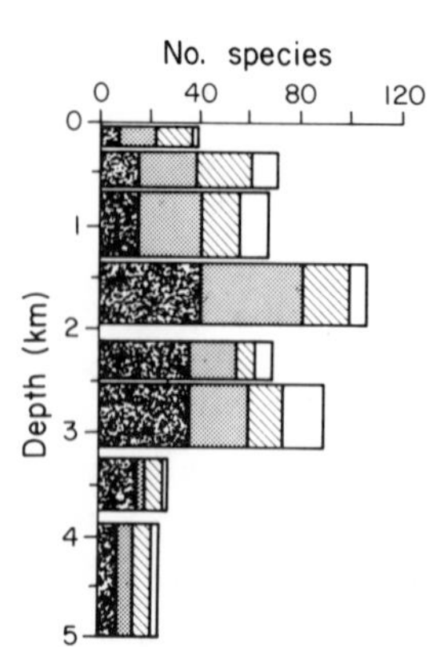

Figure 12-7. Species richness in the sea floor off New England, at different depths. The taxa are, from left to right, echinoderms (irregular stippling), fish (dotted), decapods (diagonal lines), and other groups (blank). The collections were done by otter trawls during several cruises, and all the data are pooled. Adapted from Haedrich et al. (1980).

explain taxonomic diversity. The scheme nonetheless remains as the best developed argument available and an attractive context in which to consider the evolutionary and ecological context of species diversity.

12.32 Factors Affecting Diversity on a Local Scale

Questions about trends in diversity of biotic assemblages over evolutionary time and on regional geographical scales have not yet been resolved, in spite of sustained efforts. On the other hand, work on the local, short-term control of diversity has been more successful. The local diversity of a community can be affected over relatively short periods of time by at least four kinds of factors: (a) the concentrations of deleterious substances or physiologically severe conditions in the environment, (b) the abundance of key resources, (c) the abundance of key consumers or disturbances, and (d) specific features of the local environment.

12.321 Deleterious Substances or Conditions

Few species survive in environments that contain large quantities of deleterious substances. These substances include natural and man-made compounds, such as sulfide, ammonium, or allelochemicals, and petroleum and chlorinated hydrocarbons. The effects of intense levels of pollution on assemblages of organisms are clear cut [for example, in the case of oil in marshes, Sanders et al. (1980), and Figs. 12-3, 12-4]. Despite the widespread nature of marine pollution little is known about the effects of low-level, chronic contamination by petroleum and polychlorinated biphenyls now found dispersed everywhere in the sea. Since few marine environments remain uncontaminated, deleterious substances may play a role in determining the diversity of many marine systems.

12.322 Abundance of Key Resources

A low supply of an essential resource may be insufficient to maintain many species, and thus diversity may be low. Lowered diversity of benthic polychaetes and bivalves is related to reduced oxygen concentrations in upwelling areas off southwest Africa, for example (Fig. 12-8, left). This should bring to mind the adversity-selected species discussed in Section 3.5.

Low as well as high abundance of resources may lower diversity. The very low supply of food for benthic invertebrates at depths below 3,000 m may be responsible for the lower diversity at such depths (Rex, 1981). It may be that there has to be a certain supply of food resources before much partitioning among sympatric species is feasible. On the other hand, very large amounts of resources may also lower diversity as, for example, in diatoms and higher plants in experimentally enriched salt marsh plots

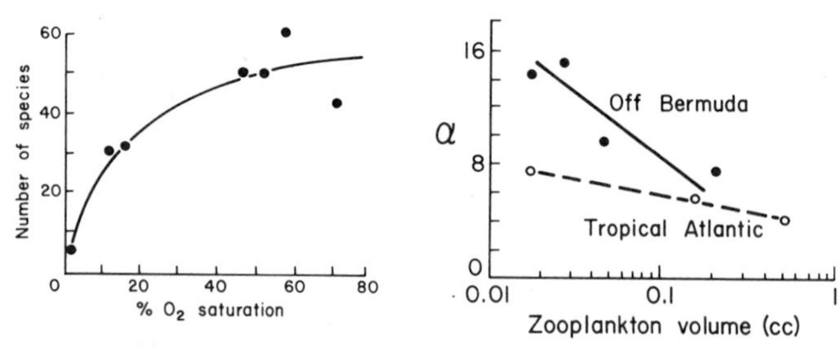

Figure 12-8. Left: Number of species of polychaetes and bivalves in a transect of benthic stations off Walvis Bay, Namibia, and percentage saturation of oxygen in the water overlying the sediments. Adapted from Sanders (1969). Right: Diversity (alpha, the index derived from the logarithmic series) of zooplankton in two areas, versus abundance of zooplankton. The volume of zooplankton between the two data sets cannot be compared because different nets and sampling methods were used. Adapted from Sutcliffe (1960). © Ecological Society of America, reprinted by permission.

(Table 12-1). In this case, in the presence of excess resources, one opportunistic (*r* or exploitatively selected) species proliferates. This also happens in phytoplankton communities of upwellings, where a few species become dominant, and diversity is lowered (Margalef, 1978). The response of diversity to resource abundance seems to be, therefore, a humped curve, with peak diversity at intermediate resource abundance.

There have been speculations as to the relation of diversity and productivity. This relation should also follow a humped pattern. Unfortunately, the examples available usually only span the lower or upper slopes of the curve, and have led to diverging conclusions in regard to the productivity hypothesis (Sanders, 1968; Huston, 1979). For example, the diversity of zooplankton near Bermuda and in the tropical Atlantic decreases the more zooplankton there are (Fig. 12-8, right). Since standing crops of zooplankton are correlated to productivity (cf. Section 8.31), higher production is correlated to lower diversity. Presumably there is a left-most portion of this relation, missing in Figure 12-8, where diversity increases as production increases.

12.323 Abundance of Consumers and Disturbances

We have already discussed (Section 8.211) experiments by Lubchenco (1978) where she manipulated the density of grazing snails on tidal pools and on the rock surface of a rocky shore. These experimental changes in density of the herbivores resulted in alteration to the diversity of the macroalgae. In the tidal pools with few snails (Fig. 12-9 top) *Enteromorpha* outcompeted other algal species, and diversity is therefore low. At intermediate densities of snails, the abundance of *Enteromorpha* and the

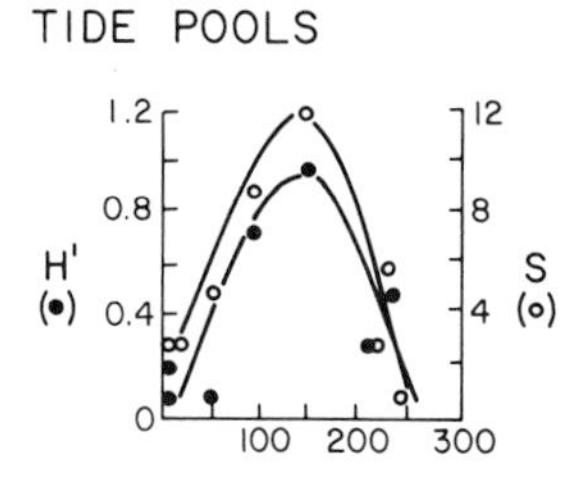

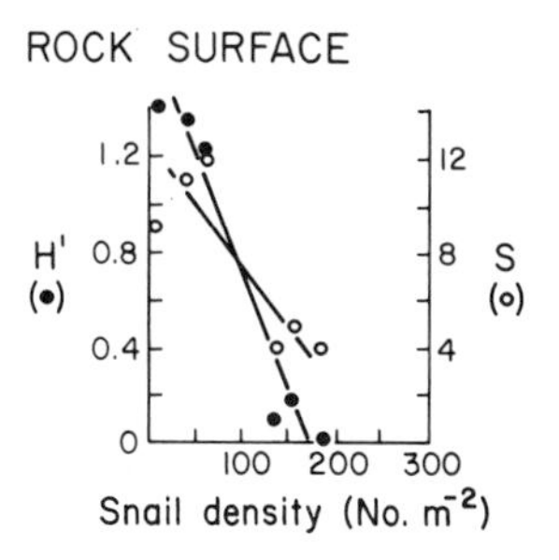

Figure 12-9. Effect of density of the grazing snail *Littorina littorea* on the diversity of algae in two habitats of the rocky shore of New England. H' is the Shannon-Weaver diversity index and S is the number of species. Adapted from Lubchenco (1978). University of Chicago, reprinted by permission.

few other preferred food species is reduced; this prevents competitive exclusion by *Enteromorpha* so that many algal species, both ephemeral and perennial, can coexist. At high densities of snails all edible algal species are consumed, leaving low diversity stands consisting mainly of the inedible *Chondrus*. Similar results, with a dome-shaped response of species richness to increased herbivore pressure, are found in cases of sea urchins feeding on macroalgae (Paine and Vadas, 1969), the sea urchin *Diadema antillarum* feeding on algal mats (Carpenter, 1981), and in damsel fish feeding on coral reef algae (Hixon and Brostoff, 1983).

The tide pool results (Fig. 12-9, top) reflect the combined impact of competition and selective grazing. This is not the case on the exposed rock surface. Disturbances such as ice damage in winter can be frequent enough in the rock surface (Fig. 12-9, bottom) that the competitive dominance of the perennials is repeatedly thwarted. If snails are absent, diversity is high, since at least 14 species of ephemeral macroalgae grow on the disturbed patches and on the perennials. As snail density increases, the ephemeral species decrease, since they are preferred food of *Littorina*. Since only the few species of perennial macroalgae remain, the canopy is one of low diversity (Fig. 12-9, bottom).

Note in Figure 12-9 that the information theory index H' provides a similar depiction as species richness. In this, and other cases, the use of the more complex index does not furnish any new information.

A more concrete example of the role of physical disturbances on species richness is provided by a study on the species growing on boulders on Californian beaches. Boulders of different size are turned over in a frequency proportional to their weight. An examination of the diversity of boulders of various sizes (Table 12-2) thus provides a gradient of distur-

Table 12-2. Species Richness (Average Number of Species ± Standard Error) of Macroalgae on Boulders Whose Size Resulted in Different Frequencies of Disturbance[a]

	Incidence of disturbance		
	Frequent	Intermediate	Seldom
Nov. 1975	1.7 ± 0.18	3.3 ± 0.28	2.5 ± 0.25
May 1976	1.9 ± 0.19	4.3 ± 0.34	3.5 ± 0.26
Oct. 1976	1.9 ± 0.14	3.4 ± 0.4	2.3 ± 0.18
May 1977	1.4 ± 0.16	3.6 ± 0.2	3.2 ± 0.21

[a] The size classes are grouped into three frequency of disturbance classes: frequent, that rock size that required less than 49 Newtons (N) to move horizontally; intermediate, 50–294 N; seldom, more than 294 N. From Sousa (1979). © Ecological Society of America, reprinted by permission.

bance frequency or intensity, since the overturning of boulders damages the attached organisms. Small boulders are populated by few species, mainly opportunistic early colonizer species, such as the alga *Ulva* and the barnacle *Chthamalus*. The largest, least disturbed boulders tend to be covered by a late colonizer, the red alga *Gigartina canaliculata*. The surface of boulders disturbed of an intermediate frequency are covered by a combination of early and late colonizers and also show bare areas. The differences in flora are not due to the size of the boulders, since small rocks experimentally attached to the sea floor supported an array of late colonizing red algal species, and eventually *G. canaliculata* became dominant. In boulders where disturbance was frequent, only the few quick-growing opportunists are present; where disturbances seldom occur, the competitive dominants exclude other species. Maximum diversity is found at intermediate disturbance frequencies.

Thus, it is clear that not just grazers, but any agent—such as predators or disturbances—that prevents competitive dominant species from monopolizing resources may affect diversity. In rocky shores predaceous sea stars are often such "keystone" species and their presence may increase diversity (Paine, 1966), and wave action can also play a key role.

In coral reefs predators and storms may remove live tissue from coral skeletons, and the percentage of live coral cores can be taken as an index of disturbance. Species richness has a dome-shaped response to increases in such disturbances (Fig. 12-10). Much as in the case of grazers, some disturbance or predation prevents competitive dominance and leads to

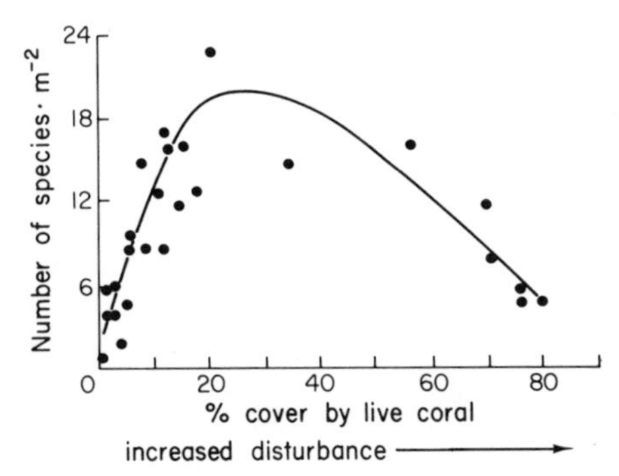

Figure 12-10. Species richness in coral reef of Heron Island, Queensland, versus the percentage of the area covered by live corals. The values are numbers of species of corals in samples of 1 m^2 in surface area. The area covered by live corals is an index of the degree to which predators and storms have been active, since starfish and other predators feed largely by scraping the live tissues off the coral skeletons, and storms break and damage corals. Adapted from data of Joseph Connell.

increased diversity.* Beyond a certain predation pressure or disturbance frequency only the few less desirable prey or tolerant species remain, and species richness is low.

12.324 Specific Features of the Local Environment

Much of what constrains within-habitat diversity has to do with the physical nature of the habitat (MacArthur, 1965). In soft sediments the activity of certain animals can provide just such a within-habitat structure (Fig. 12-11). Sea cucumbers (*Molpadia oolitica*) live head down in the mud of Cape Cod Bay, and ingest sediment particles. Feces are expelled out the anus onto the water-sediment surface and form a cone-shaped mound whose surface is considerably more consolidated than the flocculent, unconsolidated sediment surrounding the fecal mound. The mounds are colonized by tube-building polychaetes (*Euchone incolor, Ninoe nigripes*, and *Spio limicola*). These tubes in turn make it possible for the caprellid amphipod *Aeginina longicornis* and the bivalve *Thyasira gouldi* to be present. Thus the microtopography provided by the feeding of *Molpadia* increases species richness.

Other features of the habitat may also affect diversity. The particular geometry of coral heads may attract differing groups of species. Even more exotic mechanisms may be important, such as the increased diver-

* This applies to the sum of predators in the community but it may not be true for all species of predators. For example, the crown-of-thorns starfish (*Acanthaster planci*) feeds preferentially on rarer species of corals, and its feeding thus fosters the growth of common, fast-growing species. *Acanthaster* could thus lower species richness (Glynn, 1974, 1976).

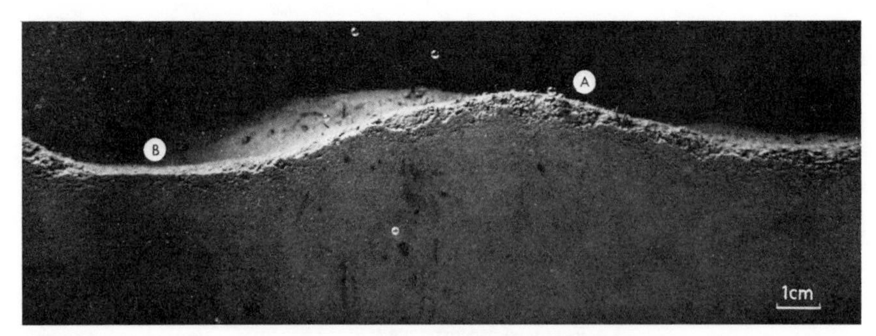

Figure 12-11. Photograph of a profile of surface sediments in Cape Cod Bay, Massachusetts. Defecation by the sea cucumber *Molpadia olitica* produces the relatively stable fecal cones (A) populated by a variety of species of suspension feeders. The cones provide habitat heterogeneity and increase the diversity compared to that of the more unconsolidated mud (B). From Rhoads and Young (1971).

sity of coral reef fish associated with the vicinity of territories of the cleaner wrasse (Table 12-3). The wrasse removes parasites off other fish, and this is apparently an important enough service to attract many individuals of many other fish species. In fact, where cleaner wrasses were absent, 26 of the 49 fish species found in the reef were not found and an additional 12 species were less abundant.

12.4 Integration of Factors Affecting Diversity and Some Consequences

Given that there are enough resources available, the diversity of communities seems, from the examples above, to be a balance between a tendency toward competitive exclusion and the prevention of this tendency by disturbances, either due to consumers, natural accidents, or pollution. Increased resources favor species with rapid growth rates and these may become dominant enough to reduce diversity; the dominance will increase up to the level allowed by the disturbances or consumers. The way in which a consumer affects the diversity of its food items may depend on its selectivity. These trends can be altered where toxic materials eliminate certain species or where particular features of the habitat play a significant role. More extensive expositions of similar conclusions are given by Connell (1978), Pielou (1975), and Huston (1979). These simple statements have not yet been developed to the extent that they can account for the notable geographical gradients in diversity with which we started this chapter. The above conclusions, however, are based on actual

Table 12-3. Effect of Presence and Absence of the Cleaning Wrasse (*Labroides dimidiatus*) on Fish Diversity on a Red Sea Coral Reef[a]

	Number of species per habitat	Number of individuals per transect	Number of species absent	Number of species with reduced abundance
Labroides absent	0–6	1–10	26	12
Labroides present	7–31	18–310	3	8

[a] From Slobodkin and Fishelson (1974). © University of Chicago, reprinted by permission.

results and provide testable predictions (Huston, 1979) that may result in eventual explanations of the observed gradients.

Even if concrete facts as to how diversity is determined within a habitat are available and reasonable ways to measure diversity are used, there still remains the matter as to its biological significance. No one has found an answer as to what it means for an organism to live in a high- or low-diversity environment. Clearly the ratio of intra- to interspecific encounters in each of these two kinds of environments may vary (Lloyd, 1967) but there is no evidence as to the significance of this.

Another area of ambiguity is the much-discussed relation of diversity and stability of the assemblage of species. Ecological wisdom followed Charles Elton and others who held that a community with many species would be more stable than one with few species. The more diverse system would be less likely to disappear due to some haphazard event since it would be more likely that some of the species present would tolerate the disturbance. A more diverse system would also be better able to adjust to changes in resources, since each component species is bound to possess slightly different characteristics. The community as a whole could thus deal with a wider variety of changes.

In the early 1970s, a spate of theoretical publications showed, using simulation models, that in fact stability was reduced in diverse systems (Gardner and Ashby, 1970; Smith, 1972; Hubbell, 1973; May, 1973; among others), only to be followed by others that showed that by changing the models diversity could increase stability (Jeffries, 1974; De Angelis, 1975; Gilpin, 1975; and others). In part, the problem is in the definition of stability, a very vague concept that has been variously defined and interpreted (Orians, 1975; Harrison, 1979). A second problem is the difficulty in formulating testable hypotheses due to the ambiguity of the concept and the often tautological nature of discussions of stability—"this community is stable because it is not perturbed by this change." The result is that there has been a great deal of modeling and reasoning and little experimental verification, a situation appropriately described by McNaughton (1977), who thinks that "the marked instability of attitudes regarding diversity-stability relationships in ecosystems arises from a low diversity of empirical tests of the hypothesis."

A few attempts at experimental evaluation of parts of the diversity-stability idea have taken place, and have come up with disagreeing conclusions (Abele and Walters, 1979; Smedes and Hurd, 1981) or failed to show meaningful relationships (Hairston et al., 1968). Studies on a few terrestrial environments show that diversity and stability increase together (McNaughton, 1977), as claimed by the earliest speculations. By and large there has been, despite much interest, little progress in recent years, and certainly no study documenting causal relationships, largely due to the difficulty in formulating testable hypotheses.

Chapter 13

Spatial Heterogeneity: Structure Over Space

13.1 Scales of Patchiness

In general, organisms are not distributed uniformly over space; consider for example, the distribution of primary production over the world's oceans (Fig. 2-30). Moreover, such heterogeneity occurs at all spatial scales. We can use the size of a patch of higher abundance or the distances among such patches as a way to assess the scale of heterogeneity. There are significantly higher or lower rates of production in distances of 10^5 to 10^8 m within any linear transect in Figure 2-30. Spatial variability is also pervasive at smaller distances: the variation in concentration of dissolved CO_2 in surface waters of the Gulf of Maine is evidence of spatially heterogeneous biological activity on a scale of 10^4–10^5 m, while the variability in concentration of chlorophyll in St. Margaret's Bay, Nova Scotia, varies on a scale of about 10^4 m (Figs. 13-1a,b). Measurements of numbers of zooplankton m^{-2} off the California coast show patches of tens of meters (Fig. 13-2c); a careful examination of spatial variability of phytoplankton production shows significant patches in a scale of 10^{-1} m, most clearly in samples from the Gulf Stream (Fig. 13-1d).

Such variability over space, especially on the smaller scales, has made field measurements notoriously variable and has long been considered as one of the chief nuisances of environmental research. Field measurements of densities of many kinds of organisms, for example, usually have associated variations of 10–30% of the mean. Much of this variability is due to unrecognized patchiness and it is only recently we have become more aware that such patchiness is the signature of some very powerful biological and physical processes. Although we are at a very primitive stage in our understanding of spatial distribution, future studies of spatial

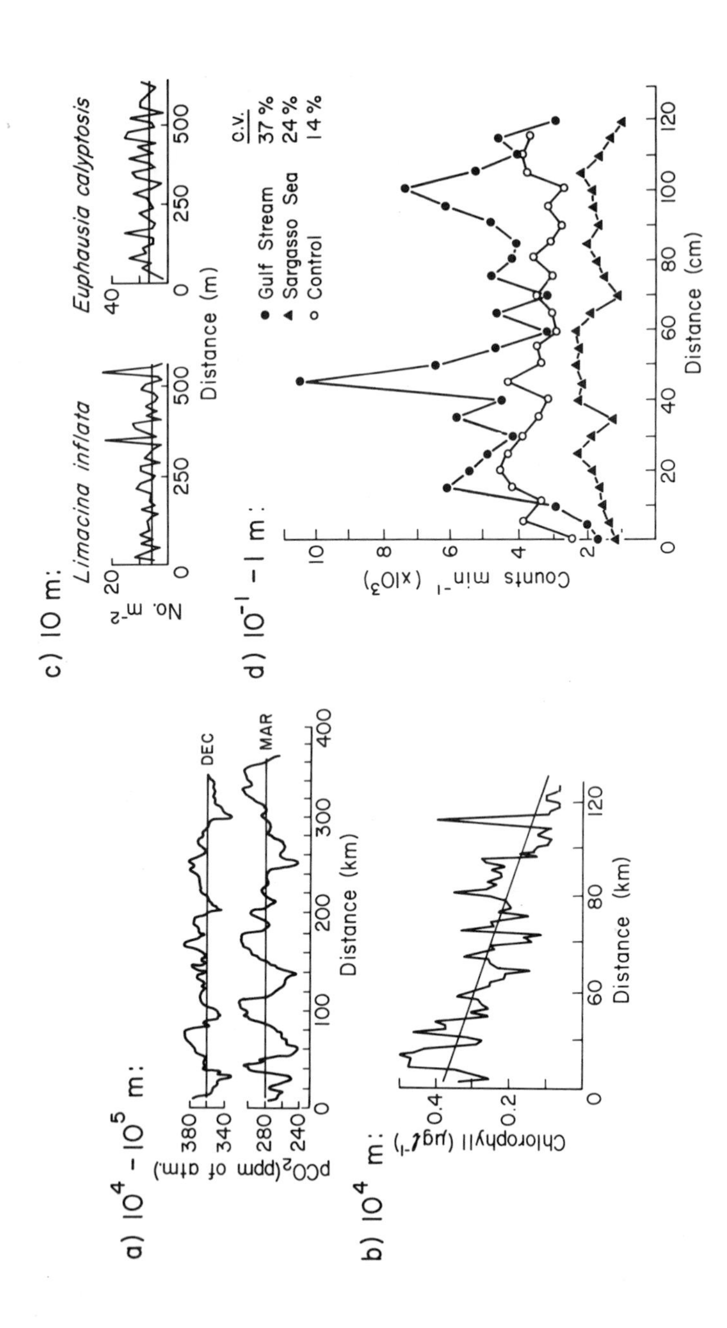

a) $10^4 - 10^5$ m:
pCO_2(ppm of atm.)
380
340
280
240
DEC
MAR
0
100
200
300
400
Distance (km)
b) 10^4 m:
Chlorophyll (µgl^{-1})
0.4
0.2
0
60
80
120
Distance (km)
c) 10 m:
Limacina inflata
Euphausia calyptosis
No. m^{-2}
20
40
0
250
500
Distance (m)
d) 10^{-1} – 1 m:
Counts min^{-1} ($x10^3$)
10
8
6
4
2
● Gulf Stream
▲ Sargasso Sea
○ Control
C.V.
37 %
24 %
14 %
0
20
40
60
80
100
120
Distance (cm)

Figure 13-1. Patchiness on approximate scales of 10^5 to 10^{-1} m for abundance and activity of plankton. (a) Concentrations of CO_2 in surface waters for transects in the Gulf of Maine. Adapted from Teal and Kanwisher (1966). (b) Chlorophyll concentrations at 10 m along a transect in St. Margaret's Bay, Nova Scotia. The straight line is the regression of chlorophyll on distance. Adapted from Platt et al. (1970). Reprinted by permission of the Canadian Journal of Fisheries and Aquatic Sciences. (c) Abundance of a pteropod mollusc and an euphausiid in near-surface tows off California. Medians of entire data shown by the horizontal line. Adapted from Wiebe (1970). (d) Radioactive counts in a ^{14}C uptake experiment with phytoplankton in the Atlantic. The control consisted of shelf water that was thoroughly stirred to homogenize the distribution of cells. The height along the y-axis depends on level of productivity and is not relevant here; the important thing to notice is the variability over space. A measure of this variability is given by the coefficient of variation (C.V. = coefficient of variation = variance/mean). Data from Cassie (1962a).

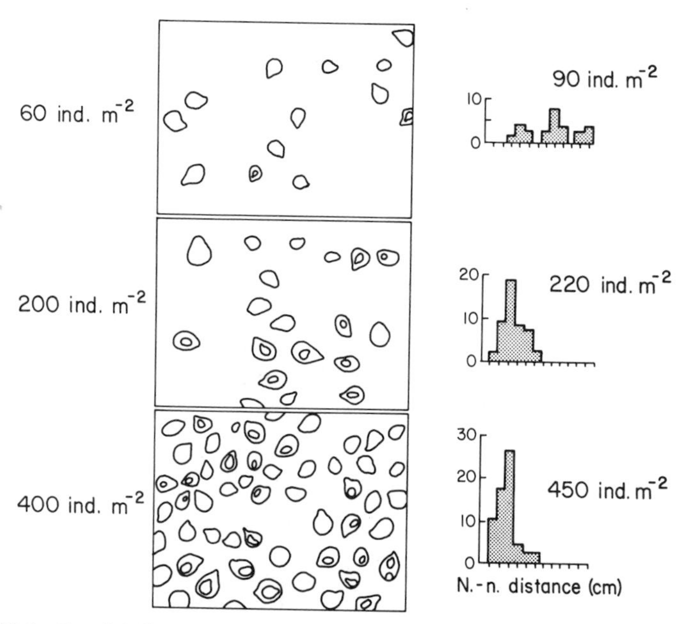

Figure 13-2. Spatial distributions of limpet (*Patella cochlear*) on a South African rocky shore. Left panels contain plan views of the distribution of limpets, which are shown as the tear-shaped objects. Small limpets settle on large limpets since territories on the rock surface are defended by adults, so that young limpets are prevented from settling on bare rock. Histograms on right show the frequency of distribution of nearest-neighbor distances (excluding those on the backs of others) for three densities of limpets approximating the densities of the diagram in the panels on the left. Adapted from Branch (1975a).

distributions may provide insight as to how the structure of marine ecosystems is determined, as well as provide a way to understand the variability of ecological measurements.

Spatial structure is evidenced by a wide variety of phenomena. We have already dealt with examples of vertical zonation within the intertidal zone (Section 9.2). In Section 2.112 we examined the mechanisms that provide a marked vertical structure to the distribution of phytoplankton. In Section 14.2 we discuss the patterns of vertical migration that give zooplankton communities a time-varying but pervasive spatial structure in the vertical dimension. In this chapter we concentrate on the scale of spatial variations, on how spatial heterogeneity is described, and what some of the sources of spatial heterogeneity may be. Lastly, we will consider some consequences of spatial variability.

13.2 Description of Spatial Distributions

Patchiness can be described in terms of the distances that separate units of the populations being studied over a continuous space, or by assessing the occurrence of units of the population within discrete spatial units such as quadrats, tidal pools, or samples. Methods for both of these situations are included in Cassie (1962) and Poole (1974).

13.21 Dispersion Patterns Over Continuous Space

Spatial distributions are most easily studied in two-dimensional surfaces such as rock surfaces. The distribution of intertidal surface-dwelling limpets provide examples of various spatial patterns (Fig. 13-2, left panels). Measurements of nearest neighbor distances* are often compiled into frequency distributions (Fig. 13-2, right panels). The mean ($\bar{x}$) and variance (s^2) of these distributions are not the same in these three examples (Fig. 13-2, right panels). The ratio $s^2/\bar{x}$ can be used as an indicator of the kind of spatial distribution displayed by the population. In randomly dispersed populations, $s^2/\bar{x} = 1$. When spacing is fairly uniform, as in Figure 13-2, bottom right, $s^2/x < 1$. When spacing is very variable, and clusters of individuals are found, $s^2/x > 1$. This latter condition is generally the most common of all distributions, and is variously referred to as aggregated or contagiously distributed.

* Distributions of nearest neighbor distances are usually calculated by compiling distances on a plane from only one of the four possible 90° quadrants; this makes it possible to include distances between individuals situated in different aggregations. Otherwise, nearest neighbors would always be within aggregation and the distribution of nearest neighbor distances would not include any reflection of the distribution of patches, an important aspect of spatial structure.

Clark and Evans (1954) used nearest neighbor distances to calculate an index, R, that also can distinguish among the three kinds of spatial distribution,

$$R = \frac{\sum_{i=1}^{n} r_i}{N} 2\sqrt{p}. \tag{13-1}$$

In this expression, r_i is the distance between an organism and its nearest neighbor, where there are n such measurements. N is the number of organisms, and p is the density of the population (individuals/unit area). Values of this index between 0 and 1 indicate that the population is aggregated, and the more aggregated, the closer to 0. Values of the index between 1 and 2.15 correspond to uniform distributions, while a value of 1 indicates a randomly spaced population. Values of R calculated for a limpet population (Fig. 13-3) show that at low density individuals are aggregated, while at high densities, as we surmise from Figure 13-2, limpets are closely and uniformly spaced. Such a pattern may be a feature of species whose major limiting factor is space.

Many other populations share the tendency to aggregate when their density is low and be either uniform or randomly spaced at high densities (Hairston, 1959). It is possible, however, for rare and abundant populations to show any of the spatial distributions. For example, judging by s^2 to $\bar{x}$ ratios, most species of crustaceans, molluscs, and echinoderms in the deep sea off California have mostly random spatial distributions, with some occasionally aggregated species (Jumars, 1976). The most dense populations tended to show the most aggregation. In the example of limpets (Fig. 13-2) it just so happened that increases in density were correlated to changes from aggregated to uniform distributions; the actual response is apparently species specific.

The size of the sample or quadrant needs to be considered carefully since the choice of sample size can artificially create one or another distribution. Consider a population distributed as in Figure 13-4, and sampled using quadrats of three sizes. An array of the largest quadrats provide a more or less uniform number of individuals, and misses the smaller-

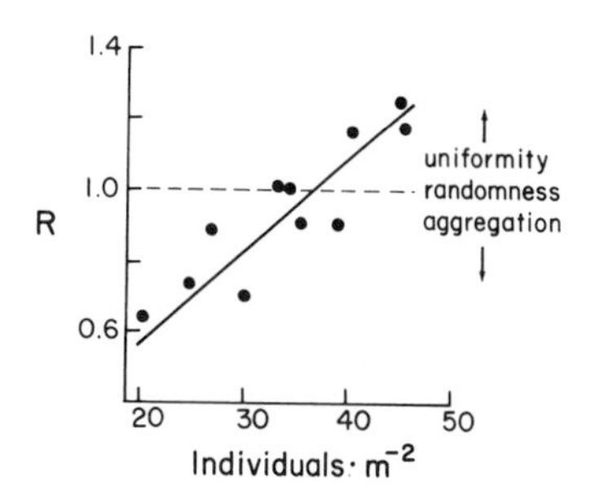

Figure 13-3. Relation of aggregation (R is the Clark–Evans index) to density of *Patella longicosta*. The horizontal dashed line shows the value of R that indicates a random distribution. Adapted from Branch (1975a).

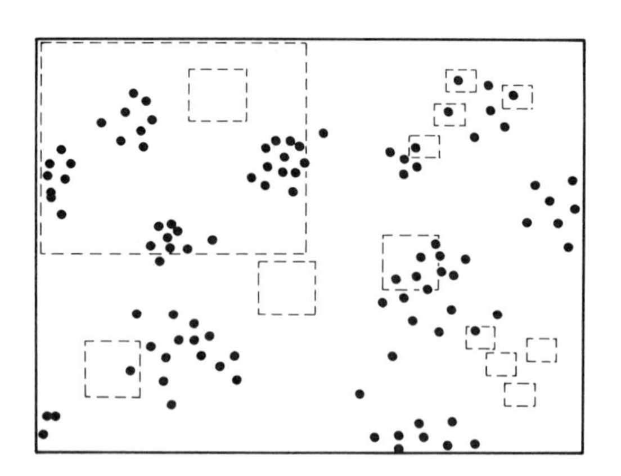

Figure 13-4. Hypothetical distribution of a population over space. The dashed squares indicate the changing effect of the use of different sample sizes.

scale patchy nature of the distribution. A collection of individuals obtained by use of the intermediate-size quadrats shows the aggregated distribution quite well, while the smaller quadrats either provide a very uniform distribution or show that the species is very rare, depending on where the quadrats are placed.

13.22 Dispersion Patterns in Discrete Units

The frequency of occurrence of individuals in discrete units or in samples from an environment can be described in a fashion similar to that used for nearest-neighbor distances. Figure 13-5 shows a generalized scheme of the frequency distribution associated with uniform, random, and aggregated populations. There are well-established statistical descriptions of these distributions (Cassie, 1962; Poole, 1974). The simplest assumption that can be made is that any point in space is equally likely to be occupied by an organism. The expected frequency distribution of samples with 0, 1, 2, 3, etc., individuals is given by successive terms of the binomial series $(q + p)^k$. In this expression k is the maximum number of individuals in sample, p is the probability of occurrence of an organism in a sample, and $q = 1 - p$. For the binomial distribution, the mean $\mu = kp$, and the variance $= \sigma^2 = kpq = \mu - [\mu^2/k]$. If k is very large, $\sigma^2 \simeq \mu$. Such a frequency distribution, where each organism is equally likely to appear in a sample, and the number in the sample is very small compared to the total population, is called a Poisson distribution, a special case of the binomial, and both are random distributions. Based on the Poisson relationship, $\sigma^2 = \mu$.

Two other kinds of distributions—uniform and aggregated—can be distinguished by comparison of values of R or by using σ^2/x (Fig. 13-5, bottom). An example of the latter procedure is possible with data on the occurrence of a planktonic mysid in samples taken by net tows (Fig. 13-6). In this particular case, increased abundance is correlated to increased aggregation, while less dense populations tend to be randomly distrib-

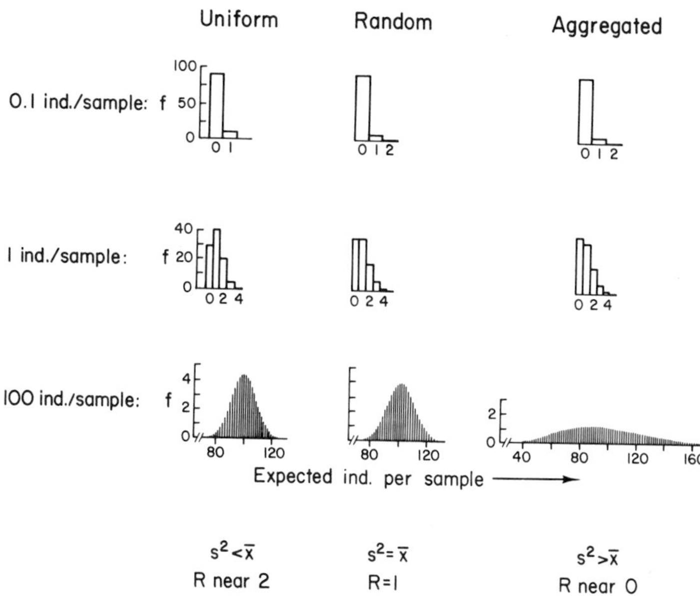

Figure 13-5. Generalized frequency distributions for uniform, random, and aggregated distribution of occurrence in discrete units of sampling. Each distribution is shown as it would look with mean number of 0.1, 1, and 100 individuals per sample. The relative size of variance and mean, and the Clark–Evans index of aggregation (R), two criteria that can be used to identify each of the three distributions, are shown at the foot of the figure. Adapted from Cassie (1962).

uted. This pattern seems to be typical for plankton (Barnes and Marshall, 1951; Clutter, 1969), and contrasts with the pattern shown by benthic or sessile animals on two-dimensional surfaces (cf. Fig. 13-3). In plankton, it seems that uniform distributions are rare: there are few points where the variance is smaller than the mean (Fig. 13-6). The ability to disperse by swimming or water motion probably enables most plankton to avoid crowding and uniform distributions. Plankton populations are seldom so dense that they cannot swim or disperse away from each other. We might thus expect that competition might be less intense in the water column or soft sediments than on the surface of intertidal rocks (cf. Chapter 9). This may be a basic difference of life in a dilute three-dimensional medium as opposed to an often crowded two-dimensional substrate.

The abundance of a population has a significant impact on spatial arrangements and on our ability to detect various spatial patterns. Only when density is relatively high is it possible to see the differences between uniform and aggregated distribution. At low densities (Fig 13-5, top row), all three distributions are highly skewed and it is very difficult to distinguish one distribution from another. Of course, rarity in the samples may be due to the smallness of the sampling unit, and an increase in the larger number or size of samples may allow the identification of the spatial arrangement.

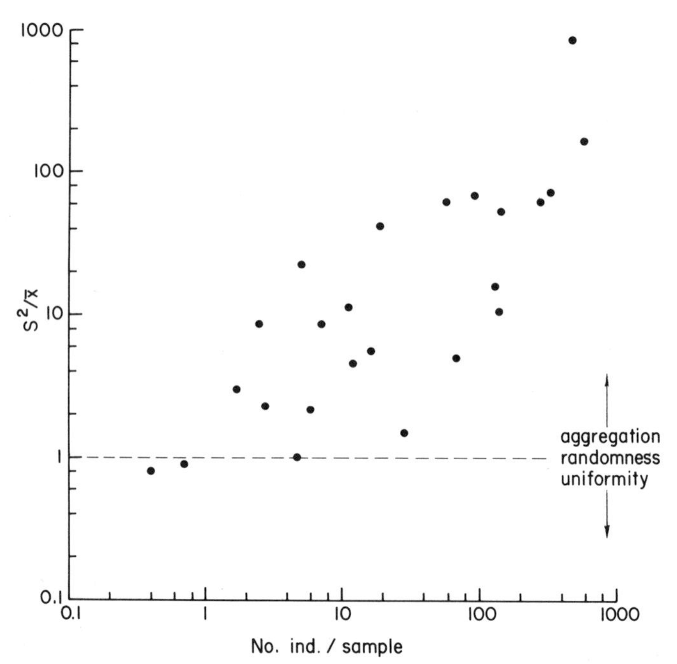

Figure 13-6. Index of aggregation ($s^2/\bar{x}$) for a variety of zooplankton samples containing the mysid *Metamysidopsis elongata*. Data from Clutter (1969).

It is clear from the foregoing that the ratio of variance to mean partly depends on the number of individuals per sample rather than being due solely to the spatial distribution. The aggregation of Figure 13-6 could be a function of density, for example. It may be common for plankton data to move diagonally across Figure 13-5, from "uniform" distributions at low densities to "aggregated" distributions at high densities. If it is of interest to avoid the effect of density, coefficients of aggregation that are not affected by abundance may be used (Cole, 1946; Pielou, 1977).

13.3 Sources of Patchiness

13.31 Social and Reproductive Mechanisms

13.311 Associations of Individuals in Mobile Groups

Many marine organisms form schools, clusters, or swarms (Fig. 13-7) including fish, euphausids, mysids, and copepods, among other taxa. These patches are due to social and reproductive behavior. The social nature of such aggregations is evident by the cohesion of movement of individuals within the clusters and by the uniform distances among members of a swarm or school.

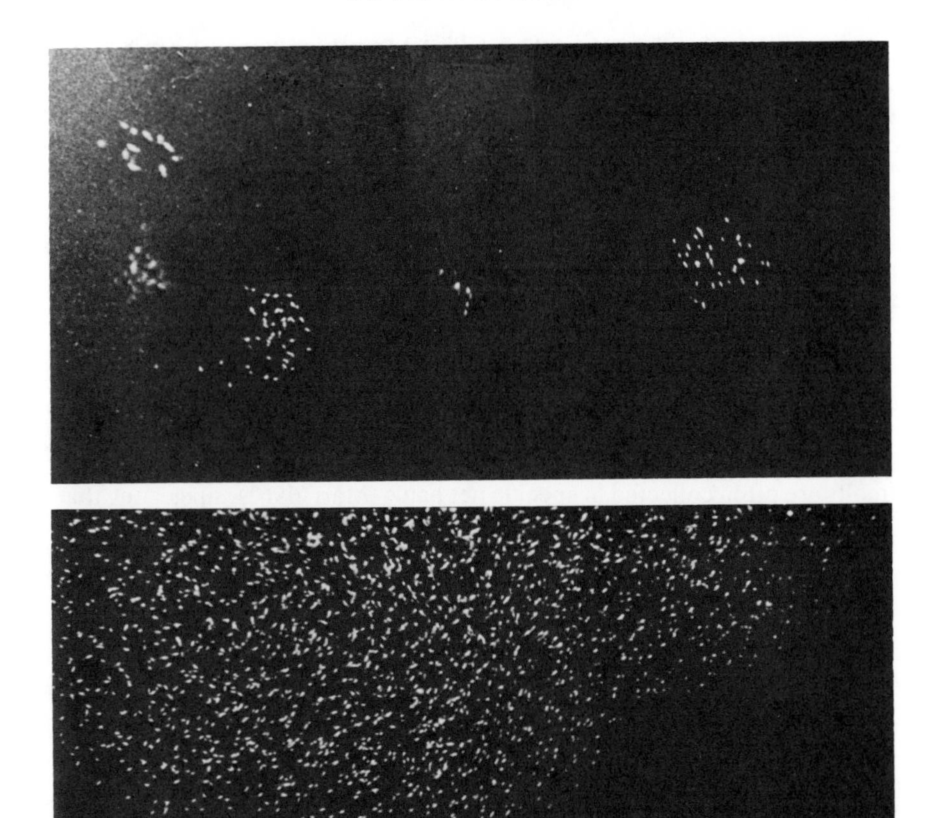

Figure 13-7. Schools or swarms of the mysid *Acanthomysis* sp. photographed at a depth of about 6 m in the Gulf of California. Length of each specimen is about 8 mm. From Clutter (1969).

Patches of the krill *Euphausia superba* move "...almost as if there was some leader in command..." (Hardy and Gunther, 1935). The shape and size of clusters are very fluid but there is a strong tendency for social plankton to reconstitute clusters even after disturbances (Kamamura, 1974; Clutter, 1969). There is clearly a strong impetus to remain in the swarm, as shown by motion pictures of swarms of pelagic mysids (Clutter, 1969). Individuals that lag behind their swarm make rapid leaping movements to regain their place in the school, and even overcompensate for their temporary isolation by swimming well into the swarm on their return. Smaller groups of mysids coalesce with larger groups if they happen to be less than about 25 cm away from the larger swarm.

Underwater observations and motion pictures of mysid swarms show that, in spite of great variability in the size of the swarm (13,000–32,000 individuals), the average nearest neighbor distance is quite constant (1–4 cm) and is maintained at less than 4 mysid lengths (Clutter, 1969).* Similar small distances are maintained by schooling fish and euphausiids. *Euphausia pacifica* swarms with an average distance of 1-2 cm among individuals (Komaki, 1967), while the density of *E. superba* can be 1 individual cm^3 (Marr, 1962). The distribution of intraswarm distances is one of the few instances of uniform spatial spacing in planktonic populations. This spatial pattern is of course destroyed when samples are taken with nets.

Some aggregations are due to reproductive phenomena, not necessarily involving social behavior. For example, eggs and larvae may be patchy due to the occurrence of reproduction in one area. Such aggregations may last for considerable time. Copepod and mysid (Clutter, 1969) clusters are made up primarily of individuals at specific stages in the life history or at similar stages of sexual maturity. In *E. superba* high numbers of cast molts are often collected in specific net hauls (Marr, 1962), suggesting that molting of individuals within a cluster is sychronized. Such observations indicate that clusters are often made up of individuals whose growth and reproduction are to some extent linked.

The sensory mechanisms used to maintain motile swarms or schools are not clear. In mysids, visual means are probably important, since eyes are well developed and swarming is lessened at night (Clutter, 1969). Since copulation takes place at night, however, other senses, including mechano- and chemoreception must also operate. Social mechanisms operate over small spatial scales, perhaps a centimeter to a meter for larger zooplankton; the reproductive patterns, however, can synchronize populations and create patches over larger spatial scales, up to diameters of perhaps 10^3 km (Haury et al., 1978).

13.312 Spacing of Individuals in Sessile Clusters

Territoriality is another social mechanism that affects the distribution of individuals over space; in this case, behavioral mechanisms set the distances among individuals within a cluster or aggregation. We have already discussed territorial spacing in shorebirds (Chapter 5) and in limpets (Chapter 4 and Fig. 13-2). Territorial defense by individuals is a proximate mechanism that provides for some ultimate need, such as sufficient supply of food or suitable breeding area. The garibaldi, a pomacentrid fish common off the California coast, offers a good example of a population whose spatial distribution is dominated by territoriality. The population clusters on areas of the sea bottom that are appropriate as a habitat (Clarke, 1970). Territorial behavior then more or less uniformly spaces

* We saw earlier that mysid distributions tended toward aggregation at higher densities (Fig. 13-6). The process of sampling collects swarms or parts of swarms so that the individuals appear aggregated. This is a matter of size of the sample unit, as discussed in reference to Figure 13-4.

the individuals within the dense cluster (Fig. 13-8). Outside the cluster the density of the garibaldi population is very reduced, so that on a larger scale the fish are patchily distributed.

The spacing is primarily defined by aggresive behavior of nesting males, who seldom leave their territory. Male garibaldi have unique nest-building behavior in which a thick growth of red algae is somehow slowly fostered in a 15–40 cm diameter patch within their territory. This patch is the nest, made more visible during the nesting season when the males clear away all vegetation from a band 5–11 cm wide around the nest. These algal nests require 2 or more years to culture and are used for several years, quite unlike other fish whose nests are abandoned after breeding. Thus the investment in a garibaldi nest is very high, and nests are defended throughout the year. Any nests that become vacant are taken over by other males at any time of the year. Because continuous defense of the nest is needed, each territory must also be large enough to provide food and shelter. In the case of garibaldi, both food abundance and reproductive needs define spatial distribution. Successful reproduction apparently requires some minimum space (van der Assem, 1967); in the garibaldi, the minimal average distance among nests is about 1.9 m.

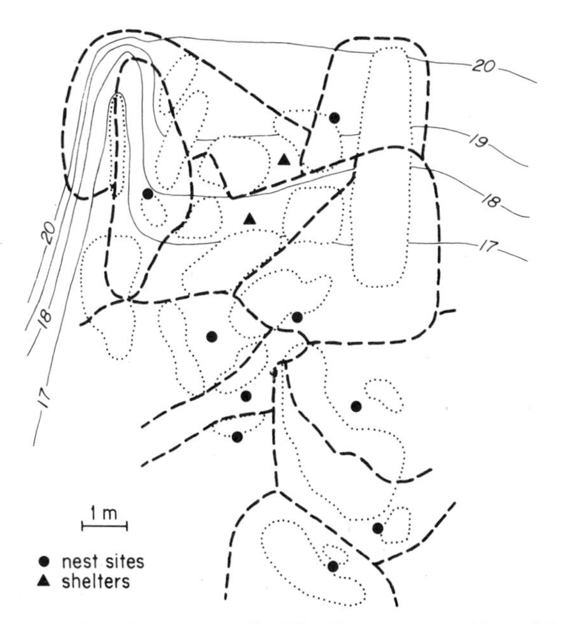

Figure 13-8. Territories of eleven garibaldi (*Hypsypops rubicunda*) off La Jolla, California. Nine males had nests within their territory (dashed lines). Depth contours (thin lines) are in meters; rocks protruding from the sea bottom are indicated by the dotted lines. Adapted from Clarke (1970). © Ecological Society of America, reprinted by permission.

13.32 Mechanisms Related to Consumption of Resources

Nutrient uptake, grazing, and predation are very likely to be involved in determining spatial distributions, but we are a long way from understanding their effect in the sea. We have already mentioned current ideas on the potential importance of small nutrient patches for phytoplankton in oligotrophic water in Chapter 2.

There has been considerable theoretical work in relation to patchiness in plankton. Kierstead and Slobodkin (1953) derived a relationship between the size of a patch of phytoplankton and the rate of growth of the algae required to counter the losses of cells by diffusion out of the patch. Their model shows that patches smaller than 10 km could not be maintained against turbulent diffusion. Modification of the model by Steele (1974) brought the critical patch size down to a few kilometers, but we know that spatial heterogeneity smaller than this scale exists (Fig. 13-1). The addition of an Ivlev equation (Section 5.21) to include grazing in the Kierstead–Slobodkin model results in the repeated appearance of patches of all sizes (Wroblewski et al., 1975). Simulation studies using models thus suggest that the fate of phytoplankton patches depends on growth rates of the phytoplankton, grazing rates by zooplankton, spatial distribution of grazers, limiting nutrients, and turbulent diffusion.

13.33 Physical Mechanisms

The properties and motion of seawater are the most important mechanisms affecting spatial distribution of plankton in most spatial scales. In scales less than 100 m, for example, the distribution of chlorophyll seems effectively determined by turbulence (Platt, 1972), but often other physical phenomena affect spatial distributions.

Zooplankton may aggregate in spatial scales of tens of meters due to local movements of water in near-shore waters. The density of zooplankton in the lee of a point in the Queensland coast reaches up to 40 times the density in adjoining water (Alldredge and Hamner, 1980). The actual mechanism by which this aggregation is accomplished is not known.

One of the better-known physical causes of patchiness is Langmuir circulation. Wind blowing over the surface of the sea causes the surface water to flow in Langmuir cells (Pollard, 1977). Langmuir cells extend up to 400 m in length, and can be several meters in diameter. These cells consist of water moving in horizontal helical paths in the direction of the wind. Wind of about 3 m sec^{-1} is required to form Langmuir cells, but higher wind speeds may disrupt them. Many parallel spirals are usually formed, and adjoining spirals roll in opposite directions, so that there are alternate parallel lines of convergence and divergence along a field of Langmuir cells. Such flow patterns can cause accumulations of floating particles on the sea surface in the areas of convergence between cells and also may affect the spatial distribution of plankton in the water within the spirals (Stavn, 1971; Evans and Taylor, 1980).

There are many other larger-scale physical mechanisms that can be important (Haury et al., 1978). Intrusion of water masses from the English Channel or from Dutch coastal waters determines whether either *Oikopleura dioica* or *Fritillaria borealis* are present in large patches of the southern North Sea (Wyatt, 1973). The impact of such movements of large parcels of water on the distribution and abundance of organisms is not unusual. Over half the variation in abundance of copepods in certain areas of the North Pacific may be due to variation in such advective transport (Wickett, 1967).

Another example of physical processes that operate at a very large scale are the consequences of large rings that spin off meanders of major western boundary currents such as the Gulf Stream (Wiebe, 1982), the Kuroshio (Tomosada, 1978), and the East Australia Current (Nilson and Cresswell, 1981). In the northwest Atlantic rings can contain a core of warm or cold water, depending on whether a portion of continental slope or Sargasso Sea water has been pinched off during the formation of the meander and ring. About five to eight cold core rings are formed per year, and each may reach a diameter of 250 km and depths of over 1000 m (Lai and Richardson, 1977). Rings last for 2, perhaps 3 years. There may be 10–15 rings in the northwest Sargasso at any one time, covering a substantial part of the area (Richardson, 1976). Warm core rings are fewer and smaller. The formation of rings thus inserts large patches of nutrient-rich, cold waters of the contintental slope into the Sargasso Sea and depauperate, warm water into the slope water. In addition, the organisms within the rings can be very different from those in the surrounding waters (Wiebe et al., 1976a), even in the case of mobile species such as fish. In warm-core rings in the Tasman Sea, for example, warm-water species dominate the fish fauna, while the cold water surrounding the rings holds a different array of cold water fish species (Brandt, 1981). Warm and cold core rings are therefore a major feature affecting the spatial distribution of organisms. Their study may clarify many questions of zoogeographical distribution as well as contribute to understanding heat transfer and productivity of the North Atlantic.

The very largest scales of patchiness such as shown in Figure 2-30 are also primarily due to hydrography, as discussed in Chapter 2. Nutrients in large measure determine regional primary production and are made available by the transport and mixing of water masses. It is therefore clear that the physics of water is behind much of the spatial structure of planktonic distributions.

13.34 Statistical Processes

Statistical properties of biological processes can generate nonrandom distributions. Suppose the settling of larval limpets on each quadrat has a random (Poisson) distribution. If survival of the settled limpets within the quadrat is also random, the resulting spatial pattern of the surviving limpets will be aggregated and will have a distribution called Neyman's type A (Poole, 1974), even though both processes that contributed to it were random. If survival is instead distributed logarithmically, another common aggregated distribution, the negative binomial results. There are also other statistical ways to generate negative binomial distributions (Cassie, 1962). Since the distribution may then be of statistical rather than have a direct ecological origin, presence of a specific distribution should be regarded at best as an empirical description rather than as evidence of the action of specific processes.

13.35 Artifacts of Measurement

We measure spatial patterns through procedures or methods that act as a filter that often imposes its own constraints on the data. This should be clear from our discussion of the effect of quadrat and sample size on the resulting frequency distributions. This is true of all the different areas of marine ecology that we have examined, in fact of all of science. The problem is especially obvious in dealing with the statistical descriptions of spatial arrangement as, for example, in a study carried out by Haury et al. (1979).

In Cape Cod Bay, Massachusetts, tidally generated internal waves (waves moving within the ocean rather than at the surface) with a period of 6–8 min and a wavelength of about 300 m propagate over a shallow

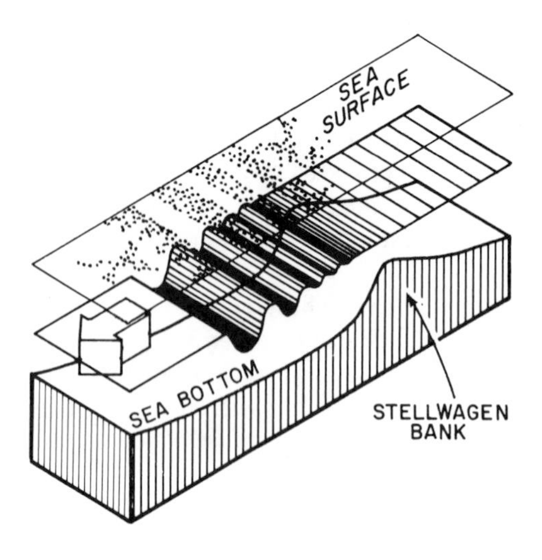

Figure 13-9. Schematic diagram of internal waves generated in the lee of Stellwagen Bank, Cape Cod Bay. While the waves can be quite steep at depth, there are only minor surface expressions. Adapted from Haury et al. (1979). © Macmillan Journal Ltd., reprinted by permission.

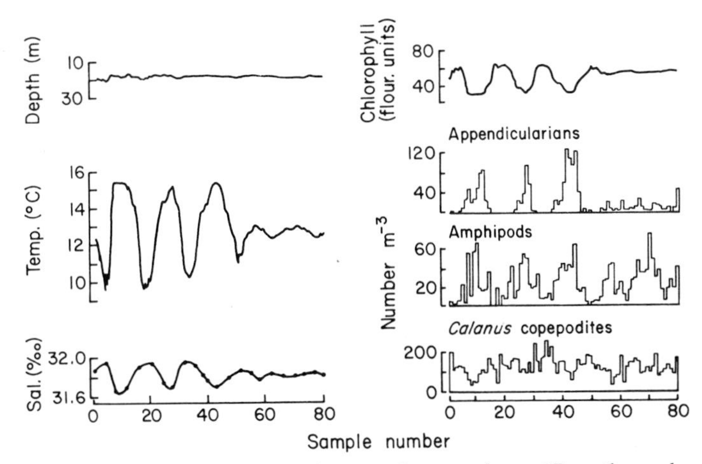

Figure 13-10. Data from a horizontal tow taken at about 17 m through a wave packet such as shown in Figure 13-9. Samples were taken every 14.6 m, and the total length of the tow was 1,155 m. Adapted from Haury et al. (1979). © Macmillan Journal Ltd, reprinted by permission.

bank (Fig. 13-9). Such internal waves are important physical mechanisms that may produce patchiness at intermediate (10–100 m) spatial scales (Haury et al., 1978). Data collected by net tows* at a depth of about 17.5 m reveal some effects of these waves (Fig. 13-10). The pattern of the internal waves is depicted by the excursions recorded in temperature and salinity. Deeper water was colder and saltier (Fig. 13-10, left). At sample number 20, for example, there was a wave peak that brought deeper water up. On the right side of Figure 13-10 are biological data collected at the same time. The deeper water contained more chlorophyll, fewer appendicularians and amphipods, and more copepods. The results of these tows are an excellent example of the ambiguous nature of patchiness and the potentially artifactual effects of the sampling procedure.

Consider the data on the right-hand side of Figure 13-10. These transects could be taken to be excellent examples of patchiness; they are quite comparable to transects such as shown in Figure 13-1. Should we consider the high densities true patches or are they just artifacts due to the collection of a transect at a uniform depth of 17.5 m? Further, are the

* These samples were collected with a modified Longhurst–Hardy plankton recorder (LHPR). This device enables the collection of a series of samples as it is towed through the water. Ribbons of mesh are rolled across the cod end of the net so that the plankton from different sections of the tow are caught on successive portions of the mesh. New versions of the LHPR reduce difficulties of net avoidance by fast-swimming copepods, extrusion of specimens through the mesh and the hang up of specimens along the net (Haury et al., 1976). A number of other instruments were used to measure temperature, salinity, and chlorophyll (Haury et al., 1979).

measured patches meaningful to organisms, that is, do organisms (other than the ones in Fig. 13-10) cross the very clear boundaries between colder and warmer water? If the species present always remain in their own water mass, the patches do not have much biological significance. On the other hand, if a whale, for instance, were feeding while swimming horizontally at 17 m, prey would be patchy in abundance.

The nature of "patches," in this and many instances, depends on the point of view and scale of measurement. This is a pervasive problem that needs careful attention. The physics of the water masses and the *in situ* behavior of the organisms are of particular importance.

13.4 Ecological Consequences of Patchiness

The pervasive patchy distribution of both resources and consumers has a variety of ecological consequences. Some aspects particularly linked to material in earlier chapters merit some note.

Feeding. We have already discussed the aggregative response of predators to patches of high prey abundance. This may occur at every scale; on small scales, distance among prey items is less than that among isolated individuals, so feeding in patches is more efficient for the consumer. For example, carp feeding on the benthos of the Volga River only achieve their full rations when they fed on highly aggregated food items (Ivlev, 1961). Laboratory populations of *Calanus pacificus* reach sizes comparable to those in the field only when densities of phytoplankton higher than natural concentrations are available (Anderson et al., 1972). As predators move about their environment, they often remain to feed in patches where prey items occur at high densities, and so accumulate in places where food is more abundant. Sperm whales, for example, gather in this fashion when feeding on patches of food thousands of miles long (Cushing, 1981). Notice that aggregations of feeding predators are not the same as predators hunting while in schools. Eggers (1976) has shown using simulation studies that hunting in schools may reduce the prey consumption per predator, except when prey densities are very high or the prey are patchily distributed. Since the latter situation is very common, we find that many predators—tuna, swordfish, porpoises—do feed while in schools.

Reduction of risk of being eaten. The behavior that leads to aggregation into schools and the maintenance of uniform distances within a school seem designed to avoid risk of predation (Brock and Riffenburgh, 1960; Vine, 1971). The chances of one individual being eaten are reduced by such spatial distributions, even though the school may provide predators with a patch of high density.

Demographic effects. Copepods in an oceanic water column feed on patchily distributed food particles, so they must necessarily encounter areas where food items are scarce. If *Centropages* and *Acartia* fail to find a patch of food within 3-5 days there is considerable mortality (Fig. 13-11, top). *Calanus* is far more tolerant of periods without feeding, while *Pseudocalanus* show a remarkable ability to sustain periods of starvation. The survival patterns of these genera are clearly adapted to quite different spatial distributions of food. The reproductive effort of copepods is also related to food distribution. If copepods are reared at various intervals of food deprivation (Fig. 13-11, bottom) there is a reduction in egg production in *Acartia* and *Centropages* at longer intervals without food. *Pseudocalanus*, on the other hand, sustain intervals of up to 17 days without food with no consequent decrease in fecundity compared to continuously fed controls (Dagg, 1977). We may speculate that demographic patterns exhibited by different species may therefore vary depending on the patchy nature of the environment of each species.

Habitat use and selection. Since different patches will differ in what they offer as resources, individual animals will show patch or habitat selectivity. We have already seen this when discussing the aggrega-

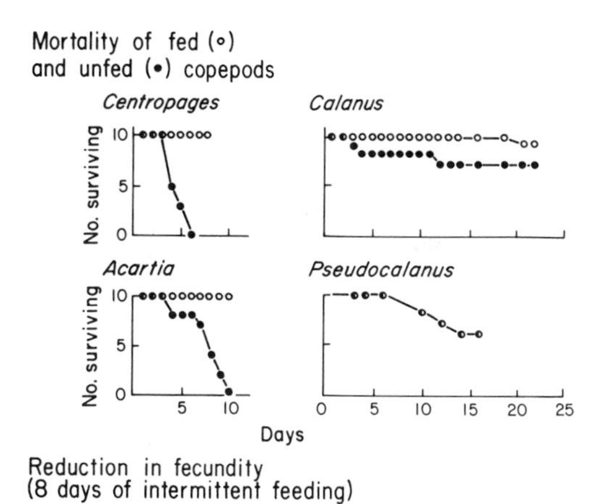

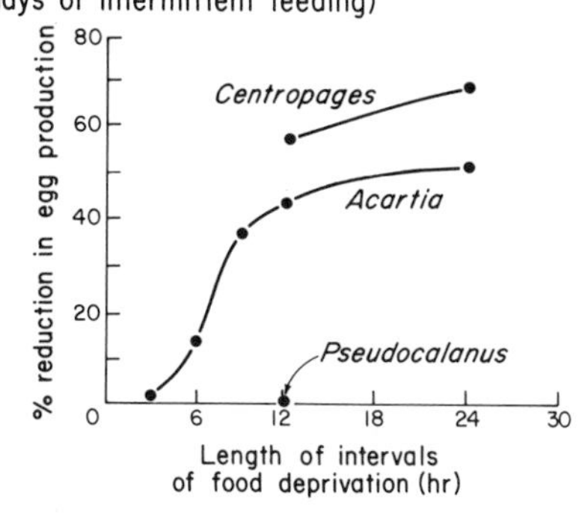

Figure 13-11. Demographic consequences of intermittent feeding in copepods. Top: Mortality of fed and unfed copepods in the laboratory over time. Bottom: Reduction in fecundity of copepods under feeding regimes where excess food was offered intermittently and the interval of food deprivation varied. Adapted from Dagg (1977).

tion response to prey density and in the discussion of searching time. Both an active choice of habitat and a simple response to resource abundance can bring about such "habitat selection."

Social organization. In species where resources are allocated by interference competition, the most favorable of patches in a habitat are occupied first. This leads to a uniform distribution of individuals within the favorable areas. If there are more individuals than can fit in the choice habitats, less favorable areas are used. In the garibaldi, the nesting males take up the best habitats; other fish, including females, have less desirable areas and there is a far less intense defense and attachment to these poorer territories. In many other cases, the areas of habitat needed for nesting and obtaining food differ widely. In terns, for example, as we saw in Chapter 5, foraging areas are necessarily far larger than nesting areas; the limited amount of suitable nesting habitat required by the population forces a colonial type of social organization. Depending on the spatial distributions of the patches suitable for a species, other forms of social behavior may occur. In whales (Chapter 9), tuna, and in other animals, long-distance migration of groups to winter feeding grounds is common. These organisms form pods, herds, or flocks that migrate from one area to another.

Dispersal. Dispersal patterns are frequently bimodal (Wiens 1976), with many individuals moving short distances within a favorable patch, and some moving long distances from one patch to another. Species that specialize in rapid colonization of newly available habitats, often called "fugitive" species (Chapter 15), show propensities for long-distance movements. Both the distributions of patches of suitable habitat and the frequency of change in a habitat affect the need or advantages of long-distance dispersal.

The above list of ecological consequences does not exhaust the possibilities; rather it serves to show the fundamental implications of spatial heterogeneity for almost every aspect of ecological organization. Awareness of the importance of patchiness goes back into the history of ecology, but concerted work on this topic is relatively recent. Many critical questions remain to be answered, such as: what is the effect of patchiness on consumption and growth? How does patchiness affect the ability of a consumer to limit its food resource? Does patchiness prevent a consumer from exterminating prey? How do biological mechanisms interact with the powerful effects of physical mechanisms?

It is to an extent paradoxical that in a section of this book, where we are supposedly dealing with the community and ecosystem levels, most of the examples have involved populations or individuals. Our knowledge of spatial distribution needs to be carried further, and we need to ask questions at the level of the community. For instance, are some environments likely to contain more species distributed in aggregated fashion? As communities age, are randomly distributed species replaced by aggregated or uniformly distributed populations? Does spatial arrangement reflect resource availability? These are among the many difficult questions raised by the existence of spatial heterogeneity. Bright young ecologists will no doubt find ways to answer them experimentally in the future.

Chapter 14

The Structure of Marine Communities Over Time: Long- and Short-Term Changes

14.1 Long- and Short-Term Changes

Conditions and organisms change on marine environments at a wide range of scales, from minutes to thousands of years (Haury et al., 1978). Even at great depths near the bottom where temperature and the chemical milieu are rather constant, there are pulses of materials raining down from the seasonal cycles of production near the surface (cf. Section 10.1121). Presumably this is why seasonal breeding takes place even in benthic organisms in the otherwise very constant deep sea (Grassle and Sanders, 1973; Rokop, 1974, 1977). The magnitude of production at the surface also varies from year to year, so that bottom fauna are exposed to variation over years as well as months.

The relative time scale of changes needs to be expressed in relation to the life span of organisms. Organisms generally respond to environmental changes that are shorter than one generation time by using behavioral (hiding, inactivity) or physiological mechanisms (dormancy, resting stages, tolerance). One of the most well-documented responses to short-term changes is the daily vertical migration of organisms through water or sediment columns. Although only one of many short-term phenomena, vertical migration is so prominent that it is given separate treatment below.

Longer-term changes in conditions, in time scale of months, years, and longer are also pervasive and prompt significant changes in communities of organisms. Seasonal cycles are one such kind of change, and are discussed below. Other less obviously periodic changes such as variations in wind direction and strength, surface temperature, and salinity (Fig. 14-1) are also common. Other phenomena that change structure of marine com-

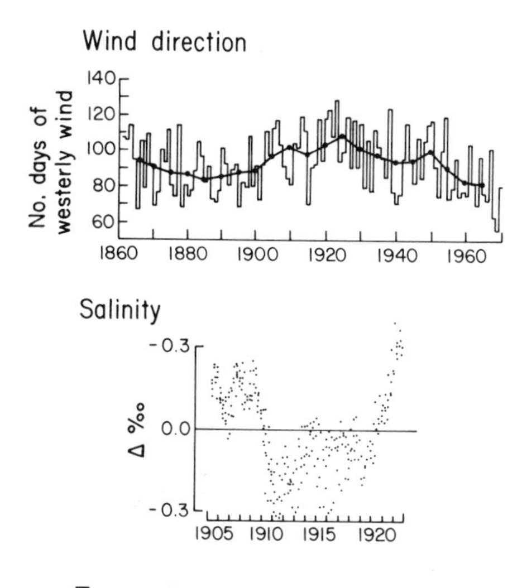

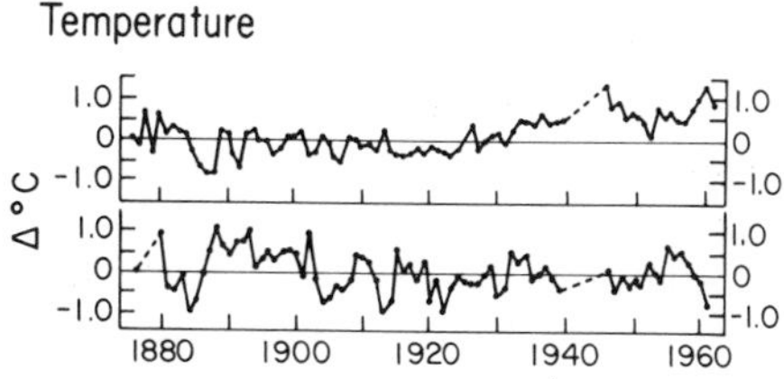

Figure 14-1. Long-term trends in wind direction, salinity, and surface water temperature near Great Britain. Top: The number of days of westerly wind over the British Isles, 1865–1963. The dots show 10-year means plotted at 5-year intervals. Middle: Salinity anomalies in the Western English Channel, 1905–1920, expressed as a deviation from the average. Bottom: Temperature anomalies of surface seawater at two stations off Britain, 1880–1960. Adapted from Cushing (1975).

munities on different time scales are discussed in Naylor and Hartnoll (1979). It is difficult to identify concretely how such changes affect marine communities, but there are pronounced changes in the abundances of many species that are related to large-scale, long-term variations in climate and hydrography.

The waters around the British Isles are among the best and longest studied of any marine environment. During the 1920s the cold-water, open-ocean community in the English Channel (dominated by the chaetognath *Sagitta elegans* and herring) was replaced by a warmer-water assemblage (dominated by *S. setosa* and another clupeid fish, the pilchard). Accumulations of nutrients in surface waters during the winter were lower, spring blooms of phytoplankton were accordingly reduced, zooplankton and demersal fish were reduced in numbers. About 1965 the trends in the Western English Channel reversed and by 1975 there were virtually no pilchard left, and zooplankton and larvae of demersal fish increased to the levels of the 1920s. In 1979 *S. setosa* was scarce or absent and *S. elegans* was the dominant chaetognath. The transition is not complete; there are occasional short-term reversals to warmer-water faunas. The herring have not returned in large numbers; rather the mackerel has become the dominant pelagic fish. The specific mechanisms behind these major changes in the pelagic community in the English Channel are not well understood (Southward, 1980), but there are three major points to emphasize: (1) the ocean is a continuously changing environment, where changes can be fast or slow and can affect small or large areas; (2) climate and hydrography play an overwhelmingly important role in determining the major outlines of marine communities, with biological interactions specifying the details of community structure within the constraints allowed by the physical environment; (3) whatever regularities we may find in the structure of assemblages of marine species are within the context of a very changeable environment so generalizations and trends will be hard to find and define.

14.2 Daily Changes: Vertical Migration

14.21 Nature and Stimuli in Vertical Migration

In sediments, motile blue-greens, dinoflagellates, and diatoms may have daily or tidally governed excursions of a few millimeters up to the sediment surface (Gallagher and Daiber, 1974; Eaton and Simpson, 1979).

In the water column, vertical migration may involve dinoflagellates, zooplankton, or fish, and the vertical movements may span tens to many hundreds of meters. Some photosynthetic dinoflagellates migrate upward as light begins to increase at dawn, and travel distances up to 50 m. The dinoflagellates travel at speeds of $1–2 \times 10^{-2}$ cm sec^{-1}, sufficient to overcome the upwelling velocity in the Baja California coastal upwelling (Blasco, 1978) but perhaps not where upwelling is more intense. Faster currents and changes in the density of water can hinder migration of dinoflagellates. Different species of dinoflagellates have characteristic sensitivities to light intensity; this results in a different pattern of vertical migration and different temporal and vertical distribution of the species.

In zooplankton, vertical migration involves an active upward movement of most species as night approaches (Fig. 14-2). The speeds required are substantial; salps, for instance, are most abundant near the surface during the night, and may be as deep as 600–800 m or deeper by day; the swimming speeds to cover these distances must be 5–10 m min^{-1} (Wiebe et al., 1979).

Feeding by adult zooplankton usually takes place at night in the shallow layers. At dawn there is a return to the daytime depths, where the individuals idle until the next dusk (Longhurst, 1976).

In any parcel of the sea not all species of plankton migrate to the same extent or synchronously, and young stages tend to be less active migrators. Tows taken in the North Atlantic show different patterns of vertical migration in adult copepods (Fig. 14-2). One species shows a lack of migration, with only a very short downward excursion at night. A second

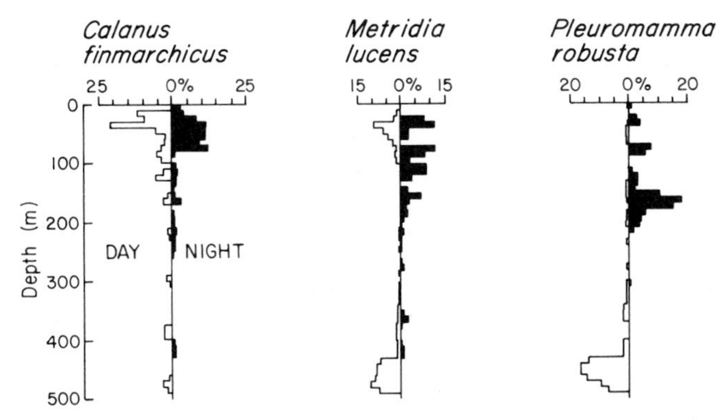

Figure 14-2. Vertical profiles of three zooplankton species (adults only) collected at night (2400) and day (1200) at Ocean Weather Station J (52°N 20°W), 13 May 1970. Data of R. Williams, adapted from Longhurst (1976).

species does migrate downward about 350 m by day but newly molted individuals remain near the surface. A third species shows a well-defined vertical movement by the entire adult population. Thus the assemblage of populations found at any one depth varies markedly over a daily interval: behavioral changes on scale of a day affect the structure of the community.

Near the surface, light is the stimulus that starts upward migrations; after dusk there is commonly a passive sinking as night brings darkness. At the next dawn there may be a reversed short period of ascent before their downward movement begins. On bright, moonlit nights these short episodes of sinking after the migrants reach the surface are reduced, indicating that light intensity does govern this behavior (Longhurst, 1976). It is not clear what cues are used for vertical migration by zooplankton located at depths below 750–1000 m, since organisms supposedly cannot detect downwelling light at such depths (Clarke and Denton, 1962) (cf. Fig. 2-3), yet 25 of 207 species collected between 600 and 1,700 m showed vertical migration cued to day–night cycles. Either organisms are extremely sensitive to some aspect of light or other cues are used.

14.22 Reasons for Vertical Migration

There must be some important advantage to the migratory behavior, since the vertical excursions frequently take the plankton between 10,000 and 50,000 times their body lengths and since for most zooplankton species it would seem preferable to feed continuously. There are a number of hypotheses about why vertical migration takes place. None is proven, but it seems likely that migration has multiple causes.

Avoidance of competition. Dinoflagellates can migrate from surface waters to deeper layers. Dinoflagellates have a relatively high light saturation curve (Fig. 2-5), so they can actively photosynthesize near the surface where other producers may be inhibited; at night they can migrate downward to deeper waters where nutrients are not as depleted as at the surface. Since the abundance of other phytoplankton may be lower in deeper water, migration may lessen competition.

Avoidance of predation. Upward migration at dusk brings zooplankton to the surface layers during the period when their visual predators are at a disadvantage. Many migrants, however, move only short distances, and spend the day at depths where there is enough light for predators to see prey. In addition, many migrants go deeper than might be required just to escape predators; last, luminescent migrants are common. On the other hand, Ohman et al. (1983) find a "reverse" migration of the copepod *Pseudocalanus* in a temperate fjord, presumably an escape response to "normally" migrating predators of *Pseudocalanus*. The "reverse" migration is not seen during periods when predators are less abundant. This suggests a strong influence of predation on migratory activity.

Horizontal transport. Vertical migration, both on a daily and on a seasonal time scale may aid a population both in maintaining its location in a very dynamic environment and in dispersal. It is often the case that different layers of water travel in different directions. Galatheid crabs transported by the California Current to tropical waters have a potential way back by migrating into an undercurrent moving in an opposite direction (Longhurst, 1968). Similar situations are found in currents of the Strait of Gibraltar (Frassetto et al., 1962), in two-layered estuaries (Bosch and Rowland Taylor, 1973), and in upwellings (Wroblewski, 1982).

Energetic advantages. Energetic expenditures by animals are strongly correlated to ambient temperatures, as we saw in Chapter 7, so that part-time residence in deeper, colder water reduces energy losses. Feeding is easier near the surface, where food particles are most abundant. Vertical migration could thus allow feeding while minimizing respiratory expenditures. Migration is not very energetically costly since swimming by zooplankton is not a very energy-demanding activity (Haury and Weihs, 1976). This is especially so in some salps, where the organisms swim actively all day long and feeding, locomotion, and respiration all are accomplished by the same motion (Wiebe et al., 1979). Growth is, however, slower at lower temperatures, so the energy bonus due to vertical migration must be enough to compensate for both locomotory expenditures and lower growth rate (McLaren, 1963). The reduced respiration due to lower temperatures (Section 7.321) may allow a reallocation of energy into growth (Section 7.431), so that copepods that migrate may grow to a larger size. Since larger animals have increased fecundities (Section 4.21), migration has a selective advantage (McLaren, 1974). The energy bonus is, in theory at least, larger at high temperatures so more migrant species are expected in tropical waters where the surface water is warm is strong contrast to deeper layers. At higher latitudes the vertical profile temperatures are more uniform and fewer vertical migrators can be expected (McLaren, 1963). These patterns seem to be true to some extent but have not been proven. The argument may hold for seasonal as well as daily vertical migrations.

Enright (1977) developed the energy-advantage idea and suggested that nocturnal grazing by copepods might provide a greater gain of energy than continuous feeding, if three conditions were met: (1) if photosynthesis is high enough so that the algal biomass accumulates appreciably from dawn to dusk; (2) if metabolism of copepods is lowered while they are in the deeper layers; and (3) if the grazing rate of the zooplankton is higher than average in returning to the surface. From earlier discussion (Sections 1.4, 6.222, 7.321, and 5.221) we know that the three conditions are reasonable. The relative importance of this intermittent feeding version of the energy-advantage hypothesis versus the predator hypothesis can be tested by noting when the zooplankton arrive at the surface. If migration is a response to escape visual predators, arrival at the surface should take place after sunset. If grazers begin feeding in surface water a few hours before sunset, they may gain a significant advantage, and the predator hypothesis would seem to be less applicable. Enright and Honneger (1977) examined the timing of arrival of copepods in surface waters off California. In one sampling series carried out during spring, when copepod density was high, they found that the onset of migration began 1–2 hr before sunset. In the summer, when predators were less numerous, migration began 1–2 hr after sunset. Thus, at different times in the season, different factors may affect vertical migration. It is unlikely that there is a simple answer to the question of why zooplankton migrate.

Short-term phenomena such as vertical migration make the concept of a plankton community a very diffuse one. There are only tenuous associations among the species present at any one depth, since some migrate vertically and some do not. Further, there are associations vertically up and down the water column, since species of deep water may feed or compete with surface species when the latter are in the deep phase of their migration cycle. Thus, where vertical migration is important, biological processes at the physiological, behavioral, and populational levels combine to make it very difficult to find community properties.

14.3 Seasonal Production Cycles

The importance of hydrography and weather on the control of marine communities is clearly illustrated in the seasonal cycles of production. We have already mentioned the importance of hydrographic processes in Chapter 2. Here we enlarge on the subject by considering the consequences of changes in hydrography over time.

Of course, not all interactions that structure the communities are physical; there are important biological processes. The biological processes, however, tend to take effect within the broad constraints provided by hydrography and weather.

The seasonal patterns of many marine communities have been described, but few examples are available in which the important controls have been clearly identified. Below we will go over the annual fluctuation in major elements of marine communities. The interpretation of the causes of the seasonal cycles given below should be taken with skepticism, since they are largely based on speculative correlations that can easily be incorrect. Nevertheless, such seasonal cycles provide a view of the impact of extrinsic and of internal controls not easily available in other ways.

14.31 Phytoplankton

There are several patterns of seasonal cycles in the plankton, but the identification of the kind of cycle is not as important as is the insight that the changes may provide as to how planktonic assemblages are governed.

14.311 Temperate Seas

The early development of marine research centers took place in the coastal areas of temperate regions, so that naturally enough annual seasonal cycles in plankton were first described for areas such as the English Channel. During winter the nutrients regenerated by mineralization, resuspended from the bottom or transported by mixing of richer deeper water accumulate in surface waters. The accumulated nutrients are not depleted by algal uptake because the absolute amount of light is low and the depth to which surface waters mix exceeds the critical depth, so algae are often too deep for photosynthesis to be very active (Sverdrup, 1953).

We have earlier (Section 2.11) mentioned the compensation depth (D_c, where the rate of photosynthesis of a phytoplankton cell (P_c) equals its rate of respiration (R_c). Above D_c, P_c exceeds R_c, and below D_c, P_c is smaller than R_c. Cells will be transported above and below D_c, and they will experience an average compensation light intensity (I_c). The critical depth (D_{cr}) is the point at which the average light intensity in the water column equals the average compensation light intensity. When the critical depth is a large value it can be calculated (Parsons et al., 1977) as $D_{cr} = 0.5\, I_0/I_c k$, where k is the extinction coefficient and I_0 is solar radiation. The importance of the critical depth is that if it is less than the depth to which surface waters are mixed, no net production in the water column takes place, since P_w is smaller than R_w, where P_w and R_w are the production and respiration in the water column. If the critical depth is greater than the depth of mixing, net production takes place in the water column (P_w exceeds R_w) and phytoplankton growth occurs.

In the early spring the critical depth increases in proportion to the increased light intensity (Fig. 14-3, top). The depth of mixing decreases as wind strength decreases during that same period, and there may be some warming of near-surface water that increases stratification. The result of all this is that more light is available, and the critical depth falls below the depth of mixing of surface waters (Fig. 14-3, top). The algae are therefore not carried below the critical depth and the spring bloom ensues (Fig. 14-3, bottom). The interaction between critical depth and the mixed layer depth starts off a bloom in temperate seas, and perhaps ends growth of phytoplankton in the fall.

The increase in phytoplankton in early spring provides more food for the few grazers in the water column, and the zooplankton respond by increasing consumption and reproduction—the functional and numerical responses to prey density. Many species of zooplankton overwinter in deep water in subadult stages. In the spring there are usually too few

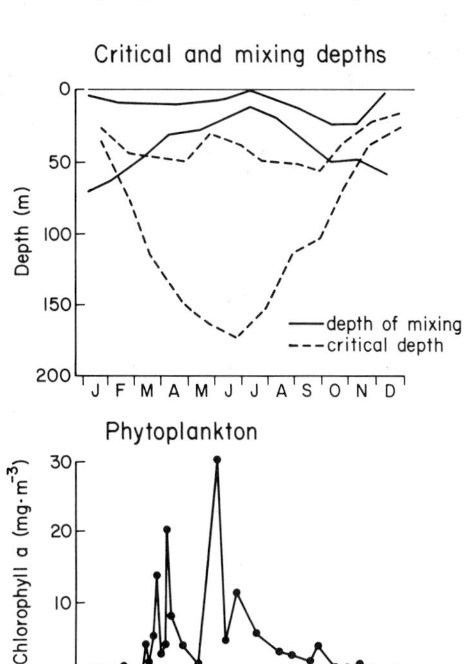

Figure 14-3. Top: Maximum and minimum depths (indicated by upper and lower lines) of the mixed layer and of the critical depths in the Strait of Georgia, west of 124°W. Bottom: Seasonal cycle of chlorophyll in nearby Departure Bay. Adapted from Parsons and Le Brasseur (1968).

individuals for their functional response to increased density of algae to prevent further growth of the phytoplankton. The rate of growth of the zooplankton population (the numerical response) involves a time lag, as pointed out in Section 5.3. Growth and reproduction also involve time lags and are temperature dependent, as shown in Chapter 7, and in early spring water temperatures are low. Since phytoplankton populations are light limited rather than temperature limited during spring, they grow at very fast rates, perhaps doubling or more each day. Under spring temperatures, it is difficult for the zooplankton to expand their numerical abundance and consumption rate fast enough to limit the increase in phytoplankton. The feeding, reproductive, and migratory rates of zooplankton just are not fast enough to produce enough mouths to restrict growth of a food supply that can double every day.

At some point in spring or early summer the algae deplete essential nutrients, principally nitrogen, from the mixed layer, and so the rate of phytoplankton growth slows down. At this time of year, recycled nitrogen is the major source of nitrogen for phytoplankton and the phytoplankton may be deficient in nutrients (Yentsch et al., 1977). Since water temperatures are higher, grazers can be more active, and build up their numbers. Because the rate of growth of algae is slower than earlier in the season, it becomes possible for grazers to remove phytoplankton cells faster than they are produced. The result is a midseason decrease in phytoplankton standing crops. One example of this changed impact of grazing through the year is provided by a study in a Swedish fjord. Bamstedt (1981) found that zooplankton only consumed 3–9% of net primary production early in the growth season, while later on, 60% of the production was consumed daily.

Depletion of nutrients in surface waters may persist through the summer if the thermally stratified water column that prevents the vertical mixing of deeper, nutrient-rich water with surface, nutrient-poor water persists. There may be intermittent periods of mixing that give rise to short bursts of algal production (Figs. 14-3, 14-4) during the summer.

Juvenile fish are most abundant in the English Channel in spring, near the time of the spring bloom (Fig. 14-4, bottom). The herbivorous young clupeids peak first, followed by other fishes. Juvenile fish grow in size and become more effective predators on zooplankton.* These fish may so decrease biomass of zooplankton as to prevent the zooplankton from entirely removing all algal cells in summer. In the late summer when the

* Recall from Section 4.4 that competition for food was evident in young fish so they must limit their food supply. Other predators of copepods such as chaetognaths may also restrict zooplankton grazing. *Noctiluca* are algal feeders but can also effectively eat eggs of copepods such as *Acartia* (Ogawa and Nakahara, 1979). In the Sea of Japan maximum densities of such predators coincide with low peaks of copepod abundance. The abundance of adult fish, as we might expect from Chapter 9, seem unrelated to events lower down in the food web.

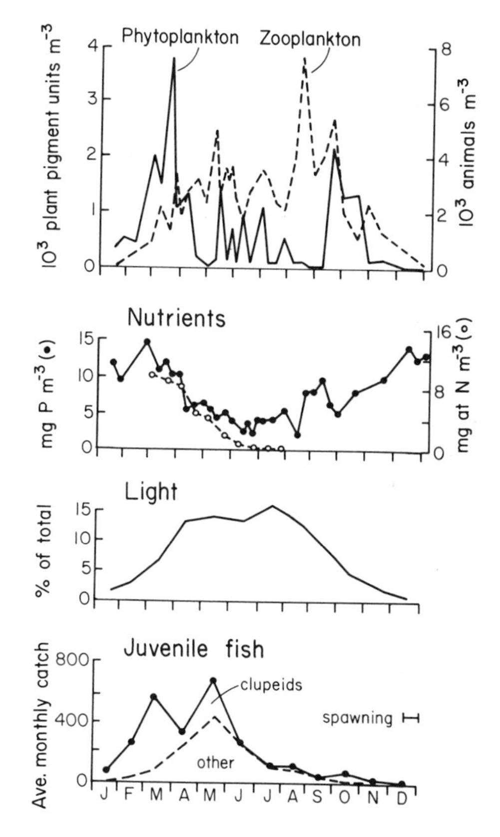

Figure 14-4. Seasonal changes in top 0–45 cm of the English Channel. Light is shown as the percentage of the total annual radiation that reached the sea surface in each month. Adapted from Harvey et al. (1935). Nitrate values for North Sea, adapted from Steele and Henderson (1977). Clupeid spawning occurs in December. Juvenile fish are the average number of fish caught on a 2-hr oblique haul with a 2-m ring trawl. Adapted from Russell (1935).

juvenile fish either migrate to deeper waters or move elsewhere, zooplankton numbers may increase, and this, combined with depletion of nutrients, may result in the lowest phytoplankton densities of the year.

After September, either because of shortage of food or because of physiological triggers cued by the shorter photoperiod, most grazers begin to die or migrate downward, eventually to become relatively inactive in deeper waters.

With cooling temperatures, or perhaps because of increased meteorological disturbances in early fall, mixing occurs and nutrient concentrations in surface water increase. Even though light intensity is reduced in the early fall, there may still be enough for phytoplankton growth. Again, whatever grazers are still present fail to control the growth of algae and hence a fall bloom of phytoplankton may take place. Soon, however, light levels become low or the mixed layer deepens due to the breakdown of stratification of the water, and the seasonal cycle of production ends, with nutrients accumulating once again in the surface layers.

This seasonal cycle, with some variation, can be seen in many temperate parts of the sea, for example, off Nova Scotia (Platt, 1971) and in the Mediterranean (Steemann Nielsen, 1975). Waters of higher latitudes have somewhat later spring blooms than lower latitudes, and the intensity of the fall versus the spring bloom also varies, depending on local circumstances.

In shallow coastal areas or where the vertical stratification is easily disrupted, nutrients may continue to be delivered to the photic zone after the spring bloom, and growth rates of phytoplankton usually continue at a high level through the summer (Fig. 14-5). Grazers seem unable to catch up to the algae and the phytoplankton may be limited by low light most of the year. Production in such shallow temperate areas seems to follow the seasonal pattern of light intensity through the year.

The seasonal dynamics of plankton at temperate latitudes seem therefore determined by light, the depth of vertical mixing, nutrient supply, and grazing pressure. Local combinations of these variables result in specific seasonal idiosyncracies for each area. How variable these patterns can be is seen in the surveys of continuous plankton recorders carried by commercial vessels in the North Atlantic (Colebrook, 1979). A mixture of the bimodal and unimodal cycles is the rule (Fig. 14-6). In open ocean areas, the warming and stratification of the water column often precede the spring bloom. In these areas, down-mixing of cells to subcritical depths is important. In coastal or shallow areas (B1, C4, D1, D2, E8, E9 in Fig. 14-6) or where haloclines or thermoclines prevent vertical

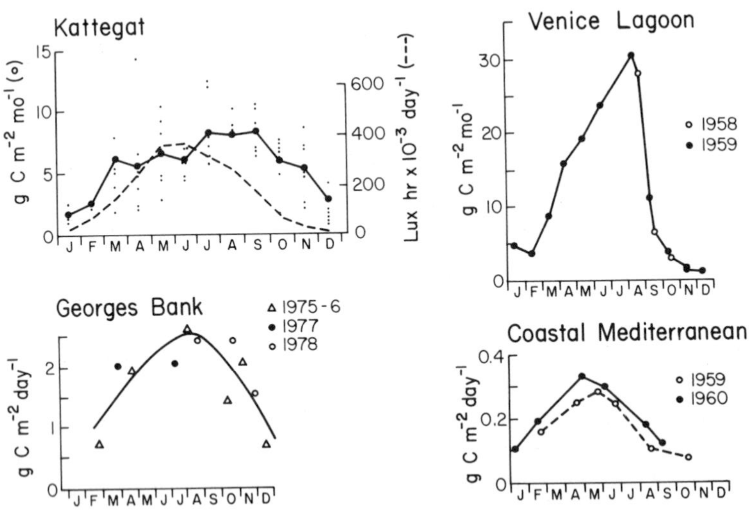

Figure 14-5. Seasonal cycle of primary production in shallow temperate waters. Top left: Gross production in Danish coastal water and light intensity in Copenhagen (Steemann Nielsen, 1975). Top right: Net production in the lagoon of Venice (Vatova, 1961). Bottom left: Production in Georges Bank, an open sea shallow area. Average standard deviation of each point is 1 g C m^{-2} day^{-1} (Cohen et al., 1981). Bottom right: Production off the city of Villafranche. From Brouardel and Rinch in Steemann Nielsen (1975).

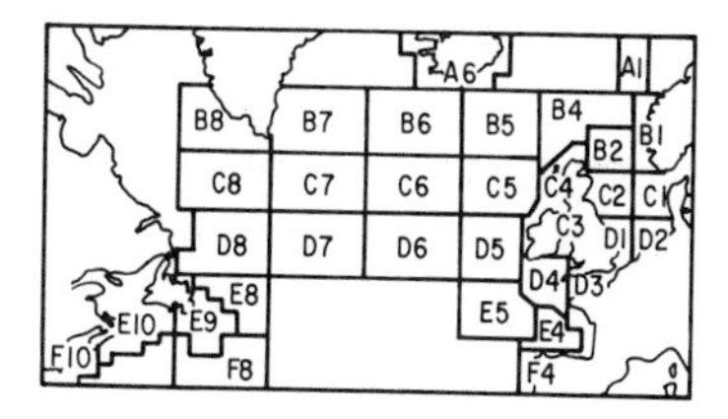

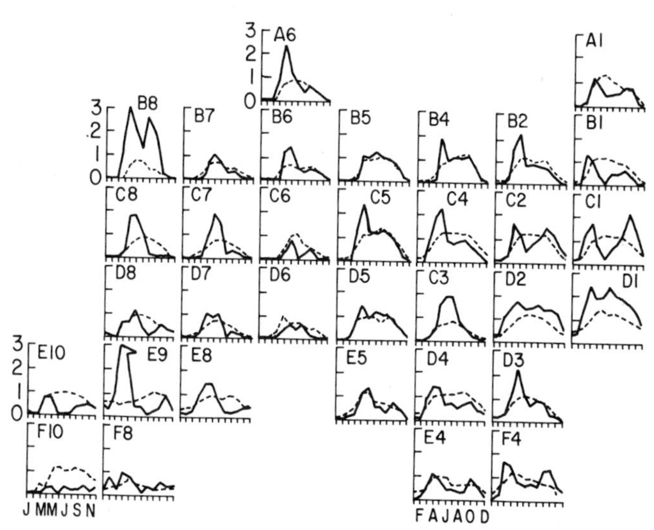

Figure 14-6. Seasonal variation (bottom) in abundance of phytoplankton (whole lines) and zooplankton (copepods) (dashed lines) in different areas of the North Atlantic (top). Phytoplankton are in arbitrary units of greenness and copepods in log means of numbers per sample. From Colebrook (1979).

mixing, for example, area B8 (compare to the deeper B7), the blooms often occur before the warming of the surface waters; the shallow depth prevents the loss of cells from the photic zone. Substantial overwintering stocks of copepods can, however, damp out the spring bloom (B1). The marked fall peak in F4 may be due to upwelling of rich water during July–September.

The point that needs emphasis here is not the identity of the type of seasonal pattern but rather that it is the interactions of a relatively few processes that control the structure of the plankton community over time. Steele (1974a) and others have incorporated mathematical expressions of the few processes we have mentioned (mixing, grazing, predation) plus nutrient uptake by phytoplankton, nutrient regeneration, assimilation, and respiration by zooplankton, into a general model of the seasonal cycle of production in the North Sea. The model is a quantitative expression of the system of controls that we have been discussing and describes the observed annual cycles reasonably well. Others, as discussed in Chapter

6, have elaborated Steele's models, and have proposed some alternative mechanisms. Further work is needed, but the predictions of the models are close enough to real values to provide some assurance that the major controls of plankton communities are understood.

14.312 Estuaries

Estuarine seasonal cycles show effects of the mechanisms described in the last section, but the nutrients are borne by seasonally varying river flow rather than provided by seasonal mixing.

In Narragansett Bay, Rhode Island, the Providence River and sewage treatment plants provide a variable but high-nutrient input (Fig. 14-7), with somewhat enhanced deliveries of ammonium and phosphate in the fall. Nitrogenous nutrients are depleted down the estuary through the year, especially during February–September. Dissolved nitrogen is the primary limiting factor* and is in very low concentrations by early March. Although there is a seasonal change in concentrations of phosphate and silicate, these nutrients are not depleted down the estuary.

The phytoplankton are probably light-limited in winter, as evidenced by the low chlorophyll concentration (Fig. 14-7, middle row and column) throughout the estuary. The phytoplankton and the zooplankton (Fig. 14-7, middle row, right column) both have more pronounced midseason peaks at the head than at the mouth of the estuary.

The high nutrient concentrations at the head of the estuary allow growth of phytoplankton through much of the year. There may be greater primary production and a higher zooplankton crop up the estuary during the warm months. The zooplankton may be responsible for the reduction of chlorophyll in the summer (Martin, 1968).

The high density of zooplankton at the head of the estuary supports the growth of large numbers of ctenophores, fish larvae, and adult menhaden (Fig. 14-7, bottom row). These predators feed primarily on zooplankton† and are very abundant, but it is not known whether they reduce the numbers of the grazer populations. Larger fish such as bluefish (*Pomatomus saltatrix*), striped bass (*Roccus saxatilis*), and bottlefish (*Preprilus triacanthus*) feed on menhaden and ctenophores. Not much is known about the role of these large top predators; they may be very important in structuring communities, as we saw in Chapter 9.

---

* We have seen that this is true for coastal waters in general (Section 2.23). In some estuaries, as those in the North Sea near Holland, silicates may be the limiting nutrient for algal growth at the head of the estuary (Gieskes and Kraay, 1975).

† Menhaden usually feed on phytoplankton, but in summer at the head of Narragansett Bay the dominant algae are small flagellates. These are too small for menhaden and therefore the fish switch feeding methods and feed raptorially on the abundant zooplankton (Kremer and Nixon, 1978).

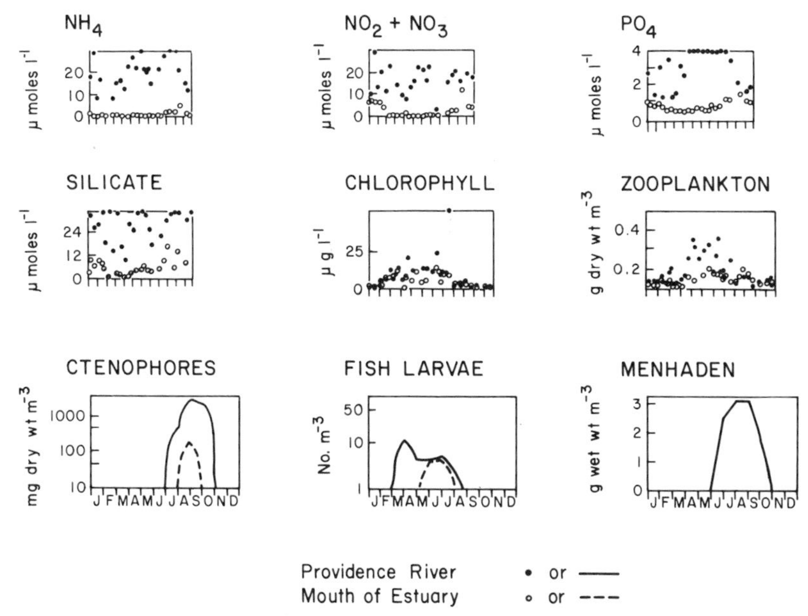

Figure 14-7. Seasonal cycles of nutrient and organisms in the head (Providence River) and mouth of the Narragansett Bay estuary. Nutrients and chlorophyll are from the surface. The copepods *Acartia tonsa* and *A. clausii* make up 95% of the zooplankton. *Mnemiopsis leiydi* is the carnivorous ctenophore. The menhaden (*Brevoortia tyrannus*) are found with various other fish species. Adapted from Kremer and Nixon (1978).

There are other estuaries where the river inputs are more strongly seasonal than in Narragansett Bay. One of these is the Nile estuary (Fig. 14-8) where the river input—shown by the sharply lowered salinity of the seawater near the discharge of the river—comes in late summer at the end of the rainy season in the Nile catchment area.* Nutrients in coastal water increase when the Nile discharges and the phytoplankton blooms quite readily. Numbers of zooplankton increase with only a slight lag behind the algal blooms. The rapidity of the numerical response of the grazers may be related to the warm temperatures. Since there are no data for nitrate, it is unclear whether grazing pressure or exhaustion of nutrients cause the sharp reduction in stock of algae during autumn, but it seems more than likely that this estuarine system, located in the very nutrient-poor Eastern basin of the Mediterranean, is virtually completely governed by nutrient supply and its exhaustion by algae.

* After the construction of the Aswan High Dam in the 1960s, this whole pattern was disrupted, and coastal fisheries reduced because of the lack of nutrient input into the nutrient-poor Eastern Mediterranean.

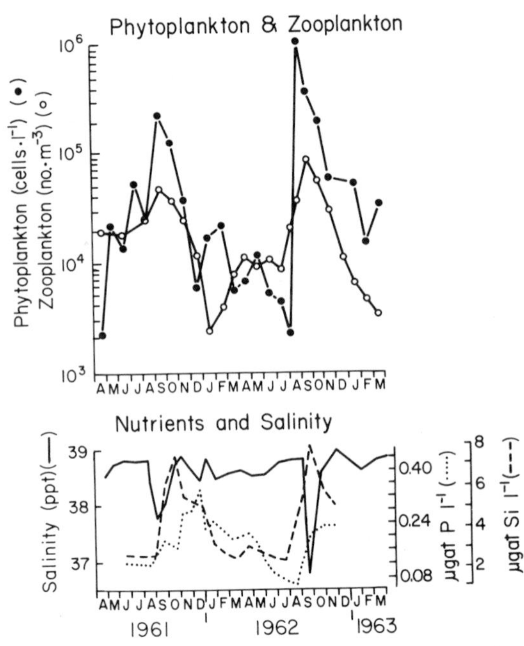

Figure 14-8. Seasonal cycle of salinity, nutrients, and production in the Nile estuary. Adapted from Aleem and Dowidar (1967).

In some estuaries and coastal waters tidal mixing may be a major mechanism providing nutrients. Puget Sound in the Pacific Northwest of the United States is a deep temperate fjord with some estuarine properties. Nutrients are regenerated by the benthos and also advected horizontally, and tidally induced vertical advection and turbulence furnish phytoplankton with adequate amounts of nutrients. Production during the winter and spring is determined primarily by light, maintained at a low level by particulates in the water (Winter et al., 1975). The grazers are probably ineffective as a mechanism limiting numbers of algae. The seasonal cycles are quite irregular, determined mainly by hydrological and weather events.

Stratification due to vertical differences in salinity may prevent mixing of phytoplankton in estuaries below the critical depth. This may allow an earlier onset of spring blooms in estuaries than in the open ocean (Parsons and LeBrasseur, 1968).

The specific mechanisms and timing by which light, nutrients and grazing interact and result in the local phenomena may differ, but the major variables are near-universal. Although at first sight estuaries may have seemed very distinctive environments, the seasonal cycles are determined by the same kinds of limiting factors that are preeminent elsewhere in the sea.

14.313 Upwelling Areas

Areas of upwelling have been studied intensively in recent years, but there are few descriptions of seasonal production cycles. In the Gulf of Panama (Fig. 14-9) the offshore winds during January–May create upwelling of cold deeper waters, new nutrients are brought into the photic zone, and blooms of phytoplankton occur. The high production increases the numbers of particles in the water enough to reduce the depth of the photic zone, even though at the latitude of the Gulf of Panama light intensity is high throughout the year.

In some areas of upwelling such as the Gulf of Panama and off Peru, the advection of cold rich water may last for several months. In the Gulf of Guinea the upwelling season is short-lived and seasonal fluctuations in the production of phytoplankton are more intense (Fig. 14-10, top). Phytoplankton densities are very variable during the upwelling season, probably due to the marked spatial heterogeneity that seems characteristic of

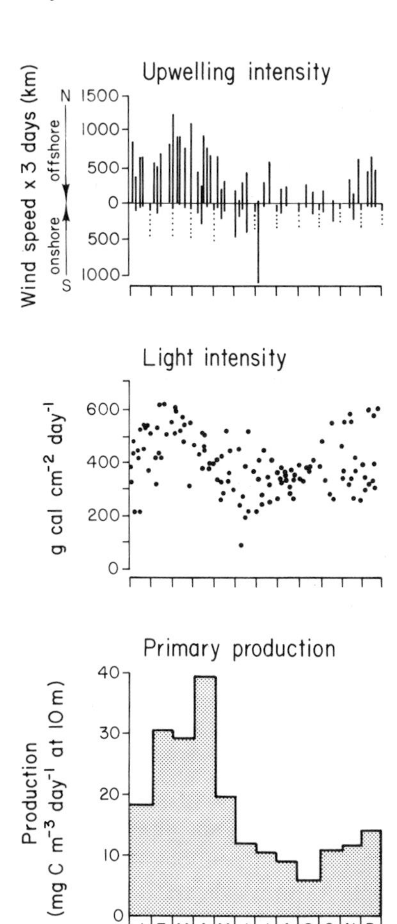

Figure 14-9. Seasonal cycle in upwelling in Gulf of Panama. Top: Average intensity of upwelling, measured by wind-speed for three day intervals, onshore and offshore, for 1954–1967. Middle: Light intensity through the year, 1955–1957. Bottom: Rate of primary production. Adapted from Forsbergh (1963) and Smayda (1966).

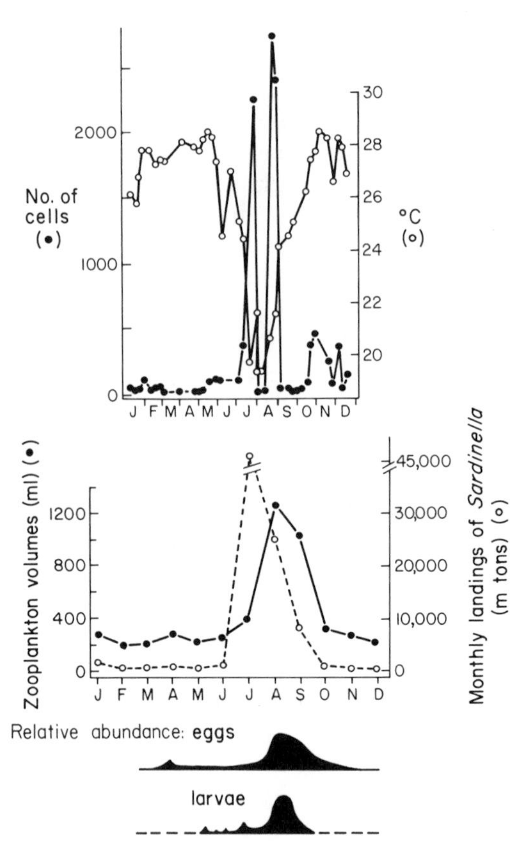

Figure 14-10. Seasonal cycle in upwelling of Gulf of Guinea, Ghana. Top: Temperature of the water and phytoplankton abundance (units are number of cells per sample taken). Note the sharp rise in phytoplankton during the time when nutrient-rich cooler water is upwelled. Bottom: Zooplankton (60–80% *Calanoids carinatus*) abundance and landings of the sardine (*Sardinella aurita*). Adapted from Houghton and Mensah (1978).

upwellings (Fig. 14-11). Small adjoining patches of upwelled water may contain phytoplankton at a different stage of the bloom or succession.

In the Gulf of Guinea the zooplankton are dominated by a herbivorous calanoid copepod and its abundance is low most of the year (Fig. 14-10, middle). As water temperature drops during the brief upwelling season, juvenile copepods migrate or are advected shoreward, feed on the abundant organisms near the surface, molt into adults, and reproduce. This aggregational and numerical response to food density may be capable of reducing phytoplankton abundance to some extent, since there is a small increase in abundance of phytoplankton in October–November, when there is no upwelling (Fig. 14-10, top) but the copepods have migrated offshore. In the upwelling off Peru, relatively large copepods (*Centropages brachiatus*, *Eucalanus inermis*, and *Calanus chilensis*) consume 0.5–4.7% of the primary production (Dagg et al., 1982). Smaller copepods (*Paracalanus* sp.) are estimated from lab experiments to consume about 33% of the primary production (Paffenhöfer et al., 1982). Other small species may add to the impact of grazers. It thus seems that grazing rates are not high enough to control phytoplankton blooms in upwellings, since the producers more than double per day.

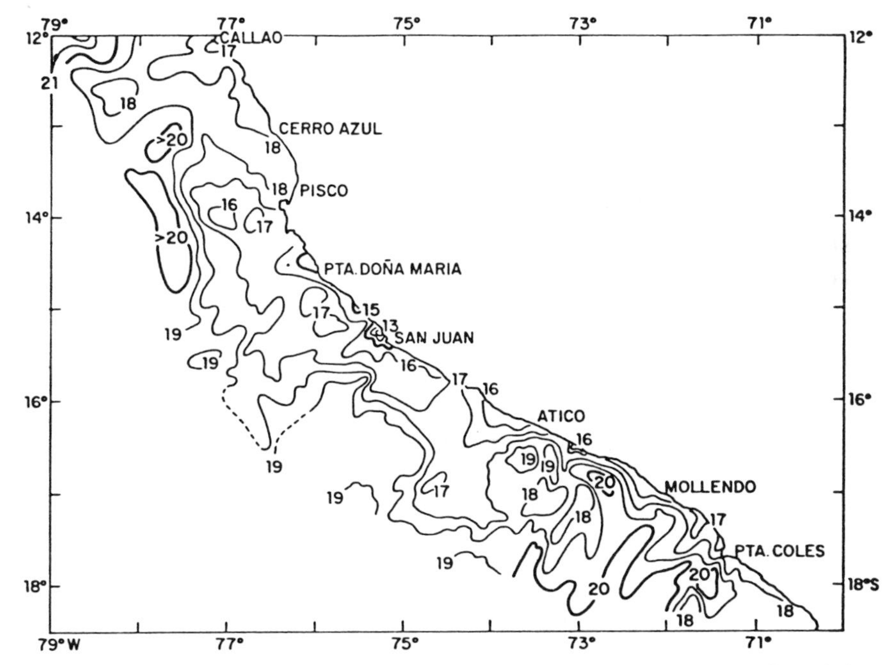

Figure 14-11. Sea-surface temperatures (°C), 28–30 May 1974, in a section of the Peruvian coastal upwelling. From Zuta et al. (1978).

The clupeid *Sardinella aurita* is the most important economic fish in the Ghanaian fishery in the Gulf of Guinea, making up to 45% of the total landings of marine fish. This sardine feeds on zooplankton, is highly mobile, and shows a marked aggregation response to food density, even over considerable distances (Fig. 14-10, middle). The fish find upwelled water and spawn there (Fig. 14-10, bottom), where the larvae have available supplies of suitably sized organisms to feed on. Perhaps predation by sardines is responsible for the decrease in zooplankton in August–October. Reduced algal food and migration to deeper, colder water are a more likely explanation for the lowered densities of zooplankton during this period.

The seasonal pattern of upwelling communities, then, seems largely governed by the intensity and timing of nutrient inputs, with the potential for some influence of young fish on the grazers.

14.314 Polar Seas

At high latitudes the pattern of seasonal growth of phytoplankton is characteristically unimodal. Light is very low during the winter, and increases in illumination in spring lead to sharp spring blooms both in the Arctic and in the Antarctic (Fig. 14-12). The farther away a station is from the pole, the earlier the peak, showing the effect of earlier onset of increasing light

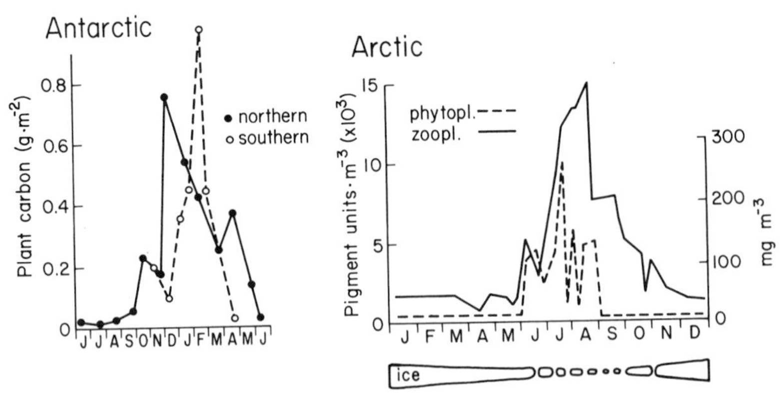

Figure 14-12. Seasonal cycles in the plankton of polar waters. Left: Stocks of phytoplankton in Antarctic waters; northern refers to stations nearer the Antarctic convergence while southern refers to stations nearer the continent. Adapted from Currie (1964) and Hart (1942). Right: Seasonal changes in copepods and phytoplankton (both net and membrane-filtered) in the upper 50 m of Scoresby Sound, East Greenland (1950–1951). Adapted from Digby (1953).

intensity away from the pole (Fig. 14-12, left). The growth of algae at the start of the season can begin even while the ice still covers the water column (Fig. 14-12, right), since growth can occur at quite low light intensities.*

In the Southern Ocean abundant nutrients are provided much of the year by the strong Antarctic Divergence (cf. Section 2.4, Figs. 2-30 and 2-32) near the continent. Farther north the effect of the upwelled water tapers off, especially near the Antarctic Convergence (Balech et al., 1968) and the seasonal peaks are less pronounced.

In the Greenland study there are no nutrient data available to assess whether nutrient depletion takes place in summer (Fig. 14-12, right). There is, however, a surprisingly rapid and early response by the zooplankton. Perhaps in this instance grazers are able to overtake the growth of phytoplankton and cause the late summer decrease in algal stock. The early increase in abundance of zooplankton may not be carried out by a numerical response through reproduction (which would have a longer lag in time) but to an aggregative response, with individual nauplii that overwinter at depth (Digby, 1954) coming to surface water to feed on the summer bloom, and quickly growing and reproducing.

14.315 Subarctic Pacific

Unlike all other areas we have examined so far, there is no obvious spring bloom in the stock of algae in the Subarctic Pacific (Fig. 14-13, top). If we exclude the coastal zones, there is a remarkably uniform amount of chlo-

* In the North Sea, for example, only 0.03 g cal cm^{-2} min^{-1} are needed to start the spring bloom (Gieskes and Kraay, 1975).

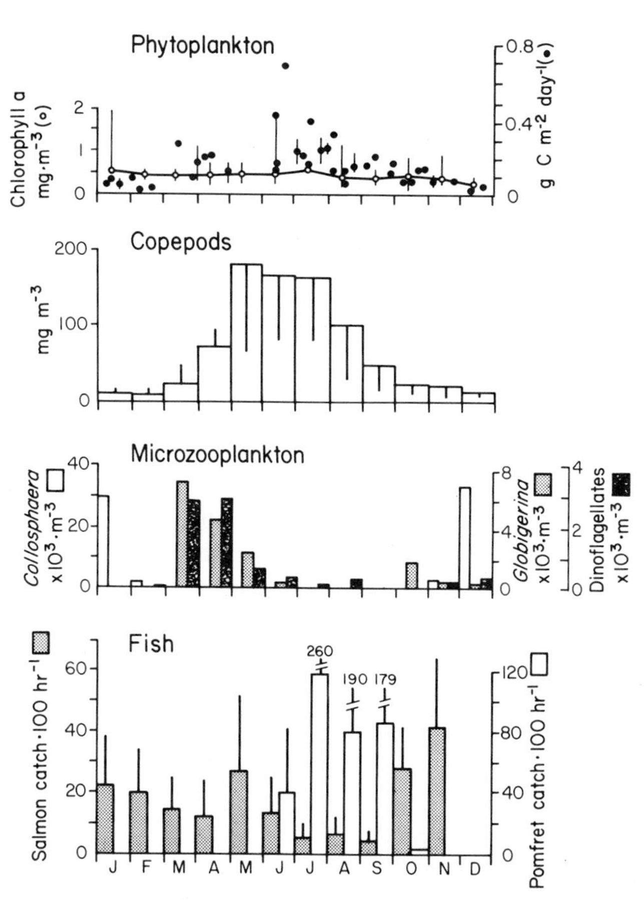

Figure 14-13. Seasonal cycle in the Subarctic Pacific, Ocean Station P, 50°N 145°W. Top: Standing crop of chlorophyll and primary production (0–150 m) through the year. Adapted from Le Brasseur (1965), Parsons and Le Brasseur (1968), and from data of G. C. Anderson. Second row: Abundance of zooplankton in net hauls, 1963–1976. Copepods are the dominant taxon, but weight includes chaetognaths, euphausiids, amphipods, and medusae. Third row: Abundance of protozoans through the year (1966–1967) in the upper 100m at Ocean Station P. Data from Le Brasseur and Kennedy (1972). Bottom: Number of fish per hour of long line fishing off Station P. Salmon include various species; pomfret is *Brama japonica*. Data from Fulton (1978).

rophyll in oceanic water from 40°N to the Bering Sea throughout the year. The values of organic carbon range from 5 to 15 μg liter^{-1}, a very low amount. There is, however, a seasonal peak of production in midsummer that is seldom expressed as an increase in stock of algae (Fig. 14-13, top). Nutrient concentrations are high throughout, even up to 15 μM nitrate-N. Even in September–November there may be 6–7 μM* nitrate-N in sur-

* These are high values compared to other oceanic waters (cf. Section 11.221, Fig. 2-18, and Table 2-3).

face waters. It seems therefore, that the low and constant algal stock is not due to low nutrient levels or to low light.

The zooplankton are dominated by two species of very large copepods, *Neocalanus plumchrus* (up to 5 mm in length) and *N. cristatus* (up to 10 mm in length). These two species make up 70–80% of the zooplankton in stock in the summer and overwinter as adults at depth (Fulton, 1973). Much of the subarctic Pacific shows a halocline at about 100–120 m (Dodimead et al., 1963) that keeps the upper mixed layer shallow and maintains algae near the surface, where the cells can use the low light intensity during winter to photosynthesize. This halocline is a key feature, because it allows young *N. plumchrus* and *N. cristatus* to find concentrated enough food even in winter. If light is too low during winter and early spring to allow much algal production, the copepods may switch to feed on the numerous microzooplankton (Fig. 14-13, third graph from top) (40–200 μm in length) in the surface water (Le Brasseur and Kennedy, 1972). Eggs are produced by *Neocalanus* from fall to spring and the young copepods come to the surface in very large numbers from early spring to late fall (Fulton, 1973). Thus there are copepods available in the surface waters throughout the year, ready to take advantage of any spurt in algal growth. These young copepods grow fast (developmental response) and since large size is correlated to higher fecundity (cf. Section 4.23), the copepods may also have a rapid numerical response.

N. plumchrus can effectively graze on particles 2–20 μm in length (B. Frost and J. Runge, unpublished data), a range that extends to a remarkably small size range for such a large copepod.* The performance of *N. plumchrus* can be contrasted with that of *Calanus pacificus*, a much smaller copepod, which does not feed when food is smaller than 3.9–5 μm. *N. plumchrus* generally filters at low rates, but can increase feeding markedly when exposed to even small increases in abundance of algae (the functional response) (B. Frost, unpublished data), another adaptation to make effective use of occasional patches of high-density phytoplankton.

Thus, features of hydrography, life history pattern, feeding behavior, and morphology of feeding appendages allow *N. plumchrus* to exploit

---

* We have seen that current views on how copepods feed suggest that passive sieving through appendages is less likely than previously thought (Chapter 6). In spite of this, intersetule distances may still have some role in feeding on very small particles. *N. plumchrus* and *N. cristatus* have relatively large feeding appendages with very closely spaced setules, perhaps an adaptation for a diet of the small cells usually found in surface waters of the subarctic Pacific. *N. plumchrus* has intersetule distances of 1.5–7.2 μm. In spite of the very large size of this copepod, the 1.5 μm is the smallest such distance for the many species surveyed by Heinrich (1963). For comparison, here are ranges of intersetule distances for a few other suspension feeding herbivores: *N. cristatus*: 1.8–9.6; *Undinula darwini*: 3.0–8.4; *Pseudocalanus elongatus*: 1.8–7.2; *Eutideus giesbrechti*: 4.2–6.6. Raptorial species, probably mainly predaceous, include *Eucheta marina*: 16.8–19.2; *Metridia pacifica*: 2.4–21.6; *Labidocera acutifrons*: 12.0–27.6; *Acartia longiremis*: 9.6–19.2.

readily a food resource in low supply and of small size, and thus to make use of whatever increase in food density occurs. The result is that in the subarctic Pacific grazers may be responsible for maintaining the very low levels of phytoplankton standing stock constant through the year (B. Frost and C. Miller, personal communication). This demonstrates that given the right circumstances, especially hydrography, timing, and a level of phytoplankton production that is not extremely high, herbivorous zooplankton *can* limit phytoplankton abundance.

The importance of the right circumstances is emphasized by a comparison to the effect of these same *Neocalanus* species in the Bering Sea to the results related above. Production by phytoplankton is considerably higher in the Bering Sea during the spring bloom (0.6–2.1 g C m^{-2} day^{-1} in oceanic and outer shelf areas, 0.7–4.1 g C m^{-2} day^{-1} in coastal waters) (Dagg et al., 1982). The zooplankton graze only 18% and 6–25% of more coastal and more open-water phytoplankton production, respectively. This seems insufficient to control the bloom of phytoplankton in the Bering Sea waters, much as we have seen in several other pelagic systems, and in contrast to what seems to be the case in the North Central Pacific. The latter seems to be an exceptional situation, rather different from other marine environments.

Fish predation may reduce zooplankton biomass in the latter half of the year (Fig. 14-13, second row). Although the fish data are not very substantial (Fig. 14-13, fourth row), there is an apparent increase of some fish synchronously with the decrease in copepods. There are estimates by R. J. Le Brasseur that roughly one fifth of the spring production of herbivores may be consumed annually by salmon maturing at sea. Sanger (1972), however, claims that salmon take less than 4% of the annual zooplankton production. There are other predators or zooplankton, including carnivorous zooplankton, medusae, squid, myctophid fish, juvenile salmon, pomfret, and baleen whales. If all these predators were considered, it seems possible that the zooplankton decline could be due to predation and not just to migration or sinking. At present this is mere speculation.

14.316 Subtropical Seas

The Sargasso Sea is a subtropical gyre in the Atlantic, consisting of a lens of 36.5‰ salinity and 18°C temperature, with a thickness of about 500 m in the center. Below this lens lies a permanent thermocline. There is a seasonal thermocline at about 100 m in depth that may disappear in winter north of Bermuda.

The intensity of light remains high year-round and the amounts of particulate and dissolved organic matter in Sargasso Sea water are extremely low, so that the photic zone extends to about 100 m or more throughout the year.

Nutrient concentrations are low throughout the year in the Sargasso Sea. The supply of nutrients to the euphotic zone is severely limited because of the pronounced vertical stratification. Nutrients, episodically brought to the surface but in low amounts, are used by the phytoplankton as quickly as they are made available. This nutrient regime creates the large area of low primary production in the Atlantic north of the Equator (cf. Fig. 2-30). In the top 100 m, phosphate ranges from 0.02 to 0.16 μg atom P liter^{-1}, nitrite plus nitrate from undetectable to 1.8 μg atoms N liter^{-1}, and silicate from 0.3 to 1.8 μg atoms Si liter^{-1}. Most of the measurements lie to the low end of these ranges, but because of the fast rate of turnover and regeneration of these nutrients, the concentration of any nutrient gives little idea of rates of use by algae, as discussed in Chapter 2.

During winter the surface waters of the northern Sargasso may be briefly mixed down to 400 m, and thus nutrients reach their highest values (1–2 μg atoms N liter^{-1}, 0.1–0.2 μg atoms P liter^{-1}). The "spring" bloom starts in December or January (Fig. 14-14). As stratification is quickly reestablished, the algae are maintained in a shallow surface layer, and the relatively high nutrients and high light prompt a vigorous bloom. Rates of production can be high (up to 2 g C m^{-2} day^{-1}) (cf. Section 1.4), but the nutrients are quickly exhausted and the bloom is short-lived.

After the single bloom, net production by phytoplankton is low and irregular for the remainder of the year (Fig. 14-14). The zooplankton also have a single seasonal peak, more or less synchronous with the phytoplankton. There are some zooplankton always present in surface waters so that they can respond quickly and effectively to the increased food supply. Since the nutrients are so quickly used by algae, the periods of high growth rate are short and the algae do not escape control by the grazers. Rough estimates of consumption of algae by the grazers show

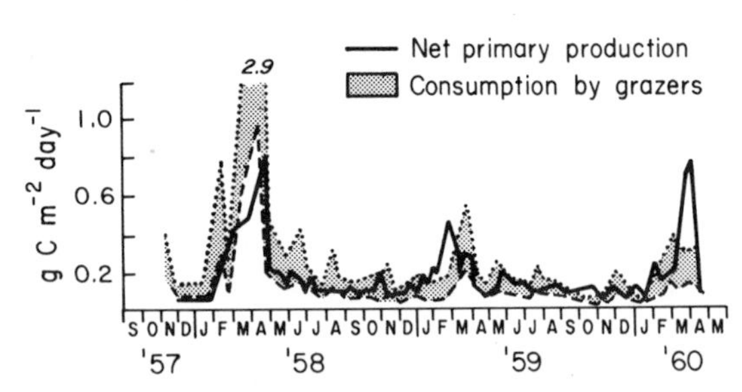

Figure 14-14. Seasonal variation in net primary production measured by ^{14}C in the subtropical Sargasso Sea and rough estimates of upper and lower limits of consumption by grazers. The latter are calculated based on standing crops of zooplankton between 0 and 500 m (dashed line) and 0 and 2,000 m (dotted line). The zooplankton were assumed to be herbivorous, and average feeding and respiration rates were used to obtain the consumption rates. Adapted from Menzel and Ryther (1961).

that virtually throughout the year the demand brackets the production rate by algae (Fig. 14-14). In contrast to coastal environments, almost 100% of the primary production goes to the grazers (Table 8-1). The potential impact of grazers on phytoplankton is large, to the extent that Sheldon et al. (1972) calculated that a 10% decrease of grazing in the Sargasso Sea for just a few days would allow growth equal to the spring bloom if sufficient nutrients were available. Nutrients are so seldom available that virtually the entire crop of phytoplankton is consumed by grazers yet consumption by grazers may not be what controls producers. In fact, the dependence of production on nutrients regenerated by the zooplankton (cf. Section 11.233) is such that probably the key role of animals is in regeneration of nutrients. There is thus a close relationship between producers and consumers mediated by regeneration of nutrients and grazing by consumers.

In the subtropical North Pacific there is a very extensive nutrient-poor gyre. This region is very homogenous over space (McGowan, 1977; Hayward and McGowan, 1979), and there is very little seasonal change (Fig. 14-15). Both phyto- and zooplankton vary hardly at all over the year. The microplankton (ranging in size from 2 to 200 μm) varied less than twofold through five cruises over 2 years (Beers et al., 1982). The North Pacific gyre resembles the Sargasso Sea, but the spring blooms are even less prominent, and the abundance of other taxa is equally unchanging (Sharp et al., 1980; McGowan and Howard, 1978). This is another pelagic system in which there is a close coupling between phyto- and zooplankton. The very low concentrations of nitrogen limit phytoplankton (cf. Section 2.2333), and zooplankton both graze the algal biomass and through excretion regenerate the nitrogen that largely supports growth of the phytoplankton.

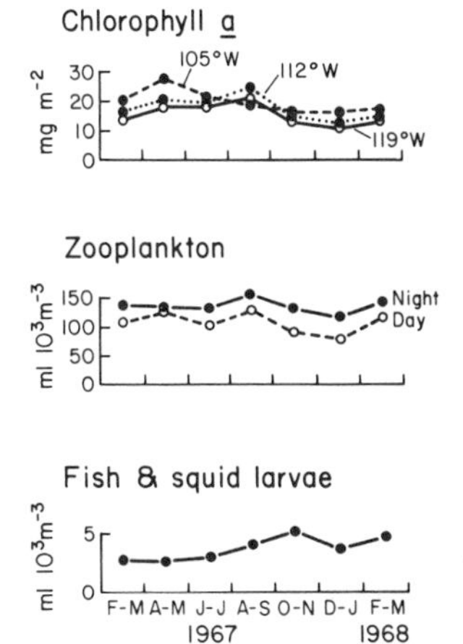

Figure 14-15. Seasonal variation in stocks in the Eastern Tropical Pacific. Chlorophyll is integrated from 0 to 150 m, fauna were collected from the top 200 m. Fish and squid larvae are 1–10 cm in length. Other large zooplankton were measured but are not included here. Adapted from Blackburn et al. (1970).

14.317 Tropical Marine Waters

In coastal tropical waters of the Great Barrier Reef there are also low and constant nutrients and very low variability in stocks of phytoplankton through the year (Orr, 1933; Marshall, 1933; Russell and Coleman, 1934; Russell, 1934). A fluctuation between 1×10^3 and 4×10^3 cells $liter^{-1}$ is very modest compared to the changes of several orders of magnitude characteristic of cycles in coastal temperate waters. Dinoflagellate densities are virtually constant through the year.

The zooplankton fauna of the Great Barrier Reef is very diverse and there is a modest seasonal pattern, peaking in late summer. The increase in zooplankton comes when algae are very scarce; perhaps the zooplankton feed on the numerous detrital particles or mucilage released by the corals. In any case, the density of zooplankton is very much lower than that in temperate waters and only varies about threefold, again a very small change compared to fluctuations in temperature marine environments. The corals themselves are of course perennial and are active through the year, primarily as producers (cf. Section 10.1232) through their zooxanthellae, with nitrogen needs provided by predation by the host coral. There are few data on the seasonal changes in coral activity, which are probably slight.

In tropical areas, as in the subtropical gyres, there is a lack of or only slight seasonality of the plankton. In the open-ocean pelagic communities, there is a remarkable lack of nutrients, primarily due to stratification of the water column and resulting isolation of the photic zone from nutrient-rich water. Whatever blooms take place are soon ended by the reduced supply of nutrients and the grazers are able to harvest algae almost as quickly as they grow. The herbivores may restrict plant growth and hence may also restrict their own abundance, although the evidence is circumstantial. Carnivorous animals may be rather unimportant in limiting prey populations in tropical waters, at least as far as the production cycle in pelagic communities is concerned, but there is little concrete information on this topic.

14.32 Large Attached Algae

The seasonal control of production cycles of macroalgae depends on how hydrography supplies nutrients and on light. Where there are seasonal variations in nutrient abundance, there is a rapid incorporation of nitrogen as nitrate becomes more available in the water column during early winter (Fig. 14-16, top right). The increase in the internal pool of nitrate does not immediately lead to peak growth rates. Maximum growth does not take place until after light intensity increases later in the spring (Fig. 14-16, top left), at which time the rate of photosynthesis increases (Fig. 14-16, bottom left). In kelp and other macroalgae the ability to respond physiologi-

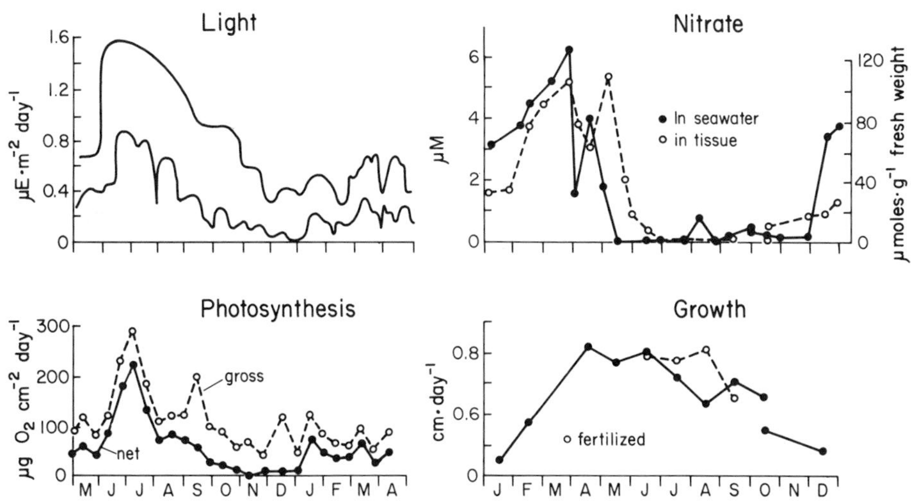

Figure 14-16. Seasonal cycle in growth and photosynthesis in the kelp *Laminaria longicruris* on the Nova Scotia coast. Left: Maximum and minimum light intensities (top) and rate of photosynthesis (bottom) during the year. Adapted from Hatcher et al. (1977). Right: Nitrate content in seawater and in tissues of kelp (top) and growth, as cm day^{-1} of blade elongation (bottom) in a kelp forest growing in water 18 m deep. Fertilization experiments done at site 9 m in depth; the growth rate of unfertilized kelp of 9 m was similar to that at 18 m. Adapted from Chapman and Craigie (1977).

cally to increased light intensity is most pronounced in spring (Brinkhuis, 1977, 1977a). Kelp can, however, grow during winter (Hatcher et al., 1977; Chapman and Craigie, 1978) by using reserves of carbohydrates stored during the previous growing season.

The decrease in growth after spring is primarily due to the reduction of nitrate in the water and in the tissues (Fig. 14-16, top right). The importance of nitrate is evidenced by the sustained growth of kelp in plots fertilized with nitrate (Fig. 14-16, bottom right). Eventually light is again reduced in early autumn and this decreases both growth and photosynthesis. In sites where upwellings or other mechanisms provide high concentrations of nutrients year around, light may limit growth of kelp (Gagné et al., 1982). In such sites the role of internal storage of carbon and nitrogen compounds is less important.

Salinity, temperature, and photoperiod may also have some influence on seasonal growth of macroalgae (Hanisak, 1979) but these factors act primarily by interacting with the primary factors of light and nutrients. The ability of macroalgae to store reserves gives them an advantage over phytoplankton, in that growth can take place during early winter and photosynthetic tissues made ready to make early use of the increased light of late winter. The seasonal pattern of growth of dominant macroalgae affects the occurrence and production of other macroalgal species over the course of a year through shading.

14.33 Vascular Plants

Most communities dominated by vascular plants—salt marshes, mangroves, eelgrass beds—have other associated producers, including epiphytes and benthic microalgae. In salt marshes of New England, growth of the predominant plant, *Spartina alterniflora*, begins in April, and the growth rate in spring is more closely related to temperature rather than light intensity (Figs. 14-17a,b). Similar correlations have been described for eelgrass beds (Harrison and Mann, 1975).

The nutrient supply determines the amount of growth achieved during the year, since more biomass accrues where *S. alterniflora* is experimentally fertilized (Fig. 14-17a). The nitrogen content of above-ground tissues is built up in early spring from stores in the perennial below-ground plant parts,* and decreases through the summer (Fig. 8-5, top). Potential demand for nitrogen by the plants exceeds actual uptake, since even though there is an abundant supply of ammonium nitrogen in the interstitial water of marsh sediment throughout the year (Fig. 14-17c), uptake of nitrogen is inhibited under anoxic conditions (Howes et al., 1981, Mendelssohn et al., 1981). The balance of root and bacterial activity determines the equilibrium between oxidized and reduced forms of sulfur as well as the redox of the marsh sediment (Howes et al., 1981). As temperatures increase, the activity of sulfate-reducing bacteria (cf. Section 11.3) produces sulfide and reduces sediments. The activity of live roots is also temperature dependent, and their release of O_2 or organic oxidants oxidizes the sulfide in the sediment. High temperatures in midsummer increase respiration and since light intensity falls during late summer (Fig. 14-17b), photosynthesis may not be able to compensate for respiratory losses of CO_2. Growth may therefore stop, and tissues may senesce and release dissolved organic carbon. Below the sediments this DOC stimulates activity of fermenters and sulfate-reducing bacteria and the latter release sulfide (cf. Section 11.3). The reduction in below-ground tissues also lowers the oxidative activity of plant roots, so the sediment becomes more anaerobic. This may further reduce uptake of the abundant ammonium and perhaps end growth of the plants for the season. The more nitrogen present in the tissues, the longer senescence of the grass is delayed, as seen by the contrast in fertilized and control grasses (Fig. 14-17a). Grasses richer in nitrogen last longer in the autumn and are eventually killed by freezing temperatures.

Grazers in salt marshes generally consume small amounts of vascular

* Below-ground plant parts store materials over winter and can exceed the weight and production of above-ground parts in marsh plants throughout the year (Valiela et al., 1976; Good et al., 1982; Kistritz et al., 1983).

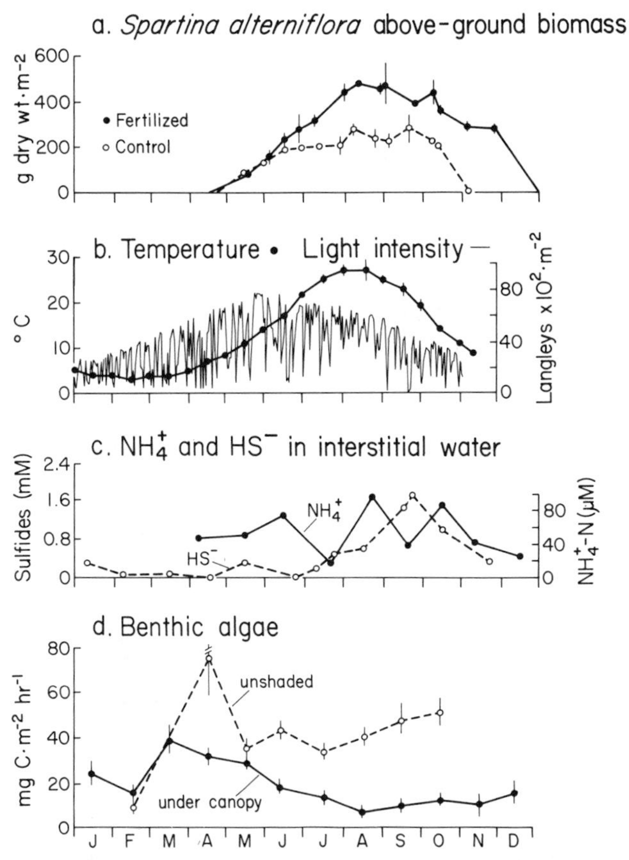

Figure 14-17. Seasonal events in a New England salt marsh. (a) Above-ground biomass of *Spartina alterniflora* in fertilized and control plots. (b) Water temperatures and light intensity throughout the year. Changes in temperature in sediments lag a few weeks behind water temperatures. (c) Concentration of nitrogen in tissues of *S. alterniflora*. From Vince and Teal (1981). © Ecological Society of America, reprinted by permission.

plants, except in certain places where dense flocks of overwintering geese accumulate (cf. Table 8-1). In general, herbivores are therefore not very important in the seasonal cycle nor in structuring salt marsh food webs; the latter are primarily detrital.

14.34 Benthic Single-Celled Algae

In shallow water there are layers of single-celled algae that grow on the sediment surface. An example of the seasonal cycle of such benthic algae is provided by salt marsh algae. Benthic algae on the surface of salt marsh sediments bloom in spring when light increases (Fig. 14-17d), but as the grasses form a canopy, the shading reduces production of benthic algae through the entire summer and fall.* This can be demonstrated by the increased production of benthic algae that follows experimental removal of the grass canopy (Fig. 14-17d). Activity of grazers on salt marsh algae increases during the warmer months (May September), so their impact is delayed until after the early spring peak of benthic algae (March–April). The reduction in algal production after the April peak is likely due to grazers, since during warmer months dense mats of benthic microalgae form where grazers are experimentally excluded. The accumulation of high densities of algal cells on a two-dimensional surface may make the benthic algae more susceptible to grazer control than the widely dispersed phytoplankton.

14.35 Control of Seasonal Cycles

In most cases that we have examined, control of seasonal cycles in the water column is by a sequence of factors. Light usually initiates the cycle as it increases in spring. The actual start may be mediated by the depth of the mixed layer relative to the photic zone. The rate and peak of primary production are then determined by the supply of available nutrients. In most cases, grazers are not abundant enough to play much of a role until the rate of primary production slows down due to depletion of nutrients within the mixed layers; then grazing rates are more capable of reducing phytoplanktonic populations.

The densities of grazers often are lower in the autumn and winter. Perhaps predation or other mortality factors are responsible, but this is not known. Many pelagic zooplankton move to deeper waters during winter, so the lower abundance may not be due to mortality. The reason for overwintering behavior is not clear. The lack of effect of pelagic fish and other predators on grazers agrees with the conclusion of Chapter 9, where we found that such predators seem unlikely to control prey populations.

Where nutrients are continually renewed, such as in shallow areas, the cycle seems determined primarily by light intensity. Seasonal inputs of nutrients as occur in estuaries and upwellings can prompt sudden and short-lived bursts of production that can be traced all the way up the planktonic food web.

Where zooplankton can maintain a dense enough population over the less productive part of the year and where phytoplankton growth is not very high—subtropical gyres, subarctic Pacific, perhaps the Arctic—the grazers may be able to crop algal blooms fast enough to restrict them.

* In eelgrass there are also important shading relationships between higher plants and epiphytes, but in this case it is the epiphytes that reduce eelgrass production as they grow over the surface of eelgrass leaves (McRoy and McMillan, 1977).

Where light is plentiful through the year, as in tropical and subtropical seas, blooms may be started by small amounts of nutrients made available by mixing. The amplitude of oscillations of the phytoplankton are damped, and zooplankton grazing rates are comparable to primary production rates. The whole community is closely adjusted, and any increase in resources is quickly consumed.

The seasonal cycles of production in attached macrophytes is driven principally by changes in light intensity and nutrient supply. In vascular plants, where nutrient uptake is by roots, the redox of the sediment also mediates growth rates.

Consideration of the seasonal cycles discussed emphasizes that communities in different marine environments are structured each in its own peculiar fashion. There are generalities, however, in that a few major types of biotic relationships are invariably present, and that the conditions constraining the biological relationships are provided by physically driven forces.

Chapter 15

The Structure of Marine Communities Over Time: Colonization and Succession

15.1 Introduction

In the last chapter we were concerned with quantitative diurnal and seasonal changes of major components of marine communities. There is a third kind of change over time, largely in composition of species, that takes place when a new resource—such as newly upwelled water or recently cleared rock surface—is made available. This is the more or less orderly sequence of replacement of some species by others that occurs during colonization or succession. The concept of succession was developed in terrestrial ecosystems, and its characteristics have been summarized by Odum (1969). Many of the properties usually associated with succession are tautological (Peters, 1976). We will examine the most useful, less tautological properties applicable to marine systems.

There are two apparently different types of observations in marine systems that have been discussed in terms of succession. First, there are recognizable sequences of taxa such as recorded in marine plankton at different stages in the production cycle (Margalef, 1967; Revelante and Gilmartin, 1976; Johnston, 1963a); the sequence of species is repeated more or less each time light triggers a seasonal production cycle or each time that nutrients are renewed in the photic zone during the summer. Such sequences may take place over months or days, and are occurring during the seasonal patterns described in Chapter 14. Second, there are recurring sequences of appearance of organisms invading and colonizing new or disturbed surfaces. This latter type of sequence of species is closer to what terrestrial ecologists call succession.*

---

* In terrestrial ecology primary succession is the sequence of species that colonize a new substrate; secondary succession takes place on previously inhabited surfaces that have been cleared of biota by some disturbance.

Although at first glance these two phenomena might appear to be rather different, the processes involved in each are similar. After a resource becomes available, species with certain properties replace others with a different set of properties. It has been difficult to ascertain the mechanisms that bring about the replacement of some species by others, since many physical, chemical, and biological variables are potentially involved (Lund, 1966), and no one study has addressed all the possibilities.

Succession deals principally with the qualitative makeup of the assemblage of organisms and is based largely on complex biological interactions among species. To understand the dynamics of succession and colonization it is necessary, therefore, to focus on biological properties of the taxa making up the sequences of species.

We now first turn to examples of succession in several marine environments to get some feeling for how the process takes place; then we will examine generalities about succession that emerge from the case histories.

15.2 Case Histories of Marine Succession

15.21 Succession in Plankton

15.211 Stages of Succession in Plankton

Many specific examples of succession in phytoplankton have been described (Johnston, 1963a,b; Ignatiades, 1969; Smayda, 1966; Sakshaug and Myklestad, 1973; Smayda, 1980). Although the species composition in different parts of the sea varies, some general properties of the species at different stages in a succession sequence have been proposed by Margalef (1967a) based on his experience in Spanish coastal waters, the Mediterranean, and the Caribbean (Table 15-1).

The proposed initial stage of succession occurs in newly upwelled nutrient-rich water. During early succession, there is high photosynthetic activity per cell, and the photoplankton grow very actively with large amounts of photosynthetic pigments per cell and high contents of chlorophyll (absorption at 665 nm) compared to carotenes (absorption at 430 nm). The cells divide rapidly, and cell concentrations are high. The principal taxa are diatoms (*Chaetocerus, Thalassiosira, Skeletonema*, and small flagellates.)

Nutrients in the water diminish during the intermediate stage of succession, due to uptake by the phytoplankton. The number of algal species increases while the abundance of most species is reduced by one or two orders of magnitude, so that there is a significant increase in the diversity of planktonic algae relative to the earlier stage (Table 15-1 and Fig. 15-1).

Table 15-1. Generalized Properties of Phytoplankton Species in Three Stages of Succession[a]

	Nutrient-rich water	Intermediate stage	Nutrient-poor water
Amount of photosynthetic pigments cell^{-1}	1–15%	0.2–0.6	Small
D_{430}/D_{665}[b]	2.5–3.5	3–4	4
Division rate	1 div day^{-1}	Intermediate	1 div wk^{-1}
Density (cells l^{-1})	10^6–10^7	2×10^4–10^5	Less than 10^4
Species richness	Low	Low to high	High to low
Size of phytoplankton cells	Smaller	Medium	Larger
Surface/volume($\mu m^2/\mu m^3$)	1	0.2–0.5	Lower
Mobility	Low	Intermediate	High
Resistance to antimetabolites[c]	Low	Intermediate	High
Ease of culture	Grow easily	Less easy to maintain	Hard to maintain

[a] The stages are based primarily on observations in the fjord-like Ria de Vigo, Spain, supplemented with information from the Caribbean and Mediterranean. Derived from the scheme of Margalef (1962, 1967), modified by Guillard and Kilham (1977).
[b] This ratio shows the spectrophotometric absorption (D) by carotenes (at about 430 nm) and chlorophyll *a* (at about 665 nm.)
[c] From Johnston (1963a,b). Results of work on inhibition of growth by synthetic metabolites similar to those found naturally released by phytoplankton.

Larger diatoms become dominant, characteristically species of *Chaetocerus* that form chains and have long, stiff bristles.

By the late stage of planktonic succession, the supply of dissolved nutrients is very low; nutrients have been taken up by organisms or lost from the photic zone by sinking of particles. Further additions of nutrients from deeper layers are prevented as long as stratification of the water column is maintained. Flagellated forms become relatively abundant, perhaps benefiting from their ability to move to patches of water with higher nutrient content. Many species of diatoms form resting spores and settle out of the surface layers. Cell densities are generally low in the third stage, about 20–100 cells liter^{-1}. Species of *Rhizosolenia, Hemiaulus*, and *Mastogloia* are characteristic of the third stage, although they may have been present during the previous stage. These species are very tolerant of low nutrient conditions and persist in the depauperate stratified water column. The mobile, large dinoflagellates prominent in late succession tend to have slow growth rates. Many of the dinoflagellate species are almost devoid of pigments and may exist heterotrophically. The slow growing algae release proportionally high amounts of DOC (cf. Chapter 10), so the relative amounts of DOC in water may be higher and heterotrophs may be more important during the last phase of succession.

Figure 15-1. Diversity (d = number of species − 1/ln number of phytoplankton) in stations in surface waters of the Ria de Vigo, Spain, 1955. The samples were ordered as to sequence in succession based on taxonomic composition. Adapted from Margalef (1958).

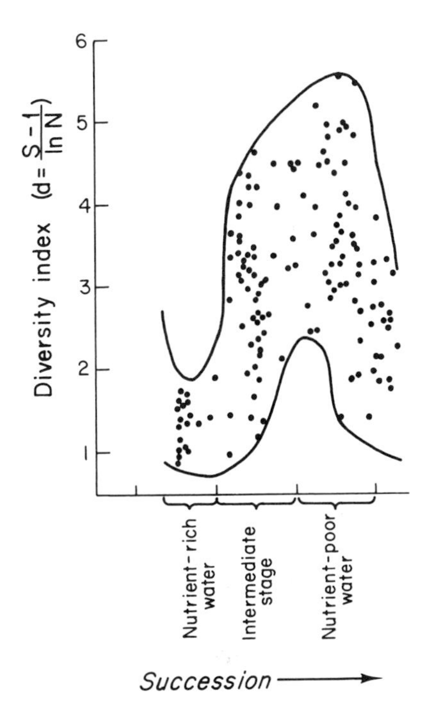

Many dinoflagellates and coccolithophorids found in the latter stages of succession are very ornate. The complex extensions of the cell surface may have several benefits (Smayda, 1970): (1) they may enlarge surface area—some protuberances have chloroplasts, and may also take up nutrients; (2) they may be involved in orientation or steering, as in the diatom *Rhizosolenia setigera*, where spines orient the largest diameter horizontally, so that sinking is slowed down; (3) they may provide rotation to the cells—the spines of *Chaetocerus* prompt a slow rotational movement during sinking, as do the slanted transverse girdles and flagella of dinoflagellates. This rotation may break up microgradients of nutrients around the cells so that uptake sites are exposed to new water during sinking. These properties seem appropriate for life in the stratified, nutrient-poor environment typical of the last stage of succession.

In addition to complex morphological adaptations, phytoplankton of late-succession assemblages tend to show complex interspecific relationships. Some of the relatively few diatom species that remain in the third stage contain *Richelia intracellularis*, a cyanobacterium suspected of fixing nitrogen. Modest (up to 2.5×10^4 liter^{-1}) blooms of diatoms occur when the blue-green is most active (Venrick, 1974). Other diatoms of depauperate waters are often found attached to tintinnid protozoans (Cupp, 1943). This association may help the diatoms by preventing sinking, providing mobility, and maintaining the diatoms in waters rich in nutrients excreted by the protozoan. Under oceanic conditions reminiscent of the summer stratification in water near shore, *Oscillatoria*, a blue-green, is a common member of the flora, and may fix a substantial amount of new nitrogen which is subsequently released into the water and used by other phytoplankton (Carpenter and Price, 1976).

15.212 Mechanisms of Replacement in Phytoplankton Succession

The scheme of succession depicted in Table 15-1 suggests a number of mechanisms that might be responsible for the replacements of some phytoplankton species by others. These mechanisms include changes in cell size—with consequent effects on nutrient uptake, sinking, and grazing—and chemical interaction among species.

15.2121 Cell Size

One of the key variables included in Margalef's scheme of phytoplankton succession (Table 15-1) is cell size, which presumably increases through succession. Such changes in cell size may be functionally important since larger surface-to-volume ratio may favor higher nutrient uptake rates, change sinking rates, and affect susceptibility to consumers.

The phytoplankton referred to in Table 15-1 are principally microphytoplankton (20–200 μm). In these size classes—primarily diatoms—it may be that larger cells are more often found in late succession. Recent work shows, however, that nanoplankton (2–20 μm in diameter) carry out most of phytoplankton production (Malone, 1971a,b; Sheldon et al., 1972, 1973; Malone et al., 1979; among others) and become more prominent late in succession. When nutrients are renewed (early succession) the nanoplankton account for 5–20% of the chlorophyll (Hallagraeff, 1981). Under stratified summer conditions in the water column (late succession), 50–99% of the chlorophyll is due to nanoplankton (Hallagraeff, 1981).

There are geographical parallels to succession. Nanoplankton (coccolithophorids, naked dinoflagellates, monads) are also usually dominant in oceanic waters (cf. Chapter 1) while larger forms (diatoms and dinoflagellates) dominate the phytoplankton in shelf and coastal waters (Malone, 1980). Similar trends in size are found in oligotrophic and eutrophic freshwaters (Malone, 1980). The importance of very small (0.5 μm in diameter) blue-greens (*Synechoccus sp.*) needs to be determined, but they seem to be more important in depauperate waters. In rich coastal waters *Synechoccus* could be responsible for 7–8% of the primary production, while in the nutrient-poor northern Sargasso Sea they may be involved in 15–40% of the total primary production (Waterbury et al., 1980).

Thus, recent work on the role of the smaller cells does not agree with the size trends of Margalef's scheme. It seems therefore necessary to review the factors that may change the relative size of phytoplankton during succession. The size distribution of phytoplankton present in the photic zone at any one time may depend on the relation of cell size and uptake of nutrients, relative rates of sinking, and grazing.

Consider nutrients first. The argument behind the generalization about cell size in Table 15-1 is as follows. Small cells are in theory best adapted to use and grow at high concentrations of nutrients because of the larger ratio of surface to volume compared to larger cells. Smaller cells have relatively more uptake sites per unit biomass, so their rates of nutrient uptake, and hence division and production rate, should be faster (Fig. 15-2). Recall, however, that smaller phytoplankton tended to have lower nutrient uptake saturation coefficients than larger phytoplankton (Fig. 2-14). In addition, Banse (1982) reviews the data on size dependence of growth rate and finds only slight effects of size. There is, therefore, no consensus as to the relation of cell size, nutrient use, and growth. Moreover, small size is a drawback, as we have seen for animals in Chapter 10, since respiratory rates are higher on a per unit weight basis for small than for larger phytoplankton (Fig. 15-3). It is argued that this results in higher growth efficiency for larger species, and therefore, when resources are scarce, large species have a competitive advantage.

Laws (1975) modeled size distribution of phytoplankton in various parts of the ocean using a more complex version of the above argument (including light, depth of the mixed layer, and other features) and his simulation was in fair agreement to field data. There is, however, also disagreement about these arguments. First, growth efficiency is not convincingly size dependent (Banse, 1976). Second, Parsons and Takahashi (1973) calculated, on the basis of data such as found in Figure 2-14, that larger cells would be found when high concentrations of nutrients were available, in contrast to the trend argued above. Hecky and Kilham (1974) pointed out that there is such broad variation in K_s (cf. Table 2-3) within one species that generalizations as to size dependence of nutrient uptake, growth, and respiration by algae may be premature. Parsons and Takahashi (1974) countered with evidence from Eppley et al. (1969), who

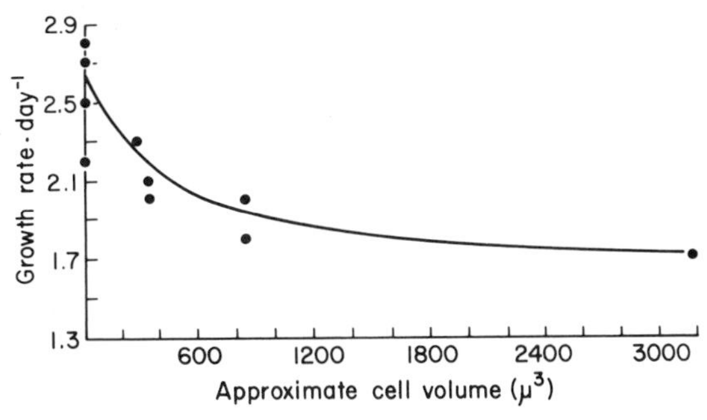

Figure 15-2. Relationships of growth and production rates to cell size in microalgae. Data from Ben-Amotz and Gilboa (1980).

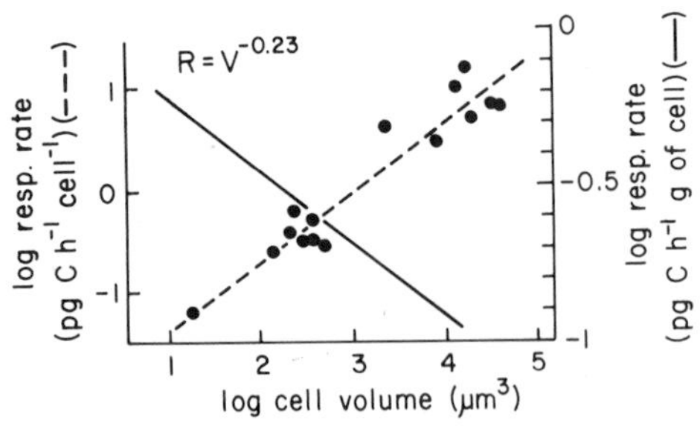

Figure 15-3. Cell and weight-specific respiration rates for several species of phytoplankton of varying size. Adapted from Laws (1975). © Ecological Society of America, reprinted by permission.

showed that small (5 μm) oceanic species of diatoms and coccolithophorids tended to have low K_s values for nitrate, less than 0.8 μmole liter^{-1}, while coastal diatoms and flagellates had higher values, typically 1–10 μg atoms liter^{-1} (cf. Fig. 2-14). It seems, though, that even if there is a pattern of size distribution of phytoplankton in nutrient-rich and nutrient-poor environments, it is not possible to make a clear *a priori* explanation based on differential uptake of nutrients and by metabolic relations due to cell size. There is a need for fresh and incisive experimentation on this topic.

Sinking of cells in the water column differentially affects cells of different sizes, since larger particles sink faster (Smayda, 1970; Vinogradova, 1977; and Chapter 10). If there were no turbulent mixing or other mechanisms to maintain cells of the larger species in the photic zone, over time, a phytoplankton community would progressively lose the larger species. In nature, physical mechanisms (upwelling, Langmuir circulation) reduce sinking rates (Semina, 1972), and physiological mechanisms provide buoyancy (Smayda, 1970). Buoyancy is improved when cells are healthy and actively growing, perhaps because the lipids in cells are high (Eppley et al., 1967).

Sinking is a specially pressing problem for phytoplankton in stratified water columns because of the stable hydrography and low nutrients. We have already mentioned some morphological adaptations of larger cells, adaptations that slow sinking of the larger species. Dinoflagellates must be favored in such circumstances because their motility allows control of their vertical position (Hasle, 1950, 1954). In fact, dinoflagellates are often most abundant where stratification is more marked (Blasco, 1971).

Grazing may also change the size distribution of phytoplankton (Frost, 1980; Malone, 1980) due to size selectivity of suspension feeders. The larger size and mobility of herbivores in upwelling systems (Table 2-4) are well suited to search for patches of high density of phytoplankton of adequate size. Sardines in the Peruvian Upwelling shift size of prey eaten depending of what is available (DeMendiola, 1971) but prefer larger diatoms to small flagellates. Grazing by sardines could therefore release flagellates from competition and lead to a bloom of these in spite of nutrient conditions favoring diatoms.

Invertebrate grazers can markedly change the size (and species) composition of phytoplankton by size-selective feeding (Fig. 15-4). In tank experiments, very high densities of grazers reduced abundance of diatoms, and the phytoplankton assemblage became dominated by small flagellates. Densities approaching 900 copepods m^{-3} may be common in rich estuaries (Olsson and Ölundh, 1974), and may occur in coastal waters (Carpenter et al., 1974, Anraku, 1964) but are seldom seen in oceanic waters. As we saw in Chapters 5, 8, and 14, the impact of grazing depends on density and feeding rate of the grazers relative to growth rate of algae. Thus grazing is likely to be most important as an agent of succession in stratified conditions where the phytoplankton is not growing very rapidly.

The relative abundance of small-celled forms in oceanic environments may mean that small grazers such as ciliates are the principal grazers that might influence the abundance of some species over others. The generation times of ciliates are 1–2 days (Beers and Stewart, 1971; Verity and Stoecker, in press), which is in the range of the doubling times of phytoplankton. Ciliates thus could show an effective numerical response, since lags between increases in their nanoplankton food and ciliate numbers would be short (Malone, 1971). Copepods can also feed on the smaller

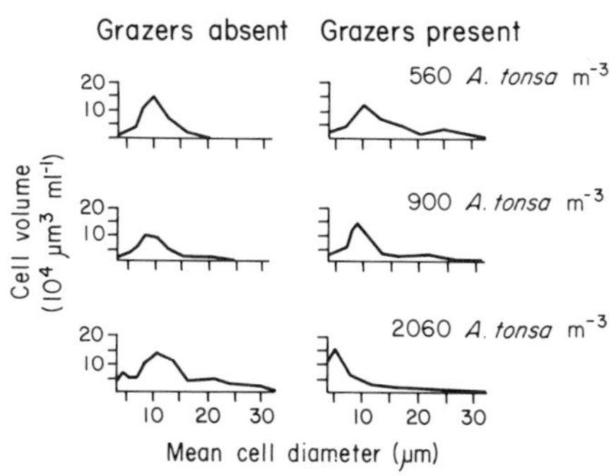

Figure 15-4. Effect of grazing by the copepod *Acartia tonsa* on natural phytoplankton from Vineyard Sound, Massachusetts. The experiments consisted of introducing 100 *Acartia tonsa* into 1,000-liter cylinders set outdoors and continuously supplied with flow-through seawater. Twenty percent of the water was renewed daily. The particle size distributions were recorded at 16 days, when the copepod density had reached 500 m^{-3}, and at 23 and 38 days, when there were about 900 and 2,060 m^{-3}, respectively. The dominant diatom in Vineyard Sound is *Skeletonema costatum* (average diameter 10 μm), and microflagellates are about 5 μm in diameter. Adapted from Ryther and Sanders (1980).

nanoplankton when microphytoplankton are scarce. Thus fluctuations in the abundance of nanoplankton may be more closely damped by grazers than those of microplankton. Since in oceanic (or late successional stages) nanoplankton are more important than microplankton, such assemblages of species may be subject to closer coupling between producer and consumer than richer coastal (early succession) assemblages. This is what we found when considering seasonal cycles of coastal and depauperate oceanic environments in Chapter 14.

The generalizations of Table 15-1 in regard to the role of cell size in succession are thus not completely supported by recent evidence. Nutrient uptake, sinking, and grazing—especially the latter two—may influence size (and species composition) of the phytoplankton, but we do not yet understand the relative importance of the various factors. We also are unable to explain the shifts in cell size or species composition during succession. These changes may not be very closely determined by any one variable; it may be that many factors are at work and that there is a great element of chance affecting the species composition of phytoplankton assemblages.

15.2122 Chemical Interactions Among Species

Replacements of species typical of the first stage, such as the diatom *Skeletonema costatum*, may be mediated by exocrine substances released by species that come later in the sequence (Pratt, 1966). In Narragansett Bay, Rhode Island, the red tide flagellate *Olisthodiscus luteus* is abundant following blooms of *S. costatum*; when the flagellate density decreases, *S. costatum* increases again. Water previously inhabited by high densities of *O. luteus* inhibits growth of the diatom, evidence that substances released by *O. luteus* may prevent its replacement by the diatom. Similarly, high densities of *O. luteus* reduce growth and increase mortality of the two tintinnid grazers (Verity and Stoecker, in press).

Other examples reviewed in Johnston (1963a) also suggest that replacement of one species by another may be mediated by antimetabolites. Furthermore, it seems that species of early succession are relatively susceptible to antimetabolites. Species characteristics of late succession may be those that have evolved tolerance to such chemical interactions. Resistance to antimetabolites (Table 15-1) may be especially developed in large dinoflagellates of stratified, nutrient-poor conditions, many of which are themselves toxic. Study of the cell membranes of such species may reveal that dinoflagellates are better able than diatoms to regulate transport of such molecules.

It is not clear just what is the target organism for the very powerful dinoflagellate toxins; the toxins do not always inhibit invertebrate grazers such as shellfish, although they may deter vertebrate grazers (White, 1981). These compounds may be anticompetitive (allelopathic) devices. Allelopathic relations are characteristic of situations where resources such as water (on land) or surfaces (in marine hard-bottom, coral reefs, etc.) are in short supply.

While resistance to metabolites may increase, the nutritional dependence of species of late succession on each other apparently increases. Early succession species are easily cultured; given basic macronutrients, they can easily manufacture their required compounds. On the other hand, it is notoriously hard to culture many species characteristic of late succession. These seem to require rather complex media, suggesting that they depend on other organisms to provide needed compounds or precursors. This emphasizes that in the late stages of the sequence there are many complex biological relationships among the species.

The specialization associated with the complex biotic interactions may result in greater susceptibility to novel chemical stresses in phytoplankton typical of late succession. Phytoplankton from nutrient-poor water (comparable to late succession) are more susceptible to man-made pollutants (Fisher, 1977). Phytoplankton adapted for environments that require nutrient uptake at very low concentrations may make a cell more susceptible to exotic pollutants. Further research is needed to confirm or disprove these observations, but they do suggest that a phytoplankton assemblage growing under nutrient-poor conditions, a situation characteristic of late succession, is rather susceptible to toxic substances.

15.213 Geographical and Successional Parallels

We have already hinted at parallels between the stages of succession (the early–late gradient) and geographical gradients such as the temperate–tropical, coastal–oceanic gradients of phytoplankton assemblages. Semina (1972) concluded that in the Pacific Ocean the largest microplankton cells are found in the quietest water (weak ascending or descending motion). Cell size tended to be small in areas of intense upwelling. This parallels Margalef's ideas to changes of microplankton cell size in early and late succession (Table 15-1). In other studies the species most frequently occurring in spring blooms are often also found in eutrophic coastal and temperate waters. Species of late stages of succession are more often found in oligotrophic, oceanic, or tropical situations. The reason of course is that hydrographic conditions in coastal areas and during mixing events provide the hydrodynamic and nutrient conditions leading to growth of phytoplankton. Similarly, midsummer stratification is similar to oceanic conditions and hence phytoplankton in these situations may be similar.

The above parallels once again show the workings of a few structuring variables—nutrients, light, grazing—mediated by hydrography on a wide variety of phenomena and aspects of marine ecosystems. Focus on these processes is likely to provide the insight to understand eventually in a general way how plankton communities are structured. At present, we only have some notions as to how these processes work.

15.22 Succession in the Rocky Intertidal Shore

There are many studies on rocky shores that describe the development of the community of sessile organisms attached to the rock surface. We will focus on a study designed specifically to describe succession and to evaluate just how species replacement takes place. This study describes the colonization and establishment of algae and animals on the top surface of intertidal cobblestones in certain parts of the southern coast of California (Table 15-2).

Species typical of early succession on these cobblestones are generally cosmopolitan, produce many flagellated motile spores for much of the year, settle densely, and grow quickly. *Ulva* spp. and *Enteromorpha*, both green algae, are representative of pioneer species, and occupy much of the space made available by disturbances (Table 15-2). Pioneers do not persist for at least two reasons. First, they suffer greater mortality from physical disturbances such as desiccation than do later species (Fig. 15-5). Second, pioneer species such as *Ulva* are preferred foods of grazers (Fig. 8-3 and Sousa, 1979a).* Over time both these mechanisms reduce the amount of space occupied by the pioneer species.

In the absence of disturbances or grazing, *Ulva* spp. do retard colonization by other species, since they are good interference competitors for space. When *Ulva* spp. are removed, whether experimentally (Fig. 15-6, top) or naturally, the species of the later successional stages increase in abundance. These taxa of later stages are often much more seasonal in their release of young stages, produce fewer young, settle more sparsely, and grow relatively slowly. *Ulva* spp. can outgrow them, and only when *Ulva* spp. are removed by grazers, or killed by a stress, does succession go on.

Once other species are established, they in turn inhibit the settlement and growth of additional species. For example, removal of *Gigartina leptorhynchos* and *Gelidium coulteri* (Fig. 15-6, bottom) leads initially to an increase of other species on the newly opened space. *Ulva* spp. may first become abundant, but as time goes on they are replaced by late-succession species. In all these examples, each species can inhibit the appearance of species characteristic of later stages in the succession.

Where late succession species were removed, *Ulva* spp. temporarily took over because they grew so much faster than any other plant on the cobblestones. The slow-growing late-succession species can tolerate the presence of *Ulva* spp., and eventually, as they grow, they replace the early colonists, perhaps through shading or by surviving grazing pressure.

---

* This seems typical of species of early succession. In terrestrial situations, palatability of plants of early succession is also greater than that of species typical of the later stages (Cates and Orians, 1975). Pioneer plants appear not to invest much effort in chemical or morphological defenses against herbivores.

Table 15-2. Species Composition of Early, Middle, and Late Successional Communities on Boulders in an Intertidal Boulder Field in California[a]

Species	Successional stage		
	Early	Middle	Late
Ulva spp. (green alga)	73.1 ± 15.6	10.8 ± 8.4	1
Chthamalus fissus (barnacle)	18.1 ± 14.7	4.3 ± 7.3	1
Gigartina leptorhynchos	1	15.8 ± 10.4	3.5 ± 5.6
Gelidium coulteri	1	4.9 ± 5.7	2.4 ± 4.8
Gigartina canaliculata	1	47.1 ± 19.0	91.8 ± 6.2
Centroceras clavulatum	1	3.7 ± 6.2	1
Anthopleura elegantissima (sea anemone)		1	1
Corallina vancouveriensis		1	1
Gastroclonium coulteri		1	1
Laurencia pacifica		1	
Porphyra perforata		1	
Rhodoglossum affine		1	
Bare area	5.4 ± 4.8	9.7 ± 6.9	1.9 ± 2.8
Number of species	6	12	10
Diversity ($e^{H'}$)[a]	1.6 ± 0.5	4.1 ± 1.3	1.5 ± 0.3

[a] Species are red algae except where indicated. Values are mean ± std. dev. of percent cover by each species on 30 boulders. Adapted from Sousa (1979, 1980).
[b] Includes only species with >1% cover. H' is the Shannon–Weaver diversity index.

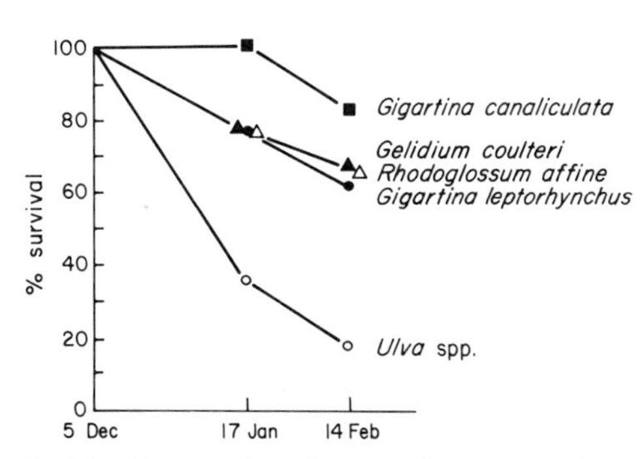

Figure 15-5. Survival for five species of macroalgae over a 2-month period during a period of harsh conditions (low tide in afternoon with the algae exposed to dessication). Thirty plants of each species were tagged on 5 December 1976. Adapted from Sousa (1979). © Ecological Society of America, reprinted by permission.

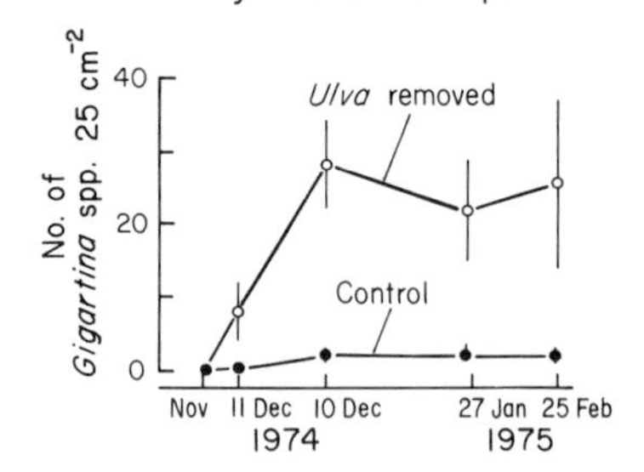

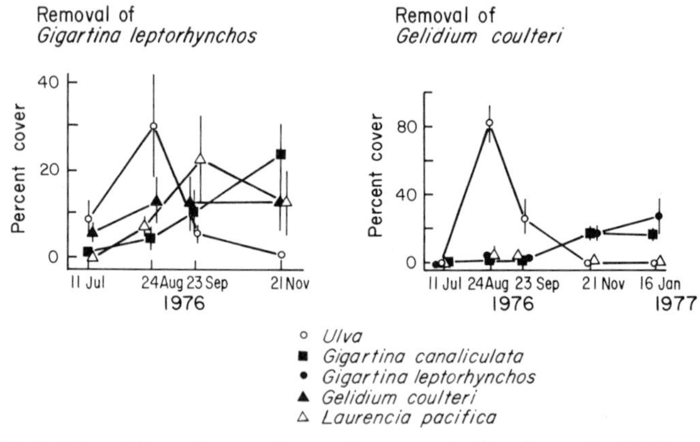

Figure 15-6. The effect of experimental removal of early and middle-succession species. Removal was repeated at each sampling interval. The controls for the experiments with *Gigartina leptorhynchos* and *Gelidium coulteri* showed no significant changes and are not shown here. Adapted from Sousa (1979). © Ecological Society of America, reprinted by permission.

In addition to desiccation and grazing, there is another important stress on the plant community of cobblestones: during episodes of strong wave action, the cobblestones may be overturned. The frequency of disturbance (cf. Section 12-3) also affects whether early- or late-succession species are present on the boulders. The boulders that are most frequently turned over only hold a few opportunistic species. Disturbance therefore halts succession. If rocks are not disturbed, middle- and late-succession species colonize the surface, and in time, *Gigartina canaliculata*, the competitive dominant, monopolizes the surface of the boulder (Sousa, 1979a).

Disturbances are important in many habitats. Strong wave action, for example, can provide open spaces in intertidal rocky shores dominated by mussels (Paine and Levin, 1981). The disturbed patches devoid of mussels vary in size from the space occupied by one mussel to 38 m^2. The rate at which these patches appear is enough to turn over the mussel population in several years, so although at any one time the open patches occupy less than 5% of the rock surface, patches are quantitatively important over several years. Small patches disappear rapidly due to movement of mussels, but large patches (on the order of square meters in area) require settlement by new individuals. In such patches succession takes place. A major portion of space is occupied by a rapid growing red alga, *Porphyra pseudolanceolata*. Herbivore limpets prevent this species from completely covering all the space during the first year or so. A herbivore-resistant red alga, *Corallina vancouveriensis*, grows during the second year. Other species of various and unknown relationships continue the succession process. Thus the patches produced by disturbance in rocky shores are an essential feature of the community.

15.23 Colonization of Hard Substrates by Fouling Communities

Ship bottoms, outfall pipes, offshore oil derricks, and any object that people place in the sea are soon colonized by organisms. This fouling of hard surfaces has had great economic interest and has been the subject of numerous studies. The species that make up these communities are those that would settle on any hard surface subtidally; fouling communities therefore cover a sizable proportion of the coastal sea bottom.

Microbial invasion is the initial process (Mitchell, 1978) that corresponds to terrestrial primary succession. Within 24 hr a new substrate may be invaded by short, rod-shaped bacteria; within 2–4 days a mixed population of bacteria and detritus particles is enmeshed in a polymeric matrix of insoluble, high-molecular-weight polysaccharides (Dempsey, 1981). By 4–7 days there may be a thick film of bacteria mixed with pennate diatoms and within 2 weeks protozoa are grazing on the microflora. The polysaccharides and glycoprotein secreted by microbes may be chemically involved in prompting settlement of larger plants and animals (Kirchman et al., 1982), but there may also be just a sticky effect of the film, since where bacteria have formed a film the ''settlement'' of inanimate latex particles is enhanced (Di Salvo and Daniels, 1978).

The development of the invertebrate community is rather unpredictable, principally because the availability of larvae, and hence larval recruitment varies markedly within and between years (Fig. 15-7). The initial composition of the assemblage is therefore variable. In the coast of North Carolina new patches of bare surface are provided by feeding of the sea urchin (*Arbacia puntulata*). Larvae of the bryozoan *Schizoporella unicornis* colonize these bare patches, and once established, the adult bryozoans can to some extent prevent the settlement of other species. The larvae of the tunicate *Styela plicata*, however, are good competitors for space, and slowly settle on the bryozoan patches, replacing them after 2

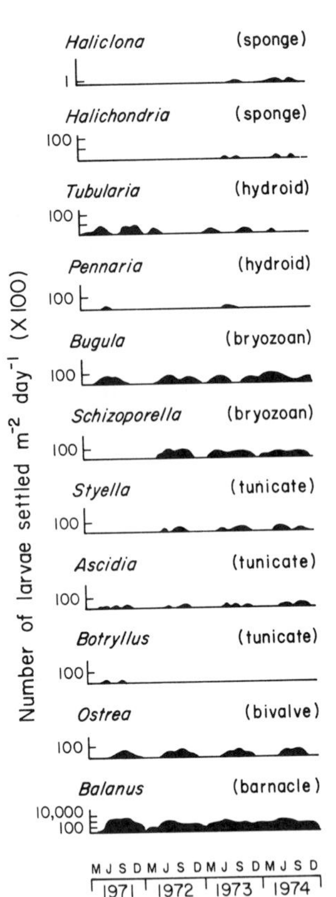

Figure 15-7. Larval recruitment over time for the 11 most abundant taxa in the fouling community in North Carolina. Adapted from Sutherland and Karlson (1977). © Ecological Society of America, reprinted by permission.

years or so. The adults of the tunicate are also good interference competitors and prevent other species from colonizing (Sutherland, 1978), at least for the life span of the specimens present on a surface.

In experiments where adult bryozoans were removed from plates (Fig. 15-8, middle) the colonization of the plate surface by *Styela* was hastened; when the tunicate was removed, the results were very much as in the untouched control plates: tunicate larvae settled on and overgrew the bryozoans (Fig. 15-8, bottom).

These and other examples from fouling communities demonstrate that inhibition of one species by another in a colonizing or successional sequence is a common phenomenon.

In another study of a fouling community on the coast of Delaware, mussels (*Mytilus edulis*) settled more abundantly on panels previously occupied by tunicates or hydroids than on bare panels (Dean and Hurd, 1980). Thus, in contrast to the many examples of inhibition reviewed above, the presence of early-succession species facilitated settlement by a later species. Once established, *Mytilus edulis* became dominant and resisted invasion by any further competitors for space, as it does in the rocky intertidal shore.

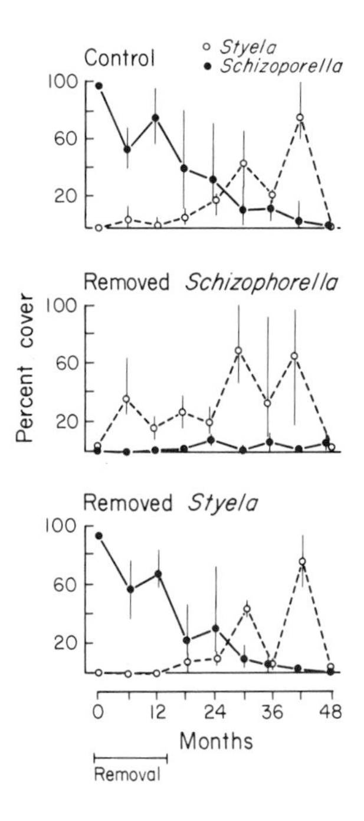

Figure 15-8. Mean ± range of abundance of an encrusting bryozoan (*Schizoporella unicornis*) and a solitary tunicate (*Styela plicata*) on experimental ceramic plates (232 cm^2) from December 1972 to December 1976. Removal of *Styela* or *Schizoporella* took place during the initial 15 months of the experiment. Adapted from Sutherland (1978). © Ecological Society of America, reprinted by permission.

In rich temperate coastal waters, where there are many larvae available to occupy surfaces, bare surfaces are nearly filled after a few months (Schoener and Schoener, 1981). Colonial species expand rapidly by asexual reproduction and cover most of the space. Solitary species settle also, but the space they occupy depends on the rate of growth of individuals. After about a year, however, the solitary species manage to cover most of the space.

The pattern of settlement in nutrient-poor tropical coastal waters is apparently different (Jackson, 1977). Colonization of bare surfaces is slow, perhaps due to low numbers of larvae available to settle, and slow growth due to low food supply. Even after 6 months half the surfaces may still be bare. Solitary species predominate early in the colonization and are replaced by colonial species through lateral overgrowth.

15.24 Species Replacement in Coral Reefs

The marvelous complexity of coral reefs (Fig. 1-9), with myriads of species vying for space has long attracted attention. These very diverse communities are continually changing, with multiple small-scale adjustments where one species with a small competitive advantage gains space or position over another while perhaps losing space to a third (Porter, 1974; Jackson and Buss, 1975; Connell, 1978).

The changes in allocation of space over time are largely under biological control. Filamentous algae grow on bare surfaces such as dead corals.

Corals and other invertebrates, once they are well established, can actively prevent fast-growing filamentous algae from growing and taking over surfaces. The filamentous algae, however, are a preferred food of herbivorous fish, and grazing also controls the growth of the algae and other early colonizers (Birkeland, 1977); sometimes territorial behavior of some fish species prevents herbivorous fish species from consuming algae within territories (Vine, 1974). The reef fish and urchins do not eat the young corals, and may in fact, enhance recruitment of corals by reducing the biomass of competitors for space (Birkeland, 1977; Sammarco, 1980). Once established, the corals (the late-succession species in this environment) dominate the community because they are good interference competitors as mature colonies or may even feed on newly settled larvae of other species. There is thus a rich pattern of interrelations that determines how far succession goes.

Any agent that removes or kills a patch of corals—storms, predators—can restart the cycle of replacements, and depending on what larvae are available at the time, a very different assemblage of species may colonize the patch. Thus succession is not simply a predetermined, linear sequence of replacements as portrayed in the classical idea of succession. The combination of species present in an area reflects the history of different patches within the area. After a patch is disturbed, the available larvae and environmental conditions in that patch may prompt the development of any of several assemblages that may retain their species composition over considerable time, as shown by Sutherland (1974) in fouling communities. A new disturbance either allows the previous species to return or causes further replacement with new species. The term "disturbance" is sufficiently vague to include any catastrophe, change in nutrient supply, appearance of new consumers or superior competitors, etc. This idea of a community as a patchwork of small parcels of slightly different history of disturbance is not just applicable to marine systems but is also emerging as the current view of the development of communities in terrestrial situations (Loucks, 1970; Denslow, 1980).

We have seen earlier that plankton assemblages have peak species diversities at intermediate to late stages in a successional sequence (Table 15-1, Fig. 15-1). The drop in the last stage is probably due to some competitive exclusion or lack of resources. This is also probably true of coral reefs (cf. Section 12-3).

In diverse communities such as coral reefs the vast assemblage of species has a long geological history that allows survival of these species as long as the vagaries of the environment fall within their past evolutionary experience. Disturbances that occur in reefs are slight enough to allow the development of the finely adjusted biological detail characteristic of coral reefs. These communities, as we also found in the case of plankton, are especially susceptible to many of the novel perturbations brought about by human activities—contamination by petroleum, heavy metals, agricultural chemicals, and disturbance of sediments such as dredging—since these disturbances are mostly new to the species making up coral reefs. This contrasts with other marine environments such as salt marshes, where the climatic and geochemical settings naturally provide some exposure to heavy metals, eutrophic water, and severe changes in salinity and temperature. Salt marsh organisms, as a consequence, are somewhat more tolerant of many kinds of contamination and disturbance than coral reef organisms.

15.25 Colonization of Soft Sediments

15.251 Characteristics and Effects of Early and Late Colonists

Physical disturbance of shallow marine bottom sediments is a common event. Tidal surges, storms, pollution, and predation can defaunate patches of soft bottoms. Dumping of spoils from dredging operations is another source of disturbance. The latter is a pressing problem in coastal zone management and has led to interest in the study of consequences of disturbance for bottom faunas.

The appearance of bare sediment almost invariably prompts massive invasion by a few opportunist species, especially polychaetes (Grassle and Grassle, 1974; Rhoads et al., 1978; Sanders et al., 1980; Thistle, 1981). The recolonization of a benthic habitat after abatement of a pollution source illustrates the process (Fig. 15-9). After the initial increase in density and subsequent crash or replacement of the few early species, there is an increase in diversity, when many of the less opportunistic species become relatively more prominent. Subsequent to this peak diversity, there may be some competitive exclusion, with resulting lower diversities, as we have seen in other environments. Over subsequent periods

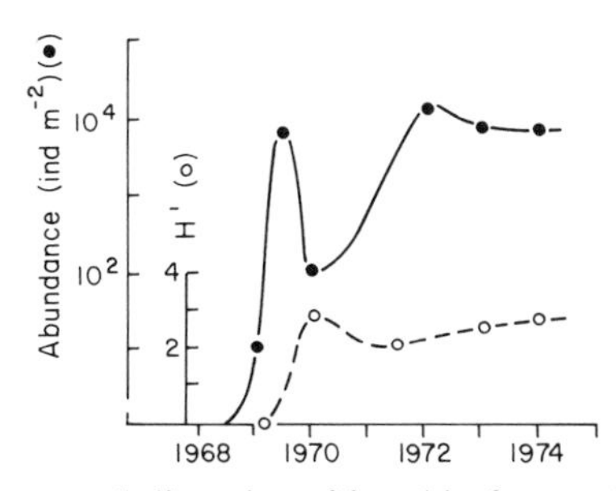

Figure 15-9. Abundance and diversity of benthic fauna after abatement of pulp mill pollution in Saltkallefjord, Sweden, allowed the recovery of natural communities. The peak in abundance is due to the early colonizers. Adapted from Pearson and Rosenberg (1978). © Pergamon Press Ltd., reprinted by permission.

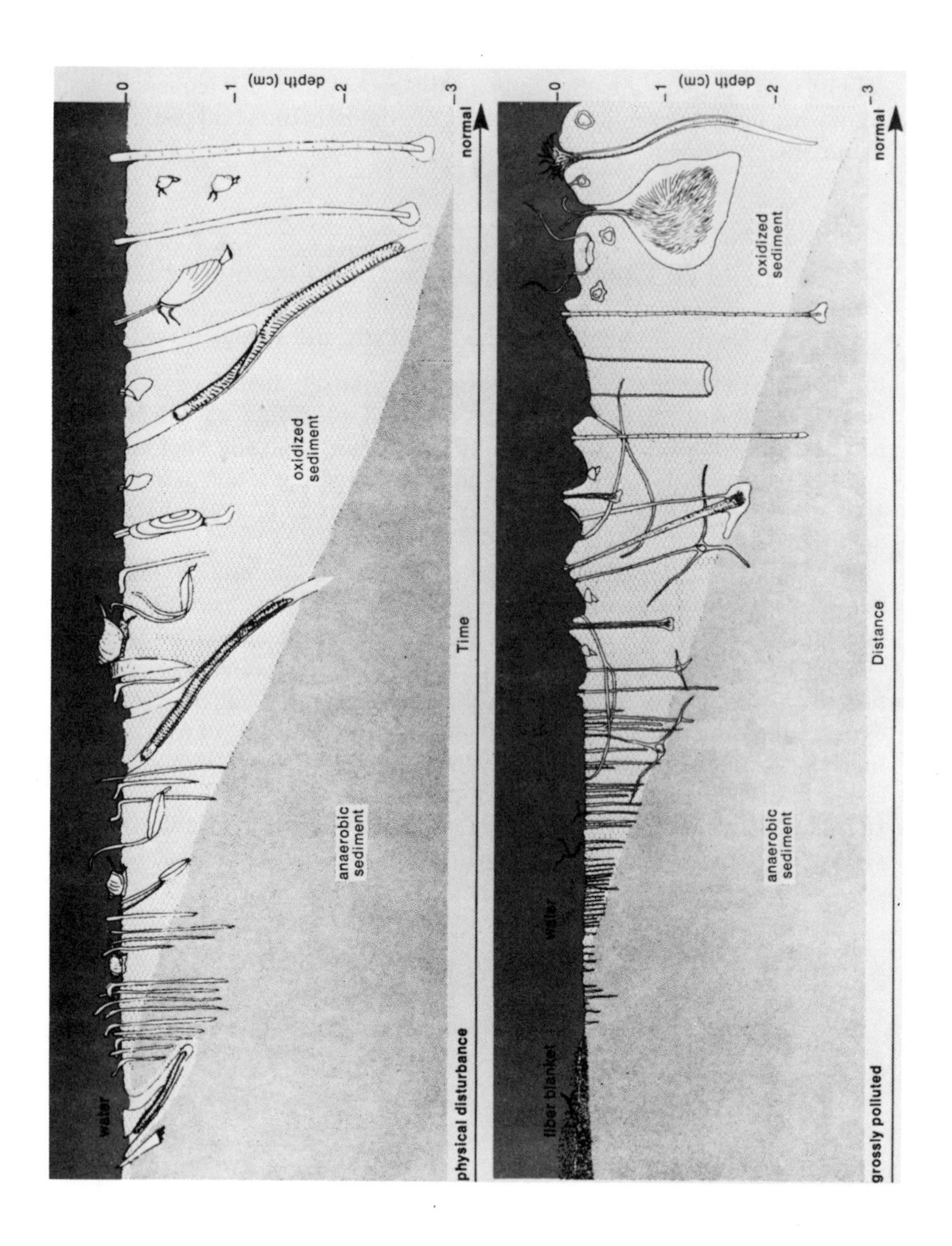
water
0
−1
−2
−3
depth (cm)
oxidized sediment
anaerobic sediment
physical disturbance
Time
normal
fiber blanket
water
0
−1
−2
−3
depth (cm)
oxidized sediment
anaerobic sediment
grossly polluted
Distance
normal

diversity may change further depending on season, vagaries of weather, and availability of larvae (Sanders et al., 1980).

Rhoads et al. (1978) studied the colonization of soft sediments by placing boxes of azoic sediment on the sea bottom and subsequently monitoring faunal changes. They also followed colonization of areas where dredge spoils were dumped. In both experiments bare muds were invaded by larvae.

The first colonizers were mostly small-sized (Fig. 15-10) species that had fast growth rates, reached very high densities, and achieved high production rates (Table 15-3). They also suffered high mortalities, so these populations turned over quickly. The high densities achieved by the opportunist species lead to depletion of resources and diminish the carrying capacity of their environment (Grassle and Grassle, 1974). The polychaete *Capitella capitata*,* is a typical opportunist. The abundance of this worm quickly increased in a box of mud placed on the bottom of salt marsh creeks and crashed after a month or so after reaching densities of over 400,000/m^2. While the polychaetes were crashing in some boxes, there was growth to about 250,000/m^2 in new boxes, so the decline in density must have been due to density-dependent depletion of food or to the accumulation of toxic substances. The former seems more likely, since at high densities there was a reduction of the size of brooding females and of number of eggs per female (cf. Chapters 4 and 7).

The early colonizer species tend to feed on particles suspended above the bottom or on recently sedimented particles. They live very near the surface (Fig. 15-10) and may form dense aggregations of tubes. Since the depth of the tubes is slight, pumping of seawater only affects the geochemical properties of a very shallow layer of the sediment. Pioneer species are easily accessible to predatory fish and crustaceans since they

* *C. capitata* has subsequently been found to consist of a group of five sibling species (Grassle and Grassle, 1976).

Figure 15-10. Idealized scheme of sequence of species invading bare sediment such as newly dumped dredge spoil (top). There is a close parallel to the gradient over space away from a grossly polluted site (chronic release of fiber from a cellulose factory) (bottom). Immediately after distubance or close to the source of pollution a few species of abundant, small and productive polychaetes are found. These are followed, either over time or space, by suspension-feeding or surface deposit-feeding molluscs. The latter are replaced by large, slow-grazing species that live deeper in the mud, feed on buried deposits, and oxidize the sediment by their activities. The close equivalence of space and time is another example of the same principle we saw in the similarity of the early–late gradient of planktonic succession and coastal–oceanic gradients. These comparisons point out the commonality of the processes involved in organizing communities. From Rhoads et al. (1978) and Pearson and Rosenberg (1976).

Table 15-3. Life History Data for Representative Species Colonizing Experimental Sites Where Either Trays of Defaunated Mud Were Provided or Where Dumping of Dredging Spoils Provided Opportunities for Colonization.[a]

	Days to peak abundance	Maximum abundance (m^{-2})	Size (mg ind^{-1})	Generations yr^{-1}	Estimated production rate (g m^{-2} day^{-1})	Mortality rate
Early colonizer						
Streblospio benedicti (polychaete)	10	420,000	0.15–50	3–4	0.57	High
Capitella capitata (polychaete)	29–50	80,000	0.15–50	5–8	0.27	High
Ampelisca abdita (amphipod)	29–50	10,000	0.5–1	2	0.06	High
Owenia fusiformis (polychaete)	—	—	0.5–1	—	—	High
Mulinia lateralis (clam)	—	—	2–10	1–2	—	High
Intermediate colonizers						
Nucula annulata (clam)	50	3,700	5–10	1–2?	0.12	Moderate
Tellina agilis (clam)	80	1,400	about 5	1?	0.04	Moderate
Pitar morrhuana (clam)	—	—	about 10	1	—	Moderate
Late colonizer						
Nepthys incisa (polychaete)	86	220	30–70	1	0.03	Low
Ensis directus (clam)	175	30	100–300	1	0.01	Low
Nassarius trivittatus (snail)	50–223	—	3–10	2	0.01	Low

[a] Experiments carried out in Long Island Sound. Adapted from Rhoads et al. (1978).

live so near the surface, and this is in part why their mortality rates are high (Table 15-3).

The later colonizers are less abundant, larger, slower growing, and less productive (Table 15-3) than the earlier species. The late colonizers also suffer lower mortalities since they tend to live buried in sediment (Fig. 15-10) and are harder for predators to find. Late colonizers are typically deposit feeders whose burrowing and pumping of seawater through long, often U-shaped tubes (Fig. 15-10) change the chemical and physical condition of the sediments to a considerable depth (20 cm) (Aller, 1980). Pumping of seawater into tubes aerates the sediments and transports chemicals and particles. The movement, ingestion, and defecation of sediment by deposit feeders increase the fluffiness of sediments and up to a point foster growth of microbes on the particles.* The animals can then feed on the microbial biomass produced by this microbial "gardening" (Yingst and Rhoads, 1980; Hylleberg, 1975; Woodin, 1977).

The surfaces of soft sediment are often covered by fecal pellets, produced by both deposit and suspension feeders. New fecal pellets are less laden with microbes or microbial films than uneaten sediment particles and are readily colonized by microbes. Defecation by benthic animals may therefore maintain relatively high microbial activity.

Water moving over the sea floor can transport surface sediments. The roughness, cohesiveness, and specific gravity of the sediment surface affect transport and are largely determined by the organisms of the benthos. There are numerous mechanisms by which animal activity can affect the integrity of the surface layer of sediment (Rhoads and Boyer, in press; Rhoads et al., 1978a; Jumars et al., 1981; Taghon et al., 1978; Grant et al., 1982).

It is difficult to predict whether the presence of pioneer or late-colonizing species will stabilize or destabilize surface sediments. Rhoads and Boyer (in press) predict that although at low densities of pioneer species sediment surfaces may be destabilized, high population densities build up very quickly and increase the stability of surface sediments by stimulating microbial films and building tubes. On the other hand, late-succession species may tend to destabilize sediment because their activity aggregates sediments, and their feeding may deplete microbial populations. These activities make the surface sediments fluffier or more fluid and easier to disrupt. These speculations need further scrutiny, but if they are true, the late-colonizing species are very important in that their activity sets the stage for disturbance and transport of the unconsolidated surface of the sediment in which they live.

The animal–sediment interaction and its effect on sediment transport and stabilization have significant applied interest. Sediment transport is obviously important in navigation, maintenance of harbors, and erosion

* The result of all these activities is often referred to as "bioturbation."

of shorelines. In addition, if pollutants are present, transport of particles along the sea floor is often accompanied by the movement of pollutant substances, since many are strongly adsorbed to particles. Particles bearing adsorbed pollutants would tend to be trapped in sediments populated by pioneer species, while further transport is more likely in sites at a later stage of succession.

The change in the oxidation state of sediment brought about by colonization and establishment of benthic fauna (Fig. 15-10) also changes nutrient dynamics (Aller, 1980). For nitrogen, the increased oxidation fosters a change from an ammonium-dominated to a nitrate-dominated phase.* Decay rates increase, and fungi may degrade some of the more refractory organic compounds. Nutrients in interstitial water are incorporated into microbial cells and are otherwise immobilized. In the case of phosphate, the more oxidized the sediment, the less soluble (and removable) the form of phosphorus (Section 11.13). For nitrogen, nitrate is not easily adsorbed, and is therefore easily transported, and the more aerobic the sediment the more that nitrogen is available for transport. Ammonium, however, can also diffuse or be pumped out of sediments (Section 11.23). It is therefore not clear which stage of succession leads to greater or lesser release of dissolved inorganic nitrogen or phosphorus. Quantitative measurements of nutrient regeneration in sediments inhabited by pioneer and late-succession assemblages will be useful.

15.252 Mechanisms of Species Replacement in Soft Sediments

The rate of species replacement in soft sediments depends on interspecific mechanisms. The species present may retard succession, as when newly settled larvae are eaten by established species (Feller et al., 1979; Wilson, 1980). On the other hand, surface-dwelling suspension feeders typical of early succession may be buried when sediments are bioturbated by late-coming deposit feeders that burrow, ingest, and defecate sediment (Rhoads and Young, 1971; Brenchley, 1981).

There is also evidence that in soft marine sediments the presence of many species facilitates rather than inhibits the settlement of other species. Experimental manipulations in a mud flat in Puget Sound show that the presence of tube-building polychaetes increases the recruitment of other polychaetes, bivalves, and oligochaetes (Gallagher et al., in press). The mechanism of facilitation may be related to the presence of tubes;

* This contrasts with some results in terrestrial environments, where it is believed that the succession moves from a nitrate-based to an ammonium-based nitrogen economy, since as succession proceeds, nitrogen may be supplied more and more by internal decay and recycling (mineralization of organic nitrogen) (Rice and Pancholy, 1972; Bormann and Likens, 1979). Such trends are not always found in terrestrial ecosystems (Vitousek et al., 1982).

artificial tubes cause aggregation of polychaetes and tanaids. Tubes change the properties of flow over the sediment surface in some favorable way, although the specific benefits provided by the tubes remain to be identified.

The presence of some species is neither inhibited nor facilitated by other species. The abundance of the polychaete *Pygospio elegans*, for example, in a mudflat in Puget Sound is unrelated to the abundance of other species (Gallager et al., in press). This species tolerates other species, and its rates of growth and reproduction are set by other factors.

15.26 Sequence of Species in Salt Marshes

New England salt marshes are mosaics of short and tall forms of *Spartina alterniflora*, bare patches, *Salicornia europaea, Spartina patens*, and *Distichlis spicata* (Fig. 15-11). A series of long-term experiments have provided a sketch of how the balance among the vegetation types is struck (Valiela et al., in press). In particular, the colonization of patches in salt marshes by plants illustrates mechanisms of replacement and the nonlinear relationships of replacement within disturbed patches.

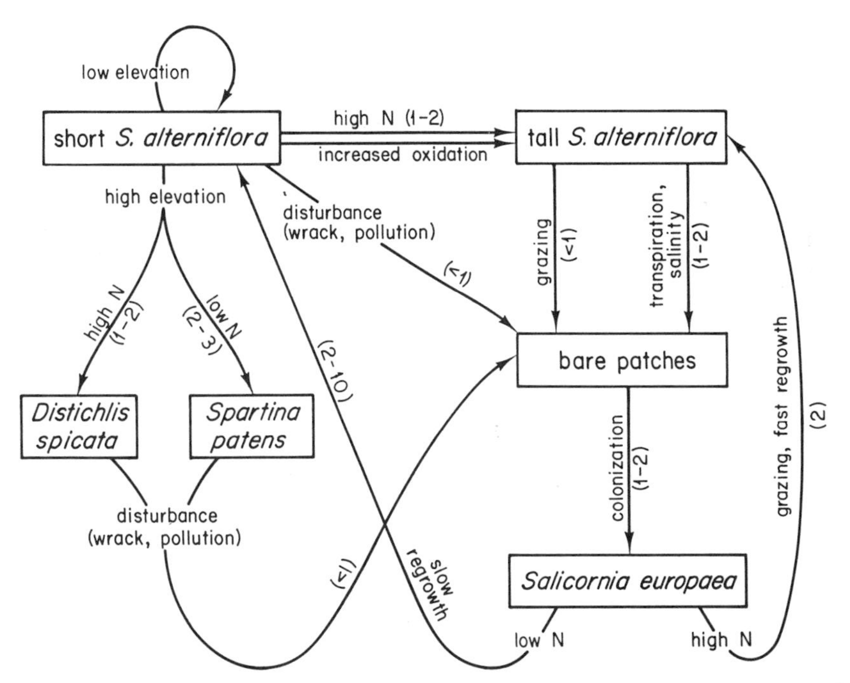

Figure 15-11. Elements (in boxes) and transformations (arrows) of the patches of salt marsh vegetation. The numbers in parentheses are duration of the transition in years. From Valiela et al. (in press).

Added supply of nitrogen (eutrophication), or increased oxidation of sediments due to increased drainage both foster conversion of patches of short to tall *S. alterniflora* within 1–2 years. The increased biomass of *S. alterniflora* could then increase salinity of the sediments, since more biomass leads to increased transpiration of fresh water. In relatively more eutrophic patches the tissues of *S. alterniflora* become enriched in nitrogen and are more readily attacked by grazers (cf. Section 6.322). Both increased salinity (Haines and Dunn, 1976) and grazer pressure (R. Buchsbaum and C. Cogswell, unpublished data) can thus lead to death of patches of *S. alterniflora* within one growing season or so.

The resulting bare patches in salt marsh sediments are colonized by an opportunistic plant, the glasswort *Salicornia europaea* within 1–2 years. This species is a poor competitor and does not do well except in bare patches. If it turns out that the sediment of the patch newly colonized by *S. europaea* is rich in nitrogen, the plants are eventually heavily attacked by herbivores and there is little regrowth of the glasswort. The high nutrients also foster the regrowth of tall *S. alterniflora*, either by new seedlings or by lateral vegetative growth. If the patch happens to have a lower nitrogen supply, the effect of grazers is less obvious, and slower regrowth of short *S. alterniflora* takes place.

If the patch in question is located at a low elevation within the intertidal range, the short form of *S. alterniflora* remains. If the patch is located at a higher elevation, above mean tide level, the short *S. alterniflora* is replaced by other species of plants. The replacement mechanism may be differential nutrient uptake. When nitrogen supply is high, *Distichlis spicata* takes over, while *Spartina patens* does so when nitrogen supply is lower.

Disturbances also create bare patches. Pollution events, such as an oil slick entering the marsh, or more commonly, ice-rafted sediments or layers of stranded sea wrack (Hartman et al., in press) will kill patches of vegetation. Such disturbances might occur both in low and high tidal elevations and kill patches of vegetation in a short time. High sites tend to collect more wrack, so are more disturbed that low sites. These bare patches are then colonized by a variety of marsh plants. Disturbance disrupts dominance by the competitively superior *Spartina* grasses, and the number of plant species (18) is greater in higher reaches of the marsh than in lower sites (8 species of plants). This is much like the situation described for coral reefs and forests in Chapter 12.

The succession of species on a particular patch of salt marsh is therefore not necessarily a predetermined, linear sequence [see den Hartog (1977) for a similar description for seagrass beds]. The sequence of events is affected by a variety of factors. "Regressive" steps are possible, where some change prompts reversion to other assemblages. Such changes can be quite local, and add to the patchiness and between-habitat diversity within a salt marsh, and probably to all marine environments.

15.3 Interrelation Among Communities or Ecosystems at Different Stages of Succession

In nature no parcel exists by itself. There are linkages among adjoining environments. Margalef (1968) has pointed out that if the adjoining parcels are in a very different state of succession, the exchanges between the two will be asymmetric. Consider an early-succession assemblage composed of fast-growing organisms and few defenses against consumers. These pioneer organisms initially have a surfeit of resources, which they exploit inefficiently. They also have demographies suited to sustain high mortalities. In contrast, late-succession assemblages are in theory slow growing, have many defenses against consumers, are unable to sustain high mortalities, and are efficient in use of resources. In late succession, resources are in short supply, and whatever is made available is consumed by some species.

If two assemblages, one in an early, the other in a late stage of succession, exist next to each other, it does not seem far-fetched for the species of the late succession to venture into the adjoining early succession parcel, since there they would have the opportunity to exploit resources that are far more freely available and consumable. There are no data for such a phenomenon for adjoining parcels in early and late stages of succession within a given community, but we can consider adjoining pairs of communities in ecosystems, one resembling our description of early succession, and another more similar to late succession situations.

We have already discussed examples of such pairs of communities. These include the sedimentation of particles from the plankton which supports the benthos. There are other examples. Benthic fish of shallow waters often feed on zooplankton in the water column. The blacksmith (*Chromis punctipennis*), an abundant pomacentrid fish, lives in rocky reefs off California, feeds on zooplankton in the water column during the day, and shelters at night in rocky crevices on the bottom. The feeding pattern of just this one species results in the importation of 8 g C m^{-2} $year^{1}$ of feces (with a C/N ratio of 8) to the nocturnal shelter on the bottom (Bray et al., 1981). Such feeding behavior is also common in other species, and in fish and zooplankton of coral reefs (Porter and Porter, 1977), where, as we have seen, food availability is typically very low.

Another example is provided by salt marshes and adjoining coastal waters. Salt marshes are very productive systems with abundant dissolved nutrients, energy-rich reduced compounds, particulate matter, and very high densities of potential prey organisms. There are asymmetric flows of nutrients and particulates to less productive and less eutrophic coastal waters (cf. Section 11.24). The annual exports of dissolved nitrogen from marshes to coastal water, for example, can in some places amount to over 40% of the amount of dissolved nitrogen in the near-shore water column (Valiela, 1984). Further, fish of the deeper coastal waters exploit salt marshes and related tidal creeks in at least two ways: (a) coastal fishes use salt marshes as nurseries for their young (Werme, 1981), since the early life stages require relatively larger food rations and so are often found in salt marshes where food is more available; and (b) larger fishes enter tidal marsh creeks during high tides to feed on the dominant prey that live within the marsh.

There are thus asymmetric flows of resources from the loosely organized plankton to the more tightly organized benthos, and from eutrophic productive marshes to the less rich coastal water column.

15.4 Generalized Properties of Succession in Marine Environments

Succession is included in the last chapter because, as seen in the discussion above, this process is the composite result of the many processes described in earlier chapters. The various examples of marine succession we have just seen provide fairly consistent patterns. Certain species have evolved colonizing characteristics and they appear first in a successional sequence. These species usually have a relatively small size, fast growth and production rates, and tolerate high mortality rates. Other properties mentioned in earlier chapters include ability to produce large numbers of young, good dispersal powers, and long-term survival of eggs, larvae, or propagules in dormant states. These properties are typical of what we have earlier called *r* or exploitatively selected species.

The diversity of the assemblage of species in an early stage of succession is low. Most early-succession species do not greatly modify the environment in which they live, except perhaps in regard to depletion of their resources as a result of their great reproductive ability.*

There are several possible general types of interactions among species that determine whether or not replacement takes place (Connell and Slatyer, 1977), and at what rates it is accomplished if it does occur.

> Depletion. The use of a resource by a species may prevent its continued presence; other species better suited to exploit resources in very short supply may then become dominant. This loose sort of interaction is prevalent in planktonic and benthic environments. Early colonizers may have lower assimilation efficiencies, or take up nutrients best at high concentrations. They therefore may be outcompeted by

---

* The terms opportunistic, fugitive, weedy, or pioneer are often applied to these species. The reproductive strategies involved were discussed in Chapter 4, where we also pointed out the *r*-*K* continuum was not a complete statement. Species adapted to live in very stressful conditions should also be included.

late-succession species. It would be of interest to compare assimilation and growth efficiency in early- and late-succession species and relate these physiological variables to demographic properties.

Tolerance. Some species that appear late in succession do so simply because their life history takes a long time to complete. Some late species may tolerate whatever the early colonizers do and, after inconspicuous development of the immature stages, may become dominant and outcompete pioneer species. We found examples of this in soft sediments, although it is probably unusual in nature. This is perhaps an odd entry here, since it essentially refers to a lack of interaction among species, at least during the early stages.

Facilitation. Some species of an early stage modify the environment so as to facilitate the appearance of species characteristic of late succession. Examples of this are bacterial films on hard substrates or certain activities of benthic animals that dramatically change the chemistry of sediments in mudflats so as to facilitate settlement of other species.

Inhibition. Earlier colonists may inhibit the appearance of later species. This delays succession, and may produce alternate assemblages that last for the life span of the species present, as we have seen in fouling communities. Many examples of this process are available in the environments surveyed above, where larvae of other or the same species are consumed, damaged, or expelled by the gauntlet of feeding appendages brandished by the animals already present on a substrate. The inhibitory interaction is frequently found in nature, especially in benthic situations, but it is only a temporary delay of succession. Open space for further colonization will still be made available by disturbances and by mortality of the early colonizers. The latter is especially applicable because early colonizers tend to be more palatable and have shorter lives, and so turn over faster than later species. This can lead to a slow change toward an assemblage of species characteristic of late succession.

Removal. Species of late succession may simply consume, overgrow, or disturb earlier species, and replace them in water, sediments, or surfaces. We have examples of this in coral reefs, rocky shores, and soft sediments.

Allelopathy. Early colonists may be replaced by a late-colonizer that can release ectocrine substances. Such substances could inhibit growth of a competing, earlier species. We have seen evidence of this in the plankton, in microbial communities, and in coral reefs.

Several of these mechanisms probably co-occur in most communities. It would be of interest to know if the preponderance of one or another of the mechanisms of succession depends on the type of community or affects the way the community is structured.

Early colonizers tend to be rapid growing, productive, *r* (or exploitatively) selected species. Early colonizers tend to be more palatable as food for consumers—relatively free of chemical deterrents, spines, and other protections. It is no accident that the major crops of agriculture are grasses, typical early succession species of terrestrial ecosystems. In the sea also the more productive fisheries are also those characteristic of early succession stages. The classic examples are clupeid fisheries in upwelling areas, where the harvest is only one trophic step from the primary producers and few species of consumers share in the harvest.

As succession proceeds, the species—in most cases—tend to be larger, grow more slowly, be less productive, and have more complex morphology and special requirements, as we saw in the plankton and benthos. Such species use resources in short supply, and to point this out we have called them *K* or saturation selected species.

As time goes on, more species accrue, and diversity increases. Spatial heterogeneity may also increase during succession (as discussed in Chapter 13), since more patches accumulate through time due to disturbances. The mosaic of heterogeneous patches containing a variety of species replaces the more homogeneous assemblages of early succession. In the latter stages of succession, diversity may decrease if disturbances are infrequent enough so that competitive dominants can exclude some species. The diversity and patchiness of the environment reflect a balance between the advantages of specialization and the degree of disturbance in the environment.

As succession proceeds, the complexity of biological interactions increases, featuring chemical inhibition, symbiotic, and behavioral mechanisms in addition to predation and competition. The activity of late-succession species often markedly changes their environment, with notable alteration of the chemistry and physics of the environment. In the water column many substances are either taken up or released by organisms, and as pointed out in Chapter 11, much of the nutrient supply in late succession may be regenerated by zooplankton. In sediments the bioturbation by benthic animals markedly changes the biogeochemistry and physics of the substrate. The marked effect of animals on soft sediments demonstrate how effectively late-succession assemblages may modify the properties of their environment (Aller, 1980). The domination of nutrient exchanges and sediment transport by animal activity are important but not well-enough known consequences of succession.

Any exploitation of a community by consumers from late-succession patches or by man tends to halt or reverse the trends of succession (cf. Fig. 15-10, bottom). The more intense the degree of pollution or exploitation, the more succession is held back. At severe levels of contamination or exploitation, only a few of the more weedy species survive.

The complex and fragile biological interactions that govern communities of advanced succession make such assemblages of species more susceptible to many man-made alterations. For example, if the density of species of late-succession deposit feeders is reduced below the number needed to maintain the oxidized state of soft sediments, the whole benthic community may be altered. The fragility of late-succession assemblages derives from two features (1) late-succession species tend to markedly change their environment and (2) late-succession species are generally more susceptible to novel stresses. The management of environments that contain assemblages of late succession needs to consider these features.

Succession is seldom a predetermined, orderly replacement of one species by another. Rather, the assemblage of species may shift to any of a number of alternate states—as we have seen in the rocky shore, fouling, and salt marsh examples. The shifts from one to another state may be due to competition and predation among the species that happen to be present, and to external factors such as nutrient supply or disturbances. The shifts may be recursive, as happens when a plankton community is exposed once again to nutrient rich water: early-succession species reappear.

A habitat is a mosaic of patches, each of which may be at a different stage of succession—either because of different age of the patch or because the species that were available for recruitment differed. This provides a remarkable degree of heterogeneity in the spatial distribution and species competition of most habitats. The degree to which disturbances create new patches is responsible for the conservation of this heterogeneity. Where disturbances (predation by large mobile species, pollution, wave action) are very infrequent, competitive and predatory interactions within the assemblage of species present take over. In some environments, the species composition is determined mainly by predation—as in the rocky intertidal shore or kelp beds. In others, predation may be less important—as in fouling communities. In still others, predation may be important but the role of competition is not yet clear—as in communities of soft sediments.

There may be interactions among patches that are in different stages in succession. Generally, these interactions are asymmetrical, with species of the later stage exploiting resources in the earlier patch. This is just another way to describe where colonizers come from; the "colonization" can be short term, though, as in the coastal fish that only feed in a salt marsh during high tide and return to deeper waters at low tide.

Succession integrates most of the topics covered in this book and reflects the processes that structure communities over time. Every exploitation of marine organisms or management of marine environments by people is necessarily done in the context of the stage of development of the community in question. Clearly, the mechanisms that govern communities are loose compared to, say, the tight organization due to endocrine regulation within individual organisms. Nevertheless, there are properties of species characteristic of different stages of community organization that can affect what we do to or what we get from any marine ecosystem.

References

Abele, L. G. 1976. Comparative species richness in fluctuating and constant environments: Coral-associated decapod crustaceans. Science 192:461–463.

Abele, L. G. 1982. Biogeography. Pp. 241–304 in The Biology of Crustacea, Vol. 1. Academic.

Abele, L. G., and K. Walters. 1979. Marine benthic diversity: A critique and alternative explanation. J. Biogeogr. 6:115–126.

Adams, B. B. 1978. The Feeding Strategies of Coastal and Open Ocean Pelagic Copepods. Ph.D. Thesis, Boston University.

Adams, J. A., and J. H. Steele. 1966. Shipboard experiments on the feeding of *Calanus finmarchicus* (Gunnerus). Pp. 19–25 in H. Barnes (ed.), Some Contemporary Studies in Marine Science. George Allen and Unwin.

Agassiz, L. 1888. Three cruises of the BLAKE. Bull. Mus. Comp. Zool. 14: 1–314.

Akre, B. G., and D. M. Johnson. 1979. Switching and sigmoid functional response curves by damselfly naiads with alternative prey available. J. Anim. Ecol. 48:703–720.

Alcaraz, M., G. -A. Paffenhöfer, and J. R. Strickler. 1980. Catching the algae: A first account of visual observations of filter feeding calanoid copepods. Amer. Soc. Limnol. Oceanogr. Spec. Symp. 3:241–248.

Aleem, A. A., and N. Dowidar. 1967. Phytoplankton production in relation to nutrients along the Egyptian Mediterranean coast. Pp. 305–327 in Proc. Int. Conf. Trop. Oceanography 1965. Univ. Miami.

Alexander, M. 1977. Introduction to Soil Microbiology. 2nd Ed. Wiley.

Allan, J. D., T. G. Kinsey, and M. C. James. 1976. Abundance and production of copepods in the Rhode River subestuary of Chesapeake Bay. Ches. Sci. 17: 86–92.

Alldredge, A. L., and W. H. Hamner. 1980. Recurring aggregation of zooplankton by a tidal current. Estuar. Coast. Mar. Sci. 10:31–37.

Allee, W. C. 1931. Animal Aggregations: A Study in General Sociology. Univ. of Chicago.

Aller, R. C. 1980. Diagenetic processes near the sediment-water interface of Long Island Sound. I: Decomposition and nutrient element geochemistry (S, N, P). Pp. 237–344 in B. Salzman (ed.), Physics and Chemistry of Estuaries: Studies in Long Island Sound. Advances in Geophysics, Vol. 22. Academic.

Aller, R. C. 1980a. Relationship of tube-dwelling benthos with sediment and overlying water chemistry. Pp. 285–308 in K. R. Tenore and B. C. Coull (eds.), Marine Benthic Dynamics. Univ. South Carolina.

Altman, P. L., and D. S. Dittmer. 1972. Biology Data Book. 2nd Ed. Vol. 1. Federat. Amer. Soc. Exp. Biol.

Ambler, J. W., and B. W. Frost. 1974. The feeding behavior of a predatory planktonic copepod, *Tortanus discaudatus*. Limnol. Oceanogr. 19:446–451.

Andersen, H. T. (ed.). 1969. The Biology of Marine Mammals. Academic.

Anderson, D. M., and F. M. Morel. 1978. Copper sensitivity of *Gonyaulax tamarensis*. Limnol. Oceanogr. 23:283–295.

Anderson, G. C., B. W. Frost, and W. K. Peterson. 1972. On the vertical distribution of zooplankton in relation to chlorophyll concentrations. Pp. 341–345 in A. Y. Takenouti (ed.), Biological Oceanography of the Northern North Pacific Ocean. Idemitsu Shoten.

Anderson, G. C., and R. P. Zeutschel. 1970. Release of dissolved organic matter by marine phytoplankton in coastal and offshore areas of the Northeast Pacific Ocean. Limnol. Oceanogr. 15:402–407.

Anderson, G. R. V., A. H. Ehrlich, P. R. Ehrlich, D. J. Roughgarden, B. C. Russell, and F. H. Talbot. 1981. The community structure of coral reef fishes. Amer. Nat. 117:476–495.

Andreae, M. O., and H. Raemdonck. 1983. Dimethyl sulfide in the surface ocean and the marine atmosphere: A global view. Science 221:744–747.

Andrews, P., and P. J. L. Williams. 1971. Heterotrophic utilization of dissolved organic compounds in the sea. III. J. Mar. Biol. Assoc. U.K. 51:111–125.

Anger, K. 1975. On the influence of sewage pollution on inshore benthic communities in the South of Kiel Bay. Part 2. Quantitative studies on community structure. Helgol. Wiss. Meeresunters. 27:408–438.

Ankar, S. 1977. The soft bottom ecosystem of the Northern Baltic proper with special reference to the macrofauna. Contr. Asko. Lab. 19:1–62.

Anraku, M. 1964. Influence of the Cape Cod Bay Canal on the hydrography and on copepods in Buzzards Bay and Cape Cod Bay, Massachusetts. I. Hydrography and distribution of copepods. Limnol. Oceanogr. 9:46–60.

Antia, N. J., C. D. McAllister, T. R. Parsons, K. Stephens, and J. D. H. Strickland. 1963. Further measurements of primary production using a large-volume plastic sphere. Limnol. Oceanogr. 8:166–183.

Arnold, D. E. 1971. Ingestion, assimilation, survival, and reproduction by *Daphnia pulex* fed seven species of blue-green algae. Limnol. Oceanogr. 16:906–920.

Arnold, K. E., and S. N. Murray. 1980. Relationships between irradiance and photosynthesis for marine benthic green algae (Chlorophyta) of differing morphologies. J. Exp. Mar. Biol. Ecol. 43:183–192.

Ashmole, N.P. 1968. Body size, prey size, and ecological segregation in five sympatric tropical terns (Aves: Laridae). Syst. Zool. 17:292–304.

Atkinson, L. P., and F. A. Richards. 1967. The occurrence and distribution of methane in the marine environment. Deep-Sea Res. 14:673–684.

Azam, F., and R. E. Hodson. 1977. Dissolved ATP in the sea and its utilization by marine bacteria. Nature 267:696–697.

Azam, F., T. Fenchel, J. G. Field, J. S. Gray, L. A. Meyer-Reil, and F. Thingstad. 1983. The ecological role of water-column microbes in the sea. Mar. Ecol. Progr. Ser. 10:257–263.

Aziz, S. A. A., and D. B. Nedwell. 1979. Microbial nitrogen transformations in the salt marsh environment. Pp. 385–398 in R. L. Jefferies and A. J. Davy (eds.), Ecological Processes in Coastal Environments. Blackwell.

Baird, D., and H. Milne. 1981. Energy flow in the Ythan estuary, Aberdeenshire, Scotland. Estuar. Coast. Shelf Sci. 13:455–472.

Baker, M. C. 1974. Foraging behavior of blackbellied plovers (*Pluvialis squatarola*). Ecology 55:162–167.

Bakus, G. J. 1981. Chemical defense mechanisms on the Great Barrier Reef, Australia. Science 211:497–499.

Balderston, W. L., and W. J. Payne. 1976. Inhibition of methanogenesis in salt marsh sediments and whole-cell suspensions of methanogenic bacteria by nitrogen oxides. Appl. Environ. Microbiol. 32:264–269.

Balech, E., S. Z. El-Sayed, G. Hasle, M. Neushul, and J. S. Zaneveld. 1968. Primary productivity and benthic marine algae of the Antarctic and Subantarctic. Antarctic Mag. Folio Series 10:1–12.

Bamstedt, U. 1981. Seasonal energy requirements of macrozooplankton from Kosterfjorden, Western Sweden. Kiel. Meeresforsch. 5:140–152.

Bannister, R. C. A. 1978. Changes in plaice stocks and plaice fisheries in the North Sea. Rapp. Proc.-Verb. Reun. Cons. Int. Explor. Mer 172:86–101.

Banse, K. 1974. The nitrogen-to-phosphorus ratio in the photic zone of the sea and the elemental composition of the plankton. Deep-Sea Res. 21:767–771.

Banse, K. 1976. Rates of growth, respiration and photosynthesis of unicellular algae as related to cell size—a review. J. Phycol. 12:135–140.

Banse, K. 1977. Determining the carbon to chlorophyll ratio of natural phytoplankton. Mar. Biol. 41:199–212.

Banse, K. 1979. On weight dependence of net growth efficiency and specific respiration rates among field populations of invertebrates. Oecologia 38:111–126.

Banse, K. 1980. Microzooplankton interference with ATP estimates of plankton biomass. J. Plankton Res. 2:235–238.

Banse, K. 1982. Cell volumes, maximal growth rates of unicellular algae and ciliates, and the role of ciliates in the marine pelagial. Limnol. Oceanogr. 27:1059–1071.

Banse, K., and S. Mosher. 1980. Adult body mass and annual production biomass relationships of field populations. Ecol. Monogr. 50:355–379.

Banse, K., F. H. Nichols, and D. R. May. 1971. Oxygen consumption by the seabed. III. On the role of the macrofauna at three stations. Vie et Milieu. I (Suppl. 22):31–52.

Barber, R. T. 1966. Interaction of bubbles and bacteria in the formation of organic aggregates in sea water. Nature 211:257–258.

Barber, R. T., R. C. Dugdale, J. J. MacIsaac, and R. L. Smith. 1971. Variations in phytoplankton growth associated with the source and conditioning of upwelling water. Invest. Pesq. 35:171–193.

Barber, R. T., and J. H. Ryther. 1969. Organic chelators: Factors affecting primary production in the Cromwell Current upwelling. J. Exp. Mar. Biol. Ecol. 3:191–199.

Barnes, H., and S. M. Marshall. 1951. On the variability of replicate plankton samples and some applications of "contagious" series to the statistical distribution of catches over restricted periods. J. Mar. Biol. Assoc. U.K. 30:233–263.

Barnes, R. D. 1980. Invertebrate Zoology. 4th Ed. Saunders.

Barsdate, J. J., R. T. Prentki, and T. Fenchel. 1974. Phosphorus cycle of model ecosystems: Significance for decomposer food chains and effect of bacterial grazers. Oikos 25:239–251.

Barsdate, R. J., and V. Alexander. 1975. The nitrogen balance of arctic tundra: Pathways, rates, and environmental implications. J. Env. Qual. 4:111–117.

Bayliss, D. E. 1982. Switching by *Lepsiella vinosa* (Gastropoda) in South Australian mangrove. Oecologia 54:212–226.

Baylor, E. R., and W. H. Sutcliffe. 1963. Dissolved organic matter in seawater as a source of particulate food. Limnol. Oceanogr. 8:369–371.

Bayne, B. L. 1973. Physiological changes in *Mytilus edulis* L. induced by temperature and nutritive stress. J. Mar. Biol. Assoc. U.K. 53:39–58.

Bayne, B. L., and C. M. Worral. 1980. Growth and production of mussels *Mytilus edulis* from two populations. Mar. Ecol. Prog. Ser. 3:328–328.

Baxter, I. G. 1959. Fecundities of winter-spring and summer-autumn herring spawners. J. Cons. Perm. Int. Exp. Mer 25:73–80.

Beddington, J. R. 1975. Mutual interference between parasites or predators and its effect on searching efficiency. J. Anim. Ecol. 44:331–340.

Beddington, J. R., M. P. Hassell, and J. H. Lawton. 1976. The components of arthropod predation. II. The predator rate of increase. J. Anim. Ecol. 45:165–186.

Beers, J. R., F. M. H. Reid, and G. L. Stewart. 1982. Seasonal abundance of the microplankton populations in the North Pacific central gyre. Deep-Sea Res. 29:227–245.

Beers, J. R., D. M. Steven, and J. B. Lewis. 1968. Primary productivity in the Caribbean Sea off Jamaica and the Tropical North Atlantic off Barbados. Bull. Mar. Sci. 18:86–104.

Beers, J. R., and G. L. Stewart. 1971. Microzooplankters in the plankton communities of the upper waters of the eastern tropical Pacific. Deep-Sea Res. 18:861–884.

Begon, M., and M. Mortimer. 1981. Population Ecology: A Unified Study of Animals and Plants. Sinauer.

Belehradek, J. 1935. Temperature and Living Matter. Borntraeger, Berlin (Protoplasma—Monogr. 8).

Bell, R. H. 1970. The use of the herb layer by grazing ungulates in the Serengeti. Pp. 111–124 in A. Watson (ed.), Animal Populations in Relation to Their Food Resources. Blackwell.

Bell, S. S. 1980. Meiofauna-macrofauna interactions in a high salt marsh habitat. Ecol. Monogr. 50:487–505.

Ben-Amotz, A., and A. Gilboa. 1980. Cryopreservation of marine unicellular algae. I. A survey of algae with regard to size, culture age, photosynthetic activity and chlorophyll-to-cell ratio. Mar. Ecol. Progr. Ser. 2:157–161.

Berkes, F. 1977. Production of the euphausiid crustacean *Thysanoessa raschii* in the Gulf of St. Lawrence. J. Fish. Res. Bd. Canada 34:443–446.

Bernstein, B. B., B. E. Williams, and K. H. Mann. 1981. The role of behavioral responses to predation in modifying urchins' (*Strongylocentrotus droebachiensis*) destructive grazing and seasonal foraging patterns. Mar. Biol. 63:39–46.

Bertalanffy, L. von. 1957. Quantitative laws in metabolism and growth. Quant. Rev. Biol. 32:217–231.

Beverton, R. J. H. 1962. Long-term dynamics of certain North Sea fish populations. Pp. 242–259 in E. D. LeCren and M. W. Holdgate (eds.), The Exploitation of Natural Animal Populations. Blackwell.

Biggs, D. C. 1977. Respiration and ammonium excretion by open ocean gelatinous zooplankton. Limnol. Oceanogr. 22:108–117.

Billen, G. 1975. Nitrification in the Scheldt estuary (Belgium and the Netherlands). Estuar. Coast. Mar. Sci. 3:79–89.

Birkeland, C. 1977. The importance of rate of biomass accumulation in early successional stages of benthic communities to the survival of coral recruits. Pp. 15–21 in Proc. 3rd Int. Coral Reef Symp., Vol. 1. Univ. of Miami, Florida.

Birkhead, T. R. 1977. The effect of habitat and density on breeding success in the common guillemot (*Uria aalge*). J. Anim. Ecol. 46:751–764.

Bishop, J. K. B. 1981. Particle sources and sinks from C-FATE (composition flux and transfer experiments) results. Pp. 9–12 in R. F. Anderson and M. P. Bacon (eds.), Sediment Trap Intercomparison Experiment. Woods Hole Oceanographic Institution Tech. Memo No. 1–81.

Bishop, J. K. B., R. W. Collier, D. R. Kettens, and J. M. Edmond. 1980. The chemistry, biology, and vertical flux of particulate matter from the upper 1500 m of the Panama Basin. Deep-Sea Res. 27A:615–640.

Bishop, J. K. B., J. M. Edmond, D. R. Ketten, M. P. Bacon, and W. B. Silker. 1977. The chemistry, biology and vertical flux of particulate matter from the upper 400 m of the equatorial Atlantic Ocean. Deep-Sea Res. 24:511–548.

Bjorndal, K. A. 1980. Nutrition and grazing behavior of the green turtle *Chelonia mydans*. Mar. Biol. 56:147–154.

Black, C. C. 1971. Ecological implications of dividing plants into groups with distinct photosynthetic capacities. Adv. Ecol. Res. 7:87–114.

Black, C. C. 1973. Photosynthetic carbon fixation in relation to net CO_2 uptake. Ann. Rev. Plant Physiol. 24:253–286.

Blackburn, M. 1973. Regressions between biological oceanographic measurements in the eastern tropical Pacific and their significance to ecological efficiency. Limnol. Oceanogr. 18:552–563.

Blackburn, M., R. M. Laurs, R. W. Owen, and B. Zeitzschel. 1970. Seasonal and areal changes in standing stocks of phytoplankton, zooplankton, and micronekton in the eastern tropical Pacific. Mar. Biol. 7:14–31.

Blasco, D. 1971. Composición y distribución del fitoplancton en la region del afloramiento de las costas peruanas. Invest. Pesq. 35:61–112.

Blasco, D. 1978. Observations on the diel migration of marine dinoflagellates off the Baja California Coast. Mar. Biol. 46:41–47.

Blegvad, H. 1925. Continued studies on the quantity of fish food in the sea bottom. Rep. Danish Biol. Stat. 31:25–56.

Blomqvist, S., and L. Håkanson. 1981. A review on sediment traps in aquatic environments. Arch. Hydrobiol. 91:101–132.

Bloom, S. A. 1981. Similarity indices in community studies: Potential pitfalls. Mar. Ecol. Progr. Ser. 5:125–128.

Bohmsack, J. A., and F. H. Talbot. 1980. Species packing by reef fishes on Australian and Caribbean reefs: An experimental approach. Bull. Mar. Sci. 30:710–723.

Boje, R., and M. Tomczak (eds.). 1978. Upwelling Ecosystems. Springer-Verlag.

Bokuniewicz, H. 1980. Groundwater seepage into Great South Bay, New York. Est. Coast. Mar. Sci. 10:437–444.

Bormann, F. H., and G. E. Likens. 1979. Pattern and Process in a Forested Ecosystem. Springer-Verlag.

Bormann, F. H., G. E. Likens, and J. M. Melillo. 1977. Nitrogen budget for an aggrading Northern hardwood ecosystem. Science 196:981–983.

Bosch, H. F., and W. Rowland Taylor. 1973. Diurnal vertical migration in an estuarine cladoceran, *Podon polyphemoides*, in the Chesapeake Bay. Mar. Biol. 19:172–181.

Bougis, P. 1976. Marine Plankton Ecology. North-Holland.

Bowen, S. H. 1980. Detrital non-protein amino acids are the key to rapid growth of *Tilapia* in Lake Valencia, Venezuela. Science 207:1216–1218.

Bowman, M. J. 1977. Nutrient distribution and transport in Long Island Sound. Estuar. Coast. Mar. Sci. 5:531–548.

Branch, G. M. 1975. Intraspecific competition in *Patella cochlear* Born. J. Anim. Ecol. 44:263–281.

Branch, G. M. 1975a. Mechanisms reducing intraspecific competition in *Patella* spp.: Migration, differentiation, and territorial behaviour. J. Anim. Ecol. 44:575–600.

Branch, G. M. 1976. Interspecific competition experienced by South African *Patella* species. J. Anim. Ecol. 45:507–529.

Brandt, S. B. 1981. Effects of a warm-core eddy on fish distributions in the Tasman Sea off East Australia. Mar. Ecol. Progr. Ser. 6:19–33.

Braudel, F. 1981. The Structures of Everyday Life. Vol. I, The Limits of the Possible. Harper and Row.

Brawn, V. M. 1969. Feeding behavior of cod (*Gadus morhua*). J. Fish. Res. Bd. Canada 26:583–596.

Bray, J. R. 1961. Measurement of leaf utilization as an index of minimal level of primary consumption. Oikos 12:70–74.

Bray, J. R. 1964. Primary consumption in three forest canopies. Ecology 45: 165–167.

Bray, J. T., O. P. Bricker, and B. N. Troup. 1973. Phosphate in interstitial waters of anoxic sediments: Oxidation effects during sampling procedure. Science 180:1362–1364.

Bray, R. N., A. C. Miller, and G. G. Geesey. 1981. The fish connection: A trophic link between planktonic and rocky reef communities? Science 214:204–205.

Breen, P. A., and K. H. Mann. 1976. Changing lobster abundance and the destruction of kelp beds by sea urchins. Mar. Biol. 34:137–142.

Brenchley, G. A. 1981. Disturbance and community structure: An experimental study of bioturbation in marine soft-bottom sediments. J. Mar. Res. 39:767–790.

Brett, J. R., V. E. Shelbourn, and C. T. Shoop. 1969. Growth rate and body composition of fingerling sockeye salmon (*Onchorynchus verka*) in relation to temperature and ration size. J. Fish. Res. Bd. Canada 26:2363–2394.

Briggs, K. B., K. R. Tenore, and R. B. Hanson. 1979. The role of microfauna in detrital utilization by the polychaete, *Nereis succinea* (Frey and Leuckart). J. Exp. Mar. Biol. Ecol. 36:225–234.

Brinkhuis, B. H. 1977. Comparisons of salt marsh fucoid production estimated from three different indices. J. Phycol. 13:328–335.

Brinkhuis, B. H. 1977a. Seasonal variations in salt marsh macroalgae photosynthesis. I. *Ascophyllum nodosum* ecad *scorpioides*. Mar. Biol. 44:165–175.

Brinkhuis, B. H. 1977b. Seasonal variations in salt marsh macroalgae photosynthesis. II. *Fucus vesiculosus* and *Ulva lactuca*. Mar. Biol. 44:177–186.

Brock, T. D. 1971. Microbial growth rates in nature. Bacteriol. Rev. 35:39–58.

Brock, V., and R. Riffenburgh. 1960. Fish schooling: A possible factor in reducing predation. J. Cons. Intern. Explor. Mer 25:307–317.

Brocksen, R. W., G. E. Davis, and C. E. Warren. 1970. Analysis of trophic processes on the basis of density-dependent function. Pp. 468–498 in J. H. Steele (ed.), Marine Food Chains, Univ. California.

Broecker, W. S. 1974. Chemical Oceanography. Harcourt, Brace Jovanovich.

Brooks, J. L., and S. I. Dodson. 1965. Predation, body size, and composition of plankton. Science 150:28–35.

Brousseau, D. J. 1978. Spawning cycle, fecundity, and recruitment in a population of soft-shell clam, *Mya arenaria*, from Cape Ann, Massachusetts. Fish. Bull. 76:155–166.

Brousseau, D. J. 1978a. Population dynamics of the soft-shell clam *Mya arenaria*. Mar. Biol. 50:63–71.

Brousseau, D. J. 1979. Analysis of growth rate in *Mya arenaria* using the von Bertalanffy equation. Mar. Biol. 51:221–227.

Brown, C. M., and B. Johnson. 1977. Inorganic nitrogen assimilation in aquatic microorganisms. Adv. Aquatic Microbiol. 1:49–114.

Brown, E. J., R. F. Harris, and J. F. Koonce. 1978. Kinetics of phosphate uptake by aquatic microorganisms: Deviations from a simple Michaelis-Menten equation. Limnol. Oceanogr. 23:26–34.

Brown, W. L., Jr., and E. O. Wilson. 1956. Character displacement. Syst. Zool. 5:49–64.

Bryan, J. R., J. R. Riley, and P. J. LeB. Williams. 1976. A Winkler procedure for making precise measurements of oxygen concentration for productivity and related studies. J. Exp. Mar. Biol. Ecol. 21:191–197.

Bryant, D. M. 1979. Effects of prey density and site character on estuary usage by overwintering waders (Charadrii). Estuar. Coast. Mar. Sci. 9:369–384.

Brylinsky, M. 1977. Release of dissolved organic matter by some marine macrophytes. Mar. Biol. 39:213–220.

Buchsbaum, R., I. Valiela, and J. M. Teal. 1982. Grazing by Canada geese and related aspects of the chemistry of salt marsh grass. Colonial Waterbirds 4:126–131.

Buesa, R. J. 1975. Population biomass and metabolic rates of marine angiosperms on the northwestern Cuban shelf. Aquat. Bot. 1:11–23.

Bunt, J. S. 1975. Primary productivity of marine ecosystems. Pp. 169–184 in H. Lieth and R. H. Whittaker (eds.), Primary Productivity of the Biosphere. Springer-Verlag.

Bunt, J. S., K. G. Boto, and G. Boto. 1979. A survey method for estimating potential levels of mangrove forest primary production. Mar. Biol. 52:123–128.

Bunt, J. S., and C. C. Lee. 1970. Seasonal primary production in Antarctic sea ice at McMurdo Sound in 1967. J. Mar. Res. 28:304–320.

Burd, A. C. 1965. Growth and recruitment in the herring of the southern North Sea. Fish. Invest. Ser. 2 23:1–42.

Burd, A. C. 1978. Long term changes in North Sea herring stocks. Rapp. Proc.-Verb. Reun. Cons. Int. Explor. Mer 172:137–153.

Burke, M. V., and K. H. Mann. 1974. Productivity and production: Biomass ratios of bivalve and gastropod populations in an eastern Canadian estuary. J. Fish. Res. Bd. Canada 31:167–177.

Burris, J. E. 1980. Respiration and photorespiration in marine algae. Pp. 411–432 in P. G. Falkowski (ed.), Primary Productivity in the Sea. Plenum.

Buss, L. W. 1979. Bryozoan overgrowth interactions—the interdependence of competition for space and food. Nature 281:475–477.

Butler, E. I., E. D. S. Corner, and S. M. Marshall. 1970. On the nutrition and metabolism of zooplankton. VII. Seasonal survey of nitrogen and phosphorus excretion by *Calanus* in the Clyde Sea area. J. Mar. Biol. Assoc. U.K. 50:525–560.

Calow, P. 1977. Conversion efficiencies in heterotrophic organisms. Biol. Rev. 52:385–409.

Calow, P. 1977a. Ecology, evolution, and energetics: A study in metabolic adaptation. Adv. Ecol. Res. 10:1–60.

Calow, P., and C. R. Fletcher. 1976. A new radiotracer technique involving ^{14}C and ^{61}Cr, for estimating the assimilation efficiencies of aquatic, primary consumers. Oecologia 9:155–170.

Cammen, L. M. 1980. The significance of microbial carbon in the nutrition of the deposit feeding polychaete *Nereis succinea*. Mar. Biol. 61:9–20.

Cammen, L., P. Rublee, and J. Hobbie. 1978. The significance of microbial carbon in the nutrition of the polychaete *Nereis succinea* and other aquatic deposit feeders. Univ. of North Carolina Sea Grant Publ. UNC-SG-78–12.

Cannon, H. G. 1928. On the feeding mechanisms of the copepods, *Calanus finmarchicus* and *Diaptomus gracilis*. Brit. J. Exp. Biol. 6:131–144.

Caperon, J., and J. Meyer. 1972. Nitrogen-limited growth of marine phytoplankton II. Uptake kinetics and their role in nutrient limited growth of phytoplankton. Deep-Sea Res. 19:619–632.

Capone, D. G., R. S. Oremland, and B. F. Taylor. 1977. Significance of N_2 fixation to the production of *Thalassia testudinum* communities. In H. B. Stewart, Jr. (ed.), Cooperative Investigations of the Caribbean and Adjacent Regions, Vol. 2. FAO, Rome.

Cappenberg, T. E. 1974. Interrelations between sulfate-reducing and methane-producing bacteria in the bottom deposits of a freshwater lake. II. Inhibition experiments. Ant. v. Leenwenhoek 40:297–306.

Cappenberg, T. E. 1975. A study of mixed continuous cultures of sulfate-reducing and methane-producing bacteria. Microb. Ecol. 2:60–72.

Capriulo, G. M., and E. J. Carpenter. 1980. Grazing by 35 to 202 μm microzooplankton in Long Island Sound. Mar. Biol. 56:319–326.

Capuzzo, J. M., and B. A. Lancaster. 1979. Larval development in the American lobster: Changes in metabolic activity and the O:N ratio. Canad. J. Zool. 57:1845–1848.

Carignan, R., and J. Kalff. 1980. Phosphorus sources for aquatic weeds: Water or sediments? Science 207:987–989.

Caron, D. A., P. G. Davis, L. P. Madin, and J. McN. Sieburth. 1982. Heterotrophic bacteria and bacterivorous protozoa in oceanic aggregates. Science 218:795–797.

Carpenter, E. J. 1973. Nitrogen fixation by *Oscillatoria* (*Trichodesmium*) *thiebautii* in the southwestern Sargasso Sea. Deep-Sea Res. 20:285–288.

Carpenter, E. J., and J. L. Culliney. 1975. Nitrogen fixation in marine shipworms. Science 187:551–552.

Carpenter, E. J., and R. R. L. Guillard. 1971. Interspecific differences in nitrate half-saturation constants for three species of marine phytoplankton. Ecology 52:183–185.

Carpenter, E. J., G. R. Harbison, L. P. Madin, N. R. Swanberg, D. C. Biggs, E. M. Hulburt, V. L. McAlister, and J. J. McCarthy. 1977. *Rhizosolenia* mats. Limnol. Oceanogr. 22:739–741.

Carpenter, E. J., and J. S. Lively. 1980. Review of estimates of algal growth using ^{14}C tracer techniques. Pp. 161–178, in P. G. Folkowski (ed.), Primary Productivity in the Sea. Plenum.

Carpenter, E. J., and J. J. McCarthy. 1975. Nitrogen fixation and uptake of combined nitrogenous nutrients by *Oscillatoria* (*Trichodesmium*) *thiebautii* in the western Sargasso Sea. Limnol. Oceanogr. 20:389–401.

Carpenter, E. J., B. B. Peek, and S. J. Anderson. 1974. Survival of copepods passing through a nuclear power station on Northeastern Long Island Sound, U.S.A. Mar. Biol. 24:49–55.

Carpenter, E. J., and C. C. Price IV. 1976. Marine *Oscillatoria* (*Trichodesmium*): Explanation for aerobic nitrogen fixation without heterocysts. Science 191:1278–1280.

Carpenter, E. J., and C. C. Price IV. 1977. Nitrogen fixation, distribution and production of *Oscillatoria* (*Trichodesmium*) spp. in the Western Sargasso and Caribbean Sea. Limnol. Oceanogr. 22:60–72.

Carpenter, E. J., C. D. Van Raalte, and I. Valiela. 1978. Nitrogen fixation by algae in a Massachusetts salt marsh. Limnol. Oceanogr. 23:318–327.

Carpenter, R. C. 1981. Grazing by *Diadema antillarum* (Philippi) and its effects on the benthic algal community. J. Mar. Res. 39:749–765.

Cassie, R. M. 1954. Some uses of probability paper in the analysis of size frequency distributions. Austral. J. Mar. Freshw. Res. 5:513–522.

Cassie, R. M. 1962. Frequency distribution models in the ecology of plankton and other organisms. J. Anim. Ecol. 31:65–92.

Cassie, R. M. 1962a. Microdistribution and other error components of ^{14}C primary production estimates. Limnol. Oceanogr. 7:121–130.

Castenholtz, R. W. 1961. The effect of grazing on marine littoral diatom populations. Ecology 42:783–794.

Caswell, H. 1978. Predator mediated coexistence: A non-equilibrium model. Amer. Nat. 112:127–154.

Caswell, H., F. C. Reed, S. N. Stephenson, and P. A. Werner. 1973. Photosynthetic pathways and selective herbivory: A hypothesis. Amer. Nat. 107:465–480.

Cates, R. G., and G. H. Orians. 1975. Successional status and the palatability of plants to generalized herbivores. Ecology 56:410–418.

Caughley, G. 1966. Mortality patterns in mammals. Ecology 47:906–918.

Cauwet, G. 1978. Organic chemistry of seawater particulates: Concepts and developments. Oceanol. Acta 1:99–105.

Cavanaugh, C. M., S. L. Gardiner, M. L. Jones, H. W. Jannasch, and J. B. Waterberg. 1981. Procaryotic cells in the hydrothermal vent tubeworm *Riftia pachyptila* Jones: Possible chemoautotrophic symbionts. Science 213:340–342.

Cederwell, H. 1977. Annual macrofauna production in a soft bottom in the Northern Baltic Proper. Pp. 155–164 in B. F. Keegan, P. O'Leidigh, and P. J. Braden (eds.), Biology of Benthic Organisms. Pergamon.

Chamberlain, W., and J. Shapiro. 1973. Phosphate measurements in natural waters—A critique. Pp. 355–366. in E. J. Griffith, A. Beeton, J. M. Spencer, and D. R. Mitchell (eds.), Environmental Phosphorus Handbook. Wiley.

Chapman, A. R. O. 1981. Stability of sea urchin dominated barren grounds following destructive grazing of kelp in St. Margaret's Bay, Southern Canada. Mar. Biol. 62:307–311.

Chapman, A. R. O., and J. S. Craigie. 1977. Seasonal growth in *Laminaria longicruris*: Relations with dissolved inorganic nutrients and internal reserves of nitrogen. Mar. Biol. 40:197–205.

Chapman, A. R.O., and J. S. Craigie. 1978. Seasonal growth in *Laminaria longicruris*: Relation with reserve carbohydrate storage and production. Mar. Biol. 46:208–213.

Chapman, V. J. 1960. Salt marshes and Salt Deserts of the World. L. Hill Ltd., Interscience.

Chapman, V. J. (ed.). 1977. Wet Coastal Ecosystems. Elsevier.

Charnov, E. L. 1976. Optimal foraging: Attack strategy of a mantid. Amer. Nat. 110:141–151.

Charnov, E. L., G. H. Orians, and K. Hyatt. 1976. Ecological implications of resource depression. Amer. Nat. 110:247–259.

Cheng, L. (ed.). 1976. Marine Insects. North Holland.

Chervin, M. B. 1978. Assimilation of particulate organic carbon by estuarine and coastal copepods. Mar. Biol. 49:265–275.

Chmyr, V. D. 1967. Radiocarbon method for determining zooplankton production in natural populations (In Russian.). Dokl. Acad. Nauk. SSSR 173:201–203.

Chock, J. S., and A. C. Mathieson. 1976. Ecological studies of the salt marsh ecad *scorpioides* (Hornemann) Hauck of *Ascophyllum nodosum* (L.) Lectolis. J. Exp. Mar. Biol. Ecol. 23:171–190.

Choi, C. I. 1972. Primary production and release of DOC in the Western North Atlantic Ocean. Deep-Sea Res. 19:731–735.

Clark, A. 1980. A reappraisal of the concept of metabolic cold adaptation in polar marine invertebrates. Biol. J. Linn. Soc. 14:77–92.

Clark, L. R., P. W. Geier, R. D. Hughes, and R. F. Morris. 1967. The Ecology of Insect Populations in Theory and Practice. Methuen.

Clark, P. J., and F. C. Evans. 1954. Distance to nearest neighbor as a measure of spatial relationships in populations. Ecology 45:445–453.

Clarke, G. L. 1946. Dynamics of production in a marine area. Ecol. Monogr. 16:323–335.

Clarke, G. L., and E. Denton. 1962. Light and animal life in the sea. Pp. 456–468 in M. N. Hill (ed.) The Sea, Vol. 1. Wiley.

Clarke, T. A. 1970. Territorial behavior and population dynamics of a pomacentrid fish, the garibaldi, *Hypsypops rubicunda*. Ecol. Monogr. 40: 189–212.

Claypool, G. E., and I. R. Kaplan. 1974. The origin and distribution of methane in marine sediments. Pp. 99–139 in I. R. Kaplan (ed.), Natural Gases in Marine Sediments. Plenum.

Clayton, R. K. 1971. Light and Living Matter. Vol. 2: The Biological Part. McGraw-Hill.
Clendenning, K. A., T. E. Brown, and H. C. Eyster. 1956. Comparative studies of photosynthesis in *Nostoc muscorum* and *Chlorella pyrenoidosa*. Canad. J. Bot. 34:943–966.
Clutter, R. I. 1969. The microdistribution and social behavior of some pelagic mysid shrimps. J. Exp. Mar. Biol. Ecol. 3:125–155.
Cock, M. J. W. 1978. The assessment of preference. J. Anim. Ecol. 47:805–816.
Cohen, E. B., M. D. Grosslein, M. P. Sissenwine, F. Steimle, and W. R. Wright. 1981. An energy budget of Georges Bank. In M. C. Mercer (ed.), Multispecies Approaches to Fisheries Management Advise. Canad. Spec. Publ. Fish. Aquatic Sci. No. 58.
Cohen, Y., W. E. Krumbein, M. Goldberg, and M. Shilo. 1977. Solar Lake (Sinai) 1. Physical and chemical limnology. Limnol. Oceanogr. 22:597–608.
Cohen, Y., W. E. Krumbein, and M. Shilo. 1977a. Solar Lake, Sinai. 2. Distribution of photosynthetic microorganisms and primary production. Limnol. Oceanogr. 22:609–620.
Cohen, Y., W. E. Krumbein, and M. Shilo. 1977b. Solar Lake, Sinai. 3. Bacterial distribution and production. Limnol. Oceanogr. 22:621–634.
Cole, J. J. 1982. Interactions between bacteria and algae in aquatic ecosystems. Ann. Rev. Ecol. Syst. 13:291–314.
Cole, L. C. 1946. A study of the cryptozoa of an Illinois woodland. Ecol. Monogr. 16:49–86.
Cole, L. C. 1954. The population consequences of life history phenomena. Quart. Rev. Biol. 29:103–137.
Cole, L. C. 1965. Dynamics of animal population growth. J. Chron. Dis. 18: 1095–1108.
Colebrook, J. M. 1979. Continuous plankton records: Seasonal cycles of phytoplankton and copepods in the North Atlantic Ocean and the North Sea. Mar. Biol. 51:23–32.
Collos, Y., and G. Slawyk. 1980. Nitrogen uptake and assimilation by marine phytoplankton. Pp. 195–211 in P. G. Falkowski (ed.), Primary Productivity in the Sea. Plenum.
Colwell, R. K., and D. J. Futuyma. 1971. On the measurement of niche breadth and overlap. Ecology 52:567–576.
Comins, H. N., and M. P. Hassell. 1979. The dynamics of optimally foraging predators and parasitoids. J. Anim. Ecol. 48:335–351.
Connell, J. H. 1961. The influence of interspecific competition and other factors on the distribution of the barnacle *Chthamalus stellatus*. Ecology 42:133–146.
Connell, J. H. 1970. A predator-prey system in the marine intertidal region. I. *Balanus glandula* and several predatory species of *Thais*. Ecol. Monogr. 40:49–78.
Connell, J. H. 1972. Community interactions on marine rocky intertidal shores. Ann. Rev. Ecol. Syst. 3:169–192.
Connell, J. H. 1978. Diversity in tropical rain forests and coral reefs. Science 199:1302–1310.
Connell, J. H. 1980. Diversity and the coevolution of competitors, or the ghost of competition past. Oikos 35:131–138.

Connell, J. H., and R. E. Slatyer. 1977. Mechanisms of succession in natural communities and their role in community stability and organization. Amer. Nat. 111:1119–1144.

Conover, R. J. 1966. Factors affecting the assimilation of organic matter by zooplankton and the question of superfluous feeding. Limnol. Oceanogr. 11:346–354.

Conover, R. J. 1974. Production in marine planktonic communities. Pp. 119–163 in Proc. 1st Inter. Cong. of Ecology, The Hague 1974. Center for Agric. Publish. and Documentation.

Conover, R. J. 1978. Transformation of organic matter. Pp. 221–499 in O. Kinne (ed.), Marine Ecology, Vol. 4. Wiley.

Conover, R. J., and C. M. Lalli. 1974. Feeding and growth in *Clione limacina* (Phipps), a pteropod mollusc. II. Assimilation, metabolism, and growth efficiency. J. Exp. Mar. Biol. Ecol. 16:131–154.

Conroy, J. W. H. 1975. Recent increases in penguin populations in Antarctica and the subantarctic. Pp. 321–336 in B. Stonehouse (ed.), The Biology of Penguins. Univ. Park.

Conway, H. L., and T. E. Whitledge. 1979. Distribution, fluxes and biological utilization of inorganic nitrogen during a spring bloom in the New York Bight. J. Mar. Res. 37:657–668.

Cook, R. M., and B. J. Cockrell. 1978. Predator ingestion rate and its bearing on feeding time and the theory of optimal diets. J. Anim. Ecol. 47:529–547.

Cooper, W. E. 1965. Dynamics and predation of a natural population of a freshwater amphipod, *Hyalella azteca*. Ecol. Monogr. 35:377–394.

Copping, A. E., and C. J. Lorenzen. 1980. Carbon budget of a marine phytoplankton-herbivore system with carbon-14 as a tracer. Limnol. Oceanogr. 25:873–882.

Corkett, C. J., and I. A. McLaren. 1969. Egg production and oil storage by the copepod *Pseudocalanus* in the laboratory. J. Exp. Mar. Biol. Ecol. 3:90–105.

Corner, E. D. S., and A. G. Davies. 1971. Plankton as a factor in the nitrogen and phosphorus cycles in the sea. Adv. Mar. Biol. 9:101–204.

Corner, E. D. S., R. N. Head, and C. C. Kilvington. 1972. On the nutrition and metabolism of zooplankton. VIII. The grazing of *Biddulphia* cells by *Calanus helgolandicus*. J. Mar. Biol. Assoc. U.K. 52:847–861.

Corner, E. D. S., and B. S. Newell. 1967. On the nutrition and metabolism of zooplankton. IV. The forms of nitrogen excreted by *Calanus*. J. Mar. Biol. Assoc. U.K. 47:113–120.

Cosgrove, D. J. 1977. Microbial transformations in the phosphorus cycle. Adv. Microb. Ecol. 1:95–134.

Cosper, T. C., and M. R. Reeve. 1975. Digestive efficiency of the chaetognath *Sagitta hispida* Conant. J. Exp. Mar. Biol. Ecol. 17:33–38.

Costopulos, J. J., G. C. Stephens, and S. H. Wright. 1979. Uptake of amino acids by marine polychaetes. Biol. Bull. 157:434–444.

Cowles, T. J. 1979. The feeding response of copepods from the Peru upwelling system: Food size selection. J. Mar. Res. 37:601–622.

Craig, H. 1971. The deep metabolism: Oxygen consumption in abyssal ocean water. J. Geophys. Res. 76:5078–5086.

Craigie, J. S., and J. McLachlan. 1964. Excretion of coloured ultraviolet absorbing substances by marine algae. Canad. J. Bot. 42:23–33.

Crawley, M. J. 1975. The numerical responses of insect predators to changes in prey density. J. Anim. Ecol. 44:877–892.

Creese, R. G. 1980. An analysis of distribution and abundance on populations of the high-shore limpet, *Notoacmea petterdi* (Tenison-Woods). Oecologia 45:252–260.

Crisp, D. J. 1975. Secondary productivity in the sea. Pp. 71–90 in D. E. Reichle, J. F. Franklyn, and D. W. Goodall (eds.), Productivity of World Ecosystems. Natl. Acad. Sci. Washington, D.C.

Crossland, C. J., D. J. Narnes, T. Cox, and M. Devereux. 1980. Compartmentation and turnover of organic carbon in the staghorn coral *Acropora formosa*. Mar. Biol. 59:181–187.

Cummins, K. W., and J. C. Wuycheck. 1971. Caloric equivalents for investigations in ecological energetics. Mitt. Int. Verein Limnol. 18:2–158.

Cupp, E. E. 1943. Marine plankton diatoms of the west coast of North America. Bull. Scripps Inst. Oceanogr. Techn. Ser. 5:1–238.

Curl, H., Jr., and L. F. Small. 1965. Variations in photosynthetic assimilation ratios in natural, marine phytoplankton communities. Limnol. Oceanogr. 10 (Suppl.):R67-R73.

Currie, R. I. 1964. Environmental features in the ecology of Antarctic Seas. Pp. 89–94 in Biologie Antarctique. Hermann.

Cushing, D. H. 1964. The work of grazing in the sea. Pp. 207–226 in D. J. Crisp (ed.), Grazing in Terrestrial and Marine Environments. Blackwell.

Cushing, D. H. 1968. Grazing by herbivorous copepods in the sea. J. Cons. Int. Explor. Mer 32:70–82.

Cushing, D. H. 1971. Upwelling and the production of fish. Adv. Mar. Biol. 9:255–334.

Cushing, D. H. 1975. Marine Ecology and Fisheries. Cambridge Univ.

Cushing, D. H. 1978. Upper trophic levels in upwelling areas. Pp. 101–110 in R. Boje and M. Tomczak (eds.), Upwelling Ecosystems. Springer-Verlag.

Cushing, D. H. 1981. Fisheries Biology: A Study in Population Dynamics. Univ. Wisconsin.

Cushing, D. H., and H. F. Nicholson. 1966. Method of estimating algae production at sea. Nature 212:310–311.

Daan, N. 1975. Consumption and production of North Sea Cod, *Gadus morhua*: An assessment of the ecological status of the stock. Neth. J. Sea Res. 9: 24–55.

Daan, N. 1978. Changes in cod stocks and cod fisheries in the North Sea. Rapp. Proc.-Verb. Reun. Cons. Int. Explor. Mer 172:39–57.

Dagg, M. J. 1974. Loss of prey body contents during feeding by an aquatic predator. Ecology 55:903–906.

Dagg, M. J. 1977. Some effects of patchy food environments on copepods. Limnol. Oceanogr. 22:99–107.

Dagg, M. J., J. Vidal, T. E. Whitledge, R. L. Iverson, and J. J. Goering. 1982. The feeding, respiration, and excretion of zooplankton in the Bering Sea during a spring bloom. Deep-Sea Res. 29:45–63.

Davies, J. M. 1975. Energy flow through the benthos in a Scottish Sea Loch. Mar. Biol. 31:353–362.

Davis, C. 1982. Processes Controlling Zooplankton Abundance on Georges Bank. Ph.D. Thesis, Boston University.

Davis, J. H. 1940. The ecology and geologic role of mangroves in Florida. Pap. Tortuga Lab. 32:1–412.

Davis, H. C., and A. Calabrese. 1964. Combined effects of temperature and salinity on development of eggs and growth of larva of *Mercenaria mercenaria* and *Crassostrea virginica*. Fish. Bull. U.S. Wild. Serv. 63:643–655.

Dawson, E. Y. 1966. Marine Botany—An Introduction. Holt, Rinehart, and Winston.

Dayton, P. K. 1971. Competition, disturbance, and community organization: the provision and subsequent utilization of space in a rocky intertidal community. Ecol. Monogr. 41:351–389.

Dean, T. A., and L. E. Hurd. 1980. Development in an estuarine fouling community: The influence of early colonists on later animals. Oecologia 46:295–301.

De Angelis, D. L. 1975. Stability and connectance in food web models. Ecology 56:238–243.

Deason, E. E. 1980. Grazing of *Acartia hudsonica* (*A. clausi*) on *Skeletonema costatum* in Narragansett Bay (U.S.A.): Influence of food concentrations and temperature. Mar. Biol. 60:101–113.

Degens, E. T. 1970. Molecular nature of nitrogenous compounds in sea water and recent marine sediments. Pp. 77–100 in D. W. Hood (ed.), Organic Matter in Natural Waters. Univ. of Alaska.

Degens, E., and K. Mopper. 1976. Factors controlling the distribution and early diagenesis of organic material in marine sediments. Pp. 59–113 in J. P. Riley and R. Chester (eds.), Chemical Oceanography. Vol. 6, 2nd Ed. Academic.

Delwiche, C. C., and B. B. Bryan. 1976. Denitrification. Ann. Rev. Microbiol. 30:241–262.

De Manche, J. M., H. C. Curl, Jr., D. W. Lundy, and P. L. Donaghey. 1979. The rapid response of the marine diatom *Skeletonema costatum* to changes in external and internal nutrient concentrations. Mar. Biol. 53:323–333.

De Mendiola, B. R. 1971. Some observations on the feeding of the Peruvian anchoveta, *Engraulis ringens*, in two regions of the Peruvian coast. Pp. 417–440 in J. Costlow (ed.), Fertility of the Sea, Vol. II. Gordon A. Breach.

Dempsey, M. J. 1981. Marine bacterial fouling: A scanning electron microscope study. Mar. Biol. 61:305–315.

Denslow, J. S. 1980. Pattern of plant species diversity during succession under different disturbance regimes. Oecologia 46:18–21.

Derby, C. D., and J. Atema. 1982. Chemosensitivity of walking legs of the lobster *Homarus americanus*: Neurophysiological response spectrum and thresholds. J. Exp. Biol. 98:303–315.

Deuser, W. G. 1975. Reducing environments. Pp. 1–37 in J. P. Riley and G. Skirrow (eds.), Chemical Oceanography. Vol. 3, 2nd Ed. Academic.

Deuser, W. G. 1979. Marine biota, nearshore sediments, and the global carbon balance. Org. Geochem. 1:243–247.

Deuser, W. G. 1980. Carbon-13 in Black Sea water and implications for the origin of hydrogen sulfide. Science 168:1575–1577.

Deuser, W. G., E. T. Degens, G. R. Harvey, and M. Rubin. 1973. Methane in Lake Kivu: New data bearing on its origin. Science 181:51–54.

Deuser, W. G., and E. H. Ross. 1980. Seasonal changes in the flux of organic carbon to the deep Sargasso Sea. Nature 280:364–365.

Deuser, W. G., E. H. Ross, and R. F. Anderson. 1981. Seasonality in the supply of sediment to the deep Sargasso Sea and implications for the rapid transfer of water to the deep sea. Deep-Sea Res. 28A:495–505.

Dickinson, C. H., and G. J. F. Pugh. 1974. Biology of Plant Litter Decomposition. Vols. 1 and 2. Academic.

Dietrich, G. 1968. General Oceanography: An Introduction. Interscience.

Digby, P. S. B. 1953. Plankton production in Scoresby Sound, East Greenland. J. Anim. Ecol. 23:289–322.

Digby, P. S. B. 1954. The biology of the marine planktonic copepods of Scoresby Sound, East Greenland. J. Anim. Ecol. 23:298–338.

Di Salvo, L. H., and G. W. Daniels. 1978. Observations on estuarine microfouling using the scanning electron microscope. Microb. Ecol. 2:234–240.

Dixon, D. R. 1976. The energetics of growth and reproduction in a brackish water serpulid *Mercierella enigmatica* (Fauvel). Pp. 197–210 in G. Persoone and E. Jaspers (eds.), Proc. 10th European Symp. Mar. Biol. Ostend, Belgium.

Dodimead, A. J., F. Favorite, and T. Hirano. 1963. Review of the oceanography of the subarctic Pacific region. Int. North Pacific Fish. Comm. Bull. No. 13.

Doherty, P. J. 1983. Tropical territorial damselfish: Is density limited by aggression or recruitment? Ecology 64:176–190.

Donaghay, P. L., and L. F. Small. 1979. Food selection capability of the estuarine copepod *Acartia clausii*. Mar. Biol. 52:137–146.

Dowd, J. E., and D. S. Riggs. 1964. A comparison of estimates of Michaelis-Menten kinetic constants for various linear transformations. J. Biol. Chem. 240:863–869.

Doyle, R. W. 1979. Ingestion rate of selective deposit feeders in a complex mixture of particles: Testing the energy-optimization hypothesis. Limnol. Oceanogr. 24:867–874.

Dring, M. J. 1981. Chromatic adaptation of photosynthesis in benthic marine algae: An examination of its ecological significance using a theoretical model. Limnol. Oceanogr. 26:271–284.

Droop, M. R. 1968. Vitamin B_{12} and marine ecology. 4. The kinetics of uptake, growth, and inhibition in *Monochrysis lutherii*. J. Mar. Biol. Assoc. U.K. 48:689–733.

Droop. M. R. 1973. Some thoughts on nutrient limitation in algae. J. Phycol. 9:264–272.

Duce, R. A., and E. K. Duursma. 1977. Inputs of organic matter to the sea. Mar. Chem. 5:319–339.

Ducklow, H. W. 1983. Production and fate of bacteria in the oceans. Bioscience 33:494–501.

Dugdale, R. C., and J. J. Goering. 1967. Uptake of new and regenerated forms of nitrogen in primary productivity. Limnol. Oceanogr. 12:196–206.

Dugdale, R. C., D. W. Menzel, and J. H. Ryther. 1961. Nitrogen fixation in the Sargasso Sea. Deep-Sea Res. 7:298–300.

Duggins, D. O. 1980. Kelp beds and sea otters: An experimental approach. Ecology 61:447–453.

Durbin, A. G., and E. G. Durbin. 1975. Grazing rates of the Atlantic menhaden *Brevoortia tyrannus* as a function of particle size and concentration. Mar. Biol. 33:265–277.

Durbin, E. G. 1974. Studies on the autecology of the marine diatom *Thalassiosira nordenskioldii* Cleve. I. The influence of daylength, light intensity, and temperature on growth. J. Phycol. 10:220–225.

Dussart, B. M. 1965. Les differentes categories de plancton. Hydrobiologia 26:72–74.

Duxbury, A. D., B. A. Morse, and N. McGary. 1966. The Columbia River effluent and its distribution at sea, 1961–1967. Univ. of Washington, Dept. of Oceanography Tech. Rep. No. 156 (Ref. M66–31).

Eaton, J. W., and P. Simpson. 1979. Vertical migration of the intertidal dinoflagellate *Amphidinium herdmaniae* Kofoid & Swegy. Pp. 339–345 in E. Naylor and R. G. Hartwell (eds.), Cyclic Phenomena in Marine Plants and Animals. Pergamon.

Ebersole, J. P. 1980. Food density and territory size: An alternative model and a test on the reef fish *Eupomacentrus leucostictus*. Amer. Nat. 115:492–509.

Ebert, T. A. 1977. Estimating growth and mortality rates from size data. Oecologia 11:281–298.

Eckman, J. E. 1979. Small-scale patterns and processes in a soft-substratum, intertidal community. J. Mar. Res. 37:437–457.

Edden, A. C. 1971. A measure of species diversity related to the log-normal distribution of individuals among species. J. Exp. Mar. Biol. Ecol. 6:199–209.

Edmond, J. M. 1970. High precision determination of titration alkalinity and total carbon dioxide content of sea water by potentiometric titration. Deep-Sea Res. 17:737–750.

Edmondson, W. T. 1960. Reproductive rates of rotifers in natural populations. Mem. Ist. Ital. Idrobiol. 12:21–77.

Edmondson, W. T., and G. G. Winberg (eds.). 1971. A Manual on Methods for the Assessment of Secondary Productivity in Fresh Waters. Blackwell.

Edwards, G. E., and S. C. Huber. 1981. C_4 pathway. Pp. 238–278 in M. D. Hatch and N. K. Boardman (eds.), The Biochemistry of Plants. Vol. 8, Photosynthesis. Academic.

Edwards, R. R. C., J. H. Steele, and A. Trevallion. 1970. The ecology of the O-group plaice and common dabs in Loch Ewe. III. Prey-predator experiments with plaice. J. Exp. Mar. Biol. Ecol. 4:156–173.

Eggers, D. M. 1976. Theoretical effects of schooling by planktivorous fish predators on rate of prey consumption. J. Fish. Res. Bd. Canada 33: 1964–1971.

Ehrlich, P. R., and L. C. Birch. 1967. The "balance of nature" and "population control." Amer. Nat. 101:97–107.

Elliot, J. M., and W. Davison. 1975. Energy equivalents of oxygen consumption in animal energetics. Oecologia 19:195–204.

Elliot, J. M., and L. Persson. 1978. The estimation of daily rates of food consumption for fish. J. Anim. Ecol. 47:977–991.

Ellis, D. V. 1977. Pacific Salmon. Management for People. West. Geogr. Series Vol. 13. Univ. of Victoria, Dept. of Geography.

Elner, R. W., and R. N. Hughes. 1978. Energy maximization in the diet of the shore crab, *Carcinus maenas*. J. Anim. Ecol. 47:107–116.

Elton, C. 1927. Animal Ecology. MacMillan.

Elton, C., and R. S. Miller. 1954. The ecological survey of animal communities with a practical system of classifying habitats by structural characters. J. Ecol. 42:460–496.

Elster, H. J. 1954. Uber die Populations dynamik von *Eudiaptomns gracilis* Sars und *Heterocope borealis* Fischer im Bodensee-Obersee. Arch. Hydrobiol. 20 (Suppl.):546–614.

Elvin, D. W., and J. J. Gossor. 1979. The thermal regime of an intertidal *Mytilus californianus* Conrad population on the central Oregon coast. J. Exp. Mar. Biol. Ecol. 39:265–279.

Elyakova, L. A. 1972. Distribution of cellulases and chitinases in marine invertebrates. Comp. Biochem. Physiol. 43:67–70.

Emerson, S., R. Jahnke, B. Bender, P. Froelich, G. Klinkhammer, C. Bowser, and G. Setlock. 1980. Early diagenesis in sediments from the Eastern Equatorial Pacific. I. Porewater nutrient and carbonate results. Earth Planet. Sci. Lett. 49:57–80.

Emlen, J. M. 1968. Optimal choice in animals. Amer. Nat. 102:385–389.

Emlen, J. M. 1973. Ecology: An Evolutionary Approach. Addison Wesley.

Emlen, J. M., and M. G. R. Emlen. 1975. Optimal choice in diet: Test of a hypothesis. Amer. Nat. 102:385–389.

Engelmann, M. D. 1961. The role of soil arthropods in the energetics of an old field community. Ecol. Monogr. 31:221–238.

Enhalt, D. H. 1976. The atmospheric cycle of methane. Pp. 13–22 in H. G. Schlegel, O. Gottschalk, and N. Pfennig (eds.), Microbial Production and Utilization of Gases. E. Goltze K.G.

Enright, J. T. 1969. Zooplankton grazing rates estimated under field conditions. Ecology 50:1070–1078.

Enright, J. T. 1977. Diurnal vertical migration: Adaptive significance and timing. Part 1. Selective advantage: A metabolic model. Limnol. Oceanogr. 22:856–872.

Enright, J. T., and H. -W. Honegger. 1977. Diurnal vertical migration: Adaptive significance and timing. Part 2. Test of the model: Details of timing. Limnol. Oceanogr. 22:873–886.

Eppley, R. W. 1972. Temperature and phytoplankton growth in the sea. Fish. Bull. 70:1063–1085.

Eppley, R. W. 1980. Estimating phytoplankton growth rates in the central oligotrophic oceans. Pp. 231–242 in P. G. Falkowski (ed.), Primary Productivity in the Sea. Plenum.

Eppley, R. W., R. W. Holmes, and J. D. M. Strickland. 1967. Sinking rates of marine phytoplankton measured with a fluorometer. J. Exp. Mar. Biol. Ecol. 1:191–208.

Eppley, R. W., S. G. Horrigan, J. A. Fuhrman, E. R. Books, C. C. Price, and K. Sellner. 1981. Origins of dissolved organic matter in Southern California coastal waters: Experiments on the role of zooplankton. Mar. Ecol. Prog. Ser. 6:149–159.

Eppley, R. W., and B. J. Peterson. 1979. Particulate organic flux and planktonic new production in the deep ocean. Nature 282:677–680.

Eppley, R. W., E. H. Renger, E. L. Venrick, and M. M. Mullin. 1973. A study of plankton dynamics and nutrient cycling in the central gyre of the North Pacific Ocean. Limnol. Oceanogr. 18:534–551.

Eppley, R. W., J. N. Rogers, and J. J. McCarthy. 1969. Half-saturation constants for uptake of nitrate and ammonium by various phytoplankton. Limnol. Oceanogr. 14:912–920.

Eppley, R. W., and P. R. Sloan. 1965. Carbon balance experiments with marine phytoplankton. J. Fish. Res. Bd. Canada 22:1083–1097.

Eriksson, E. 1952. Composition of atmospheric precipitation. I. Nitrogen compounds. Tellus 4:215–232.

Es, F. B. van, and L. -A. Meyer-Reil. 1982. Biomass and metabolic activity of heterotrophic marine bacteria. Adv. Microb. Ecol. 6:111–170.

Estes, J. A. 1979. Exploitation of marine mammals: r-selection of K-strategists? J. Fish. Res. Bd. Canada 36:1009–1017.

Estes, J. A., R. J. Jameson, and E. B. Rhode. 1982. Activity and prey selection in the sea otter: Influence of population status on community structure. Amer. Nat. 120:242–258.

Estes, J. A., and J. F. Palmisano. 1974. Sea otters: Their role in structuring nearshore communities. Science 185:1058–1060.

Estes, J. A., N. S. Smith, and J. F. Palmisano. 1978. Sea otter predation and community organization in the western Aleutian Islands, Alaska. Ecology 59:822–833.

Estrada, M. 1974. Photosynthetic pigments and productivity in the upwelling region of northwest Africa. Thethys 6:247–260.

Evans, P. D., and K. H. Mann. 1977. Selection of prey by American lobsters (*Homarus americanus*) when offered a choice between sea urchin and crabs. J. Fish. Res. Bd. Canada 34:2203–2207.

Evans, G. T., and F. J. R. Taylor. 1980. Phytoplankton accumulation in Langmuir cells. Limnol. Oceanogr. 25:840–845.

Everson, I. 1977. Antarctic marine secondary production and the phenomenon of cold adaptation. Phil. Trans. Roy. Soc. Lond. B 279:55–66.

Farrington, J. W. 1980. An overview of the biogeochemistry of fossil fuel hydrocarbons in the marine environment. Pp. 1–22 in L. Petrakis and F. T. Weiss (eds.), Petroleum in the Marine Environment, Adv. in Chem. Ser. 185, Amer. Chem. Soc.

Farrington, J. W., and B. W. Tripp. 1977. Hydrocarbons in western North Atlantic surface sediments. Geochim. Cosmochim. Acta 41:1627–1641.

Feeny, R. 1976. Plant apparency and chemical defense. Pp. 1–40 in J. Wallace and R. Mansell (eds.), Rec. Adv. Phytochem. 10. Biochemical Interaction Between Plants and Insects. Plenum.

Feigenbaum, D. 1979. Predation on chaetognaths by typhloscolecid polychaetes: One explanation for headless specimens. J. Mar. Biol. Assoc. U.K. 59:631–633.

Feigenbaum, D., and M. R. Reeve. 1977. Prey detection in the Chaetognatha: response to a vibrating probe and experimental determination of attack distance in large aquaria. Limnol. Oceanogr. 22:1052–1058.

Feller, R. J. 1977. Life history and production of meiobenthic harpacticoid copepods in Puget Sound. Ph.D. Thesis, Univ. of Washington.

Feller, R. J., G. L. Taghon, E. D. Gallagher, G. E. Kenny, and P. A. Jumars. 1979. Immunological methods for food web analysis in a soft bottom benthic community. Mar. Biol. 54:61–74.

Fenchel, T. 1968. The ecology of marine microbenthos. II. The food of marine benthic ciliates. Ophelia 5:73–121.

Fenchel, T. 1969. The ecology of marine microbenthos. IV. Structure and function of the benthic ecosystems, its chemical and physical factors and the microfauna communities with special reference to the ciliated protozoa. Ophelia 6:1–182.

Fenchel, T. 1970. Studies on the decomposition of organic detritus derived from the turtle grass *Thalassia testudinum*. Limnol. Oceanogr. 15:14–20.

Fenchel, T. 1973. Aspects of the decomposition of sea grasses. Pp. 123–145 in C. P. McRoy and C. Helfferich (eds.), Seagrass Ecosystems. Marcel Dekker.

Fenchel, T. 1974. Intrinsic rate of natural increase; the relationship with body size. Oecologia 14:317–326.

Fenchel, T. 1975. Character displacement and coexistence in mud snails (Hydrobiidae). Oecologia 20:19–32.

Fenchel, T. 1982. Ecology of heterotrophic microflagellates. II. Bioenergetics and growth. Mar. Ecol. Prog. Ser. 8:225–231.

Fenchel, T. 1982a. Ecology of heterotrophic microflagellates. IV. Quantitative occurrence and importance as bacterial consumers. Mar. Ecol. Prog. Ser. 9:35–42.

Fenchel, T., and T. H. Blackburn. 1979. Bacteria and Mineral Cycling. Academic.

Fenchel, T., and P. Harrison. 1976. The significance of bacterial grazing and mineral cycling for the decomposition of particulate detritus. Pp. 285–299 in J. M. Anderson and A. Macfadyen (eds.), The Role of Terrestrial and Aquatic Organisms in Decomposition Processes. Blackwell.

Fenchel, T. M., and B. B. Jorgensen. 1977. Detritus food chains of aquatic ecosystems: The role of bacteria. Pp. 1–58 in M. Alexander (ed.), Advances in Microbial Ecology. Vol. 1. Plenum.

Fenchel, T., and L. H. Kofoed. 1976. Evidence for exploitative interspecific competition in mud snails (Hydrobiidae). Oikos 27:367–376.

Fenchel, T., T. Perry, and A. Thane. 1977. Anaerobiosis and symbiosis with bacteria in free-living ciliates. J. Protozool. 24:154–163.

Fenchel, T., and B. J. Straarup. 1971. Vertical distribution of photosynthetic pigments and the penetration of light in marine sediments. Oikos 22:172–182.

Fenical, W. 1982. Natural product chemistry in the marine environment. Science 215:923–928.

Ferguson, R. L., and A. V. Palumbo. 1979. Distribution of suspended bacteria in neritic waters of Long Island during stratified conditions. Limnol. Oceanogr. 24:697–705.

Ferguson, R. L., and P. Rublee. 1976. Contribution of bacteria to standing crop of coastal plankton. Limnol. Oceanog. 21:141–144.

Field, J. G. 1969. The use of the information statistic in the numerical classification of heterogeneous systems. J. Ecol. 57:565–569.

Finenko, Z. Z., and V. E. Zaika. 1970. Particulate matter and its role in the productivity of the sea. Pp. 32–45 in J. H. Steele (ed.), Marine Food Chains. Univ. California.

Fischer, A. G. 1961. Latitudinal variation in organic diversity. Amer. Sci. 49:50–74.

Fisher, N. S. 1977. On the differential sensitivity of estuarine and open-ocean diatoms to exotic chemical stress. Amer. Nat. 14:871–895.

Fisher, R. A., A. S. Corbet, and C. B. Williams. 1943. The relation between the number of species and the number of individuals in a random sample of an animal population. J. Anim. Ecol. 12:42–58.

Fisher, T. R., P. R. Carlson, and R. T. Barker. 1982. Sediment nutrient regeneration in three North Carolina estuaries. Estuar. Coast. Shelf Sci. 14:101–116.

Fitt, W. K., and R. L. Pardy. 1981. Effects of starvation, and light and dark on the energy metabolism of symbiotic and aposymbiotic sea anemones. *Anthopleura elegantissima*. Mar. Biol. 61:199–205.

Fitzwater, S. E., G. A. Knauer, and J. H. Martin. 1982. Metal contamination and its effect on primary production measurements. Limnol. Oceanogr. 27:544–551.

Fleeger, J. W., and M. A. Palmer. 1982. Secondary production of the estuarine, meiobenthic copepod *Microarthridion littorale*. Mar. Ecol. Progr. Ser. 7:157–162.

Focht, D. D., and W. Verstraete. 1977. Biochemical ecology of nitrification and denitrification. Adv. Microb. Ecol. 1:135–214.

Foerster, R. E., and W. E. Ricker. 1941. The effect of reduction of predaceous fish on survival of young sockeye salmon at Cultus Lake. J. Fish. Res. Bd. Canada 5:315–336.

Fogg, G. E., C. Nalewajko, and W. D. Watt. 1965. Extracellular products of phytoplankton photosynthesis. Proc. Roy. Soc. B 162:517–534.

Fogg, G. E., and Than-tun. 1960. Interrelations of photosynthesis and assimilation of elementary nitrogen in a blue-green alga. Proc. Roy. Soc. B 153:111–127.

Fonds, M. 1979. Laboratory observations on the influence of temperature and salinity on development of the eggs and growth of the larvae of *Solea solea*. Mar. Ecol. Prog. Ser. 1:91–99.

Fong, W., and K. H. Mann. 1980. Role of gut flora in the transfer of amino acids through a marine food chain. Canad. J. Fish. Aquat. Sci. 37:88–96.

Foreman, R. E. 1977. Benthic community modification and recovery following extensive grazing by *Strongylocentrotus droebachiensis*. Helgol. Wiss. Meeres. 30:468–484.

Forsbergh, E. D. 1963. Some relationships of meteorological, hydrographic and biological variables in the Gulf of Panama. Inter-Amer. Trop. Tuna Comm. Bull. 7:1–109.

Forster, J. R. M., and P. A. Gabbott. 1971. The assimilation of nutrients from compounded diets by the prawns *Palaemon serratus* and *Pandalus platyceros*. J. Mar. Biol. Assoc. U.K. 51:943–961.

Fournier, R. O. 1971. The transport of organic carbon to organisms living in the deep oceans. Proc. Roy. Soc. Edinburgh 73:203–211.

Fox, L. R., and W. W. Murdoch. 1978. Effects of feeding history on short-term and long-term functional responses in *Notonecta hoffmanni*. J. Anim. Ecol. 47:945–959.

Frank, P. W. 1965. Shell growth in a natural population of the snail *Tegula funebralis*. Growth 29:395–403.

Frankenberg, D., and K. L. Smith, Jr. 1967. Coprophagy in marine animals. Limnol. Oceanogr. 12:443–450.

Frassetto, R., R. H. Backus, and E. Hays. 1962. Sound scattering layers and their relation to thermal structure in the Strait of Gibralter. Deep-Sea Res. 9:69–72.

Free, C. A., J. R. Beddington, and J. H. Lawton. 1977. On the inadequacy of simple models of mutual interference for parasitism and predation. J. Anim. Ecol. 46:543–554.

Frey, B. E., and L. F. Small. 1980. Effects of micronutrients and major nutrients on natural phytoplankton populations. J. Plankt. Res. 2:1–22.

Frey, R. W., and P. B. Basan. 1978. Coastal salt marshes. Pp. 101–170 in R. A. Davis, Jr. (ed.), Coastal Sedimentary Environments. Springer-Verlag.

Friedman, M. M., and J. R. Strickler. 1975. Chemoreceptors and feeding in calanoid copepods (Arthropoda:Crustacea). Proc. Natl. Acad. Sci. 72:4185–4188.

Frier, J. O. 1979. Character displacement in *Sphaeroma* spp. (Isopoda: Crustacea). I. Field evidence. Mar. Ecol. Prog. Ser. 1:159–163.

Frier, J. O. 1979a. Character displacement in *Sphaeroma* spp. (Isopoda: Crustacea). II. Competition for space. Mar. Ecol. Prog. Ser. 1:165–168.

Froelich, P. N., G. P. Klinkhammer, M. L. Bender, N. A. Luedke, G. R. Heath, O. Hammond, B. Hartman, and V. Maynard. 1979. Early oxidation of organic matter in pelagic sediments of the eastern equatorial Atlantic: Suboxic diagenesis. Geochim. Cosmochim. Acta 43:1075–1090.

Frost, B. W. 1972. Effects of size and concentration of food particles on the feeding behavior of the marine planktonic copepod *Calanus pacificus*. Limnol. Oceanogr. 17:805–815.

Frost, B. W. 1974. Feeding processes at lower trophic levels in pelagic communities. Pp. 59–77 in C. B. Miller (ed.), The Biology of the Oceanic Pacific. Oregon State.

Frost, B. W. 1975. A threshold feeding behavior in *Calanus pacificus*. Limnol. Oceanog. 20:263–266.

Frost, B. W. 1977. Feeding behavior of *Calanus pacificus* in mixtures of food particles. Limnol. Oceanog. 22:472–491.

Frost, B. W. 1980. Grazing. Pp. 465–491 in I. Morris (ed.), The Physiological Ecology of Phytoplankton. Univ. of California.

Fuhrman, J. A., J. W. Ammerway, and F. Azam. 1980. Bacteria plankton in the coastal euphotic zone: Distribution, activity and possible relationships with phytoplankton. Mar. Biol. 60:201–207.

Fuhrman, J. A., and F. Azam. 1980. Bacterioplankton secondary production estimates for coastal water of British Columbia, Antarctica and California. Appl. Env. Microbiol. 39:1085–1095.

Fuhrman, J. A., and F. Azam. 1982. Thymidine incorporation as a measure of heterotrophic bacteria plankton production in marine surface waters: Evaluation and field results. Mar. Biol. 66:109–120.

Fuhs, G. W., S. D. Demmerle, E. Canelli, and M. Chen. 1972. Characterization of phosphorus-limited plankton algae. Pp. 113–123 in G. E. Likens (ed.), Nutrients and Eutrophication. Spec. Symp. Vol 1. Amer. Assoc. Limnol. Oceanogr. Allen.

Fuji, A. 1967. Ecological studies on the growth and food consumption of Japanese common littoral sea urchin, *Strongylocentrotus intermedius* (A. Agassiz). Mem. Fac. Fish. Hokkaido Univ. 15:83–160.

Fuji, A., and M. Hashizume. 1974. Energy budget for a Japanese common scallop, *Patinopecten yessoensis* (Jay), in Mutsu Bay. Bull. Fac. Fish. Hokkaido Univ. 25:7–19.

Fulton, J. 1973. Some aspects of the life history of *Calanus plumchrus* in the Strait of Georgia. J. Fish. Res. Bd. Canada 30:811–815.

Fulton, J. 1978. Seasonal and annual variations of net zooplankton at Ocean Station P, 1965–1976. Fish. Mar. Serv. Canada, Data Rep. No. 49.

Gagné, J. A., K. H. Mann, and A. R. O. Chapman. 1982. Seasonal patterns of growth and storage in *Laminaria longicruris* in relation to differing patterns of availability of nitrogen in the water. Mar. Biol. 69:91–101.

Gagosian, R. B., and D. H. Stuermer. 1977. The cycling of biogenic compounds and their diagenetically transformed products in seawater. Mar. Chem. 5:605–632.

Gallagher, E. D., P. A. Jumars, and D. D. Trueblood. In press. Facilitation of soft-bottom benthic succession by tube builders. Ecology.

Gallagher, J. L., and F. C. Daiber. 1974. Diel rhythms in edaphic community metabolism in a Delaware salt marsh. Ecology 54:1160–1163.

Gallopin, G. C. 1971. A generalized model of a resource population system. I, II. Oecologia 7:382–413; 414–432.

Gambell, R. 1968. Seasonal cycles and reproduction in sei whales of the southern hemisphere. Disc. Repts. 35:31–134.

Gardner, W. D. 1977. Fluxes, dynamics, and chemistry of particulates in the ocean. Ph.D. Thesis, Massachussetts Inst. of Technology/Woods Hole Oceanographic Institution.

Gardner, W. D. 1980. Sediment trap dynamics and calibration: A laboratory evaluation. J. Mar. Res. 38:17–39.

Gardner, W. D. 1980a. Field assessment of sediment traps. J. Mar. Res. 38:41–52.

Gardner, M. R., and W. R. Ashby. 1970. Connectance of large dynamic (cybernetic) systems: Critical values for stability. Nature 228:784.

Gardner, W. S., and D. W. Menzel. 1974. Phenolic aldehydes as indicators of terrestrially derived organic matter in the sea. Geochim. Cosmochim. Acta 38:813–822.

Gargas, E. 1970. Measurement of primary production, dark fixation, and vertical distribution of the microbenthic algae in the Oresund. Ophelia 8:231–253.

Garrod, D. J., and A. D. Clayden. 1972. Current biological problems in the conservation of deep-sea fishery resources. Symp. Zool. Soc. Lond. 29:161–184.

Garside, C., and T. C. Malone. 1978. Monthly oxygen and carbon budgets of the New York Bight apex. Estuar. Coast. Mar. Sci. 6:93–104.

Gaskin, D. E. 1978. Form and function in the digestive tract and associated organs in cetacea, with a consideration of metabolic rates and specific energy budgets. Oceanogr. Mar. Biol. Ann. Rev. 16:313–415.

Gaudy, R. 1974. Feeding four species of pelagic copepods under experimental conditions. Mar. Biol. 25:125–141.

Geiselman, J. A., and O. S. McConnell. 1981. Polyphenols in brown algae *Fucus versicolosus* and *Arcophyllum nodosum*: Chemical defenses against the marine herbivorous snail, *Littorina littorea*. J. Chem. Ecol. 7:1115–1133.

Gerlach, S. A. 1978. Food-chain relationships in substidal silty sand marine sediments and the role of meiofauna in stimulating bacterial productivity. Oecologia 33:55–69.

Giblin, A. 1982. Uptake and Remobilization of Heavy Metals in Salt Marshes. Ph.D. Diss. Boston University.

Giere, O. 1981. The gutless marine oligochaete *Phallodrilus leukodermatus*.

Structural studies on an aberrant tubificid associated with bacteria. Mar. Ecol. Progr. Ser. 5:353–357.
Gieskes, W. W. C., and G. W. Kraay. 1975. The phytoplankton spring bloom in Dutch coastal waters of the North Sea. Neth. J. Sea Res. 9:166–196.
Gieskes, W. W. C., G. W. Kraay, and M. A. Baars. 1979. Current ^{14}C methods for measuring primary production: Gross underestimates in oceanic waters. Neth. J. Sea Res. 13:58–78.
Gilpin, M. E. 1975. Stability of feasible predator-prey systems. Nature 254:137–139.
Glander, K. E. 1981. Feeding patterns in mantled howling monkeys. Pp. 231-259 in A. C. Kamil and T. D. Sargent (eds.), Foraging Behavior. Garland STPM.
Glenn, A. R. 1976. Production of extracellular protein by bacteria. Ann. Rev. Microbiol. 30:41–62.
Glibert, P. M. 1982. Regional studies of daily, seasonal, and size fraction variability in ammonium remineralization. Mar. Biol. 70:209–222.
Glibert, P. M., and J. C. Goldman. 1981. Rapid ammonium uptake by marine phytoplankton. Mar. Biol. Lett. 2:25–31.
Glibert, P. M., F. Lipschultz, J. J. McCarthy, and M. A. Altabet. 1982. Isotope dilution models of uptake and remineralization of ammonium by marine plankton. Limnol. Oceanogr. 27:639–650.
Glombitza, M. 1977. Highly hydroxilated phenols of the phaeophyceae. Pp. 191–204 in D. J. Faulkner and W. M. Fenicol (eds.), Marine Natural Products Chemistry. Plenum.
Glover, H. E. 1980. Assimilation numbers in cultures of marine phytoplankton. J. Plankton Res. 2:69–79.
Glynn, P. W. 1974. The impact of *Acanthaster* on corals and coral reefs in the Eastern Pacific. Environ. Conser. 1:295–304.
Glynn, P. W. 1976. Some physical and biological determinants of coral community structure in the eastern Pacific. Ecol. Monogr. 46:431–456.
Godshalk, G. L., and R. G. Wetzel. 1978. Decomposition of aquatic angiosperms. I. Dissolved components. Aquat. Bot. 5:281–300.
Godshalk, G. L., and R. G. Wetzel. 1978a. Decomposition of aquatic angiosperms. III. *Zostera marina* L. and a conceptual model of decomposition. Aquat. Bot. 5:329–354.
Goering, J. J., R. C. Dugdale, and D. W. Menzel. 1966. Estimates of *in situ* rates of nitrogen uptake by *Trichodesmium* sp. in the tropical Atlantic Ocean. Limnol. Oceanog. 11:614–620.
Goldman, C. R. 1972. The role of minor nutrients in limiting the productivity of aquatic ecosystems. Pp. 21–33 in G. E. Likens (ed.), Nutrients and Eutrophication: The Limiting Nutrient Controversy. Spec. Symp. Vol. 1. Amer. Soc. Limnol. Oceanogr. Allen.
Goldman, J. C., and E. L. Carpenter. 1974. A kinetic approach to the effect of temperature on algal growth. Limnol. Oceanog. 19:756–766.
Goldman, J. C., and J. J. McCarthy. 1978. Steady state growth and ammonium uptake of a fast-growing marine diatom. Limnol. Oceanogr. 23:695–730.
Goldman, J. C., J. J. McCarthy, and D. G. Peavey. 1979. Growth rate influence on the chemical composition of phytoplankton in oceanic waters. Nature 279: 210–215.

Goldsmith, F. B., and C. M. Harrison. 1976. Description and analysis of vegetation. In S. B. Chapman (ed.), Methods in Plant Ecology. Blackwell.

Golley, F., H. T. Odum, and R. F. Wilson. 1962. The structure and metabolism of a Puerto Rico red mangrove forest in May. Ecology 43:9–19.

Golterman, H. L. 1973. Vertical movement of phosphate in freshwater. Pp. 509–538 in E. L. Griffith, A. Beeton, J. M. Spencer, and D. T. Mitchell (eds.), Environmental Phosphorus Handbook. Wiley.

Good, R. E., N. F. Good, and B. R. Frasco. 1982. A review of primary production and decomposition dynamics of the belowground marsh component. Pp. 139–158 in V. S. Kennedy (ed.), Estuarine Comparisons. Academic.

Goodman, D. 1982. Optimal life histories, optimal notation, and the value of reproductive value. Amer. Nat. 119:803–823.

Gordon, D. C., Jr. 1971. Distribution of particulate organic carbon and nitrogen at an oceanic station in the central Pacific. Deep-Sea Res. 18:1127–1134.

Gordon, D. C., Jr. 1977. Variability of particulate organic carbon and nitrogen along the Halifax-Bermuda section. Deep-Sea Res. 24:257–270.

Gordon, D. M., P. B. Birch, and A. J. McComb. 1980. The effect of light, temperature, and salinity on photosynthetic rates of an estuarine *Cladophora*. Bot. Mar. 23:749–755.

Goreau, T. F. 1959. The ecology of Jamaican coral reefs. I. Species composition and zonation. Ecology 40:67–90.

Goss-Custard, J. D. 1969. The winter feeding ecology of the redshank *Tringa totanus*. Ibis 111:338–356.

Goss-Custard, J. D. 1970. Factors affecting the diet and feeding rate of the redshank *Tringa totanus*. Pp. 101–110 in A. Watson (ed.), Animal Populations in Relation to Their Food Resources. Blackwell.

Goss-Custard, J. D. 1981. Feeding behavior of redshank, *Tringa totanus*, and optimal foraging theory. Pp. 115–134 in A. C. Kamil and T. D. Sargent (eds.), Foraging Behavior. Garland STPM.

Gotto, J. W., and B. F. Taylor. 1976. N_2 fixation associated with decaying leaves of the red mangrove (*Rhizophora mangle*). Appl. Env. Microbiol. 31:781–783.

Govindjee. 1976. Photosynthesis. Wiley.

Govindjee, and P. R. Mohanty. 1972. Photochemical aspects of photosynthesis in blue-green algae. Pp. 171–196 in T. V. Desikachary (ed.), Taxonomy and Biology of Blue-green Algae. Center for Advanced Study in Botany. Madras.

Grant, P. R. 1972. Convergent and divergent character displacement. Biol. J. Linn. Soc. 4:39–68.

Grant, W. D., L. F. Boyer, and L. P. Sanford. 1982. The effect of biological processes on the initiation of sediment motion in non-cohesive sediment. J. Mar. Res. 40:659–677.

Grassle, J. F. 1977. Slow recolonization of deep sea sediment. Nature 265:618–619.

Grassle, J. F., and J. P. Grassle. 1974. Opportunistic life histories and genetic systems in marine benthic polychaetes. J. Mar. Res. 32:253–284.

Grassle, J. F., and J. P. Grassle. 1976. Sibling species in the marine pollution indicator *Capitella* (Polychaeta). Science 192:567–569.

Grassle, J. F., and H. L. Sanders. 1973. Life histories and the role of disturbance. Deep-Sea Res. 20:643–659.

Gray, J. S. 1966. The attractive factor of intertidal sand to *Protodrilus symbioticus*. J. Mar. Biol. Assoc. U.K. 46:627–645.

Gray, J. S. 1967. Substrate selection by the archiannelid *Protodrilus rubropharyngeus*. Helgol. Wiss. Meeres. 15:252–269.

Green, G. 1977. Ecology of toxicity in marine sponges. Mar. Biol. 40:207–215.

Green, R.H. 1971. A multivariate approach to the Hutchinsonian niche: Bivalve molluscs of central Canada. Ecology 52:543–556.

Greze, V. N. 1978. Production in animal populations. Pp. 89–114 in O. Kinne (ed.), Marine Ecology Vol. 4. Wiley.

Grice, G. D., R. P. Harris, M. R. Reeve, J. F. Heinbokel, and C. O. Davis. 1980. Large scale enclosed water-column ecosystems. An overview of Foodweb I, the final CEPEX experiment. J. Mar. Biol. Assoc. U.K. 60:401–414.

Griffith, D. 1980. Foraging costs and relative prey size. Amer. Nat. 116:743–752.

Griffith, E. J. 1973. Environmental phosphorus—An editorial. Pp. 683–698 in E. J. Griffith, A. Beeton, J. M. Spencer, and D. T. Mitchell (eds.), Environmental Phosphorus Handbook. Wiley.

Griffiths, K. J., and C. S. Holling. 1969. A competition submodel for parasites and predators. Canad. Entom. 101:785–818.

Griffiths, R. J. 1980. Filtration, respiration and assimilation in the black mussel *Choromytilus meridionalis*. Mar. Ecol. Prog. Ser. 3:63–70.

Griffiths, R. J. 1980a. Natural food availability and assimilation in the bivalve *Choromytilus meridionalis*. Mar. Ecol. Prog. Ser. 3:151–156.

Grime, J. P. 1977. Evidence for the existence of three primary strategies in plants and its relevance to ecological and evolutionary theory. Amer. Nat. 111:1169–1194.

Grøntved, J. 1960. On the productivity of microbenthos and phytoplankton in some Danish fjords. Meddr. Danm. Fish.-og Havunders. N.S. 3:55–92.

Gross, M. G. 1977. Oceanography: A View of the Earth. Int. Ed. Prentice-Hall.

Grundmanis, V., and J. W. Murray. 1977. Nitrification and denitrification in marine sediments from Puget Sound. Limnol. Oceanogr. 22:804–813.

Guillard, R. R. L., and P. Kilham. 1977. The ecology of marine planktonic diatoms. Pp. 372–469 in D. Werner (ed.), The Biology of Diatoms. Blackwell.

Gulbrandsen, R. A., and C. E. Roberson. 1973. Inorganic phosphorus in seawater. Pp. 117–140 in E. J. Griffith, A. Beeton, J. M. Spencer, and D. T. Mitchell (eds.), Environmental Phosphorus Handbook. Wiley.

Gulland, J. A. 1970. Food chain studies and some problems in world fisheries. Pp. 296–315 in J. H. Steele (ed.), Marine Food Chains. Univ. California.

Gulland, J. A. 1971. Ecological aspects of fishery research. Adv. Ecol. Res. 7:115–176.

Haedrich, R. L., G. T. Rowe, and P. T. Polloni. 1980. The megabenthic fauna in the deep sea south of New England. Mar. Biol. 57:165–179.

Hagström, A., U. Larsson, P. Hörstedt, and S. Normark. 1979. Frequency of dividing cells, a new approach to the determination of bacterial growth rates in aquatic environments. Appl. Envir. Microbiol. 37:805–812.

Haines, B. L., and E. L. Dunn. 1976. Growth and resource allocation responses of *Spartina alterniflora* Loisel. to three levels of NH_4-N, Fe, and NaCl in solution culture. Bot. Gazette 137:224–230.

Haines, E., A. Chalmers, R. Hanson, and B. Sherr. 1977. Nitrogen pools and fluxes in a Georgia salt marsh. Pp. 241–254 in M. Wiley (ed.), Estuarine Processes, Vol. II. Academic.

Hairston, N. G. 1959. Species abundance and community organization. Ecology 40:404–416.

Hairston, N. G., J. D. Allan, R. K. Colwell, D. J. Futuyma, J. Howell, M. D. Lubin, J. Mathias, and J. H. Vandermeer. 1968. The relationships between species diversity and stability: An experimental approach with protozoa and bacteria. Ecology 49:1091–1101.

Hairston, N. G., F. E. Smith, and L. B. Slobodkin. 1960. Community structure, population control, and competition. Amer. Nat. 94:421–425.

Hall, C. A. S., and R. Moll. 1975. Methods of assessing aquatic primary productivity. Pp. 19–53 in J. Lieth and R. H. Whittaker (eds.), Primary Productivity in the Biosphere. Springer-Verlag.

Hall, D. J. 1964. An experimental approach to the dynamics of a natural population of *Daphnia galeata mendotae*. Ecol. 45:94–112.

Hall, D. J., W. E. Cooper, and E. E. Werner. 1970. An experimental approach to the production dynamics and structure of freshwater animal communities. Limnol. Oceanogr. 15:839–928.

Hall, D. J., S. T. Threlkeld, C. W. Burns, and P. Crowley. 1976. The size-efficiency hypothesis and the size structure of zooplankton communities. Ann. Rev. Ecol. Syst. 7:177–208.

Hallagraeff, G. M. 1981. Seasonal study of phytoplankton pigments and species at a coastal station off Sidney: Importance of the diatoms and the nanoplankton. Mar. Biol. 61:107–118.

Halldal, P. 1974. Light and photosynthesis of different marine algal groups. Pp. 345–360 in N. G. Jerlov and E. Steemann Nielsen (eds.), Optical Aspects of Oceanography. Academic.

Hamilton, W. A. 1979. Microbial energetics and metabolism. Pp. 22–44 in J. M. Lynch and N. J. Poole (eds.), Microbial Ecology. Wiley.

Hamner, P., and W. M. Hamner. 1977. Chemosensory tracking of scent trails by the planktonic shrimp *Acetes sibogae australis*. Science 195:886–888.

Hamner, W. M., L. P. Madin, A. L. Alldredge, R. W. Gilmer, and P. P. Hamner 1975. Underwater observations of gelatinous zooplankton: Sampling problems, feeding biology, and behavior. Limnol. Oceanog. 20:907–917.

Hamre, J. 1978. The effect of recent changes in the North Sea mackerel fishery on stock and yield. Rapp. Proc.-Verb. Reun. Cons. Int. Explor. Mer 172:197–210.

Handa, N. 1977. Land sources of marine organic matter. Mar. Chem. 5:341–359.

Hanisak, M. D. 1979. Growth patterns of *Codium fragile* ssp. *tomentosoides* in response to temperature, irradiance, salinity, and nitrogen source. Mar. Biol. 50:319–332.

Hanlon, R. D. G. 1981. Influence of grazing by Collembola on the activity of senescent fungal colonies grown on media of different nutrient concentration. Oikos 36:362–367.

Hanlon, R. D. G., and J.M. Anderson. 1980. The influence of macroarthropod feeding activity on fungi and bacteria in decomposing oak leaves. Soil Biol. Biochem. 12:255–261.

Hanson, R. B., K. R. Tenore, S. Bishop, C. Chamberlain, M. M. Pamatmat, and J. Tietjen. 1981. Benthic enrichment in the Georgia Bight related to Gulf Stream intrusions and estuarine outwelling. J. Mar. Res. 39: 417–441.

Haq, S. M. 1967. Nutritional physiology of *Metridia lucens* and *M. longa* from the Gulf of Maine. Limnol. Oceanogr. 12:40–51.

Harbison, G. R., and R. W. Gilmer. 1976. The feeding rates of the pelagic tunicate *Pegea* and two other salps. Limnol. Oceanogr. 21:517–528.

Harbison, G. R., L. P. Madin, and N. R. Swanberg. 1978. On the natural history and distribution of oceanic ctenophores. Deep-Sea Res. 25:233–256.

Harbison, G. R., and V. L. McAlister. 1980. Fact and artifact in copepod feeding experiments. Limnol. Oceanogr. 25:971–981.

Harding, D., J. H. Nichols, and D. Tungate. 1978. The spawning of plaice (*Pleuronectes platessa* L.) in the southern North Sea and English Channel. Rapp. Proces-Verb. Cons. Intern. Explor. Mer 172:102–113.

Hardy, A. C., and E. R. Gunther. 1935. The plankton of the South Georgia whaling grounds and adjacent waters, 1926–27. Discovery Rep. 2:1–146.

Hargrave, B. T. 1969. Epibenthic algal production and community respiration in the sediments of Marion Lake. J. Fish. Res. Bd. Canada 26:2003–2026.

Hargrave, B. T. 1970. The utilization of benthic microflora by *Hyallela azteca* (Amphipoda). J. Anim. Ecol. 39:427–438.

Hargrave, B. T. In press. Sinking of particulate matter from the surface water of the ocean. In J. E. Hobbie and P. J. Leb. Williams (eds.), Heterotrophic Activity in the Sea. Plenum.

Hargrave, B. T., and N. M. Burns. 1979. Assessment of sediment trap collection efficiency. Limnol. Oceanogr. 24:1124–1136.

Harris, G. P. 1980. The measurement of photosynthesis in natural populations of phytoplankton. Pp. 129–187 in I. Morris (ed.), The Physiological Ecology of Phytoplankton. Univ. of California.

Harrison, G. W. 1979. Stability under environmental stress: Resistance, resilience, and variability. Amer. Nat. 113:659–669.

Harrison, P. G. 1977. Decomposition of macrophyte detritus in seawater: Effects of grazing by amphipods. Oikos 28:165–169.

Harrison, P. G. 1982. Control of microbial growth and of amphipod grazing by water-soluble compounds from leaves of *Zostera marina*. Mar. Biol. 67:225–230.

Harrison, P. G., and A. T. Chan. 1980. Inhibition of growth of microalgae and bacteria by extracts of eelgrass (*Zostera marina*) leaves. Mar. Biol. 61:21–26.

Harrison, P. G., and K. H. Mann. 1975. Chemical changes during the seasonal cycle of growth and decay in eelgrass (*Zostera marina*) on the Atlantic Coast of Canada. J. Fish. Res. Bd. Canada 32:615–621.

Harrison, W. G. 1978. Experimental measurements of nitrogen remineralization in coastal waters. Limnol. Oceanog. 23:684–694.

Harrold, C. 1982. Escape responses and prey availability in a kelp forest predator-prey system. Amer. Nat. 119:132–135.

Hart, T. J. 1942. Phytoplankton periodicity in Antarctic surface waters. Discovery Rep. 21:261–356.

Hartley, R. D., and E. C. Jones. 1977. Phenolic components and degradability of cell walls of grass and legume species. Phytochemistry 16:1531–1534.

Hartman, J., H. Caswell, and I. Valiela. In press. Effect of wrack accumulation on salt marsh vegetation. Proc. 17th European Mar. Biol. Symp.

Hartog, C. den. 1970. The Sea Grasses of the World. North-Holland.

Hartog, C. den. 1977. Structure, function, and classification in seagrass communi-

ties. Pp. 89–122 in C. P. McRoy and C. Helfferich (eds.), Seagrass Ecosystems. M. Dekker.
Harvey, G. R., W. G. Steinhauer, and J. M. Teal. 1973. Polychlorobiphenyls in North Atlantic ocean water. Science 180:643–644.
Harvey, H. W. 1945. The Chemistry and Biology of Seawater. Cambridge Univ.
Harvey, H. W. 1950. On the production of living matter in the sea off Plymouth. J. Mar. Biol. Assoc. U.K. 29:97–138.
Harvey, H. W. 1955. The Chemistry and Fertility of Sea Waters. Cambridge Univ.
Harvey, H. W., L. N. Cooper, M. V. Leborn, and F. S. Russell. 1935. Plankton production and its control. J. Mar. Biol. Assoc. U.K. 20:407–441.
Hasle, G. R. 1950. Phototactic vertical migration in marine dinoflagellates. Oikos 2:162–175.
Hasle, G. R. 1954. More on phototactic diurnal migration in marine dinoflagellates. Nytt Mag. Bot. 2:139–147.
Hassell, M. P., J. H. Lawton, and J. R. Beddington. 1976. The components of arthropod predation. I. The prey death rate. J. Anim. Ecol. 45:135–164.
Hassell, M. P., J. H. Lawton, and J. R. Beddington. 1977. Sigmoid functional responses by invertebrate predators and parasitoids. J. Anim. Ecol. 46:249–262.
Hassell. M. P., and R. M. May. 1974. Aggregation of predators and insect parasites and its effect on stability. J. Anim Ecol. 43:567–594.
Hassell, M. P., and G. C. Varley. 1969. New inductive model for insect parasites and its bearing on biological control. Nature 223:1133–1137.
Hatch, M. D., and C. R. Slack. 1970. Photosynthetic CO_2 fixation pathways. Ann. Rev. Plant Physiol. 21:141–162.
Hatcher, B. G., A. R. O. Chapman, and K. H. Mann. 1977. An annual carbon budget for the kelp *Laminaria longicruris*. Mar. Biol. 44:85–96.
Hauck, R. D., and J. M. Bremner. 1976. Use of tracers for soil and fertilizer research. Adv. Agron. 28:219–266.
Haury, L. R., M. G. Briscoe, and M. H. Orr. 1979. Tidally generated internal wave packets in Massachessetts Bay. Nature 278:312–317.
Haury, L. R., J. A. McGowan, and P. H. Wiebe. 1978. Patterns and processes in the time-space scale of plankton distributions. Pp. 277–327 in J. H. Steele (ed.), Spatial Patterns in Plankton Communities. Plenum.
Haury, L. R., and D. Weihs. 1976. Energetically efficient swimming behavior of negatively buoyant zooplankton. Limnol. Oceanogr. 21:797–803.
Haury, L. R., P. H. Wiebe, and S. H. Boyd. 1976. Longhurst-Hardy plankton recorders: Their design and use to minimize bias. Deep-Sea Res. 23:1217–1229.
Haxo, F. T., and L. R. Blinks. 1950. Photosynthetic action spectra of marine algae. J. Gen. Physiol. 33:389–422.
Hayes, F. R., and J. R. Phillips. 1958. Lake water and sediment. IV. Radiophosphorus equilibrium with mud, plants, and bacteria under oxidized and reduced conditions. Limnol. Oceanogr. 3:459–475.
Hayward, T. L., and J. A. McGowan. 1979. Pattern and structure in an oceanic zooplankton community. Amer. Zool. 1045–1055.
Head, W. D., and E. J. Carpenter. 1975. Nitrogen fixation associated with the marine macroalgae *Codium fragile*. Limnol. Oceanogr. 20:815–823.

Hecky, R. E., and P. Kilham. 1974. Environmental control of phytoplankton cell size. Limnol. Oceanogr. 19:361–366.
Hedges, J. I., and D. C. Mann. 1979. The lignin geochemistry of marine sediments from the Southern Washington Coast. Geochim. Cosmochim. Acta 43:1809–1818.
Heinle, D. R. 1969. Production of a calanoid copepod, *Acartia tonsa*, in the Patuxent River estuary. Chesap. Sci. 7:59–74.
Heinle, D. R., and D. A. Flemer. 1975. Carbon requirements of a population of the estuarine copepod *Eurytemora affinis*. Mar. Biol. 31:235–247.
Heinrich, A. K. 1962. The life histories of plankton animals and seasonal cycles of plankton communities in the oceans. J. Cons. Int. Explor. Mer 27:15–24.
Heinrich, A. K. 1963. On the filtering ability of copepods in the boreal and tropical region of the Pacific. Trudy Inst. Okeanol. 71:60–71.
Helder, W. 1974. The cycle of dissolved inorganic nitrogen compounds in the Dutch Wadden Sea. Neth. J. Sea Res. 8:154–173.
Hellebust, J. A. 1967. Excretion of organic compounds by cultured and natural populations of marine phytoplankton. Pp. 761–766 in G. H. Lauff (ed.), Estuaries. Amer. Assoc. Adv. Sci. Publ. No. 83.
Hellebust, J. A. 1974. Extracellular products. Pp. 838–863 in W. D. P. Stuart (ed.), Algal Physiology and Biochemistry. Bot. Monogr. 10. Univ. Calif.
Hemmingsen, A. M. 1960. Energy metabolism as related to body size and respiratory surfaces, and its evolution. Part II. Rep. Steno Mem. Hosp. Nord. Insulin Lab. 9:1–110.
Hempel, G. 1965. On the importance of larval survival for the population dynamics of marine food fish. Calif. Coop. Oceanic Fish. Inv. Rep. 10:13–23.
Hendrix, S. D. 1980. An evolutionary and ecological perspective of the insect fauna of ferns. Amer. Nat. 115:171–196.
Henny, C. J., and H. M. Wight. 1969. An endangered osprey population: Estimates of mortality and production. Auk 86:188–198.
Herrera, C. H. 1982. Grasses, grazers, mutualism, and coevolution: A comment. Oikos 38:254–259.
Hessler, R. R., and H. L. Sanders. 1967. Faunal diversity in the deep-sea. Deep-Sea Res. 14:65–78.
Hewer, H. R. 1964. The determination of age, sexual maturity, longevity and a life-table in the grey seal (*Halichoerus grypus*). Proc. Zool. Soc. Lond. 142:593–623.
Hibbert, C. J. 1977. Energy relations of the bivalve *Mercenaria mercenaria* on an intertidal mudflat. Mar. Biol. 44:77–84.
Hildrew, C. W., and C. R. Townsend. 1977. The influence of substrate on the functional response of *Plectrocnemia conspersa* (Curtis) larvae (Trichoptera: Polycentropodidae). Oecologia 31:21–26.
Hinga, K. R. 1979. The food requirements of whales in the southern hemisphere. Deep-Sea Res. 26A:569–577.
Hinga, K. R., J. McN. Sieburth, and G. Ross Heath. 1979. The supply and use of organic material at the deep-sea floor. J. Mar. Res. 37:557–579.
Hixon, M. A., and W. N. Brostoff. 1983. Damselfish as keystone species in reverse: Intermediate disturbance and diversity of reef algae. Science 220:511–513.
Hobbie, J. E., R. J. Daley, and J. Jasper. 1977. Use of nucleopore filters for

counting bacteria by fluorescence microscopy. Appl. Env. Microbiol. 33:1225–1228.

Hobbie, J.E., O. Holm-Hansen, T. T. Packard, L. R. Pomeroy, R. W. Sheldon, J. P. Thomas, and W. J. Wiebe. 1972. A study of the distribution and activity of microorganisms in ocean water. Limnol. Oceanogr. 17:544–555.

Hobbie, J. E., and C. Lee. 1980. Microbial production of extracellular material: Importance in benthic ecology. Pp. 341–346 in B. C. Coull and K. R. Tenore (eds.), Marine Benthic Dynamics, Univ. South Carolina.

Hobbie, J. E., and R. T. Wright. 1965. Competition between planktonic bacteria and algae for oceanic solutes. Mem. Ist. Ital. Idrobiol. 18 Suppl.:175–185.

Hobson, L. A. 1971. Relationships between particulate organic carbon and microorganisms in upwelling areas off Southwest Africa. Inv. Pesq. 35:195–208.

Hobson, L. A. 1974. Effects of interaction of irradiance, daylength, and temperature on division rates of three species of marine unicellular algae. J. Fish. Res. Bd. Canada 31:391–395.

Hoffmann, E. E., J. M. Klink, and G.-A. Paffenhöfer. 1981. Concentrations and vertical fluxes of zooplankton fed pellets on a continental shelf. Mar. Biol. 61:327–335.

Hogetsu, K., M. Hatanaka, T. Hanaoka, and T. Kawamura (eds.). 1977. Productivity of Biocenoses in Coastal Regions of Japan. JIBP Synthesis 14. Univ. Tokyo.

Holden, M. J. 1973. Are long-term sustainable fisheries for elasmobranchs possible? Rapp. Proc. -Verb. Cons. Intern. Explor. Mer 164:360–367.

Holden, M. J. 1978. Long-term changes in landings of fish from the North Sea. Rapp. Proc.-Verb. Reun. Cons. Int. Explor. Mer 172:11–26.

Holeton, G. F. 1974. Metabolic adaptations of polar fish: Fact or artifact? Physiol. Zool. 47:137–152.

Holling, C. S. 1959. Some characteristics of simple types of predation and parasitism. Canad. Entomol. 91:385–398.

Holling, C. S. 1965. The functional response of predators to prey density and its role in mimicry and population regulation. Mem. Entomol. Soc. Canada No. 45.

Holling, C. S. 1966. The functional response of invertebrate predators to prey density. Mem. Entomol. Soc. Canada 48:1–85.

Holme, N. A., and A. D. McIntyre. 1971. Methods for the Study of Marine Benthos. IBP Handbook 16. Blackwell.

Holm-Hansen, O. 1970. ATP levels in algal cells as influenced by environmental conditions. Plant Cell Physiol. 11:689–700.

Holm-Hansen, O. and C. R. Booth. 1966. The measurement of adenosine triphosphate in the ocean and its ecological significance. Limnol. Oceanog. 11:510–519.

Holm-Hansen, O. and B. Riemann. 1978. Chlorophyll a determination: Improvements in methodology. Oikos 30:438–447.

Holm-Hansen, O., S. Z. El-Sayed, G. A. Franceschini, and R. L. Cuhel. 1977. Primary production and the factors controlling phytoplankton growth in the Southern Ocean. In G. A. Llano (ed.), Adaptations Within Antarctic Ecosystems. Proc. 3rd SCAR Symp. on Antarct. Biol. Smithsonian Institution.

Honjo, S. 1976. Coccoliths: Production, transportation and sedimentation. Mar. Micropaleont. 1:65–79.

Honjo, S. 1980. Material fluxes and modes of sedimentation in the mesopelagic and bathypelagic zones. J. Mar. Res. 38:53–97.
Honjo, S., and K.O. Emery 1976. Suspended matter of eastern Asia: Scanning electron microscopy and X-ray peak analysis. Pp. 259–288 in H. Aoki and S. Iizuka (eds.), Volcanoes and Techosphere. Tokai Univ.
Honjo, S., and M. R. Roman. 1978. Marine copepod fecal pellets: Production, preservation, and sedimentation. J. Mar. Res. 36:45–57.
Hooper, F. F. 1973. Origin and fate of organic phosphorus compounds in aquatic systems. Pp. 179–202 in E. J. Griffith, A. Beeton, J. M. Spencer, and D. T. Mitchell (eds.), Environmental Phosphorus Handbook. Wiley.
Hopkinson, C. S., J. G. Gosselink, and R. T. Parrondo. 1980. Production of coastal Louisiana marsh plants calculated from phenometric techniques. Ecology 61:1091–1098.
Hopkinson, C. S., and R. L. Wetzel. 1982. In situ measurements of nutrient and oxygen fluxes in a coastal marine benthic community. Mar. Ecol. Progr. Ser. 10:29–35.
Hoppe, G. 1978. Relations between active bacteria and heterotrophic potential in the sea. Neth. J. Sea Res. 12:78–98.
Hoppe, H. A., T. Levring, and Y. Tanaka. 1979. Marine Algae in Pharmaceutical Science. Walter de Gruyter.
Horn, H. S. 1966. Measurement of "overlap" in comparative ecological studies. Amer. Nat. 100:419–424.
Horn, M. H. 1972. The amount of space available for marine and freshwater fishes. Fish. Bull. U.S. Fish. Wildlife Ser. 72:1295–1297.
Horwood, J. W., and J. D. Goss-Custard. 1977. Predation by the oystercatcher, *Haematopus ostralegus* (L.), in relation to the cockle, *Cerastoderma edule* (L.), fishery in the Burry Inlet, South Wales. J. Appl. Ecol. 14:139–158.
Houghton, R. W., and M. A. Mensah. 1978. Physical aspects and biological consequences of Ghanian coastal upwelling. Pp. 167–180 in R. Boje and M. Tomczak (eds.), Upwelling Ecosystems. Springer-Verlag.
Howarth, R. W. 1979. Pyrite: Its rapid formation in a salt marsh and its importance in ecosystem metabolism. Science 203:49–51.
Howarth, R. W., A. Giblin, J. Gale, B. J. Peterson, and G. W. Luther. 1982. Reduced sulfur compounds in the pore waters of a New England salt marsh. In R. O. Hallberg (ed.), Environmental Biogeochemistry. Ecol. Bull. 35:135–152.
Howarth, R. W., and J. M. Teal. 1979. Sulfate reduction in a New England salt marsh. Limnol. Oceanogr. 24:999–1013.
Howarth, R. W., and J. M. Teal. 1980. Energy flow in a salt marsh ecosystem: The role of reduced inorganic sulfur compounds. Amer. Nat. 116:862–872.
Howell, A. B. 1930. Aquatic Mammals. Charles Thomas.
Howes, B. W., R. W. Howarth, J. M. Teal, and I. Valiela. 1981. Oxidation-reduction potentials in a salt marsh: Spatial patterns and interactions with primary production. Limnol. Oceanogr. 26:350–360.
Howes, B. L., J. W. H. Dacey, and G. M. King. In press. Carbon flow through oxygen and sulfate reduction pathways in salt marsh sediments. Limnol. Oceanogr.
Howes, B. L., J. W. H. Dacey, and J. M. Teal. In press a. Annual carbon mineralization and belowground production of *Spartina alterniflora* in a New England salt marsh. Ecology.

Hubbell, S. P. 1973. Population and simple food webs as energy filters. II. Two-species systems. Amer. Nat. 107:122–151.

Hubold, G. 1978. Variations in growth rate and maturity of herring in the Northern North Sea in the years 1955–1973. Rapp. Proc.-Verb. Reun. Cons. Int. Explor. Mer 172:154–163.

Huffaker, C. B. 1958. Experimental studies on predation. II. Dispersion factors and predator-prey oscillations. Hilgardia 27:343–383.

Hughes, R. N. 1970. An energy budget for a tidal flat population of the bivalve *Scrobicularia plana* DaCosta. J. Anim. Ecol. 39:357–381.

Hughes, R. N. 1980. Optimal foraging theory in the marine context. Oceanogr. Mar. Biol. Ann. Rev. 18:423–481.

Hughes, R. N., and R. W. Elner. 1979. Tactics of a predator, *Carcinus maenas* and morphological responses of the prey, *Nucella lapillus*. J. Anim. Ecol. 48:65–78.

Hulburt, E. M. 1962. Phytoplankton in the Southwestern Sargasso Sea and North Equatorial Current, February, 1961. Limnol. Oceanogr. 7:307–315.

Hulburt, E. M. 1966. The distribution of phytoplankton, and its relationship to hydrography, between southern New England and Venezuela. J. Mar. Res. 24:67–81.

Hulburt, E. M., and R. R. L. Guillard. 1968. The relationship of the distribution of the diatom *Skeletonema tropicum* to temperature. Ecology 49:337–339.

Humphreys, W. F. 1979. Production and respiration in animal populations. J. Anim. Ecol. 48:427–453.

Hunter, J. R. 1966. Procedure for analysis of schooling behavior. J. Fish. Res. Bd. Canada 23:547–562.

Hunter, J. R., and G. L. Thomas. 1974. Effect of prey distribution and density on the searching and feeding behaviour of larval anchovy *Engraulis mordax*. Pp. 559–574 in J. H. S. Baxter (ed.), The Early Life History of Fish. Springer-Verlag.

Hurlbert, S. H. 1971. The non-concept of species diversity: A critique and alternative parameters. Ecology 52:577–586.

Huston, M. 1979. A general hypothesis of species diversity. Amer. Nat. 113:81–101.

Hutcheson, K. 1970. A test for comparing diversities based on the Shannon formula. J. Theor. Biol. 29:151–154.

Hutchinson, G. E. 1957. Concluding Remarks. Cold Spr. Harbor Symp. Quant. Biol. 22:415–427.

Hutchinson, G. E. 1961. The paradox of the plankton. Amer. Nat. 95:137–145.

Hutchinson, G. E. 1975. A Treatise on Limnology. Vol. III. Limnological Botany. Wiley.

Hutchinson, G. E. 1978. An Introduction to Population Ecology. Yale Univ.

Hylleberg, J. 1975. Selective feeding by *Abarenicola pacifica* with notes on *Abarenicola vagabunda* and a concept of gardening in lugworms. Ophelia 14:113–137.

Hylleberg-Kristensen, J. 1972. Carbohydrates of some marine invertebrates with notes on their food and natural occurrence of the carbohydrates studied. Mar. Biol. 14:130–142.

Ichimura, S. 1967. Environmental gradient and its relation to primary productivity in Tokyo Bay. Records Oceanogr. Works (Japan) 9:115–128.

Ignatiades, L. 1969. Annual cycle, species diversity and succession of phytoplankton in lower Saronicos Bay, Aegean Sea. Mar. Biol. 3:196–200.

Ikeda, T. 1970. Relationship between respiration rate and body size in marine phytoplankton animals as a function of the temperature of habitat. Bull. Fac. Fish. Hokkaido Univ. 21:91–112.

Ikeda, T. 1974. Nutritional ecology of marine zooplankton. Mem. Fac. Fish. Hokkaido Univ. 22:1–97.

Iles, T. D. 1967. Growth studies on North Sea herring. I. The second year's growth (I-group) of East Anglican herring, 1939–1963. J. Cons. Intern. Explor. Mer 31:56–76.

Iles, T. D. 1968. Growth studies on North Sea herring. II. O-group growth of East Anglican herring. J. Cons. Intern. Explor. Mer 32:98–116.

Inter-American Tropical Tuna Commission 1980. Annual Report of the Inter-American Tropical Tuna Commission. 1979. Pp. 1–227.

Isaacs, J. D. 1973. Potential trophic biomasses and trace-substance concentrations in unstructured marine food webs. Mar. Biol. 22:97–104.

Isaacs, J. D., and R. A. Schwartzlose. 1978. Active animals of the deep-sea floor. Sci. Amer. 233:84–91.

Ittekkot, V., U. Brookmann, W. Michaelis, and E. T. Degens. 1981. Dissolved free and combined carbohydrates during a phytoplankton bloom in the Northern North Sea. Mar. Ecol. Progr. Ser. 4:259–305.

Iturriaga, R. 1979. Bacterial activity related to sedimentary particulate matter. Mar. Biol. 55:157–169.

Iturriaga, R., and H. G. Hoppe. 1977. Observations of heterotrophic activity on photoassimilated organic matter. Mar. Biol. 40:101–108.

Ivlev, V. S. 1961. Experimental Ecology and Feeding of Fishes (D. Scott, translator). Yale Univ..

Ivleva, I. V. 1970. The influence of temperature on the transformation of matter in marine invertebrates. Pp. 96–112 in J. H. Steele (ed.), Marine Food Chains. Univ. California.

Jackson, G. A. 1980. Phytoplankton growth and zooplankton grazing in oligotrophic oceans. Nature 284:439–441.

Jackson, J. B. C. 1977. Habitat area, colonization, and development of epibenthic community structure. Pp. 349–358 in B. F. Keegan, P. O. Leidigh, and P. J. S. Boaden (eds.), Proc. 11th Eur. Mar. Biol. Symp. Pergamon.

Jackson, J. B. C. 1977a. Competition on marine hard substrata: The adaptive significance of solitary and colonial strategies. Amer. Nat. 111:743–767.

Jackson, J. B. C. 1979. Overgrowth competition between encrusting cheilostome ectoprocts in a Jamaican cryptic reef environment. J. Anim. Ecol. 48: 805–823.

Jackson, J. B. C., and L. Buss. 1975. Allelopathy and spatial competition among coral reef invertebrates. Proc. Nat. Acad. Sci. U.S.A. 72:5160–5163.

Jacobs, J. 1974. Quantitative measurements of food selection. A modification of the forage ratio and Ivlev's electivity index. Oecologia 14:413–417.

Jannasch, H. W., K. Eimhjellen, C. O. Wirsen, and A. Farmanfarmanian. 1971. Microbial degradation of organic matter in the deep sea. Science 171:672–675.

Jannasch, H. W., and C. O. Wirsen 1973. Deep-sea microorganisms: In situ response to nutrient enrichment. Science 180:641–643.

Jannasch, H. W., and C. O. Wirsen. 1979. Chemosynthetic primary production at East Pacific seafloor spreading centers. Bioscience 29:592–598.

Jannasch, H. W., and C. O. Wirsen. In press. Microbial activities in undecomposed and decomposed deep-sea water samples. Appl. Env. Microbiol.

Jassby, A., and T. Platt. 1976. Mathematical formulation of the relationships between photosythesis and light for phytoplankton. Limnol. Oceanogr. 21:540–547.

Jawed, M. 1973. Ammonia excretion by zooplankton and its significance to primary productivity during summer. Mar. Biol. 23:115–120.

Jeffrey, S. W. 1980. Algal pigment systems. Pp. 33–58 in P. G. Falkowski (ed.), Primary Productivity in the Sea. Plenum.

Jeffries, C. 1974. Qualitative stability and digraphs in model ecosystems. Ecology 56:238–243.

Jenkin, P. M. 1937. Oxygen production by the diatom *Coscinodiscus excentricus* in relation to submarine illumination in the English Channel. J. Mar. Biol. Assoc. U.K. 22:301–342.

Jenkins, W. J. 1977. Tritium-helium dating in the Sargasso Sea: A measurement of oxygen utilization. Science 196:291–292.

Jensen, A. 1973. Studies on the phytoplankton ecology of the Trondheim fjord II. Chloroplast pigments in relation to abundance and physiological state of the phytoplankton. J. Exp. Mar. Ecol. 11:137–155.

Jerlov, N. G. 1951. Optical studies of ocean water. Rep. Swed. Deep Sea Exped. 3:1–59.

Jerlov, N. G. 1968. Optical Oceanography. Elsevier.

Johannes, R. E. 1964. Phosphorus excretion and body size in marine animals: Microzooplankton and nutrient regeneration. Science 146:923–924.

Johannes, R. E. 1965. Influence of marine protozoa on nutrient regeneration. Limnol. Oceanogr. 10:434:442.

Johannes, R. E. 1980. The ecological significance of the submarine discharge of groundwater. Mar. Ecol. Prog. Ser. 3:365–373.

Johannes, R. E., S. L. Coles, and N. T. Kuenzel. 1970. The role of zooplankton in the nutrition of some scleractinian corals. Limnol. Oceanogr. 15:579–586.

Johnson, B. D., and R. S. Cooke. 1980. Organic particle and aggregate formation resulting from the dissolution of bubbles in seawater. Limnol. Oceanogr. 25:653–661.

Johnson, K. M., C. M. Burney, and J. McN. Sieburth. 1981. Enigmatic marine ecosystem metabolism measured by direct diel CO_2 and O_2 flux in conjunction with DOC release and uptake. Mar. Biol. 65:49–60.

Johnson, P. W., and J. McN. Sieburth. 1979. Chroococcoid cyanobacteria in the sea: A ubiquitous and diverse phototrophic biomass. Limnol. Oceanogr. 24:928–934.

Johnson, R. G. 1976. Conceptual models of benthic marine communities. Pp. 149–159 in T. J. Schopf (ed.), Models in Paleobiology. Freeman and Cooper.

Johnston, R. 1963a. Antimetabolites as an aid to the study of phytoplankton nutrition. J. Mar. Biol. Assoc. U.K. 43:409–425.

Johnston, R. 1963b. Seawater, the natural medium of phytoplankton. I. General features. J. Mar. Biol. Assoc. U.K. 43:427–456.

Jolley, E. T., and A. K. Jones. 1977. The interaction between *Navicula muralis* and an associated species of *Flavobacterium*. Br. Phycol. J. 12:315–328.

Jones, B. C., and G. H. Geen 1977. Food and feeding of spiny dogfish (*Squalus acanthias*) in British Columbia waters. J. Fish. Res. Bd. Canada 34: 2067–2078.

Jones, R. 1964. A review of methods of estimating population size from marking experiments. Rapp. Proc.-Verb. Cons. Int. Explor. Mer 155:202–209.
Jones, R. 1973. Density dependent regulation of the numbers of cod and haddock. Rapp. Proc.-Verb. Reun. Cons. Int. Explor. Mer 164:156–173.
Jones, R. 1978. Competition and coexistence with particular reference to gadoid fish species. Rapp. Proc.-Verb. Reun. Cons. Int. Explor. Mer 172:292–300.
Jones, R., and J. R. G. Hislop. 1978. Changes in North Sea haddock and whiting. Rapp. Proc.-Verb. Reun. Cons. Int. Explor. Mer 172:58–71.
Jones, R. C. 1973a. The stock and recruitment relation as applied to the North Sea haddock. Rapp. Proc.-Verb. Reun. Cons. Int. Explor. Mer 164:156–173.
Jordan, T. E., and I. Valiela. 1982. A nitrogen budget of the ribbed mussel, *Geukensia demissa*, and its significance in nitrogen flow in a New England salt marsh. Limnol. Oceanogr. 27:75–90.
Jørgensen, B. B. 1977. The sulfur cycle of a coastal marine sediment (Limfjorden, Denmark). Limnol. Oceanogr. 22:814–832.
Jørgensen, B. B. 1977a. Bacterial sulfate reduction within reduced microniches of oxidized marine sediments. Mar. Biol. 41:7–17.
Jørgensen, B. B. 1977b. The distribution of colorless sulfurbacteria (*Beggiatoa* spp.) in a coastal marine sediment. Mar. Biol. 41:19–28.
Jørgensen, B. B. 1980. Mineralization and the bacterial cycling of carbon, nitrogen, and sulfur in marine sediments. Pp. 239–251 in D. C. Ellwood, J. N. Hedges, M. J. Leatham, J. M. Lynch, and J. H. Slater (eds.), Contemporary Microbial Ecology. Academic.
Jørgensen, B. B. 1982. Mineralization of organic matter in seabed: The role of sulfate reduction. Nature 296:643–645.
Jørgensen, C. B. 1962. The food of filter-feeding organisms. Rapp. Proc.-Verb. Reun. Cons. Int. Explor. Mer 153:99–107.
Jørgensen, C. B. 1966. Biology of Suspension Feeding. Pergamon.
Jørgensen, C. B. 1981. Mortality, growth, and grazing import of a cohort of bivalve larvae, *Mytilus edulis*. L. Ophelia 20:185–192.
Jørgensen, N. O. G., and E. Kristensen. 1980. Uptake of amino acids by three species of *Nereis* (Annelida: Polychaeta). I. Transport kinetics and net uptake from natural concentrations. Mar. Ecol. Prog. Ser. 3: 329–340.
Jørgensen, N. O. G., and E. Kristensen. 1980a. Uptake of amino acids by three species of *Nereis* (Annelida: Polychaeta). II. Effects of anaerobiosis. Mar. Ecol. Prog. Ser. 3:341–346.
Jørgensen, S. E. (ed.). 1979. Handbook of Environmental Data and Ecological Parameters. Pergamon.
Josefson, A. B. 1982. Regulation of population size, growth, and production of a deposit feeding bivalve: A long-term field study of three deep-water populations off the Swedish west coast. J. Exp. Mar. Biol. Ecol. 59:125–150.
Jumars, P. A. 1975. Methods for measurement of community structure in deep-sea macrobenthos. Mar. Biol. 30:245–252.
Jumars, P. A. 1976. Deep-sea species diversity: Does it have a characteristic scale? J. Mar. Res. 34:217–246.
Jumars, P. A. 1978. Spatial autocorrelation with RUM (Remote Underwater Manipulator): Vertical and horizontal structure of a bathyal benthic community. Deep-Sea Res. 25:589–604.
Jumars, P. A., A. R. M. Nowell, and R. L. F. Self. 1981. A simple model of flow-sediment-organism interaction. Mar. Geol. 42:155–172.

Jurasz, C. M. and V. P. Jurasz. 1979. Feeding modes of the humpback whale, *Megaptera novaeangliae* in Southeast Alaska. Sci. Rep. Whales Res. Inst. 31:69–83.

Kabanova, Y. G. 1969. Primary production of the Northern part of the Indian Ocean. Okeanologiya 8:270–278.

Kaestner, A. 1970. Invertebrate Zoology. Vols. I-III. Translated by H. W. Levi. Interscience.

Kalle, K. 1966. The problem of the gelbstoff in the sea. Mar. Biol. Ann. Rev. 4:91–104.

Kalmijn, A. J. 1978. Electric and magnetic sensory world of sharks, skates, and rays. Pp. 507–528, in E. S. Hodgson and R. F. Mathewson (eds.), Sensory Biology of Sharks, Skates, and Rays. Office of Naval Research, Dept. of the Navy.

Kamamura, A. 1974. Food and feeding ecology in the Southern sei whale. Sci. Rep. Whales Res. Inst. 26:25–144.

Kanwisher, J. W. 1966. Photosynthesis and respiration in some seaweeds. Pp. 407–420 in H. Barnes (ed.), Some Contemporary Studies in Marine Science. Allen and Unwin.

Kanwisher, J. W., K. D. Lawson, and L. R. McCloskey 1974. An improved, self-contained polarographic dissolved oxygen probe. Limnol. Oceanogr. 19:700–704.

Kaplan, W., J. M. Teal, and I. Valiela. 1977. Denitrification in salt marsh sediments: Evidence for seasonal temperature selection among populations of denitrifiers. Microb. Ecol. 3:193–204.

Kaplan, W., I. Valiela, and J. M. Teal. 1979. Denitrification in a salt marsh ecosystem. Limnol. Oceanogr. 24:726–734.

Karl, D. M., C. D. Winn, D. C. L. Wong. 1981a. RNA synthesis as a measure of microbial growth in aquatic environments. I. Evaluation, verification, and optimization of method. Mar. Biol. 64:1–12.

Karl, D. M., C. D. Winn, D. C. L. Wong. 1981b. RNA synthesis as a measure of microbial growth in aquatic environments. II. Field application. Mar. Biol. 64:13–22.

Karl, D. M., C. O. Wirsen, and H. W. Jannasch. 1980. Deep sea primary production at the Galapagos hydrothermal vents. Science 207:1345–1347.

Kasuya, T. 1972. Growth and reproduction of *Stenella caeruleoalba* based on the age determination by means of dentinal growth layers. Sci. Rep. Whales Res. Inst. 24:57–79.

Kay, D. G., and A. E. Brafield 1973. The energy relations of the polychaete Neanthes (*Nereis*) *virens* (Sars). J. Anim. Ecol. 42:673–692.

Keating, K. I. 1978. Blue-green algal inhibition of diatom growth: Transitions from mesotrophic to eutrophic community structure. Science 199:971–973.

Keeling, C. D., R. B. Bacastow, A. E. Bainbridge, C. A. Ekdahl, Jr., P. R. Guenther, and L. S. Waterman. 1976. Atmospheric carbon dioxide variations at Mauna Loa Observatory, Hawaii. Tellus 28:537–551.

Keenleyside, M. H. A. 1955. Some aspects of the schooling behavior of fish. Behavior 8:183–248.

Kenyon, K. W. 1969. The Sea Otter in the Eastern Pacific Ocean. Govt. Printing Office.

Kerr, S. R. 1971. Analysis of laboratory experiments on growth efficiency of fishes. J. Fish. Res. Bd. Canada 28:801–808.

Kerr, S. R. 1971a. Prediction of fish growth efficiency in nature. J. Fish. Res. Bd. Canada 28:809–814.

Ketchum, B. H., and N. Corwin. 1965. The cycle of phosphorus in a plankton bloom in the Gulf of Maine. Limnol. Oceanogr. 10:R148-R161.

Kierstead, H. and L. B. Slobotkin. 1953. The size of water masses containing plankton blooms. J. Mar. Res. 12:141–147.

Kiey, J. 1973. Primary production in the Indian Ocean I. Pp. 115–126 in B. Zeitzschel and S. A. Gerlach (eds.), The Biology of the Indian Ocean. Springer-Verlag.

King, G. M. 1983. Sulfate reduction in Georgia salt marsh soils: An evaluation of pyrite formation by use of ^{35}S and ^{55}Fe tracers. Limnol. Oceanogr. 28:987–995.

King, G. M., and W. J. Wiebe. 1978. Methane release from soils of a Georgia salt marsh. Geochim. Cosmochim. Acta 42:343–348.

King, K. R., J. T. Hollibaugh, and F. Azam. 1980. Predator-prey interactions between the larvacean *Oikopleura dioica* and bacterioplankton in enclosed water columns. Mar. Biol. 56:49–57.

King, W. B. 1970. The trade wind zone oceanography pilot study. Part VII. Observation of sea birds March 1964 to June 1965. Spec. Sci. Rep. Fish. U.S. Fish Wild. Serv. 586:1–136.

Kinne, O. (ed.) 1970. Marine Ecology, Vol. I. Environmental Factors. Part I. Wiley Interscience.

Kinne, O., and G.-A. Paffenhöfer. 1965. Hydraulic structure and digestion rate as a fuction of temperature and salinity in *Clava multicornis* (Cnidaria, Hydrozoa). Helgol. Wiss. Meeres. 12:329–341.

Kinsey, D. W. 1981. Is there outwelling in Georgia? Estuaries 4:277–278.

Kiørboe, T., F. Møhlenberg, and O. Nøhr. 1981. Effect of suspended bottom material on growth and energetics in *Mytilus edulis*. Mar. Biol. 61:283–288.

Kirchman, D., S. Graham, D. Reish, and R. Mitchell. 1982. Bacteria induce settlement and metamorphosis of *Janua* (*Dexiospira*) *brasiliensis* Grube (Polychaeta: Spirorbidae). J. Exp. Mar. Biol. Ecol. 56:153–163.

Kirchman, D., S. Graham, D. Reish, and R. Mitchell. 1982a. Lectins may mediate in the settlement and metamorphosis of *Janua* (*Dexiospira*) *brasiliensis* Grube (Polychaeta: Spirorbidae). Mar. Biol. Letters 3:131–142.

Kislaliogu, M., and R. N. Gibson. 1976. Prey handling time and its importance in food selection by the 15-spined stickleback, *Spinachia spinachia* (L.). J. Exp. Mar. Biol. Ecol. 25:115–158.

Kistritz, R. U., K. J. Hall, and I. Yesaki. 1983. Productivity, detritus flux, and nutrient cycling in a *Carex lyngbyei* tidal marsh. Estuaries 6:227–236.

Klump, J. V., and C. S. Martens. 1981. Biogeochemical cycling in an organic rich coastal marine basin. II. Nutrient sediment-waters exchange processes. Geochim. Cosmochim. Acta 45:101–121.

Knauer, G. A., and J. H. Martin. 1981. Primary production and carbon-nitrogen fluxes in the upper 1,500 m of the northeast Pacific. Limnol. Oceanogr. 26:181–186.

Knauer, G. A., J. A. Martin, and K. W. Bruland. 1979. Fluxes of particulate carbon, nitrogen, and phosphorus in the upper water column of the northeast Pacific. Deep-Sea Res. 26A:97–108.

Kneib, R. T. 1981. Size-specific effects of density on the growth, fecundity, and mortality of the fish *Fundulus heteroclitus* in an intertidal salt marsh. Mar. Ecol. Progr. Ser. 6:203–212.

Koehl, M. A. R., and J. R. Strickler 1981. Copepod feeding currents: Food capture at low Reynolds number. Limnol. Oceanogr. 26:1062–1073.

Kofoed, L. H. 1975. The feeding biology of *Hydrobia ventrosa* (Montague). I. The assimilation of different components of food. J. Exp. Mar. Biol. Ecol. 19:1–9.

Kofoed, L. H. 1975a. The feeding biology of *Hydrobia ventrosa*. 2. Allocation of the components of the carbon-budget and the significance of the reactions of dissolved organic material. J. Exp. Mar. Biol. Ecol. 19:243–256.

Kohl, D. H., G. D. Shearer, and B. Commoner. 1971. Fertilizer nitrogen: Contribution to nitrate in surface waters in a corn belt watershed. Science 174:1331–1334.

Kohn, A. J. 1971. Diversity, utilization of resources, and adaptive radiation in shallow water marine invertebrates of tropical islands. Limnol. Oceanogr. 16:332–348.

Komaki, Y. 1967. On the surface swarming of euphausiid crustaceans. Pac. Sci. 21:433–448.

Koop, K., R. C. Newell, and M. J. Lucas 1982. Microbial regeneration of nutrients from the decomposition of macrophyte debris on the shore. Mar. Ecol. Prog. Ser. 9:91–96.

Kooyman, G. L., R. W. Davis, J. P. Croxall, and D. P. Costa. 1982. Diving depth and energy requirements of King penguins. Science 217:726–727.

Krebs, C. J. 1978. Ecology: The Experimental Analysis of Distribution and Abundance. 2nd Ed. Harper and Row.

Krebs, C. T. 1976. The Effects of Sewage Sludge on the Marsh Fiddler Crab, *Uca pugnax*. Ph.D. Diss., Boston Univ.

Krebs, J. R., A. I. Houston, and E. L. Charnov. 1981. Some recent developments in optimal foraging. Pp. 3–18 in A. C. Kamil and T. D. Sargent (eds.), Foraging Behavior. Garland STDM.

Kremer, J. M., and S. W. Nixon. 1978. A Coastal Marine Ecosystem. Springer-Verlag.

Kriss, A. E. 1963. Marine Microbiology (Deep Sea). Oliver and Boyd.

Krom, M. D., and R. A. Berner. 1980. Adsorption of phosphate in anoxic marine sediments. Limnol. Oceanogr. 25:797–806.

Krom, M. D., and R. A. Berner. 1980a. The diffusion coefficients of sulfate, ammonium, and phosphate ions in anoxic marine sediments. Limnol. Oceanogr. 25:327–337.

Kuenzler, E. J. 1961. Phosphorus budget of a mussel population. Limnol. Oceanogr. 6:400–415.

Kuipers, B. R., P. A. W. J. de Wilde, and F. Creutzberg. 1981. Energy in a tidal flat ecosystem. Mar. Ecol. Progr. Ser. 5:215–221.

Kurten, B. 1953. On the variation and population dynamics of fossil and recent mammal populations. Acta Zool. Fenn. 76:1–122.

Lack, D. 1947. Darwin's Finches. Cambridge Univ.

Ladle, M. 1972. Larval Simuliidae as detritus feeders in chalk streams. Mem. Ist. Ital. Idrobiol. 29 (Suppl.):429–439.

Lai, D. V., and P. L. Richardson. 1977. Distribution and movement of Gulf Stream rings. J. Phys. Oceanogr. 7:670–683.

Lam, R. K., and B. R. Frost. 1976. Model of copepod filtering responses to changes in size and concentration of food. Limnol. Oceanogr. 21:490–500.

Lampert, W. 1970. Release of dissolved organic carbon by grazing zooplankton.Limnol. Oceanogr. 23:831–834.

Lampitt, R. S. 1978. Carnivorous feeding by a small marine copepod. Limnol. Oceanogr. 23:1228–1230.

Lancelot, C. 1979. Gross excretion rates of natural marine phytoplankton and heterotrophic uptake of excreted products in the Southern North Sea, as determined by short-term kinetics. Mar. Ecol. Progr. Ser. 1:179–186.

Land, L. S., J. C. Lang, and B. N. Smith. 1975. Preliminary observations on the carbon isotopic composition of some reef coral tissues and symbiotic zooxanthellae. Limnol. Oceanogr. 20:283–287.

Landenberger, D. E. 1968. Studies on selective feeding in the Pacific starfish *Pisaster* in Southern California. Ecology 49:1002–1075.

Landry, M. R. 1976. The structure of marine ecosystems: An alternative. Mar. Biol. 35:1–7.

Landry, M. R. 1977. A review of important concepts in the trophic organization of pelagic ecosystems. Helgol. Wiss. Meeres. 30:8–17.

Landry, M. R. 1978. Predatory feeding behavior of a marine copepod, *Labidocera trispinosa*. Limnol. Oceanogr. 23:1103–1113.

Landry, M. R. 1980. Detection of prey by *Calanus pacificus*: Implications of the first antenna. Limnol. Oceanogr. 25:545–549.

Landry, M. R. 1981. Switching between herbivory and carnivory by the planktonic marine copepod *Calanus pacificus*. Mar. Biol. 65:77–82.

Landry, M. R., and R. P. Hassett. 1982. Estimating the grazing impact of marine microzooplankton. Mar. Biol. 67:283–288.

Langmuir, I. 1938. Surface motion of water induced by the wind. Science 87:119–123.

Larkin, P. A. 1977. An epitaph for the concept of maximum yield. Trans. Amer. Fish. Soc. 106:1–11.

Larsson, U., and A. Hagström. 1979. Phytoplankton exudate release as an energy source for the growth of pelagic bacteria. Mar. Biol. 52:199–206.

Lasker, R. 1966. Feeding, growth, respiration and carbon utilization of an euphausiid crustacean. J. Fish. Res. Bd. Canada 23:1291–1317.

Lasker, R. 1970. Utilization of zooplankton energy by a Pacific sardine population in the California Current. Pp. 265–284 in J. H. Steele (ed.), Marine Food Chains. Univ. California.

Lasker, R., J. B. J. Wells, and A. D. McIntyre. 1970. Growth, reproduction, respiration and carbon utilization of the sand-dwelling harpacticoid copepod, *Asellopsis intermedia*. J. Mar. Biol. Assoc. U.K. 50:147–160.

Lawlor, L. L. 1980. Structure and stability in natural and randomly constructed competitive communities. Amer. Nat. 116:394–408.

Lawrence, J. M. 1975. On the relationship between marine plants and sea urchins. Oceanogr. Mar. Biol. Ann. Rev. 13:213–286.

Laws, E. A. 1975. The importance of respiration losses in controlling the size distribution of marine phytoplankton. Ecology 56:419–426.

Laws, E. A., and D. G. Redalje. 1979. Effect of sewage enrichment on the phytoplankton population of a subtropical estuary. Pac. Sci. 33:129–144.

Laws, R. M. 1977a. Seals and whales of the Southern Ocean. Phil. Trans. Roy. Soc. Lond. B 279:81–96.

Laws, R. M. 1977b. The significance of vertebrates in the Antarctic Marine Ecosystem. Pp. 411–438 in G. A. Llano (ed.), Adaptations Within Antarctic Ecosystems. Proc. 3rd SCAR Symp. Antarct. Biol. Smithsonian Inst.

Leach, J. H. 1970. Epibenthic algal production in an intertidal mudflat. Limnol. Oceanog. 15:514–521.

Leatham, G. F., V. King, and M. A. Stahmann. 1980. In vitro protein polymerization by quinones or free radicals generated by plant or fungal oxidative enzymes. Phytopathology 70:1134–1140.

Le Brasseur, R. J. 1965. Seasonal and annual variation of net zooplankton at Ocean Station P, 1956–1964. Fish Res. Bd. Canada, MS Rep. Ser. No. 202.

Le Brasseur, R. J., and O. D. Kennedy. 1972. Microzooplankton in coastal and oceanic areas of the Pacific Subarctic water mass: A preliminary report. Pp. 355–365 in A. Y. Takenouti (ed.), Biological Oceanography of the Northern North Pacific Ocean. Idemitsu Shoten.

LeCren, E. D. 1965. Some factors regulating the size of populations of freshwater fish. Mitt. Intern. Verein. Limnol. 13:88–105.

Lee, C., and C. Cronin. 1982. The vertical flux of particulate organic nitrogen in the sea: Decomposition of amino acids in the Peru upwelling area and the equatorial Atlantic. J. Mar. Res. 40:227–251.

Lee, C., J. W. Farrington, and R. B. Gagosian 1979. Sterol geochemistry of sediments in the western North Atlantic Ocean and adjacent coastal areas. Geochim. Cosmochim. Acta 43:35–46.

Lehman, J. T. 1976. The filter feeder as an optimal forager, and the predicted shapes of feeding curves. Limnol. Oceanogr. 21:501–516.

Lehman, J. T., and D. Scavia 1982. Microscale patchiness of nutrients in plankton communities. Science 216:729–730.

Leighton, D. L. 1966. Studies of food preference in algivorous invertebrates of Southern California kelp beds. Pac. Sci. 20:104–113.

Lenz, J. 1977. On detritus as a food source for pelagic filter-feeders. Mar. Biol. 41:39–48.

Leong, R. J. H., and C. P. O'Connell. 1969. A laboratory study of particulate and filter feeding on the Northern Anchovy (*Engraulis mordax*). J. Fish. Res. Bd. Canada 26:557–582.

Leslie, P. H. 1966. The intrinsic rate of increase and overlap of successive generations in a population of guillemots (*Uria aalge* Pont.). J. Anim. Ecol. 35:291–301.

Levin, S. A., and L. A. Segel. 1976. Hypothesis for origin of planktonic patchiness. Nature 259–659.

Levins, R. 1966. The strategy of model building in population biology. Amer. Sci. 54:421–431.

Levinton, J. S. 1979. Deposit feeders, their resources, and the study of resource limitaion. Pp. 117–141 in R. J. Livingston (ed.), Ecological Processes in Coastal and Marine Systems. Plenum.

Levinton, J. S., and G. R. Lopez. 1977. A model of renewable resources and limitation of deposit-feeding benthic populations. Oecologia 31:177–190.

Lewin, R. A. (ed.). 1962. Physiology and Biochemistry of Algae. Academic.

Lewis, J. 1977. Organic Production of Coral Reefs. Biol. Rev. 52:305–347.

Lewis, J. R. 1964. The Ecology of Rocky Shores. Hodder & Stoughton (formerly The English Universities Ltd).

Li, W. K. W., D. V. Subba Rao, W. G. Harrison, J. C. Smith, J. J. Cullen, B. Irwin, and T. Platt. 1983. Autotrophic picoplankton in the tropical ocean. Science 219:292–295.

Liebig, J. 1840. Chemistry in its Application to Agriculture and Physiology. Taylor and Walton.

Lieth, H. 1975. Historical survey of primary productivity research. Pp. 7–16 in H. Lieth and R. H. Whittaker (eds.), Primary Productivity of the Biosphere. Springer-Verlag.

Lindquist, A. 1978. A century of observations on sprat in the Skagerrak and the Kattegat. Rapp. Proc.-Verb. Reun. Cons. Int. Explor. Mer 172:187–196.

Linley, E. A. S., R. C. Newell, and S. A. Bosma. 1981. Heterotrophic utilization of microalgae released during fragmentation of kelp (*Eklonia maxima* and *Laminaria pallida*). I. Development of microbial communities associated with the degredation of kelp mucilage. Mar. Ecol. Progr. Ser. 4:31–41.

Linthurst, R. A., and R. J. Reimold. 1978. An evaluation of method for estimating the net aerial primary productivity of estuarine angiosperms. J. Appl. Ecol. 15:919–931.

Livingstone, D. A. 1963. Chemical composition of rivers and lakes. Chapter G in M. Fleischer (ed.), Data of Geochemistry, 6th Ed. Geol. Survey Prof. Pap. 440-G.

Lloyd, M. 1967. Mean crowding. J. Anim. Ecol. 36:1–30.

Lockwood, S. J. 1980. Density-dependent mortality in O-group plaice (*Pleuronectes platessa* L.) populations. J. Cons. Int. Explor. Mer 39:148–153.

Lockyer, C. 1976. Body weights of some species of large whales. J. Cons. Int. Expl. Mer 36:259–273.

Loder, T. C., W. B. Lyons, S. Murray, and H. D. McGuinness. 1978. Silicate in anoxic pore waters and oxidation effects during sampling. Nature 273:373–374.

Longhurst, A. R. 1967. Diversity and trophic structure of zooplankton communities in the California Current. Deep-Sea Res. 14:393–408.

Longhurst, A. R. 1968. Distribution of the larvae of *Pleuroncodes planipes* in the California Current. Limnol. Oceanogr. 13:143–155.

Longhurst, A. R. 1976. Vertical migration. Pp. 116–137 in D. H. Cushing and J. J. Walsh (eds.), The Ecology of the Seas. Blackwell.

Longhurst, A., M. Colebrook, J. Gulland, R. Le Brasseur, C. Lorenzen, and P. Smith. 1972. The instability of ocean populations. New Scientist 1 June 1972:2–4.

Lonsdale, P. 1977. Clustering of suspension-feeding macrobenthos near abyssal hydrothermal vents at oceanic spreading centers. Deep-Sea Res. 24:857–863.

Lopez, G. R., J. S. Levinton, and L. B. Slobodkin. 1977. The effect of grazing by the detritivore *Orchestia grillus* on *Spartina* litter and its associated microbial community. Oecologia 30:111–127.

Lorenzen, C. J. 1976. Primary production in the sea. Pp. 173–185 in D. H. Cushing and J. J. Walsh (eds.), The Ecology of the Seas. Saunders, CBS College Publishing.

Lorenzen, C. J., F. R. Shuman, and J. T. Bennett. 1981. In situ calibration of a sediment trap. Limnol. Oceanogr. 26:580–585.

Lotka, A. J. 1925. Elements of Physical Biology. Williams and Wilkins. Reprinted by Dover 1956.

Lotka, A. J. 1956. Elements of Mathematical Biology. Dover.
Loucks, O. L. 1970. Evolution of diversity efficiency, and community stability. Amer. Zool. 10:17–25.
Lubchenco, J. 1978. Plant species diversity in a marine intertidal community: Importance of herbivore food preference and algal competitive ability. Amer. Nat. 112:23–39.
Lubchenco, J., and J. Cubit. 1980. Heteromorphic life history of certain marine algae as adaptations to variations in herbivory. Ecol. 6:676–687.
Lubchenco, J., and B. A. Menge 1978. Community development and persistence in a low rocky intertidal zone. Ecol. Monogr. 48:67–94.
Luecke, C., and W. J. O'Brien. 1981. Prey location volume of a planktivorous fish: A new measure of prey vulnerability. Canad. J. Fish. Aquat. Sci. 38:1264–1270.
Lugo, A. E., G. Evink, M. M. Brinsom, A. Bruce, and S. C. Snedaker. 1975. Diurnal rates of photosynthesis, respiration, and transpiration in mangrove forests of South Florida. Pp. 335–352 in F. B. Golley and E. Medina (eds.), Tropical Ecological Systems. Springer-Verlag.
Lund, J. W. G. 1966. Summation. Pp. 227–249 in C. H. Oppenheimer (ed.), Marine Biology, Vol. 2. New York Academy of Science.
Luther, G. W., III, A. Giblin, R. W. Howarth, and R. A. Ryans. 1982. Pyrite and oxidized iron mineral phases formed from pyrite oxidation in salt marsh and estuarine sediments. Geochim. Cosmochim. Acta. 46:2665–2669.
MacArthur, R. H. 1965. Pattern of species diversity. Biol. Rev. 40:510–533.
MacArthur, R. H., and E. O. Wilson. 1967. The Theory of Island Biogeography. Princeton Univ.
MacCall, A. D. 1979. Population estimates for the waning years of the Pacific sardine fishery. Calif. Coop. Fish. Invest. Rep. 20:72–82.
MacDonald, D. D. M., and T. J. Pitcher. 1979. Age-groups from size-frequency data: A versatile and efficient method of analyzing distribution mixtures. J. Fish. Res. Bd. Canada 36:987–1001.
MacGregor, J.S. 1957. Relation between fish conditions and population size in the sardine (*Sardinops caerulea*). Fish. Bull. U.S. Fish. Wildl. Serv. 60:215–230.
MacIntosh, N. A. 1970. Whales and krill in the twentieth century. Pp. 195–212 in M. W. Holdgate (ed.), Antarctic Ecology. Academic.
MacIsaac, J. J., and R. C. Dugdale. 1969. The kinetics of nitrate and ammonia uptake by natural populations of marine phytoplankton. Deep-Sea Res. 16:45–57.
MacLeod, P., and I. Valiela. 1975. The effect of density and mutual interference by a predator: A laboratory study of predation by the nudibranch *Coryphella rufibranchialis* on the hydroid *Tubularia larynx*. Hydrobiologia 47:339–346.
Majak, W., J. S. Craigie, and J. McLachlan. 1966. Photosynthesis in the Rhodophyceae. Canad. J. Bot. 44:541–549.
Malone, T. C. 1971a. The relative importance of nanoplankton and net plankton as primary producers in tropical and neritic phytoplankton communities. Limnol. Oceanogr. 16:633–639.
Malone, T. C. 1971b. The relative importance of nanoplankton and net plankton as primary producers in the California Current system. Fish. Bull. U.S. Fish Wildl. Serv. 69:799–820.

Malone, T. C. 1980. Algal size. Pp. 433–463 in I. Morris (ed.), The Physiological Ecology of Phytoplankton. Univ. California.

Malone, T. C., M. B. Chervin, and D. C. Boardman. 1979. Effects of 22μm screens on size-frequency distributions of suspended particles and biomass estimates of phytoplankton size fractions. Limnol. Oceanogr. 24:956–960.

Manahan, D. T., S. H. Wright, G. C. Stephens, and M. A. Rice. 1982. Transport of dissolved amino acids by the mussel, *Mytilus edulis*: Demonstration of net uptake from natural seawater. Science 215:1253–1255.

Mann, K. H. 1969. The dynamics of aquatic ecosystems. Adv. Ecol. Res. 6:1–81.

Mann, K. H. 1973. Seaweeds: Their productivity and strategy for growth. Science 182:975–981.

Mann, K. H. 1978. Production on the bottom of the sea. Pp. 225–250 in D. H. Cushing and J. J. Walsh (eds.), The Ecology of the Seas. Blackwell.

Mann, K. H. 1982. Ecology of Coastal Waters. Univ. California.

Marais, J. F. K. 1980. Aspects of food intake, food selection, and alimentary canal morphology of *Mugil cephalus* (Linneaus, 1958), *Liza tricuspidens* (Smith, 1935), *L. richardsoni* (Smith, 1846) and *L. dummerili* (Steindauher, 1869). J. Exp. Mar. Biol. Ecol. 44:193–209.

Margalef, R. 1951. Diversidad de especies en las comunidades naturales. Publ. Inst. Biol. Aplic. Barcelona 9:5–27.

Margalef, R. 1957. La teoría de la información en ecología. Mem. Real Acad. Cienc. Artes. (Barcelona) 33:373–449.

Margalef, R. 1958. Temporal sucession and spatial heterogeneity in phytoplankton. Pp. 323–349 in A. A. Buzzati-Traverso (ed.), Perspectives in Marine Biology. Univ. California.

Margalef, R. 1967. The food web in the pelagic environment. Helgol. Wiss. Meeres. 18:548–559.

Margalef, R. 1967a. Succession in marine populations. Adv. Front. Pl. Sci. (New Delhi) 2:137–188.

Margalef, R. 1968. Perspectives in Ecological Theory. Univ. of Chicago.

Margalef, R. 1974. Ecología. Omega.

Margalef, R. 1978. Phytoplankton communities in upwelling areas. The example of NW Africa. Oecol. Aquat. 3:97–132.

Margalef, R. 1978a. General concepts of population dynamics and food links. Pp. 617–704 in O. Kinne (ed.), Marine Ecology Vol. 4. J. Wiley & Sons, Ltd.

Margulis, L. 1981. Symbiosis in Cell Evolution: Life and Its Environment in the Early Earth. W.H. Freeman.

Marinucci, A. C., J. E. Hobbie, and J. V. K. Helfrich. 1983. Effect of litter nitrogen on decomposition and microbial biomass in *Spartina alterniflora*. Microb. Ecol. 9:23–40.

Marr, J. C. 1960. The causes of major variations in the catch of the Pacific sardine *Sardinops cerulea*. Proc. World Sci. Meet. Biol. Sardines and Related Species 3:667–791.

Marr, J. W. S. 1962. The natural history and geography of the Antarctic Krill (*Euphausia superba* Dana). Discovery Rep. 32:33–464.

Marsh, J. A., Jr., and S. V. Smith 1978. Productivity measurements of coral reefs in flowing water. Pp. 361–378 in D. R. Stoddart and R. E. Johannes (eds.), Coral Reefs: Research Methods. UNESCO.

Marshall, S. M. 1933. The production of microplankton in the Great Barrier Reef Region. Great Barrier Reef Exp. 1928–29. Sci. Repts. 2:111–157.

Marshall, S. M., and A. P. Orr. 1955. The Biology of a Marine Copepod, *Calanus finmarchicus* (Gunnerus). Oliver and Boyd.

Marshall, S. M., and A. P. Orr. 1964. Grazing by copepods in the sea. Pp. 227–238 in D. J. Crisp (ed.), Grazing in Terrestrial and Marine Environments. Blackwell.

Marshall, S. M., and A. P. Orr. 1966. Respiration and feeding in some small copepods. J. Mar. Biol. Assoc. U.K. 46:513–530.

Marsho, T. V., R. P. Burchard, and R. Fleming. 1975. Nitrogen fixation in the Rhode River estuary of Chesapeake Bay. Canad. J. Microbiol. 21:1348–1356.

Martens, C. S. 1976. Control of methane sediment-water bubble tranport by macrofaunal nitrification in Cape Lookout Bight, North Carolina. Science 192:998–1000.

Martens, C. S., and R. A. Berner. 1974. Methane production in the interstitial waters of sulfate depleted estuarine sediments. Science 185:1167–1169.

Martens, C. S., and R. A. Berner. 1979. Interstitial water chemistry of anoxic Long Island Sound sediments. I. Dissolved gases. Limnol. Oceanogr. 22:10–25.

Martin, J. H. 1968. Phytoplankton-zooplankton relationships in Narragansett Bay. Seasonal changes in zooplankton excretion rates relative to phytoplankton abundance. Limnol. Oceanogr. 13:63–71.

Martin, J. H. 1970. Phytoplankton-zooplankton relationships in Narragansett Bay. IV. The seasonal importance of grazing. Limnol. Oceanogr. 15:413–418.

Mattson, W. J., Jr. 1980. Herbivory in relation to plant nitrogen content. Ann. Rev. Ecol. Syst. 11:17–25.

Mattson, W. J., and N. D. Addy 1975. Phytophagous insects as regulators of forest primary production. Science 190:515–522.

May, R. M. 1973. Stability and Complexity in Model Ecosystems. Princeton Univ.

May, R. M. 1975. Patterns of species abundance and diversity. Pp. 81–120 in M. L. Cody and J. M. Diamond (eds.), Ecology and Evolution of Communities. Harvard Univ.

May, R. M., G. R. Conway, M. P. Hasell and J. R. E. Southwood 1974. Time delays, density dependence, and single species oscillations. J. Anim. Ecol. 43:747–770.

Mayzaud, P. 1973. Respiration and nitrogen excretion of zooplankton. II. Studies of the metabolic characteristics of starved animals. Mar. Biol. 21:19–28.

Mayzaud, P., and S. Dallot 1973. Respiration et excretion azotée du zooplancton. I. Etude des niveaux métaboliques de quelques espèces de Mediterranée occidentale. Mar. Biol. 19:307–314.

Mayzaud, P., and S. A. Poulet 1978. The importance of the time factor in the response of zooplankton to varying concentrations of naturally occurring particulate matter. Limnol. Oceanogr. 23:1144–1154.

McAllister, C. D. 1969. Aspects of estimating zooplankton production from phytoplankton production. J. Fish. Res. Bd. Canada 26:199–220.

McAllister, C. D. 1970. Zooplankton rations, phytoplankton mortality, and the estimation of marine production. Pp. 419–457, in J. H. Steele (ed.), Marine Food Chains. Univ. California.

McAllister, C. D., T. R. Parsons, K. Stephens, and J. D. H. Strickland. 1961. Measurements of primary production in coastal seawater using a large volume plastic sphere. Limnol. Oceanogr. 6:237–258.

McCall, P. L. 1977. Community patterns and adaptive strategies of the infaunal benthos of Long Island Sound. J. Mar. Res. 35:221–266.

McCarthy, J. J. 1972. The uptake of urea by natural populations of marine phytoplankton. Limnol. Oceanogr. 17:738–748.

McCarthy, J. J. 1980. Nitrogen. Pp. 191–234 in I. Morris (ed.), The Physiological Ecology of Phytoplankton. Univ. California.

McCarthy, J. J., and E. J. Carpenter. 1979. *Oscillatoria* (*Trichodesmium*) *thiebautii* (Cyanophyta) in the central north Atlantic Ocean. J. Phycol. 15:75–82.

McCarthy, J. J., and E. J. Carpenter. 1984. Nitrogen cycling in near-surface waters of the open ocean. Pp. 487–512 in E. J. Carpenter and D. G. Capone (eds.), Nitrogen in the Marine Environment. Academic.

McCarthy, J. J., and J. C. Goldman. 1979. Nitrogenous nutrition of marine phytoplankton in nutrient-depleted waters. Science 203:670–672.

McCarthy, J. J., W. R. Taylor, and M. E. Loftus. 1974. Significance of nanoplankton in the Chesapeake Bay Estuary and problems associated with the measurement of nanoplankton productivity. Mar. Biol. 24:7–16.

McGowan, J. A. 1977. What regulates pelagic community structure in the Pacific? Pp. 423–444 in N. R. Anderson and B. J. Zahuranec (eds.), Oceanic Sound Scattering Prediction. Plenum.

McGowan, J. A., and T. L. Hayward. 1978. Mixing and oceanic productivity. Deep-Sea Res. 25:771–793.

McGregor, J. S. 1957. Fecundity of the Pacific Sardine (*Sardinops caerulea*). Fish. Bull. U.S. Wildl. Serv. 17:427–449.

McFarland, W. N., and J. Prescott. 1959. Standing crop, chlorophyll content and *in situ* metabolism of a giant kelp community in Southern California. Contr. Mar. Sci. 6:110–132.

McLaren, I. A. 1963. Effects of temperature on growth of zooplankton, and the adaptive value of vertical migration. J. Fish. Res. Bd. Canada 20:685–727.

McLaren, I. A. 1965. Some relationships between temperature and egg size, body size, development rate, and fecundity of the copepod *Pseudocalanus*. Limnol. Oceanogr. 10:528–538.

McLaren, I. A. 1974. Demographic strategy of vertical migration by a marine copepod. Amer. Nat. 108:91–102.

McNaughton, S. J. 1977. Diversity and stability of ecological communities: A comment on the role of empiricism in ecology. Amer. Nat. 111:515–525.

McNeill, S. 1973. The dynamics of a population of *Leptopterna dolabrata* (Heteroptera: Miridae) in relation to its food resources. J. Anim. Ecol. 42:495–508.

McNeill, S., and T. R. E. Southwood. 1978. The role of nitrogen in the development of insect/plant relationships. Pp. 77–98 in J. B. Harborne (ed.), Biochemical Aspects of Plant and Animal Coevolution. Academic.

McRoy, C. P. 1974. Seagrass productivity: Carbon uptake experiments in eelgrass, *Zostera marina*. Aquaculture 4:131–137.

McRoy, C. P., and C. McMillan. 1977. Production ecology and physiology of sea grasses. Pp. 53–87 in C. P. McRoy and C. Helfferich (eds.), Seagrass Ecosystems. M. Dekker.

Mendelssohn, I. A., K. L. McKee, and W. H. Patrick. 1981. Oxygen deficiency in *Spartina alterniflora* roots: Metabolic adaptation to anoxia. Science 214:439–441.

Menzel, D. 1974. Primary production, dissolved and particulate organic matter. Pp. 659–678. In E. D. Goldberg (ed.), The Sea. Wiley.
Menzel, D. W., and J. J. Goering. 1966. The distribution of organic detritus in the ocean. Limnol. Oceanogr. 11:333–337.
Menzel, D. W., E. M. Hulburt, and J. R. Ryther. 1963. The effects of enriching Sargasso Sea water on the production and species composition of the phytoplankton. Deep-Sea Res. 10:209–219.
Menzel, D. W., and J. H. Ryther. 1960. The annual cycle of primary production in the Sargasso Sea off Bermuda. Deep-Sea Res. 6:351–367.
Menzel, D. W., and J. H. Ryther. 1961. Zooplankton in the Sargasso Sea off Bermuda and its relation to organic production. J. Cons. Perm. Inter. Explor. Mer 26:250–258.
Menzel, D. W., and J. H. Ryther. 1970. Distribution and cycling of organic matter in the oceans. Pp. 31–54 in D. W. Hood (ed.), Organic Matter in Natural Waters. Inst. Mar. Sci. Univ. Alaska.
Menzel, D. W., and J. P. Spaeth 1962. Occurrence of ammonia in Sargasso Sea waters and in rainwater at Bermuda. Limnol. Oceanogr. 7:159–162.
Mertz, D. B. 1970. Notes on methods used in life history studies. Pp. 4–17 in J. H. Connell, D. N. Mertz, and W. W. Murdoch (eds.), Readings in Ecology and Ecological Genetics. Harper and Row.
Meyer-Reil, L. -A. In press. Bacterial biomass and heterotrophic activity in sediments and overlying waters. In J. Hobbie and P. J. Leb. Williams (eds.), Heterotrophic Activity in the Sea. Plenum.
Meyer-Reil, L. -A., and A. Faubel. 1980. Uptake of organic matter by meiofauna organisms and interrelationships with bacteria. Marine Ecol. Progr. Ser. 3:251–256.
Miller, C. B., and B. W. Frost. 1982. Subarctic pacific ecosystem research (SUPER). EOS Trans. 63:62.
Mills, E. L. 1980. The structure and dynamics of shelf and slope ecosystems off the North East Coast of North America. Pp. 25–47 in K. R. Tenore and B. C. Coull (eds.), Marine Benthic Dynamics. Univ. South Carolina.
Mitchell, R. 1978. Mechanism of biofilm formation in seawater. Pp. 45–50 in R. H. Gray (ed.), Proc. Ocean Thermal Energy Conversion (OTEC) Biofouling and Corrosion Symp. Oct. 10–12, 1977, Seattle, Wash. U.S. Dept. of Energy and Pacific Northwest Lab.
Miyazaki, N. 1977. Growth and reproduction of *Stenella coeruleoalba* off the Pacific coast of Japan. Sci. Rep. Whales Res. Inst. 29:21–48.
Moe, R. L., and P. C. Silva 1977. Antarctic marine flora: Uniquely devoid of kelps. Science 196:1206–1208.
Moment, G. B. 1962. Reflexive selection: A possible answer to an old puzzle. Science 136:262–263.
Mootz, C. A., and C. E. Epifanio. 1974. An energy budget for *Menippe mercenaria* larvae fed *Artemia nauplii*. Biol. Bull. 146:44–55.
Morel, A. 1974. Optical properties of pure water and pure seawater. Pp. 1–24, in N. G. Jerlov and E. Steemann Nielsen (eds.), Optical Aspects of Oceanography. Academic.
Morel, F. M. M., and J. Morgan. 1972. A numerical method for computing equilibria in aqueous chemical systems. Env. Sci. Tech. 6:58–67.
Moriarty, D. J. W. 1977. Improved method using muramic acid to estimate biomass of bacteria in sediments. Oecologia 26:317–323.

Morisita, M. 1959. Measuring the dispersion of individuals and analysis of the distributional patterns. Mem. Fac. Sci. Kyushu Univ. Ser. E (Biol.) 2:215–235.

Morita, R. Y. 1980. Microbial life in the deep sea. Canad. J. Microbiol. 26:1375–1385.

Morris, I. 1980. Paths of carbon assimilated in marine phytoplankton. Pp. 139–159 in P. O. Falkowski (ed.), Primary Productivity in the Sea. Plenum.

Morris, I., and H. Glover. 1981. Physiology of photosynthesis by marine coccoid cyanobacteria—Some ecological implications. Limnol. Oceanogr. 26:957–961.

Morrison, S. J., and D. C. White. 1980. Effects of grazing by estuarine gammaridean amphipods on the microbiota of allochthonous detritus. Appl. Env. Microbiol. 40:659–671.

Muller, P. J., and E. Suess. 1979. Productivity, sedimentation rate, and sedimentary organic matter in the oceans. I. Organic carbon preservation. Deep-Sea Res. 26A:1347–1362.

Mullin, M. M. 1969. Production of zooplankton in the ocean: The present status and problems. Oceanogr. Mar. Biol. Ann. Rev. 7:293–314.

Mullin, M. M., E. F. Stewart, and F. J. Fuglister. 1975. Ingestion by planktonic grazers as a function of concentration of food. Limnol. Oceanogr. 20:259–262.

Murdoch, W. W. 1966. "Community structure, population control, and competition"—a critique. Amer. Nat. 100:219–226.

Murdoch, W. W. 1969. Switching in general predators: Experiments on predator specificity and stability of prey populations. Ecol. Monogr. 39:335–354.

Murdoch, W. W. 1971. The development response of predators to changes in prey density. Ecology 52:132–137.

Murdoch, W. W., and A. Oaten 1975. Predation and population stability. Adv. Ecol. Res. 9:1–132.

Murphy, G. I. 1967. Vital statistics of the Pacific sardine (*Sardinops caerulea*) and the population consequences. Ecology 48:731–736.

Murphy, G. I. 1968. Pattern in life history and the environment. Amer. Nat. 102:390–404.

Muscatine, L., and R. E. Marian. 1982. Dissolved inorganic nitrogen flux in symbiotic and nonsymbiotic medusae. Limnol. Oceanogr. 27:910–917.

Myers, J. P, P. G. Connors, and F. A. Pitelka 1979. Territory size in wintering sanderlings: The effects of prey abundance and intruder density. Auk 96:535–561.

Naito, Y., and M. Nishiwaki. 1972. The growth of two species of the harbour seal in the adjacent waters of Hokkaido. Sci. Rep. Whales Res. Inst. 24:127–144.

Navarro, J. M., and J. E. Winter 1982. Ingestion rate, assimilation efficiency, and energy balance in *Mytilus chilensis* in relation to body size and different algal concentrations. Mar. Biol. 67:255–266.

Naylor, E. and R. G. Hartnoll (eds.). 1979. Cyclic Phenomena in Marine Plants and Animals. Proc. 13th Europ. Mar. Biol. Symp. Pergamon.

Neushul, M. 1971. Submarine illumination in *Macrocystis* beds. Pp. 241–254 in W. J. North (ed.), The Biology of Giant Kelp Beds (*Macrocystis*) in California. J. Cramer.

Nelson-Smith, A. 1968. In J. D. Carthy and D. R. Arthur (eds.), The Biological Effects of Oil Pollution on Littoral Communities. Vol. 2, Suppl. Field Studies Council, London.

Newell, R. C., M. I. Lucas, and E. A. S. Linley. 1981. Rate of degradatıon and efficiency of conversion of phytoplankton debris by maine microorganisms. Mar. Ecol. Progr. Ser. 6:123–136.

Newell, R. C., M. I. Lucas, B. Velimisov, and L. J. Seiderer. 1980. Quantitative significance of dissolved organic losses following fragmentation of kelp (*Ecklonia maxima* and *Laminaria pollida*). Mar. Ecol. Progr. Ser. 2:45–59.

Newell, S. Y., and R. E. Hicks. 1982. Direct count estimates of fungal and bacterial biovolume in dead leaves of smooth cordgrass (*Spartina alterniflora* Loisel.). Estuaries 5:246–260.

Nicholson, A. J. 1954. An outline of the dynamics of animal populations. Austral. J. Zool. 2:9–65.

Nicotri, M. E. 1980. Factors involved in herbivore food preference. J. Exp. Mar. Biol. Ecol. 42:13–26.

Nienhuis, P. H., and E. T. van Ierland. 1978. Consumption of eelgrass, *Zostera marina*, by birds and invertebrates during the growing season in Lake Grevelinge (SW Netherlands). Neth. J. Sea Res. 12:180–194.

Nilson, C. S., and G. R. Cresswell. 1981. The formation and evolution of East Australian Current Eddies. Progr. Oceanogr. 9:133–183.

Nissenbaum, A., and I. R. Kaplan. 1972. Chemical and isotopic evidence for the *in situ* origin of marine humic substances. Limnol. Oceanogr. 17:570–582.

Nixon, S. W. 1980. Between coastal marshes and coastal waters—A review of twenty years of speculation and research on the role of salt marshes in estuarine productivity and water diverstiy. Pp. 437–525 in R. Hamilton and K. B. MacDonald (eds.), Estuarine and Wetland Processes. Plenum.

Nixon, S. W. 1981. Remineralization and nutrient cycling in coastal marine ecosystems. Pp. 111–138 in B. J. Neilson and L. E. Cronin (eds.), Estuaries and Nutrients. Humana.

Nixon, S. W., J. R. Kelly, B. N. Furnas, C. A. Oviatt, and S. S. Hale. 1980. Phosphorus regeneration and the metabolism of coastal marine bottom communities. Pp. 219–242 in K. R. Tenore and B. C. Coull (eds.), Marine Benthic Dynamics. Univ. South Carolina.

Nixon, S. W., and C. A. Oviatt. 1973. Ecology of a New England salt marsh. Ecol. Monogr. 43:463–498.

Nixon, S. W., C. A. Oviatt, and S. S. Hale. 1976. Nitrogen regeneration and the metabolism of coastal marine bottom communities. Pp. 269–283 in J. M. Anderson and A. MacFadyen (eds.), The Role of Terrestrial and Aquatic Organisms in Decomposition Processes. Blackwell.

Noy-Meir, I. 1975. Stability of grazing systems: An application of predator-prey graphs. J. Ecol. 63:459–483.

O'Brien, W. J., N. A. Slade, and G. L. Vinyard. 1976. Apparent size as the determinant of prey selection by bluegill sunfish. Ecology 57:1304–1310.

O'Connors, H. B., Jr., C. B. Wurster, C. D. Powers, D. C. Biggs, and R. G. Rowland. 1978. Polychlorinated biphenyls may alter marine trophic pathways by reducing phytoplankton size and production. Science 201:737–739.

Odum, E. P. 1969. The strategy of ecosystem development. Science 164:267–270.

Odum, E. P. 1971. Fundamentals of Ecology. 3rd Ed. Saunders.

Odum, E. P., C. E. Connell, and L. B. Davenport. 1962. Population energy flow of three primary consumer components of old-field ecosystems. Ecology 43:88–96.

Odum, H. T., and E. P. Odum. 1955. Trophic structure and productivity of a windward coral reef community on Eniwetok Atoll. Ecol. Monogr. 25:291–320.

Odum, W. E., P. W. Kirk, and J. C. Zieman. 1979. Non-protein nitrogen compounds associated with particles of vascular plant detritus. Oikos 32:363–367.

Ogawa, Y., and T. Nakahara. 1979. Interrelationships between pelagic fishes and plankton in the coastal fishing ground of the Southwestern Japan Sea. Mar. Ecol. Progr. Ser. 1:115–122.

Ogden, J. C., R. A. Brown, and N. Salesby. 1973. Grazing by the echinoid *Diadema antillarum* Philippi: Formation of halos around West Indian patch reefs. Science 182:715–717.

Ogura, N. 1972. Rate and extent of decomposition of dissolved organic matter in surface seawater. Mar. Biol. 13:89–93.

Ogura, N. 1975. Further studies on decomposition of dissolved organic matter in coastal seawater. Mar. Biol. 31:101–111.

Ohman, M. D., B. W. Frost, and E. B. Cohen. 1983. Reverse vertical migration: An escape from invertebrate predators. Science 220:1404–1407.

Olsson, I., and E. Olundh. 1974. On plankton production in Kungsbacka fjord, an estuary on the Swedish west coast. Mar. Biol. 24:17–28.

Omori, M. 1978. Zooplankton fisheries of the world: A review. Mar. Biol. 48:199–205.

Onuf, C. P, J. M. Teal, and I. Valiela. 1977. Interactions of nutrients, plant growth, and herbivory in a mangrove ecosystem. Ecology 58:514–526.

Oremland, R. S. 1975. Methane production in shallow water tropical marine sediments. Appl. Microbiol. 30:602–608.

Oremland, R. S., and B. F. Taylor. 1977. Diurnal fluctuation of O_2, N_2 and CH_4 in the rhizosphere of *Thalassia testudinum*. Limnol. Oceanogr. 22: 566–570.

Oremland, R. S., and B. F. Taylor. 1978. Sulfate reduction and methanogenesis in marine sediments. Geochim. Cosmochim. Acta 42:209–214.

Orians, G. H. 1962. Natural selection and ecological theory. Amer. Nat. 96:257–263.

Orians, G. H. 1975. Diversity, stability and maturity in natural ecosystems. Pp. 139–150 in W. H. van Dobben and R. H. Lowe-McConnell (eds.), Unifying Concepts in Ecology. A. W. Junk.

Oritsland, T. 1977. Food consumption of seals in the Antarctic pack ice. Pp 749–768 in G. A. Llano (ed.), Adaptations Within Antarctic Ecosystems. Proc. 3rd SCOR Symp. on Antarctic Biol. Smithsonian Inst.

Orr, A. P. 1933. Physical and chemical conditions in the sea in the neighbourhood of the Great Barrier Reef. Great Barrier Reef Exped. 1928–29, Sci. Reps. 2:37–86.

Osman, R. W., and R. B. Whitlach. 1978. Patterns of species diversity: Fact or artifact? Paleobiology 4:41–54.

Ostfeld, R. S. 1982. Foraging strategies and prey switching in the California sea otter. Oecologia 53:170–178.

Otsuki, A., and T. Hanya. 1972. Production of dissolved organic matter from dead green algal cells. I. Aerobic microbial decomposition. Limnol. Oceanogr. 17:248–257.

Otsuki, A., and T. Hanya. 1972a. Production of dissolved organic matter from dead green algal cells. II. Anaerobic microbial decomposition. Limnol. Oceanogr. 17:258–264.

Owen, D. F., and R. G. Wiegert. 1976. Do consumers maximize plant fitness? Oikos 27:488–492.

Owen, D. F., and R. G. Wiegert 1981. Mutualism between grasses and grazers: An evolutionary hypothesis. Oikos 36:376–378.

Paasche, E. 1966. Action spectrum of coccolith formation. Physiol. Plant. 19:770–779.

Paasche, E. 1973. Silicon and the ecology of marine plankton diatoms. I. *Thalassisira pseudonana* (*Cyclotella nana*) grows in a chemostat with silicate as the limiting nutrient. Mar. Biol. 19:117–126.

Paasche, E. 1980. Silicon. Pp. 259–284 in I. Morris (ed.), The Physiological Ecology of Phytoplankton. Univ. California.

Pacala, S. W., and J. Roughgarden. 1982. An experimental investigation of the relationship between resource partitioning and interspecific competition in two two-species insular *Anolis* lizard communities. Science 217:444–446.

Packard, T. T., D. Blasco, J. J. MacIsaac, and R. C. Dugdale. 1971. Variations in nitrate reductase activity in marine phytoplankton. Inv. Pesq. 35:209–220.

Paffenhöfer, G.-A. 1968. Nahrungsaufnahme, Stoffumsatz und Energiehaushalt desmarinen Hydroidenpolypen *Clava multicornis*. Helgol. Wiss. Meeresunters. 18:1–44.

Paffenhöfer, G.-A. 1971. Grazing and ingestion rates of nauplii, copepodids, and adults of the marine planktonic copepod *Calanus helgolandicus*. Mar. Biol. 11:286–298.

Paffenhöfer, G.-A., and S. C. Knowles. 1979. Ecological implications of fecal pellet size, production and consumption by copepods. J. Mar. Res. 37:35–49.

Paffenhöfer, G.-A., J. R. Strickler, and M. Alcaraz. 1982. Suspension-feeding by herbivorous calanoid copepods: A cinematographic study. Mar. Biol. 67:193–199.

Paine, R. T. 1966. Food web complexity and species diversity. Amer. Nat. 100:65–75.

Paine, R. T. 1969. A note on trophic complexity and community stability. Amer. Nat. 103:91–93.

Paine, R. T. 1971. The measurement and application of the calorie to ecological problems. Ann. Rev. Ecol. Syst. 2:145–164.

Paine, R. T., and S. A. Levin 1981. Intertidal landscapes: Disturbance and the dynamics of pattern. Ecol. Monogr. 51:145–178.

Paine, R. T., and R. L. Vadas 1969. The effects of grazing by sea urchins, *Strongylocentrotus* spp. on benthic algae populations. Limnol. Oceanogr. 14:710–719.

Painter, H. W. 1970. A review of the literature on inorganic nitrogen metabolism in microorganisms. Water Res. 4:393–450.

Paloheimo, J. E. 1974. Calculation of instantaneous birth rate. Limnol. Oceanogr. 19:692–694.

Pamatmat, M. M. 1968. Ecology and metabolism of a benthic community on an intertidal sandflat. Int. Rev. Ges. Hydrobiol. 53:211–298.

Pamatmat, M. M. 1977. Benthic community metabolism: A review and amendment of present status and outlook. Pp. 89–111 in B. C. Coull (ed.), Ecology of Marine Benthos. Univ. South Carolina.

Pandian, T. J. 1975. Mechanisms of heterotrophy. Pp. 61–249 in O. Kinne (ed.), Marine Ecology. Vol. II, Pt. 1. Wiley.

Paranjape, M. A. 1967. Moulting and respiration of euphausiids. J. Fish. Res. Bd. Canada 24:1229–1240.

Parker, M. 1975. Similarities between the uptake of nutrients and the ingestion of prey. Verb. Internat. Verein. Limnol. 19:56–59.

Parker, P. L., E. W. Behrens, J. A. Calder, and D. Shultz. 1972. Stable carbon isotope ratio variations in the organic carbon from Gulf of Mexico sediments. Contr. Mar. Sci. 16:139–147.

Parsons, T. R. 1963. Suspended organic matter in seawater. Progr. Oceanogr. 1:205–239.

Parsons, T. R., and R. J. Le Brasseur 1968. A discussion of some critical indices of primary and secondary production for large scale ocean surveys. Calif. Mar. Res. Comm. Calif. Coop. Oceanic Fish. Invest. Rep. 12:54–63.

Parsons, T. R., and R. J. Le Brasseur 1970. The availability of food to different trophic levels in the marine food chain. Pp. 325–343 in J. H. Steele (ed.), Marine Food Chains. Univ. California.

Parsons, T. R., R. J. LeBrasseur, and J. D. Fulton. 1967. Some observations on the dependence of zooplankton grazing on the cell size and concentration of phytoplankton blooms. J. Oceanogr. Soc. Japan 23:10–17.

Parsons, T. R., R. J. Le Brasseur, J. D. Fulton, and O. D. Kennedy. 1969. Production studies in the strait of Georgia. Part II. Secondary production under the Fraser River plume, February to May, 1967. J. Exp. Mar. Biol. Ecol. 3:39–50.

Parsons, T. R., and H. Seki. 1970. Importance and general implications of organic matter in aquatic environments. Pp. 1–27 in D. W. Hood (ed.), Organic Matter in Natural Waters. Univ. of Alaska.

Parsons, T. R., and J. D. H. Strickland. 1962. On the production of particulate organic carbon by heterotrophic processes in sea water. Deep-Sea Res. 8:211–222.

Parsons, T. R., and M. Takahashi. 1973. Environmental control of phytoplankton cell size. Limnol. Oceanogr. 28:511–515.

Parsons, T. R., and M. Takahashi. 1974. A rebuttal to the comment by Hecky and Kilham. Limnol. Oceanogr. 19:366–368.

Parsons, T. R., M. Takahashi, and B. Hargrave. 1977. Biological Oceanographic Processes. 2nd Ed. Pergamon.

Patrick, R., M. H. Hohn, and J. H. Wallace. 1954. A new method for determining the pattern of the diatom flora. Not. Naturae 259:1–12.

Paul, E. A., and R. L. Johnson. 1977. Microscopic counting and adenosine 5'-triphosphate measurements in determining microbial growth in soils. Appl. Environ. Microbiol. 34:263–269.

Payne, M. R. 1977. Growth of a fur seal population. Phil. Trans. Roy. Soc. Lond. B 279:67–79.

Pearcy, W. G. 1962. Ecology of young winter flounder in an estuary. Bull. Bingh. Oceanogr. Coll. 18:1–78.

Pearre, S., Jr. 1980. Feeding by chaetognatha: The relation of prey size to predator size in several species. Marine Ecology Progr. Ser. 3:125–134.

Pearse, J. S., and A. H. Hines. 1979. Expansion of a central California kelp forest following the mass mortality of sea urchins. Mar. Biol. 51:83–91.

Pearson, T. H., and R. Rosenberg. 1976. A comparative study of the effects on the marine environment of wastes from the coastal industries in Scotland and Sweden. Ambio 5:77–79.

Pearson, T. H., and R. Rosenberg. 1978. Macrobenthic succession in relation to organic enrichment and pollution of the marine environment. Oceanogr. Mar. Biol. Ann. Rev. 16:229–311.

Peer, D. L. 1970. Relation between biomass, productivity, and loss to predators in a population of a marine benthic polychaete, *Pectinaria hyperborea*. J. Fish. Res. Bd. Canada 27:2143–2153.

Peet, R. K. 1974. The measurement of species diversity. Ann. Rev. Ecol. Syst. 5:285–307.

Penhale, P. A., and W. O. Smith, Jr. 1977. Excretion of dissolved organic carbon by eelgrass (*Zostera marina*) and its epiphytes. Limnol. Oceanogr. 22:400–407.

Perry, M. J., and R. W. Eppley. 1981. Phosphate uptake by phytoplankton in the central North Pacific Ocean. Deep-Sea Res. 28:39–49.

Perry, M. J., M. C. Talbot, and R. S. Alberte. 1981. Photoadaption in marine phytoplankton: Response of the photosynthetic unit. Mar. Biol. 62:91–101.

Persson, L.-E. 1981. Were macrobenthic changes induced by thinning out of flatfish stocks in the Baltic proper? Ophelia 20:137–152.

Peterman, R. M. 1980. Testing for density-dependent marine survival in Pacific salmonids. Pp. 1–23 in W. J. McNeil and D. C. Himsworth (eds.), Salmonid Ecosystems of the North Pacific. Oregon State Univ.

Peterman, R. M., and M. Gatto 1978. Estimation of functional responses of predators of juvenile salmon. J. Fish. Res. Bd. Canada 35:797–808.

Peters, R. H. 1976. Tautology in evolution and ecology. Amer. Nat. 110:1–12.

Peterson, B. J. 1980. Aquatic primary productivity and the ^{14}C-CO_2 method: A history of the productivity problem. Ann. Rev. Ecol. Syst. 11:359–385.

Peterson, B. J., J. E. Hobbie, and J. F. Haney. 1978. *Daphnia* grazing on natural bacteria. Limnol. Oceanogr. 23:1039–1044.

Peterson, C. H. 1979. Predation, competitive exclusion, and diversity in the soft-sediment benthic communities of estuaries and lagoons. Pp. 233–264 in R. J. Livingston (ed.), Ecological Processes in Coastal and Marine Systems. Plenum.

Petipa, T. S. 1979. Trophic relationships in communities and the functioning of marine ecosystems: 1. Studies in trophic relationships in pelagic communities of the southern seas of the USSR and in the tropical Pacific. Pp. 237–250 in M. J. Dunbar (ed.), Marine Production Mechanisms. Cambridge Univ..

Petipa, T. S., E. V. Pavlova, and G. N. Mironov. 1970. The food web structure, utilization and transport of energy by trophic levels in the planktonic communities. Pp. 142–167 in J. H. Steele (ed.), Marine Food Chains. Univ. California.

Pfeiffer, W. J, and R. G. Wiegert. 1981. Grazers on *Spartina* and their predators. Pp. 87–112 in L. R. Pomeroy and R. G. Wiegert (eds.), The Ecology of a Salt Marsh. Springer-Verlag.

Pianka, E. R. 1966. Latitude gradients in species diversity: A review of concepts. Amer. Nat. 100:33–46.

Pianka, E. R. 1973. The structure of lizard communities. Ann. Rev. Ecol. Syst. 4:53–74.

Pickard, G. L. 1964. Descriptive Physical Oceanography: An Introduction. Pergamon.

Pielou, E. C. 1966. The measurement of diversity in different types of biological collection. J. Theor. Biol. 13:131–144.

Pielou, E. C. 1969. An Introduction to Mathematical Ecology. Wiley.

Pielou, E. C. 1975. Ecological Diversity. Wiley.

Pielou, E. C. 1977. Mathematical Ecology. Wiley.

Platt, T. 1971. The annual production by phytoplankton in St. Margaret's Bay, Nova Scotia. J. Cons. Perm. Int. Explor. Mer 33:324–333.

Platt, T. 1972. Local phytoplankton abundance and turbulence. Deep-Sea Res. 19:183–197.

Platt, T., L. M. Dickie, and R. W. Trites. 1970. Spatial heterogeneity of phytoplankton in a near-shore environment. J. Fish. Res. Bd. Canada 27:1453–1473.

Platt, T., and C. L. Gallegos. 1980. Modeling primary production. Pp. 339–362 in P. Falkowski (ed.), Primary Productivity in the Sea. Plenum.

Platt, T., and D. V. Subba Rao. 1975. Primary production of marine microphytes. Pp. 249–280 in J. P. Cooper (ed.), Photosynthesis and Productivity in Different Environments. Cambridge Univ.

Poindexter, J. S. 1981. Oligotrophy: Fast and famine existence. Adv. Microbial Ecol. 5:63–89.

Pollard, R. T. 1977. Observations and theories of Langmuir circulations and and their role in near surface mixing. Pp. 235–251 in M. Angel (ed.), A Voyage of Discovery. Pergamon.

Pomeroy, L. R. 1959. Algal productivity in salt marshes of Georgia. Limnol. Oceanogr. 4:386–397.

Pomeroy, L. R. 1960. Residence times of dissolved phosphate in natural waters. Science 131:1731–1732.

Pomeroy, L. R. 1974. The ocean's food web, a changing paradigm. Bioscience 24:499–504.

Pomeroy, L. R., and R. E. Johannes 1968. Occurrence and respiration of ultraplankton in the upper 500 m of the ocean. Deep-Sea Res. 15:381–391.

Pomeroy, L. R., H. M. Mathews, and Hong Saik Min. 1963. Excretion of phosphate and soluble organic phosphoric compounds by zooplankton. Limnol. Oceanogr. 8:50–55.

Pomeroy, L. R., and R. G. Wiegert (eds.). 1981. The Ecology of a Salt Marsh. Springer-Verlag.

Poole, R. W. 1974. An Introduction to Quantitative Ecology. McGraw-Hill.

Porter, J. W. 1972. Predation by *Acanthaster* and its effect on coral species diversity. Amer. Nat. 106:487–492.

Porter, J. W. 1974. Community structure of coral reefs on opposite sides of the isthmus of Panama. Science 186:543–545.

Porter, J. W. 1976. Autotrophy, heterotrophy, and resource partitioning in Caribbean reef-building corals. Amer. Nat. 110:731–742.

Porter, J. W., and K. G. Porter. 1977. Quantitative sampling of demersal plankton migrating from different coral reef substrates. Limnol. Oceanogr. 22:553–556.

Porter, K. G. 1973. Selective grazing and differential digestion of algae by zooplankton. Nature 244:179–180.

Porter, K. G. 1976. Enhancement of algal growth and productivity by grazing zooplankton. Science 192:1332–1334.

Postma, H., and J. W. Rommets. 1979. Dissolved and particulate organic carbon in the North Equatorial Current of the Atlantic Ocean. Neth. J. Sea Res. 13:85–98.

Potts, M., and B. A. Whitton. 1977. Nitrogen fixation by blue-green algal communities in the intertidal zone of the lagoon of Aldabra Atoll. Oecologia 27:275–283.

Poulet, S. A. 1978. Comparison between five coexisting species of marine copepods feeding on naturally occurring particulate matter. Limnol. Oceanogr. 23:1126–1143.

Poulet, S. A., and P. Marsot. 1980. Chemosensory feeding and food gathering by omnivorous marine copepods. Pp. 198–218 in W. C. Kerfoot (ed.), Evolution and Ecology of Zooplankton Communities. Spec. Symp. Vol. 3. Amer. Soc. Limnol. Oceanogr. Univ. Press of New England.

Prakash, A., M. A. Rashid, A. Jensen, and D. V. Subba Rao. 1973. Influence of humic substances on the growth of marine phytoplankton: Diatoms. Limnol. Oceanogr. 18:516–524.

Prakash, A., A. Jensen, and M. A. Rashid. 1975. Humic substances and aquatic productivity. Pp. 219–268 in D. Povoledo and H. L. Golterman (eds.), Proc. Int. Meet. Humic Substances, Nieuwersluis, 1972. PUDOC.

Pratt, D. M. 1966. Competition between *Skeletonema costatum* and *Olisthodiscus luteus* in Narragansett Bay and in culture. Limnol. Oceanogr. 11:447–455.

Preston, F. W. 1948. The commonness and rarity of species. Ecology 29:254–283.

Preston, F. W. 1962. The canonical distribution of commonness and rarity. Ecology 39:185–215.

Prezelin, B., and B. M. Sweeney. 1979. Photoadaptation of photosynthesis in two bloom-forming dinoflagellates. Pp. 101–106 in D. L. Taylor and H. H. Seliger (eds.), Toxic Dinoflagellates Blooms. Elsevier North Holland.

Price, P. W., C. E. Bouton, P. Gross, B. A. McPheron, J. N. Thompson, and A.E. Weis. 1980. Interactions among three trophic levels: Influence of plants on interactions between insect herbivores and natural enemies. Ann. Rev. Ecol. Syst. 11:41–65.

Prince, J. S. 1974. Nutrient assimilation and growth of some seaweeds in mixtures of seawater and secondary sewage treatment effluents. Aquaculture 4:69–80.

Prinslow, T., I. Valiela, and J. M. Teal. 1974. The effect of detritus and ration size on the growth of *Fundulus heteroclitus* L., a salt marsh killifish. J. Exp. Mar. Biol. Ecol. 16:1–10.

Pulliam, R. H. 1974. On the theory of optimal diets. Amer. Nat. 108:59–74.

Purcell, J. E. 1980. Influence of siphonophore behavior upon their natural diets: Evidence for aggressive mimicry. Science 209:1045–1047.

Qasim, S. Z. 1970. Some problems related to the food chain in a tropical estuary. Pp. 45–51 in J. H. Steele (ed.), Marine Food Chains. Oliver and Boyd.

Qasim, S. Z. 1979. Primary production in some tropical environments. Pp. 31–69 in M. J. Dunbar (ed.), Marine Production Mechanisms. Cambridge Univ.

Radovich, J., and A. D. McCall. 1979. A management model for the central stock of the Northern Anchovy, *Engraulis mordax*. Calif. Coop. Fish. Invest. Rep. 20:83–88.

Raitt, D. F. S. 1968. The population dynamics of the Norway pout in the North Sea. Dept. Agric. Fish. Scotland, Mar. Res. 24 pp.

Ramus, J. 1978. Seaweed anatomy and photosynthetic performance: The ecological significance of light guides, heterogeneous absorption and multiple scatter. J. Phycol. 14:352–362.

Rand, A. L. 1954. Social feeding behavior of birds. Fieldiana: Zoology 36:1–71.

Rauck, G., and J. J. Zijlstra. 1978. On the nursery aspects of the Wadden sea for some commercial fish species and possible long-term changes. Rapp. Proc. - Verb. Reun. Cons. Int. Explor. Mer 172:266–275.

Real, L. A. 1979. Ecological determinants of functional response. Ecology 60:481–485.

Reddy. K. R., and W. H. Patrick. 1976. Effect of frequent changes in aerobic and anaerobic conditions on redox potential and nitrogen loss in a flooded soil. Soil Biol. Biochem. 8:491–495.

Redfield, A. C. 1934. On the proportion of organic derivatives in sea water and their relation to the composition of plankton. Pp 176–192 James Johnston Memorial Volume. Univ. Press of Liverpool.

Redfield, A. C., B. H. Ketchum, and F. A. Richards. 1963. The influence of organisms on the composition of sea-water. Pp. 26–77 in M. N. Hill (ed.), The Sea, Vol. 2. Interscience.

Reeburgh, W. S. 1976. Methane consumption in Cariaco Trench waters and sediments. Earth Planet. Sci. Lett. 28:337–344.

Reeburgh, W. S., and D. T. Heggie. 1977. Microbial methane consumption reaction and their effect on methane distributions in freshwater and marine sediments. Limnol. Oceanogr. 22:1–9.

Rees, C. P. 1975. Competitive interactions and substratum preferences of two intertidal isopods. Mar. Biol. 30:21–25.

Reeve, M. R. 1970. The biology of Chaetognatha. I. Quantitative aspects of growth and egg production in *Sagitta hispida*. Pp. 168–189 in J. H. Steele (ed.), Marine Food Chains. Univ. California.

Reeve, M. R., T. C. Cooper, and M. A. Walter. 1975. Visual observations on the program of digestion and the production of fecal pellet in the chaetognath *Sagitta hispida* Conant. J. Exp. Mar. Biol. Ecol. 17:39–46.

Reeve, M. R., and M. A. Walter. 1972. Conditions of culture, food-size selection and the effects of temperature and salinity on the rate and generation drive in *Sagitta hispida* Conant. J. Exp. Mar. Biol. Ecol. 9:191–200.

Reeve, M. R., M. A. Walter, and T. Ikeda. 1978. Laboratory studies of ingestion and food utilization in lobate and tentaculate ctenophores. Limnol. Oceanogr. 23:740–751.

Reichle, D. E., R. A. Goldstein, R. I. Van Hook, Jr., and G. J. Dodson. 1973. Analysis of insect composition in a forest canopy. Ecology 54:1077–1084.

Reiswig, H. M. 1974. Water transport, respiration, and energetics of three tropical marine sponges. J. Exp. Mar. Biol. Ecol. 14:231–249.

Revelante, N., and M. Gilmartin. 1976. Temporal succession of phytoplankton in the Northern Adriatic. Neth. J. Sea Res. 10:377–396.

Rex, M. A. 1981. Community structure in the deep-sea benthos. Ann. Rev. Ecol. Syst. 12:331–353.

Rhee, G.-Y. 1972. Competition between an alga and an aquatic bacterium for phosphate. Limnol. Oceanogr. 17:505–514.

Rhee, G.-Y. 1978. Effects of N:P atonic ratio and nitrate limitation on algal growth, cell composition, and nitrate uptake. Limnol. Oceanogr. 23:10–25.

Rhoads, D. C., and L. F. Boyer. In press. The effects of marine benthos on physical properties of sediments: A successional perspective. In P. McCall and M. Tevesz (eds.), The Biogenic Alterations of Sediment. Plenum Geobiology Series, Vol. 2. Plenum.

Rhoads, D. C., and R. A. Lutz (eds.). 1980. Skeletal Growth of Aquatic Organisms. Plenum.

Rhoads, D. C., P. L. McCall, and J. Y. Yingst. 1978. Disturbance and production on the estuarine sea floor. Amer. Sci. 66:577–586.

Rhoads, D. C., and D. K. Young. 1971. Animal-sediment relations in Cape Cod Bay, Massachusetts. II. Reworking by *Molpadia oolitica* (Holothuroidea). Mar. Biol. 11:255–261.

Rice, D. L. 1982. The detritus nitrogen problem. New observations and perspectives from organic geochemistry. Mar. Ecol. Progr. Ser. 9:153–162.

Rice, E. L., and S. K. Pancholy. 1972. Inhibition of nitrification by climax ecosystems. Amer. J. Bot. 53:1033–1040.

Richards, F. A. 1970. The enhanced preservation of organic matter in anoxic marine environments. Pp. 399–412 in D. W. Hood (ed.), Organic Matter in Natural Waters. Inst. Mar. Sci. Alaska Occ. Publ. 1.

Richards, F. A., and W. W. Broenkow. 1971. Chemical changes, including nitrate reduction in Darwin Bay, Galapagos Archipelago, over a 2 month period, 1969. Limnol. Oceanogr. 16:758–765.

Richards, F. A. 1968. Chemical and biological factors in the marine environment. Pp. 259–303 in J. F. Brahtz (ed.), Ocean Engineering. Wiley.

Richards, S. W., D. Merriman, and L. H. Calhoun. 1963. Studies on the marine resources of southern New England. IX. The biology of the little skate, *Raja erinacea* Mitchell. Bull Bingham Oceanogr. Coll. 18:4–67.

Richardson, P. 1976. Gulf stream rings. Oceanus 19:65–68.

Richerson, P., R. Armstrong, and C. R. Goldman. 1970. Contemporaneous disequilibrium, a new hypothesis to explain the "paradox of the plankton." Proc. Nat. Acad. Sci. 67:1710–1714.

Richman, S., D. R. Heinle, and R. Huff. 1977. Grazing by adult estuarine calanoid copepods of the Chesapeake Bay. Mar. Biol. 42:69–84.

Rickard, D. T. 1975. Kinetics and mechanisms of pyrite formation at low temperature. Amer. J. Sci. 275:636–652.

Ricker, W. E. 1975. Composition and interpretation of biological statistics of fish populations. Bull. Fish. Res. Bd. Canada 191:1–382.

Ricklefs, R. E. 1979. Ecology. 2nd Ed. Chiron.

Rieper, M. 1979. Microautoradiographic studies on North Sea sediment bacteria. Mar. Ecol. Prog. Ser. 1:337–345.

Riley, G. A. 1946. Factors controlling phytoplankton populations on Georges Bank. J. Mar. Res. 6:54–73.

Riley, G. A. 1947. A theoretical analysis of the zooplankton populations of Georges Bank. J. Mar. Res. 6:104–113.

Riley, G. A. 1956. Oceanography of Long Island Sound, 1952–1954. IX. Production and utilization of organic matter. Bull. Bingham Oceanogr. Coll. 18:324–341.

Riley, G. A. 1956a. Review of the oceanography of Long Island Sound. Deep-Sea Res. 3 (suppl.):224–238.

Riley, G. A. 1963. Theory of food chain relations in the ocean. Pp. 438–463 in M. N. Hill (ed.), The Sea, Vol. 2. Interscience.

Riley, G. A. 1963b. Organic aggregates in seawater and the dynamics of their formation and utilization. Limnol. Oceanogr. 8:372–381.

Riley, J. P., and R. Chester. 1971. Introduction to Marine Chemistry. Academic.

Robertson, S. B., and B. W. Frost. 1977. Feeding by an omnivorous planktonic copepod *Aetideus divergens* Bradford. J. Exp. Mar. Biol. Ecol. 29:231–244.

Robinson, J. D., K. H. Mann, and J. A. Novitsky. 1982. Conversion of the particulate fraction of seaweed detritus to bacterial biomass. Limnol. Oceanogr. 27:1072–1079.

Rogers, D. E. 1980. Density-dependent growth of Bristol Bay sockeye salmon. Pp.257–283 in W. J. McNeil and D. C. Himsworth (eds.), Salmonid Ecosystems of the North Pacific. Oregon State Univ.

Rogers, D. J., and M. P. Hassell. 1974. General models for insect parasite and predator searching behavior: Interference. J. Anim. Ecol. 43:239–253.

Rokop, F. J. 1974. Reproductive patterns in the deep-sea benthos. Science 186:743–745.

Rokop, F. J. 1977. Seasonal reproduction in the deep sea. Mar. Biol. 43:237–246.

Roman, M. R., and P. A. Rublee. 1980. Containment effects in copepod grazing experiments: A plea to end the black box approach. Limnol. Oceanogr. 25: 982–990.

Rosenfeld, J. K. 1979. Ammonium adsorption in near shore anoxic sediments. Limnol. Oceanogr. 24:356–364.

Rosenthal, G. A., and D. H. Janzen. 1979. Herbivores: Their interaction with secondary plant metabolites. Academic.

Rosenzweig, M. L., and R. H. MacArthur. 1963. Graphic representation and stability conditions of predator-prey interactions. Amer. Nat. 97:209–223.

Roughgarden, J. 1974. Species packing and the competitive function with illustrations from coral reef fish. Theor. Pop. Biol. 5:163–186.

Rowe, G. T., C. H. Clifford, and K. L. Smith. 1977. Nutrient regeneration in sediments off Cap Blanc, Spanish Sahara. Deep-Sea Res. 24:57–63.

Rowe, G. T., and W. D. Gardner 1979. Sedimentation rates in the slope water of the Northwest Atlantic Ocean measured directly with sediment traps. J. Mar. Res. 37:581–600.

Rubenstein, D. I., and M. A. R. Koehl. 1977. The mechanisms of filter feeding. Some theoretical considerations. Amer. Nat. 111:981–994.

Rudd, J. W. M., and C. D. Taylor. 1980. Methane cycling in aquatic environments. Adv. Aquat. Microbiol. 2:77–150.

Runge, J. A. 1980. Effects of hunger and season on the feeding of *Calanus pacificus*. Limnol. Oceanogr. 25:134–145.

Russell, F. S. 1934. The zooplankton. III. A comparison of the abundance of zooplankton in the barrier reef lagoon with that of some regions in northern European waters. Great Barrier Reef Exp. 1928–29. Sci. Rep. 2:176–201.

Russell, F. S. 1935. The seasonal abundance and distribution of the pelagic young of teleostean fishes caught in the ring-trawl in offshore waters in the Plymouth area. Part II. J. Mar. Biol. Assoc. U.K. 20:147–179.

Russell, F. S., and J. S. Colman. 1934. The zooplankton. II. The composition of the zooplankton of the Barrier Reef Lagoon. Great Barrier Reef Exp. 1928–29. Sci. Rep. 2:159–201.

Russell-Hunter, W. D. 1970. Aquatic Productivity. McMillan.

Ryther, J. H. 1956. Photosynthesis in the ocean as a function of light intensity. Limnol. Oceanogr. 1:61–70.

Ryther, J. H. 1969. Photosynthesis and fish production in the sea. Science 166:72–76.

Ryther, J. H., and W. M. Dunstan. 1971. Nitrogen, phosphorus and eutrophication in the coastal marine environment. Science 171:1008–1013.

Ryther, J. H., D. W. Menzel, E. M. Hulbert, C. J. Lorenzen, and N. Corwin. 1971. The production and utilization of organic matter in Peru coastal current. Inv. Pesq. 35:43–59.

Ryther, J. H., and J. G. Sanders. 1980. Experimental evidence of zooplankton control of the species composition and size distribution of marine phytoplankton. Mar. Ecol. Progr. Ser. 3:279–283.

Ryther, J. H., and C. S. Yentsch. 1957. The estimation of phytoplankton production in the ocean from chlorophyll and light data. Limnol. Oceanogr. 2:281–286.

Saks, N. M., and E. G. Kahn. 1979. Substrate competition between a salt marsh diatom and a bacterial population. J. Phycol. 15:17–21.

Sakshaug, E., and S. Myklestad. 1973. Studies on the phytoplankton ecology of the Tronsheimsfjord. III. Dynamics of phytoplankton blooms in relation to environmental factors, biomass experiments, and parameters for the physiological state of the populations. J. Exp. Mar. Biol. Ecol. 11:157–188.

Sale, P. F. 1977. Maintenance of high diversity in coral reef fish communities. Amer. Nat. 111:337–359.

Sale, P. F. 1978. Reef fishes and other vertebrates: A comparison of social structure. Pp. 313–346 in E. S. Reese and F. J. Lighter (eds.), Contrasts in Behavior. Wiley-Interscience.

Sameoto, D. D. 1971. Life history, ecological production, and an empirical mathematical model of the population of *Sagitta elegans* in St. Margaret's Bay, Nova Scotia. J. Fish. Res. Bd. Canada 28:971–985.

Sameoto, D. P. 1973. Annual life cycle and production of the chaetognath *Sagitta elegans* in Bedford Basin, Nova Scotia. J. Fish. Res. Bd. Canada 30:333–344.

Samina, L. V. 1976. Rate and intensity of filtration in some Caspian sea bivalve molluscs. Oceanology 15:496–498.

Sammarco, P. W. 1980. *Diadema* and its relationship to coral spat mortality, grazing, competition, and biological disturbance. J. Exp. Mar. Biol. Ecol. 45:245–272.

Sanders, H. L. 1956. The biology of marine bottom communities. Bull Bingh. Oceanogr. Coll. Yale Univ. 15:344–414.

Sanders, H. L. 1968. Marine benthic diversity: A comparative study. Amer. Nat. 102:243–282.

Sanders, H. L. 1969. Benthic marine diversity and the stability-time hypothesis. Pp. 71–81 in G. M. Woodwell and H. H. Smith (eds.), Diversity and Stability in Ecological Systems. Brookhaven Symp. Biol. 22. Brookhaven Natl. Lab.

Sanders, H. L., J. F. Grassle, G. R. Hampson, L. S. Morse, S. Garner-Price, and C. J. Jones. 1980. Anatomy of an oil spill: Long term effects from the grounding of the barge *Florida* off West Falmouth, Mass. J. Mar. Res. 38:265–380.

Sand-Jensen, K. 1975. Biomass, net production and growth dynamics in an eelgrass (*Zostera marina* L.) population in Vellerup Vig, Denmark. Ophelia 14:185–201.

Sanger, G. A. 1972. Fisheries potentials and estimated biological productivity of the Subarctic Pacific region. Pp. 561–574 in A. Y. Takenouti (ed.), Biological Oceanography of The Northern North Pacific Ocean. Idemitsu Shoten.

Sansone, F. J., and C. J. Martens. 1978. Methane oxidation in Cape Lookout Bight, North Carolina. Limnol. Oceanogr. 23:349–355.

Saunders, R. L. 1963. Respiration in the Atlantic Cod. J. Fish. Res. Bd. Canada 20:373–386.

Saville, A. 1978. The growth of herring in the northwestern North Sea. Rapp. Proc.-Verb. Reun. Cons. Int. Explor. Mer 172:164–171.

Schaefer, M. B. 1957. A study of the dynamics of the fishery for yellowfin tuna in the eastern tropical Pacific Ocean. Bull. Interamer. Trop. Tuna Comm. 2:245–285.

Schaefer, M. B. 1965. The potential harvest of the sea. Trans. Amer. Fish. Soc. 94:123–128.

Schaefer, M. B. 1970. Men, birds, and anchovies in the Peru Current—dynamic interactions. Trans. Amer. Fish. Soc. 99:461–467.

Schaeffer, W. M. 1974. Optimal reproductive effort in fluctuating environments. Amer. Nat. 108:783–790.

Scheer, G. 1978. Application of phytosociologic methods. Pp. 175–196 in D. R. Stoddart and R. E. Johannes (eds.), Coral Reefs: Research Methods. UNESCO.

Scheltema, R. 1967. Relationship of temperature to the larval development of *Nassarius obsoleta* (Gastropoda). Biol. Bull 132:253–255.

Schindler, D. W. 1974. Whole-lake eutrophication experiments with phosphorus, nitrogen and carbon. Int. Ver. Theor. Angew. Limnol. Verh. 19:3221–3231.

Schindler, D. W. 1977. Evolution of phosphorus limitation in lakes. Science 195:260–267.

Schlichter, D. 1980. Adaptations of cnidarians for integumentary absorption of dissolved organic material. Rev. Can. Biol. 39:259–283.

Schoener, A., and T. W. Schoener. 1981. The dynamics of the species-area relation in marine fouling systems: 1. Biological correlates of changes in the species-area slope. Amer. Nat. 118:339–360.

Schoener, T. W. 1971. Theory of feeding strategies. Ann. Rev. Ecol. Syst. 2:369–404.

Schoener, T. W. 1974. Resource partitioning in ecological communication. Science 185: 27–39.

Schoener, T. W. 1983. Field experiments on interspecific competition. Amer.Nat. 122:240–285.

Scholander, P. F., W. Flagg, V. Walters, and L. Irving. 1953. Climatic adaptation in arctic and tropical poikilotherms. Physiol. Zool. 26:67–92.

Schottler, U. 1979. On the anaerobic metabolism of three species of *Nereis* (Annelida). Mar. Ecol. Progr. Ser. 1:249–254.

Schultz, J. C., and I. T. Baldwin. 1982. Oak leaf quality declines in response to defoliation by gypsy moth larvae. Science 217:149–151.

Schwinghamer, P. 1981. Characteristic size distributions of integral benthic communities. Canad. J. Fish. Aquat. Sci. 38:1255–1263.

Scott, J. A., N. R. French, and J. W. Leatham. 1979. Patterns of consumption in grasslands. Pp. 89–105 in N. French (ed.), Perspectives in Grasslands Ecology. Ecol. Stud. 32. Springer-Verlag.

Scranton, M. I., and P. G. Brewer. 1977. Occurrence of methane in the near-surface waters of the Western subtropical North-Atlantic. Deep-Sea Res. 24:127–138.

Scranton, M. I., and J. W. Farrington. 1977. Methane production in waters off Walvis Bay. J. Geophys. Res. 82:4947–4953.

Sebens, K. P., and K. De Riemer. 1977. Diel cycles of expansion and contraction in coral reef anthozoans. Mar. Biol. 43:247–256.

Seitzinger, S., S. Nixon, M. Pilson, and S. Burke. 1980. Denitrification and N_2O production in near-shore marine sediments. Geochim. Cosmochim. Acta 44:1053–1860.

Seki, H. 1968. Relation between production and mineralization in Aburatsubo Inlet, Japan. J. Fish. Res. Bd. Canada 25:625–637.

Sekiguchi, H., and T. Kato, 1976. Influence of *Noctiluca* predation on the *Acartia* populations in Isle Bay, coastal Japan. J. Oceanogr. Soc. Japan 32:195–198.

Self, R. F. L., and P. A. Jumars. 1978. New resource axes for deposit feeders? J. Mar. Res. 36:627–641.

Sellner, K. G. 1981. Primary productivity and the flux of dissolved organic matter in several marine environments. Mar. Biol. 65:101–112.

Semina, H. L. 1972. The size of phytoplankton cells in the Pacific Ocean. Int. Rev. Gesamten Hydrobiol. 57:177–205.

Sen Gupta, R. 1973. Nitrogen and phosphorus budgets in the Baltic Sea. Mar. Chem. 1:267–280.

Sergeant, D. E. 1973. Environment and reproduction in seals. J. Reprod. Fertil. Suppl. 19:555–561.

Shafir, A., and J. G. Field. 1980. Importance of a small carnivorous isopod in energy transfer. Mar. Ecol. Progr. Ser. 3:203–215.

Shanks, A. L., and J. D. Trent. 1980. Marine snow: Sinking rates and potential role in vertical flux. Deep-Sea Res. 27A:137–143.

Shapiro, J. 1973. Blue-green algae: Why they become dominant. Science 179:382–384.

Sharp, J. H. 1973. Size classes of organic carbon in seawater. Limnol. Oceanogr. 18:441–447.

Sharp, J. H. In press. Inputs into microbial food chains. In J. Hobbie and P. L. LeB. Williams (eds.), Heterotrophic Activity in the Sea. Plenum.

Sharp, J. H., M. J. Perry, E. H. Renger, and R. W. Eppley. 1980. Phytoplankton rate processes in the oligotrophic waters of the central North Pacific Ocean. J. Plankton Res. 2:335–353.

Shelbourn, J. E. 1962. A predator-prey size relationship for plaice larvae feeding on *Oikopleura*. J. Mar. Biol. Assoc. U.K. 42:243–252.

Shelbourn, J. E., J. R. Brett, and S. Shirata. 1973. Effect of temperature and feeding regime on the specific growth rate of sockeye salmon fry (*Oncorhynchus nerka*), with a consideration of size effect. J. Fish. Res. Bd. Canada 30:1191–1194.

Sheldon, R. W. 1979. Measurements of phytoplankton growths by particle counting. Limnol. Oceanogr. 24:760–767.

Sheldon, R. W., and S. R. Kerr. 1976. The population density of monsters in Loch Ness. Limnol. Oceanogr. 21:796–979.

Sheldon, R. W., and T. R. Parsons. 1967. A continuous size spectrum for particulate matter in the sea. J. Fish. Res. Bd. Canada 24:909–915.

Sheldon, R. W., A. Prakash, and W. H. Sutcliffe, Jr. 1972. The size distribution of particles in the ocean. Limnol. Oceanogr. 17:327–340.

Sheldon, R. W., W. H. Sutcliffe, Jr., and A. Prakash. 1973. The production of particles in the surface waters of the ocean with particular reference to the Sargasso Sea. Limnol. Oceanogr. 18:719–733.

Shukla, S. S., J. K. Syers, J. D. H. Williams, D. E. Armstrong, and R. F. Harris. 1971. Sorption of inorganic phosphate by lake sediments. Soil Sci. Soc. Amer. Proc. 35:244–249.

Shulenberger, E., and J. L. Reid. 1981 The Pacific shallow oxygen maximum, deep chlorophyll maximum, and primary productivity, reconsidered. Deep-Sea Res. 28A:901–919.

Siebers, D. 1979. Transintegumentary uptake of dissolved amino acids in the sea star *Asterias rubens*. A reassessment of its nutritional role with special reference to the significance of heterotrophic bacteria. Mar. Ecol. Progr. Ser. 1:169–177.

Sieburth, J. McN. 1968. The influence of algal antibiosis on the ecology of marine microorganisms. Pp. 63–94 in M. R. Droop and E. J. F. Wood (eds.), Advances in Microbiology of the Sea. Academic.

Sieburth, J. McN. 1969. Studies on algal substances in the sea. III. The production of extracellular organic matter by littoral marine algae. J. Exp. Mar. Biol. Ecol. 3:290–309.

Sieburth, J. McN., and J. T. Conover. 1965. *Sargassum* tannin, an antibiotic that retards fouling. Nature 208:52–53.

Sieburth, J. McN., and A. Jensen. 1969. Studies on algal substances in the sea II. The formation of gelbstoff (humic material) by exudates of Phaeophyta. J. Exp. Mar. Biol. Ecol. 3:275–289.

Sieburth, J. McN., V. Smetacek, and J. Lenz. 1978. Pelagic ecosystem structure: Heterotrophic compartments of the plankton and their relationship to plankton size fractions. Limnol. Oceanogr. 23:1256–1263.

Siegismund, H. R. 1982. Life cycle and production of *Hydrobia ventrosa* and *H. neglecta* (Mollusca: Prosobranchia). Mar. Ecol. Progr. Ser. 7:75–82.

Silliman, R. P. 1968. Interaction of food level and exploitation in experimental fish populations. Fish. Bull. U.S. Fish Wildl. Serv. 66:425–430.

Silver, M. W., and A. L. Alldredge. 1981. Bathypelagic marine snow: Vertical transport system and deep-sea algal and detrital community. J. Mar. Res. 39:510–530.

Silver, M. W., and K. W. Bruland. 1981. Differential feeding and fecal pellet organization of salps and pteropods and the possible origin of the deep-water flux and olive-green "cells." Mar. Biol. 62:263–273.

Silverman, M. P., and H. L. Ehrlich. 1964. Microbial formation and degradation of minerals. Adv. Appl. Microbiol. 6:153–206.

Silvertown, J. W. 1982. No evolved mutualisms between grasses and grazers. Oikos 38:253–254.

Simenstad, C. A., J. A. Estes, and K. W. Kenyon. 1978. Aleuts, sea otters, and alternate stable state communities. Science 200:403–411.

Simenstad, C. A., K. L. Fresh, and E. O. Salo. 1982. The role of Puget Sound and Washington coastal estuaries in the life history of Pacific salmon: An unappreciated function. Pp. 343–364 in V. S. Kennedy (ed.), Estuarine Comparisons. Academic.

Simpson, E. H. 1949. Measurement of diversity. Nature 163:688.

Sinclair, A. R. E. 1975. The resource limitation of trophic levels in tropical grassland ecosystems. J. Anim. Ecol. 44:497–520.

Skauen, D. M., N. Marshall, and R. J. Frajala. 1971. A liquid scintillation method for assaying ^{14}C labeled benthic microflora. J. Fish. Res. Bd. Canada 28:769–770.

Skirrow, G. 1975. The dissolved gases—carbon dioxide. Pp. 1–192 in J. P. Wiley and G. Skirrow (eds.), Chemical Oceanography, Vol. 2. Academic.

Skopintsev, B. A. 1972. A discussion of some views of the sizes, distribution,and composition of organic matter in deep ocean waters. Oceanology 12: 471–474.

Slater, J. H. 1979. Microbial population and community dynamics. Pp. 45–66 in J. M. Lynch and N. J. Poole (eds.), Microbial Ecology. Wiley.

Slichter, D. 1980. Adaptations of cnidarians for integumentary absorption of dissolved organic material. Rev. Can. Biol. 39:259–282.

Slobodkin, L. B., and L. Fishelson. 1974. The effect of the cleaner-fish *Labroides dimidiatus* on the point diversity of fishes on the reef front at Eilat. Amer. Nat. 108:369–376.

Slobodkin, L. B., F. E. Smith, and N. G. Hairston. 1967. Regulation in terrestrial ecosystems, and the implied balance of nature. Amer. Nat. 101:109–124.

Small, L. F., S. W. Fowles, and M. Y. Unlu. 1979. Sinking rates of natural copepod fecal pellets. Mar. Biol. 51:233–241.

Small, L. F. and J. F. Hebard. 1967. Respiration of a vertically migrating marine crustacean *Euphausia pacifica* Hansen. Limnol. Oceanogr. 12:272–280.

Smayda, T. J. 1966. A quantitative analysis of the phytoplankton of the Gulf of Panama. III. General ecological conditions and the phytoplankton dynamics at 8°45'N, 79°23'W from November 1954 to May 1957. Inter-Amer. Trop. Tuna Comm. Bull. 11:353–612.

Smayda, T. J. 1970. The suspension and sinking of phytoplankton in the sea. Oceanogr. Mar. Biol. Ann. Rev. 8:357–414.

Smayda, T. J. 1980. Phytoplankton species succession. Pp. 483–570 in I. Morris (ed.), The Physiological Ecology of Phytoplankton. Univ. California.

Smedes, G. W., and L. E. Hurd. 1981. An empirical test of community stability: Resistance of a fouling community to a biological patch-forming disturbance. Ecology 62:1561–1572.

Smith, F. E. 1972. Spatial heterogeneity, stability, and diversity in ecosystems. Trans. Conn. Acad. Arts. Sci. 44:309–335.

Smith, K. L. 1973. Respiration of a sublittoral community. Ecology 54:1065–1075.

Smith, K. L., K. A. Burns, and J. M. Teal. 1972. *In situ* respiration of benthic communities in Castle Harbor, Bermuda. Mar. Biol. 12:196–199.

Smith, K. L., Jr., and J. M. Teal. 1973a. Deep-sea benthic community respiration: An *in-situ* study at 1850m. Science 179:282–283.

Smith, K. L., Jr., and J. M. Teal. 1973b. Temperature and pressure effects on thecosomatous pteropods. Deep-Sea Res. 20:853–858.

Smith, K. L., Jr., G. A. White, and M. B. Laver. 1979. Oxygen uptake and nutrient exchange of sediments measured *in situ* using a free vehicle grab respirometer. Deep-Sea Res. 26A:337–346.

Smith, S. H. 1968. Species succession and fishery exploitation in the Great Lakes. J. Fish. Res. Bd. Canada 25:667–693.

Smith, T. J. III., and W. E. Odum. 1981. The effects of grazing by snow geese on coastal salt marshes. Ecology 62:98–106.

Smith, V. H. 1983. Low nitrogen to phosphorus ratios favor dominance by blue-green algae in lake phytoplankton. Science 221:669–671.

Smith, W. O., R. T. Barber, and S. A. Huntsman. 1977. Primary production off the Coast of Northwest Africa: Excretion of dissolved organic matter and its heterotrophic uptake. Deep-Sea Res. 24:35–47.

Solomon, M. E. 1949. The natural control of animal populations. J. Anim. Ecol. 18:1–35.

Somero, G. N., and P. W. Hochachka 1976. Biochemical adaptation to temperature. Pp. 125–190 in R. C. Newell (ed.), Adaptation to Environment. Butterworths.

Sorokin, Y. I. 1964. On the primary production and the bacterial activities in the Black Sea. J. Cons. Int. Explor. Mer 29:41–60.

Sournia, A. 1969. Cycle annual du phytoplankton, et de la production primaire dans les mers tropicales. Mar. Biol. 3:287–303.

Sournia, A. 1977. Analyse et bilan de la production primaire dans les récifs coralliens. Ann. Inst. Océanogr. 53:47–74.

Sousa, W. P. 1979. Experimental investigations of disturbance and ecological succession in a rocky intertidal algal community. Ecol. Monogr. 49:227–254.

Sousa, W. P. 1979a. Disturbance in marine intertidal boulder fields: The cross-equilibrium maintenance of species diversity. Ecology 60:1225–1239.

Sousa, W. P. 1980. The responses of a community to disturbance: The importance of successional age and species life histories. Oecologia 45:72–81.

Soutar, A., and J. D. Isaacs. 1969. History of fish populations inferred from fish scales in anaerobic sediments off California. Calif. Coop. Fish. Inv. Rep. 13:63–70.

Southward, A. J. 1980. The Western English Channel—An inconstant ecosystem? Nature 285:361–366.

Southward, A. J., and E. C. Southward. 1968. Uptake and incorporation of labelled glycine by pogonophores. Nature 218:875–876.

Southward, G. M. 1967. Growth of Pacific halibut. Rep. Int. Pac. Halibut Comm. 43:1–40.

Southwood, T. R. E. 1978. Ecological Methods. 2nd Ed. Chapman and Hall/Wiley.

Spencer, D. W. 1981. The sediment trap intercomparison experiment: Some preliminary data. Pp. 57–104 in R. F. Anderson and M. P. Bacon (eds.), Sediment Trap Intercomparison Experiment. Woods Hole Oceanographic Institutions Techn. Memo. WHOI-1–81.

Staresinic, N., G. T. Rowe, D. Shaughnessey, and A. J. Williams. 1978. Measurement of the vertical flux of particulate matter with a free drifting sediment trap. Limnol. Oceanogr. 23: 559–563.

Stavn, R. H. 1971. The horizontal-vertical distribution hypothesis: Langmuir circulations and *Daphnia* distributions. Limnol. Oceanogr. 16:453–466.

Stearns, S. C. 1976. Life-history tactics: A review of the ideas. Quart. Rev. Biol. 5:3–47.

Steele, J. H. 1974a. The Structure of Marine Ecosystems. Harvard Univ..

Steele, J. H. 1974b. Spatial heterogeneity and populations stability. Nature 248:83.

Steele, J. H. 1976. The role of predation in ecosystem models. Mar. Biol. 35:9–11.
Steele, J. H., and E. W. Henderson. 1977. Plankton patches in the Northern North Sea. Pp. 1–19 in J. H. Steele (ed.), Fisheries Mathematics. Academic.
Steele, J. H., and B. W. Frost. 1977. The structure of plankton communities. Phil. Trans. Roy. Soc. London 280:485–534.
Steemann Nielsen, E. 1975. Marine Photosynthesis with Special Emphasis on the Ecological Aspects. Amsterdam: Elsevier Oceanography Series 13.
Steemann Nielsen, E., and A. Jensen. 1957. Primary ocean production, the autotrophic production of organic matter in the oceans. Galathea Rep. 1:49.
Steemann Nielsen, E., and E. G. Jorgensen. 1968. The adaptation of plankton algae. III. With special consideration of the importance in nature. Physiol. Plant. 21:647–654.
Steemann Nielsen, E., and S. Wium-Anderson 1970. Copper ions as poison in the sea and freshwater. Mar. Biol. 6:93–97.
Stewart, M. G. 1979. Absorption of dissolved organic nutrients by marine invertebrates. Oceanogr. Mar. Biol. Ann. Rev. 17:163–192.
Stewart, M. G. 1981. Kinetics of dipeptide uptake by the mussel *Mytilus edulis*. Comp. Biochem. Physiol. 69A:311–315.
Stewart, M. G., and R. C. Dean. 1980. Uptake and utilization of amino acids by the shipworm *Bankia gouldi*. Comp. Biochem. Physiol. 66B:443–450.
Stimson, J. 1970. Territorial behavior of the owl limpet, *Lottia gigantea*. Ecology 51:113–118.
Stimson, J. 1973. The role of territory in the ecology of the intertidal limpet *Lottia gigantea* (Gray). Ecology 54:1020–1030.
Stoecker, D. 1980. Relationships between chemical defense and ecology in benthic ascidians. Mar. Ecol. Prog. Ser. 3:257–265.
Stoecker, D., R. R. L. Guillard, and R. M. Kavee. 1981. Selective predation by *Favella ehrenbergii* (Tintinnia) on and among dinoflagellates. Biol. Bull. 160:136–145.
Stonehouse, B. (ed.). 1975. The Biology of Penguins. Univ. Park.
Stout, J. D., K. R. Tate, and L. F. Molloy. 1976. Decomposition processes in New Zealand soils with particular respect to rates and pathways of plant degredation. Pp. 97–144 in J. M. Anderson and A. Macfadyen (eds.), The Role of Terrestrial and Aquatic Organisms in Decomposition Processes. Blackwell.
Strickland, J. D. H. 1965. Production of organic matter in the primary stages of the oceanic food chains. Pp. 478–610 in J. B. Riley and G. Skirrow (eds.), Chemical Oceanography. Academic.
Strickland, J. D. H., and K. H. Austin. 1960. On the forms, balance, and cycle of phosphorus observed in the coastal and oceanic waters of the Northeastern Pacific. J. Fish. Res. Bd. Canada 17:337–345.
Strickland, J. D. H., and T. R. Parsons. 1968. A practical handbook of seawater analysis. Fish. Res. Bd. Canada Bull. 167:1–311.
Strickler, J. R. 1975. Intra and interspecific information flow among planktonic copepods: Receptors. Int. Ver. Theor. Angew. Limnol. Verh. 19:2951–2958.
Strickler, J. R., and A. K. Skal 1973. Setae of the first antennae of the copepod *Cyclops scutifer* (Sars.): Their structure and importance. Proc. Nat. Acad. Sci. 70:2656–2659.
Stuart, V., M. I. Lucas, and R. C. Newell. 1981. Heterotrophic utilization of

particulate matter from the kelp *Laminaria pallida*. Mar. Ecol. Progr. Ser. 4:357–348.

Stuermer, D. H., and J. R. Payne. 1976. Investigation of seawater and terrestrial humic substances with carbon-13 and proton nuclear magnetic resonance. Geochim. Cosmochim. Acta 40:1109–1114.

Stumm, W., and J. J. Morgan. 1981. Aquatic Chemistry. 2nd Ed. Wiley.

Suess, E. 1980. Particulate organic carbon flux in the oceans—surface productivity and oxygen utilization. Nature 288:260–263.

Sunda, W., and R. R. L. Guillard. 1976. The relationship between Cu ion activity and the toxicity of Cu to phytoplankton. J. Mar. Res. 34:511–527.

Sunda, W. G., R. T. Barber, and S. A. Huntsman. 1981. Phytoplankton growth in nutrient rich seawater: Importance of copper-manganese cellular interactions. J. Mar. Res. 39:567–586.

Sunda, W. G., and J. A. M. Lewis. 1978. Effect of complexation by natural organic ligands on the toxicity of copper to a unicellular algae, *Monochrysis lutheri*. Limnol. Oceanogr. 23:870–876.

Sushchenya, L. M. 1968. Elements of energy balance in the amphipod *Orchestia bottae* Mil.-Edw. (Amphipoda-Talitroidea). Biol. Morya 15:52–70. In Russian. Fish. Res. Bd. Canada Transl. 1225.

Sushchenya. L. M. 1970. Food rations, metabolism, and growth of crustaceans. Pp. 127–141 in J. H. Steele (ed.), Marine Food Chains. Univ. California.

Sutcliffe, W. H. 1960. On the diversity of the copepod population in the Sargasso Sea of Bermuda. Ecology 41:585–587.

Sutherland, J. P. 1974. Multiple stable points in natural communities. Amer. Nat. 108:859–873.

Sutherland, J. P. 1978. Functional roles of *Schizoporella* and *Styela* in the fouling community at Beaufort, N. C. Ecology 59:257–264.

Sutherland, J. P. 1980. Dynamics of the epibenthic community on roots of the mangrove *Rhizophora mangle*, at Bahia de Buche, Venezuela. Mar. Biol. 58:75–84.

Sutherland, J. P., and R. H. Karlson. 1977. Development and stability of the fouling community at Beaufort, N.C. Ecol. Monogr. 47:425–446.

Sverdrup, H. U. 1953. On conditions for the vernal blooming of phytoplankton. J. Cons. Int. Explor. Mer 18:287–295.

Swain, T. 1979. Tannins and lignins. Pp. 657–682 in G. E. Rosenthal and D. H. Janzen (eds.), Herbivores: Their Interaction with Secondary Plant Metabolites. Academic.

Swift, M. J., O. W. Heal, and J. M. Anderson (eds.). 1979. Decomposition in Terrestrial Ecosystems. Stud. in Ecol. Vol. 5. Univ. California.

Szyper, J. P., J. Hirota, J. Caperon, and D. A. Ziemann. 1976. Nutrient regeneration by the larger net zooplankton in the southern basins of Kaneshe Bay. Pac. Sci. 30:363–372.

Taghon, G. L., R. F. Self, and P. A. Jumars. 1978. Predicting particulate selection by deposit feeders: A model and its implications. Limnol. Oceanogr. 23:752–759.

Taguchi, S. 1976. Relationship between photosynthesis and cell size of marine diatoms. J. Phycol. 12:185–189.

Takahashi, M., and S. Ichimura. 1968. Vertical distribution and organic matter production of photosynthetic sulfur bacteria in Japanese lakes. Limnol. Oceanogr. 13:644–655.

Talbot, F. H., B. C. Russell, and G. R. V. Anderson 1978. Coral reef fish communities: Unstable high diversity systems? Ecol. Monogr. 48:425–440.

Talling, D. F. 1960. Comparative laboratory and field studies of photosynthesis by a marine planktonic diatom. Limnol. Oceanogr. 5:62–77.

Taniguchi, A. 1973. Phytoplankton-zooplankton relationships in the Western Pacific Ocean and adjacent seas. Mar. Biol. 21:115–121.

Taylor, F. J. R. 1978. Problems in the development of an explicit hypothetical phylogeny of the larval eukaryotes. Biosystems 10:67–89.

Taylor, L. R., R. A. Kempton, and I. P. Woiwod. 1976. Diversity statistics and the log-series model. J. Anim. Ecol. 45:255–272.

Teal, J. M. 1962. Energy flow in the salt marsh ecosystem of Georgia. Ecology 43:614–624.

Teal, J. M., and F. G. Carey. 1967. Effects of pressure and temperature on the respiration of euphausiids. Deep-Sea Res. 14:725–733.

Teal, J. M., and J. W. Kanwisher. 1961. Gas exchange in a Georgia salt marsh. Limnol. Oceanogr. 6:388–399.

Teal, J. M., and J. W. Kanwisher. 1966. The use of pCO_2 for the calculation of biological production, with examples from waters off Massachusetts. J. Mar. Res. 24:4–14.

Teal, J. M., and J. W. Kanwisher. 1966a. Gas transport in the marsh grass, *Spartina alterniflora*. J. Exp. Bot. 17:355–361.

Teal, J. M., I. Valiela, and D. Berlo. 1979. Nitrogen fixation by rhizosphere and freeliving bacteria in salt marsh sediments. Limnol. Oceanogr. 24:126–132.

Tempel, D., and W. Westenheide. 1980. Uptake and incorporation of dissolved amino acids by interstitial turbellaria and polychaeta and their dependence on temperature and salinity. Mar. Ecol. Progr. Ser. 3:41–50.

Tenore, K. R. 1977. Growth of *Capitella capitata* cultured on various levels of detritus derived from different sources. Limnol. Oceanogr. 22:936–941.

Tenore, K. R. 1983. What controls the availability of detritus derived from vascular plants: Organic nitrogen enrichment or caloric availability? Mar. Ecol. Progr. Ser. 10:307–309.

Tenore, K. R., L. Cammen, S. E. G. Findlay, and N. Philips. 1982. Perspectives of research on detritus: Do factors controlling the availability of detritus to macroconsumers depend on its source? J. Mar. Res. 40: 473–480.

Tenore, K. R., and W. M. Dunstan. 1973. Comparison of feeding and biodeposition of three bivalves at different feed levels. Mar. Biol. 21:190–195.

Tenore, K. R., J. C. Goldman, and J. P. Clarner. 1973. The food chain dynamics of the oyster, clam, and mussel in an aquaculture food chain. J. Exp. Mar. Biol. Ecol. 12: 157–165.

Tenore, K. R., R. B. Hanson, B. E. Donseif, and C. N. Wiederhold. 1979. The effect of organic nitrogen supplement on the utilization of different sources of detritus. Limnol. Oceanogr. 84:350–355.

Tenore, K. R., and D. L. Rice. 1980. A review of trophic factors affecting secondary production of deposit feeders. Pp. 325–340 in K. R. Tenore and B. C. Coull (eds.), Marine Benthic Dynamics. Univ. of South Carolina.

Tenore, K. R., J. H. Tietjen, and J. J. Lee. 1977. Effects of meiofauna on incorporation of aged eelgrass detritus by the polychaete *Nepthys incisa*. J. Fish. Res. Bd. Canada 34:563–567.

Thistle, D. 1981. Natural physical disturbances and communities of marine soft bottoms. Mar. Ecol. Progr. Ser. 6:223–228.

Thomas, W. H. 1969. Phytoplankton nutrients enrichment experiments off Baja California and in the eastern equatorial Pacific Ocean. J. Fish. Res. Bd. Canada 26:1133–1145.

Thomas, W. H. 1979. Anomalous nutrient—chlorophyll interrelationships in the offshore eastern tropical Pacific Ocean. J. Mar. Res. 37:327–335.

Thomas, W. H., and A. N. Dodson. 1972. On nitrogen deficiency in tropical Pacific phytoplankton. II. Photosynthetic and cellular characteristics of a chemostat-grown diatoms. Limnol. Oceanogr. 17:515–523.

Thomas, W. H., E. H. Renger, and A. N. Dodson. 1971. Near surface organic nitrogen in the eastern tropical Pacific Ocean. Deep-Sea Res. 18:65–71.

Thompson, D. J. 1978. Towards a realistic predator-prey model: The effect of temperature on the functional response and life history of larvae of the damselfly, *Ischnura elegans*. J. Anim. Ecol. 47:757–767.

Thorson, G. 1957. Bottom communities (sublittoral or shallow shelf). Pp. 401–534 in J. Hedgepeth, Treatise on Marine Ecology and Paleoecology. Geol. Soc. Amer. Mem. 67.

Tietjen, J. H., and J. J. Lee. 1977. Feeding behavior of marine nematodes. Pp. 21–35 in B. C. Coull (ed.), Ecology of Marine Benthos. Univ. South Carolina.

Tiews, K. 1978. On the disappearance of bluefin tuna in the North Sea and its ecological implications for herring and mackerel. Rapp. Proc.-Verb. Cons. Intern. Explor. Mer 172:301–309.

Tijssen, S. B. 1979. Diurnal oxygen rhythms and primary production in the mixed layer of the Atlantic Ocean at 20°N. Neth. J. Sea Res. 13:79–84.

Tilman, D. 1977. Resource competition between planktonic algae: An experimental and theoretical approach. Ecology 58:338–348.

Timonin, A. G. 1969. The structure of pelagic associations. The quantitative relationship between different trophic groups of plankton in frontal zones of the tropical ocean. Oceanology 9:686–694.

Tinbergen, L. 1960. The natural control of insects in pine woods. I. Factors influencing the intensity of predation by songbirds. Arch. Neerl. Zool. 13:265–343.

Tolbert, N. E. 1974. Photorespiration. Pp. 474–504 in W. D. Stewart (ed.), Algal Physiology and Biochemistry. Univ. California.

Tomosada A. 1978. A large warm eddy detached from the Kuroshio east of Japan. Bull. Tokai. Reg. Fish. Res. Lab. 94:59–103.

Topinka, J. A., and J. V. Robbins. 1976. Effect of nitrate and ammonium enrichment on growth and nitrogen physiology in *Fucus spiralis*. Limnol. Oceanogr. 21:659–664.

Tranter, D. J., and B. S. Newell. 1963. Enrichment experiments in the Indian Ocean. Deep-Sea Res. 10:1–9.

Trench, R. K. 1974. Nutritional potentials in *Zoanthus sociatus* (Coelenterata, Anthozoa). Helgol. Wiss. Meeresunt. 26:174–216.

Trench, R. K. 1979. The cell biology of plant-animal symbiosis. Ann. Rev. Pl. Physiol. 30:485–592.

Trevallion, A. 1971. Studies on *Tellina tenuis* Da Costa. III. Aspects of general biology and energy flow. J. Exp. Mar. Biol. Ecol. 7:95–122.

Trevallion, A., R. R. C. Edwards, and J. H. Steele. 1970. Dynamics of a benthic bivalve. Pp. 285–295. in J. H. Steele (ed.), Marine Food Chains, Univ. California.

Tseytlin, B. V. 1976. Gigantism in deep-water plankton-consuming organisms.Oceanology 15:499–503.

Turner, R. E. 1978. Community plankton respiration in a salt marsh estuary and the importance of macrophytic leachates. Limnol. Oceanogr. 23:422–451.

Turner, R. E., W. W. Woo, and H. R. Jitts. 1979. Estuarine influences on a continental shelf plankton community. Science 206:218–220.

Turpin, D. H., and P. J. Harrison. 1979. Limiting nutrient patchiness and its role in phytoplankton ecology. J. Exp. Mar. Biol. Ecol. 39:151–166.

Tyler, A. V. 1970. Rates of gastric emptying in young cod. J. Fish. Res. Bd. Canada 27:1177–1189.

Ursin, E. 1973. On the prey size preferences of cod and dab. Meed. Danm. Fisk.-og Havunders. N.S. 7:85–98.

Valiela, I., B. Howes, R. Howarth, A. Giblin, K. Foreman, J. M. Teal, and J. E. Hobbie. 1982. Regulation of primary production and decomposition in a salt marsh ecosystem. Pp. 151–168 in B. Gopal, R. E. Turner, R. G. Wetzel, and D. F. Whigham (eds.), Wetlands, Ecology, and Management. Nat. Inst. of Ecol. India.

Valiela, I., L. Koumjian, T. Swain, J. M. Teal, and J. E. Hobbie. 1979. Cinnamic acid inhibition of detritus feeding. Nature 280:55–57.

Valiela, I., and J. M. Teal. 1979. The nitrogen budget of a salt marsh ecosystem. Nature 280:652–656.

Valiela, I., J. M. Teal, C. Cogswell, J. Hartman, S. Allen, R. Van Etten, and D. Goehringer. In press. Some long-term consequences of sewage contamination in salt marsh ecosystems. Proc. Conf. Uses of Wetlands in Sewage Treatment. U.S. Fish. Wildl. Serv., and E.P.A.

Valiela, I., J. M. Teal, and N. Y. Persson. 1976. Production and dynamics of experimentally-enriched salt marsh vegetation: Below-ground biomass. Limnol. Oceanogr. 21:245–252.

Valiela, I. 1983. Nitrogen in salt marsh ecosystems. Pp. 649–678 in E. J. Carpenter and D. G. Capone (eds.), Nitrogen in the Marine Environment. Academic.

Valiela, I., J. M. Teal, S. Volkmann, D. Shafer, and E. C. Carpenter. 1978. Nutrient and particulate fluxes in a salt marsh ecosystem: Tidal exchanges and inputs by precipitation and groundwater. Limnol. Oceanogr. 23:798–812.

Valley, S. L. (ed.) 1965. Handbook of Geophysics and Space Environments. McGraw Hill.

Van Blaricom, G. R. 1982. Experimental analysis of structural regulation in a marine sand community exposed to oceanic swell. Ecol. Monogr. 52:283–305.

Van der Assam, J. 1967. Territory in the three-spined stickleback *Gasterosteus aculeatus* L. Behavior Suppl. 16:1–164.

Van Raalte, C. D. 1975. Epibenthic salt marsh algae: Light and nutrient limitation. Ph.D. Thesis, Boston Univ.

Van Raalte, C. D., I. Valiela, E. J. Carpenter, and J. M. Teal. 1974. Nitrogen fixation: Presence in salt marshes and inhibition by addition of combined nitrogen. Estuar. Coast. Mar. Sci. 2:301–305.

Van Raalte, C. D., I. Valiela, and J. M. Teal. 1976. Productivity of benthic algae in experimentally fertilized salt marsh plots. Limnol. Oceanogr. 21:862–872.

Van Raalte, C. D., I. Valiela, and J. M. Teal. 1976a. The effect of fertilization on the species of salt marsh diatoms. Water Res. 10:1–4.

Van Sickle, J. 1977. Mortality rates from size distribution. Oecologia 27:311–318.

Van Someren, V. G. L., and T. H. E. Jackson. 1959. Some comments on protective resemblance amongst African Lepidoptera (Rhopalocera). J. Lepidopterists Soc. 13:121–150.

Van Valkenberg, J., K. Jones, and D. R. Heinle. 1978. A comparison by size class and volume of detritus versus phytoplankton in Chesapeake Bay. Estuar. Coast. Mar. Sci. 6:569–582.

Van Wazer, J. R. 1958. Phosphorus and Its Compounds. Vol. I. Interscience.

Van Wazer, J. R. 1973. The compounds of phosphorus. Pp. 169–178. in E. J. Griffith, A. Beeton, J. M. Spencer, and D. T. Mitchell (eds.), Environmental Phosphorus Handbook. Wiley.

Vatova, A. 1961. Primary production in the High Venice Lagoon. J. Cons. Perm. Inter. Explor. Mer 26:148–155.

Veen, J. F. de. 1978. Changes in North Sea sole stocks (*Solea solea* (L.)). Rapp. Proc.-Verb. Cons. Int. Explor. Mer 172:124–136.

Velimirov, B., J. A. Ott, and R. Novak. 1981. Microorganisms on macrophyte debris: Biodegradation and its implication for the food web. Kieler Meeresforsch. 5:333–344.

Venrick, E. L. 1974. The distribution and significance of *Richelia intracellularis* Schmidt in the North Pacific Central Gyre. Limnol. Oceanogr. 19:473–445.

Venrick, E. L., J. R. Beers, and J. F. Heinbokel. 1977. Possible consequences of containing microplankton for physiological rate measurements. J. Exp. Mar. Biol. Ecol. 26:55–76.

Verity, P. G., and D. Stoecker. In press. The effects of *Olisthodiscus luteus* Carter on the growth and abundance of tintinnids. Mar. Biol.

Vermeij, G. J., and A. D. Covich. 1978. Coevolution of freshwater gastropods and their predators. Amer. Nat. 112:833–843.

Vernberg, W. B., and B. C. Coull. 1974. Respirations of an interstitial ciliate and benthic energy relationships. Oecologia 16:259–264.

Victor, B. C. 1983. Recruitment and population dynamics of a coral reef fish. Science 219:419–420.

Vidal, J. 1980. Physioecology of zooplankton. I. Effects of phytoplankton concentration, temperature, and body size on the growth rate of *Calanus pacificus* and *Pseudocalanus* sp. Mar. Biol. 56:111–134.

Vince, S., and I. Valiela. 1973. The effects of ammonium and phosphate enrichment on chlorophyll *a*, pigment ratio, and species composition of phytoplankton of Vineyard Sound. Mar. Biol. 19:69–73.

Vince, S., I. Valiela, N. Backus, and J. M. Teal. 1976. Predation by the salt marsh killifish *Fundulus heteroclitus* (L.) in relation to prey size and habitat structure: Consequences for prey distribution and abundance. J. Exp. Mar. Biol. Ecol. 23:255–266.

Vince, S. W., I. Valiela, and J. M. Teal. 1981. An experimental study of the structure of herbivorous insect communities in a salt marsh. Ecology 62:1662–1678.

Vine, I. 1971. Risk of visual detection and pursuit by a predator and the selective advantage of flocking behavior. J. Theor. Biol. 30:405–428.

Vine, P. J. 1974. Effects of algal grazing and aggressive behavior of the fishes *Pomacentrus lividus* and *Acanthurus sohal* on coral-reef ecology. Mar. Biol. 24:131–136.

Vinogradov, A. P. 1953. The elementary chemical composition of marine organisms. Sears Found. Mar. Res. Mem. 2:1–647.

Vinogradova, L. A. 1977. Experimental determination of the gravitational sinking rate of marine planktonic algae. Oceanology 17:458–461.

Vinogradov, M. E. 1970. Vertical Distribution of the Oceanic Zooplankton. Israel Program for Scientific Translations.

Vinogradov, M. E. 1972. Vertical stratification of zooplankton in the Kurile-Kamchatka trench. Pp. 333–340 in A. Y. Takenouti (ed.), Biological Oceanography of the Northern North Pacific Ocean. Idemitsu Shoten.

Vinogradov, M. Y., I. V. Kulina, L. P. Lebedea, and E. A. Shushkina. 1977. Variation with depth of the functional characteristics of a plankton community in the equatorial upwelling region of the Pacific Ocean. Oceanology 17:345–355.

Virnstein, R. W. 1977. The importance of predation by crabs and fishes on benthic infauna in the Chesapeake Bay. Ecology 58:1199–1217.

Vitousek, P. M., J. R. Gosz, C. C. Grier, J. M. Melillo, and W. A. Reiners. 1982. A comparative analysis of potential nitrification and nitrate mobility in forest ecosystems. Ecol. Monogr. 52:155–177.

Volkmann-Rocco, B. and G. Fava. 1969. Two sibling species of *Tisbe* (Copepoda, Harpacticoidea): *Tisbe reluctans* and *T. persimilis* n. sp. Mar. Biol. 3:159–164.

Volterra, V. 1926. Variazioni e fluttuazione del numero d'individui in specie animali conviventi. Mem. Acad. Lincei Roma 2:31–113.

Von Arx, W. S. 1962. An Introduction to Physical Oceanography. Addison- Wesley.

Vooys, C. G. N. de. 1979. Primary production in aquatic environments. Pp. 259–252 in B. Bolin, E. T. Degens, S. Kempe, and P. Ketner (eds.), The Global Carbon Cycle. Wiley.

Wakeham, S. G., and J. W. Farrington. 1980. Hydrocarbons in contemporary aquatic sediments. Pp. 3–32 in R. A. Baker (ed.), Contaminants and Sediments. Ann Arbor. Sci. Publ.

Walsh, J. J. 1975. A spatial simulation model of the Peru upwelling ecosystem. Deep-Sea Res. 22:201–246.

Walsh, J. J. 1976. Models of the sea. Pp. 388–446 in D. H. Cushing and J. J. Walsh (eds.), The Ecology of the Seas. Blackwell.

Walsh, J. J., J. C. Kelley, T. E. Whitledge, J. J. MacIsaac, and S. A. Huntsman. 1974. Spin-up of the Baja California upwelling ecosystem. Limnol. Oceanogr. 19:553–572.

Warren, C. E., and G. E. Davis. 1967. Laboratory studies on the feeding, bioenergetics, and growth of fish. Pp. 175–214 in S. D. Gerking (ed.), The Biological Basis of Freshwater Fish Production. Blackwell.

Warwick, R. M. 1980. Population dynamics and secondary production of benthos. Pp. 1–24 in K. R. Tenore and B. C. Coull (eds.), Marine Benthic Dynamics. Univ. of South Carolina.

Warwick, R. M., and R. Price. 1975. Macrofauna production in an estuarine mudflat. J. Mar. Biol. Assoc. U.K. 55:1–18.

Warwick, R. M., and R. Price. 1979. Ecological and metabolic studies on free-living nematodes from an estuarine mudflat. Estuar. Coast. Mar. Sci. 9:257–271.

Waterbury, J. B., S. W. Watson, and F. Valois. 1980. Preliminary assessment of the importance of *Synechoccus* spp. as oceanic primary producers. Pp. 516–517 in P. G. Falkowski (ed.), Primary Productivity in the Sea. Plenum.

Watt, W. D., and F. R. Hayes. 1963. Tracer study of the phosphorus cycle in seawater. Limnol. Oceanogr. 8:276–285.

Welch, H. E. 1968. Relationships between assimilation efficiencies and growth efficiencies for aquatic consumers. Ecology 49:755–759.

Wellington, G. M. 1982a. An experimental analysis of the effects of light and zooplankton on coral formation. Oecologia 52:311–320.

Wellington, G. M. 1982b. Depth zonation of corals in the Gulf of Panama: control and facilitation by resident reef fishes. Ecol. Monogr. 52:223–241.

Werme, C. 1981. Resource Partitioning in a Salt Marsh Fish Community. Ph.D. Diss., Boston Univ.

Werner, E. E. 1977. Species packing and niche complementarity in three sunfishes. Amer. Nat. 111:553–578.

Werner, E. E., and D. J. Hall. 1976. Niche shifts in sunfishes: Experimental evidence and significance. Science 191:404–406.

Wetzel, R. G. 1975. Limnology. Saunders.

Wharton, W. G., and K. H. Mann. 1981. Relationship between destructive grazing by the sea urchin, *Strongylocentrotus droebachiensis*, and the abundance of American lobster, *Homarus americanus*, on the Atlantic Coast of Nova Scotia. Can. J. Fish. Aquat. Sci. 38:1339–1349.

Wheeler, P. A. 1979. Uptake of methylamine (an ammonium analogue) by *Macrocystis pyrifera* (Phaeophyta). J. Phycol. 15:12–17.

Wheeler, P. A. 1980. Use of methylammonium as an ammonium analogue in nitrogen transport and assimilation studies with *Cyclotella cryptica* (Bacillariophyceae). J. Phycol. 16:328–334.

Wheeler, P. A., and J. A. Hellebust. 1981. Uptake and concentration of alkylamines by a marine diatom. Plant Physiol. 67:367–372.

White, A. W. 1981. Sensitivity of marine fishes to toxins from red-tide dinoflagellate *Gonyaulax excavata* and implication for fish kills. Mar. Biol. 65:255–260.

Whitledge, T. E. 1978. Regeneration of nitrogen by zooplankton and fish in the Northwest Africa and Peru upwelling ecosystems. Pp. 90–110 in R. Boje and M. Tomczak (eds.), Upwelling Ecosystems. Springer-Verlag.

Whitney, D. M., A. G. Chalmers, E. B. Haines, R. B. Hanson, L. R. Pomeroy, and B. Sherr. 1981. The cycles of nitrogen and phosphorus. Pp. 163–182 in L. R. Pomeroy and R. G. Wiegert (eds.), The Ecology of a Salt Marsh. Springer-Verlag.

Whittaker, R. H. 1972. Evolution and measurement of species diversity. Taxon 21:213–251.

Whittaker, R. H. 1973 (ed.). Ordination and Classification of Communities. Junk. b.v.-Publishers.

Whittaker, R. H., and D. Goodman. 1979. Classifying species according to their demographic strategy. I. Population fluctuations and environmental heterogeneity. Amer. Nat. 113:185–200.

Whittaker, R. H., and G. E. Likens 1975. The biosphere and man. Pp. 305–328 in H. Lieth and R. H. Whittaker (eds.), Primary Productivity of the Biosphere. Springer-Verlag.

Whittle, K. J. 1977. Marine organisms and their contribution to organic matter in the oceans. Mar. Chem. 5:381–411.

Wickett, W. P. 1967. Ekman transport and zooplankton concentrations in the North Pacific Ocean. J. Fish. Res. Bd. Canada 24:581–594.

Wiebe, P. H. 1970. Small-scale spatial distribution in oceanic zooplankton. Limnol. Oceanogr. 15:205–217.

Wiebe, P. H. 1982. Rings of the Gulf Stream. Scientific Amer. 246:60–70.

Wiebe, P. H., S. H. Boyd, and C. Winget. 1976. Particulate matter sinking to the deep sea floor at 2000m in the Tongue of the Ocean, Bahamas, with a description of a new sedimentation trap. J. Mar. Res. 34:341–354.

Wiebe, P. H., E. M. Hulbert, E. J. Carpenter, A. E. Jahn, G. P. Knapp, S. H. Boyd, P. B. Ortner, and J. L. Cox. 1976a. Gulf stream cold core rings: Large scale interaction sites for open ocean plankton communities. Deep-Sea Res. 23:695–710.

Wiebe, P. H., L. P. Madin, L. R. Haury, G. R. Harbison, L. M. Philbin. 1979. Diel vertical migration by *Salpa aspera* and its potential for large scale particulate organic transport ot the deep-sea. Mar. Biol. 53:249–256.

Wiebe, W. T., and D. F. Smith. 1977. Direct measurement of dissolved organic carbon release by phytoplankton and incorporation by microheterotrophs. Mar. Biol. 42:213–223.

Wiegert, R. G., and F. C. Evans. 1964. Primary production and disappearance of dead vegetation on an old field in southeastern Michigan. Ecology 45:49–63.

Wiegert, R. G., and F. C. Evans. 1967. Investigations of secondary productivity in grasslands. Pp. 499–518 in K. Petrusewicz (ed.), Secondary Productivity of Terrestrial Ecosystems. Polish Acad. Sci.

Wiegert, R. G., and D. F. Owen. 1971. Trophic structure, available resources, and population density in terrestrial vs aquatic ecosystems. J. Theoret. Biol. 30:69–81.

Wiens, J. A. 1976. Population responses to patchy environments. Ann. Rev. Ecol. Syst. 7:81–120.

Williams, D. McB. 1980. Dynamics of the ponacentrid community on small patch reefs in One Tree Lagoon (Great Barrier Reef). Bull. Mar. Sci. 30:159–170.

Williams, P. J. Le B. 1975. Biological and chemical dissolved organic material in sea water. Pp. 301–364 in G. Riley and O. Skirrow (eds.), Chemical Oceanography, Vol.2. 2nd Ed.

Williams, P. J. Le B. 1981. Incorporation of microheterotrophic processes into the classical paradigm of the planktonic food web. Kieler Meeresforsch. Suppl. 5:1–28.

Williams, P. J. Le B., R. T. C. Raine, and J. R. Bryan. 1979. Agreement between the ^{14}C and oxygen method of measuring phytoplankton production, measurement of the photosynthetic quotient. Oceanol. Acta 2:411–416.

Williams, R. B., and M. B. Murdoch. 1972. Compartmental analysis of the production of *Juncus roemerianus* in a North Carolina salt marsh. Chesap. Sci. 13:69–79.

Williams, R. B., M. N. Murdoch, and L. K. Thomas. 1968. Standing crop and importance of zooplankton in a system of shallow estuaries. Chesap. Sci. 9:42–51.

Williams, W. T., G. N. Lance, L. J. Webb, J. G. Tracey, and M. B. Dale. 1969.

Studies in the numerical analysis of complex rain-forest communities. III. The analysis of successional data. J. Ecol. 57:515–535.

Wilson, E. O., and W. H. Bossert 1971. A Primer of Population Biology. Sinauer.

Wilson, W. H. 1980. A laboratory investigation of the effects of a terebellid polychaete on the survivorship of nereid polychaete larvae. J. Exp. Mar. Biol. Ecol. 46:73–80.

Wiltse, W. I., K. H. Foreman, J. M. Teal, and I. Valiela. In press. Importance of predators and food resources to the macrobenthos of salt marsh creeks. J. Mar. Res.

Winberg, G. G. 1971. Methods for the Estimates of Production of Aquatic Animals. Academic.

Winter, D. F., K. Banse, and G. C. Anderson. 1975. The dynamics of Phytoplankton blooms in Puget Sound, a fjord in the Northwestern United States. Mar. Biol. 29:139–176.

Winter, J. E. 1978. A review of the knowledge of suspension-feeding in lamellibranchiate bivalves, with special reference to artificial aquaculture systems. Aquaculture 13:1–33.

Winters, G. H. 1976. Recruitment mechanisms of Southern Gulf of St. Lawrence Atlantic herring (*Clupea harengus harengus*). J. Fish. Res. Bd. Canada 33:1751–1763.

Wishner, K. F. 1980. The biomass of the deep-sea benthopelagic plankton. Deep-Sea Res. 27A:203–216.

Wohlschlag, D. E. 1964. Respiratory metabolism and ecological characteristics of some fishes in McMurdo Sound, Antarctica. Antarctic Res. Ser. 1:33–62.

Wolff, T. 1976. Utilization of seagrass in the deep sea. Aquat. Bot. 2:161–174.

Womersley, H. B. S., and R. E. Norris. 1959. A free-floating marine red algae. Nature 184:828–829.

Wood, E. J. F., W. E. Odum, and J. C. Zieman. 1967. Influence of seagrasses on the productivity of coastal lagoons. Pp. 495–502 in Lagunas Costeras, un Symposio. Mem. Simp. Intern. Lagunas Costeras. UNAM-UNESCO, Mexico D.F.

Woodin, S. A. 1977. Algal "gardening" behavior by nereid polychaetes: Effects on soft-bottom community structure. Mar. Biol. 44:39–42.

Woodin, S. A. 1978. Refuges, disturbance, and community structure: A marine soft bottom example. Ecology 59:274–284.

Wright, R. T. In press. Dynamics of pools of dissolved organic carbon. In J. E. Hobbie and P. J. Le B. Williams (eds.), Heterotrophic Activity in the Sea. Plenum.

Wright, R. T., and J. E. Hobbie. 1965. The uptake of organic solutes in lake water. Limnol. Oceanogr. 10:22–28.

Wroblewski, J. A. 1982. Interaction of currents and vertical migration in maintaining *Calanus marshallae* in the Oregon upwelling zone—A simulation. Deep-Sea Res. 29:665–686.

Wroblewski, J. S., and J. J. O'Brien. 1976. A spatial model of phytoplankton patchiness. Mar. Biol. 35:161–175.

Wroblewski, J. S., J. J. O'Brien, and T. Platt. 1975. On the physical and biological scales of phytoplankton patchiness in the ocean. Mem. Soc. R. Sci. Liege 7:43–57.

Wyatt, T. 1973. The biology of *Oikopleura dioica* and *Fritillaria borealis* in the Southern Bight. Mar. Biol. 22:137–158.

Wyrtki, K. 1975. El Niño: The dynamic response of the Equatorial Pacific Ocean to atmospheric forcing. J. Phys. Oceanogr. 5:572–584.

Yamaguchi, M. 1975. Estimating growth parameters from growth rate dates. Oecologia 20:321–332.

Yellin, M. B., C. R. Agegian and J. S. Pearse. 1977. Ecological benchmarks in the Santa Cruz County kelp forests before the re-establishment of sea otters. Center for Coastal Marine Studies. Univ. Calif., Santa Cruz.

Yentsch, C. S. 1967. The measurement of chloroplastic pigments—thirty year progress? Pp. 255–270 in H. L. Golterman and R. J. Clymo (eds.), Chemical Environment in the Aquatic Habitat. N.V. Noord-Hollandsche Uitgevers Maatschappij.

Yentsch, C. S. 1980. Light attenuation and phytoplankton photosynthesis. Pp. 95–127 in I. Morris (ed.), The Physiological Ecology of Phytoplankton. Univ. California.

Yentsch, C. M., C. S. Yentsch, and L. R. Strube. 1977. Variations in ammonium enhancement, an indication of nitrogen deficiency in New England coastal phytoplankton populations. J. Mar. Res. 35:537–555.

Yingst, J. Y. 1976. The utilization of organic matter in shallow marine sediments by an epibentic deposit-feeding holothurian. J. Exp. Mar. Biol. Ecol. 23:55–69.

Yingst, J. Y., and D. C. Rhoads. 1980. The role of bioturbation in the enhancement of bacterial growth rates in marine sediments. Pp. 407–421 in K. R. Tenore and B. C. Coull (eds.), Marine Benthic Dynamics. Univ. South Carolina.

Young, D. K. 1971. Effects of infauna on the sediment and seston of a subtidal environment. Vie Mil. Suppl. 22:557–571.

Zach, R., and J. N. M. Smith. 1981. Optimal foraging in wild birds? Pp. 95–110 in A. C. Kamil and T. D. Sargent (eds.), Foraging Behavior. Garland STPM.

Zaika, V. E., and N. P. Makarova. 1979. Specific production of free-living marine nematodes. Mar. Ecol. Progr. Sci. 1:153–158.

Zaret, T. M., and T. R. Paine. 1973. Species introduction in a tropical lake. Science 182:449–455.

Zeitzschel, B. (ed.) 1973. The Biology of the Indian Ocean. Springer-Verlag.

Zeitzschel, B. 1980. Sediment-water interaction in nutrient dynamics. Pp. 195–218 in K. R. Tenore and B. C. Coull (eds.), Marine Benthic Dynamics. Univ. South Carolina.

Zenkevitch, L. 1963. Biology of the Seas of the USSR. Interscience.

Zuberer, D. A., and W. S. Silver 1978. Biological dinitrogen fixation (acetylene reduction) associated with Florida mangroves. Appl. Env. Microbiol. 35:567–575.

Zuta, S., J. Rivera, and A. Bustamante. 1978. Hydrologic aspects of the main upwelling areas off Peru. Pp. 235–257 in R. Boje and M. Tomczak (eds.), Upwelling Ecosystems. Springer-Verlag.

Index

A

B

D

F

G

O

P

W

Y

Z

Springer Advanced Texts in Life Sciences
Series Editor: David E. Reichle

Environmental Instrumentation
By Leo J. Fritschen and Lloyd W. Gay

Biophysical Ecology
By David M. Gates

Forest Succession: Concepts and Application
Edited by Darrel C. West, Herman H. Shugart, and Daniel B. Botkin

Marine Ecological Processes
By Ivan Valiela